Principles of
ANALOG
ELECTRONICS

Giovanni Saggio

CRC Press
Taylor & Francis Group
Boca Raton   London   New York

CRC Press is an imprint of the
Taylor & Francis Group, an **informa** business

CRC Press
Taylor & Francis Group
6000 Broken Sound Parkway NW, Suite 300
Boca Raton, FL 33487-2742

© 2014 by Taylor & Francis Group, LLC
CRC Press is an imprint of Taylor & Francis Group, an Informa business

No claim to original U.S. Government works

Printed on acid-free paper
Version Date: 20131212

International Standard Book Number-13: 978-1-4665-8201-9 (Hardback)

Library of Congress Cataloging-in-Publication Data

Saggio, Giovanni.
  Principles of analog electronics / Giovanni Saggio. -- 1st edition.
    pages cm
  Includes bibliographical references and index.
  ISBN 978-1-4665-8201-9 (hardback)
  1. Analog electronic systems. 2. Electronic circuit design. I. Title.

  TK7867.S 2014
  621.3815--dc23   2013018107

Visit the Taylor & Francis Web site at
http://www.taylorandfrancis.com

and the CRC Press Web site at
http://www.crcpress.com

We live in a world that does not tolerate the slightest grammatical mistake but indulges in the complete ignorance in scientific matters. Could this strange disparity perhaps be caused by the tumultuous scientific developments in the past few decades? Fortunately, there is a silver lining: for one who knows science, he or she potentially has the key to success in his or her hands.

Anyone who really knows electronics can design new ideas.

Contents

Contents

Contents

Contents

Contents

Contents

Contents

Contents

Contents

Contents

Contents

Preface

Anyone who knows electronics can create new ideas, and this book explores that possibility by focusing on analog electronics. This is because in the real world, signals are mostly analog, spanning continuously varying values, so that circuits interfaced with the physical world have an analog nature, to process analog signals.

The fascinating area of analog "grows" to overlap fundamental areas (fields, circuits, signals and systems, and semiconductors), here morphed into a self-consistent comprehensive book. This approach leads to a text that captures the big picture, while still providing the necessary details, to reduce knowledge fragmentation and to improve learning outcomes.

The presentation is accurate and clear, including appropriate examples and detailed explanations of the behavior of real electronic circuits.

The text is "humanized" not only because the important theorems and laws are treated, but also because we look at the people who fundamentally contributed to those. Curiosities (How did Google get its name? Why is it difficult to locate crickets in a field? Why does the violin bridge have that strange shape?, etc.), observations, and real life application-oriented examples (electrocardiogram instrumentation, active noise-canceling headphones, USB-powered charger, etc.) are provided to contribute to a practical approach.

The first part of the book includes a clear and thorough presentation of the mathematical (Chapter 1), physical (Chapter 2), and chemical concepts (Chapter 3) that are essential to understanding the principles of operation of electronic devices. This may be particularly useful for students with a limited background in basic matters who want to take a serious approach to electronics.

The circuit approach is detailed with models (Chapters 4, 5, 10) and main theorems (Chapter 6).

Passive and active electronic devices are described and analyzed, with specific reference to the fundamental filters (Chapter 7), and to the most common Si-based components such as diodes, BJTs, and MOSFETs (Chapters 4, 8). Semiconductor devices are then used to design electronic circuits, such as rectifiers, power suppliers, clamper and clipper circuits (Chapter 9). The main topologies of amplifiers based on BJTs (Chapter 11), and on MOSFETs (Chapter 12), along with their variants and improvements (Chapters 13, 14), are also discussed. Relevant or curious circuit applications are analyzed as well (Chapter 15).

At the end of each chapter, helpful summaries are provided, with key points, jargon, terms, and exercises with solutions also included.

Practical tables, often missing in many books on electronics, are included here to illustrate the coding schemes necessary to recognize commercial passive and active components.

MATLAB® is a high-level language and interactive environment for numerical computation, visualization, and programming. MATLAB® is a registered trademark of The MathWorks Inc. For product information, please contact:

The MathWorks, Inc.
3 Apple Hill Drive
Natick, MA 01760-2098 USA
Tel: 508 647 7000
Fax: 508-647-7001
E-mail: info@mathworks.com
Web: www.mathworks.com

Author

Giovanni Saggio earned a masters degree in electronics engineering and a PhD in microelectronics and telecommunication engineering at the University of Rome Tor Vergata.

He was offered one fellowship by Texas Instruments and two fellowships by the Italian National Research Council (CNR). He developed research on nanoelectronics at Glasgow University (Ultra Small Structure Lab), and on electronic devices at Cambridge University (Cavendish Lab), and Oxford University (Rutherford Appleton Lab). He has been a designer, planner, and director of electrical installations and security systems, as well as hydraulic systems.

Professor Saggio is currently a researcher and aggregate professor at the University of Rome Tor Vergata (Italy), where he holds chairs in electronics at the engineering faculty (Departments of Information, Automation, Mathematics, Biomedical Engineering, Master of Sound, and Master of CBRN Protection) and at the medical faculty (Departments of Neurophysiology, Cardiovascular Medicine, Orthopedics, and Audiology).

He has been working on problems concerning electronic noise, SAWs, and electronic sensors, and more recently, his research activity concerns the field of biotechnology.

Professor Saggio has been project leader of research for the Italian Space Agency (ASI), for the avionic service of the Italian Defence Department (Armaereo), and for the Italian Workers' Compensation Authority (INAIL). He is currently a member of Italian Space Biomedicine, and the founder and manager of HITEG (Health Involved Technical Engineering Group).

Professor Saggio has authored or co-authored more than 100 scientific publications for conferences and international journals, four patents, several book chapters, and is the sole author of three books (in Italian): *Basi di Elettronica* (three editions), *Applicazioni di Elettronica di Base*, and *Elettronica Analogica Fondamentale* (UniversItalia ed.).

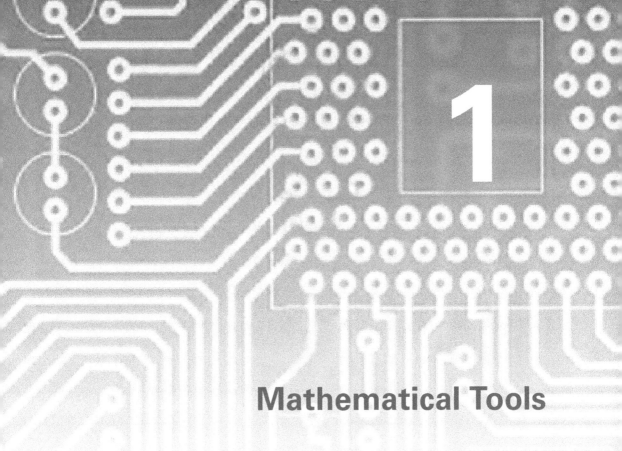

Mathematical Tools

Electronics is based on fundamental mathematical tools that are summarized here for the convenience of the reader.

1.1 MULTIPLES AND SUBMULTIPLES

The modern metric system of measurement adopts prefixes to form *multiples* and *submultiples* of *SI units* (from the French *le Système International d'Unités, International System of Units*). The *SI* is founded on base units (*meter, kilogram, second, ampere, kelvin, mole, candela*) and base independent quantities (*length, mass, time, current, thermodynamic temperature, amount of substance, luminous intensity*). The prefixes used to form multiples and submultiples of *SI* units were established at the *Conférence Générale des Poids et Mésures* (*CGPM* in 1960 with later revisions) and are shown in Table 1.1.

In this way, for example, considering the *length* for which the unit of measurement is the *meter*, the kilo*meter* expresses 1000 *meters*, the micro*meter* expresses 0.000001 *meter*, and so on.

Observations

Kilogram is the only SI unit with a prefix as part of its name and symbol.

SI prefix symbols are capitalized for multipliers 10^6 and larger, and lowercase for multipliers 10^3 and smaller.

TABLE 1.1
Prefixes Used to Form Multiples and Submultiples of SI Units

Multiplicator		Symbol	Prefix
10^{24}	1000000000000000000000000	Y	yotta
10^{21}	1000000000000000000000	Z	zetta
10^{18}	1000000000000000000	E	exa
10^{15}	1000000000000000	P	peta
10^{12}	1000000000000	T	tera
10^{9}	1000000000	G	giga
10^{6}	1000000	M	mega
10^{3}	1000	k	kilo
10^{2}	100	h	hecto
10^{1}	10	da	deka
	1		
10^{-1}	0.1	d	deci
10^{-2}	0.01	c	centi
10^{-3}	0.001	m	milli
10^{-6}	0.000001	μ	micro
10^{-9}	0.000000001	n	nano
10^{-12}	0.000000000001	p	pico
10^{-15}	0.000000000000001	f	femto
10^{-18}	0.000000000000000001	a	atto
10^{-21}	0.000000000000000000001	z	zepto
10^{-24}	0.000000000000000000000001	y	yocto

Curiosities

The large number 10^{100} is expressed by the term *googol*, from which the name of the famous search engine google.com was derived. The creators of the famous search engine wanted this as an indicator of the immense capacity to index a huge amount of websites.

Today we can describe any number with a huge value and a lot of numbers have proper names, say for example a *million* or a *billion*. In ancient Greece, the largest number with a proper name was 10,000: *myriad*. The word comes from myrios (meaning "innumerable") and today we still use it with a different meaning. However, Archimedes (Archimedes of Siracusa, a Greek mathematician, astronomer, physicist, and engineer, 287–212 B.C.) was not limited by the fact that the largest number having a proper name was "only" 10,000. Instead, he was able to express the number "one followed by one hundred million of billions of zeros," meaning the myriads (ten thousands) of myriads, equal to 100 million, on rows and columns of an enormous chart. He counted until he reached the gigantic number that he called "a myriad of myriads of the myriad-myriadth row by the myriad-myriadth column."

Every word can have a prefix, which usually we ignore. As an example the word "television" uses "tele," which has Greek origin and means "far," so "television" stands for something like "far sight." Think about "telephone," "telegraph," "telefax," "telemetry," "telenovelas," "telecom," "telehealth," "telescope," "telework," etc.

The age of the universe is on the order of 4×10^{17} *seconds*, and the part observable by man has dimensions on the order of 10^{24} *km*.

1.2 PERIODIC WAVEFORMS

Waves describe the phenomenon of propagation that occurs both in time and space. For example, when a coin is thrown in a fountain, a wave will be created rippling outwards from the coin that spreads over a certain time and covers a certain area.

These waves are essential in many fields of science and engineering (physics, chemistry, electrical engineering, electronics, etc.), because it is possible to associate and convey information with their variations. When a wave is used for this purpose, it represents a *signal* and depends on the physical quantity that is associated with it. There are signals of *light, sound, pressure, electricity*, and so on.

In the real world, signals are mostly *analog*, namely they have a continuous nature and span continuously varying values. That is why circuits that interface with the physical world are of an *analog* nature.

The waves can be *periodic*, that is, they repeat themselves at regular intervals, and can be described by a *mathematical function*, which relates two (or more) variables. Figure 1.1 describes a specific type of wave x that propagates in time t over space s.

The duration time of a single cycle is called a *period*, indicated with the letter T and measured in seconds (s). The distance travelled by the wave in a cycle is called a *wavelength*, indicated with the letter λ and measured in meters (m).

Observation

No function can be really considered to be strictly *periodic*. In fact, the periodicity assumes a recurrence at regular intervals but *without* limits in time, existing from $t_0 = -\infty$ to $t_1 = +\infty$. Clearly each real function begins and ends, and is practically a *piecewise-defined* function:

$$f(t) \begin{cases} = 0, & if \ t \leq t_0 \\ \neq 0 \ and \ periodic, & if \ t_0 < t \leq t_1 \\ = 0, & if \ t > t_1 \end{cases}$$

It is a common practice to work with piecewise-defined periodic functions as if they were really strictly periodic.

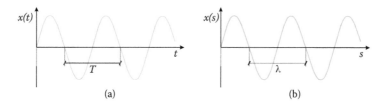

(a)   (b)

FIGURE 1.1   Periodic waveform that propagates (a) in time and (b) in space.

It is well known that the rate of change of variation of distance Δs as a function of a time interval Δt defines an *average velocity* $v(t)$ equal to

$$v(t) = \frac{\Delta s}{\Delta t}$$

which can be rewritten for *instantaneous velocity* $v_i(t)$ as the limit of the average velocity

$$v_i(t) = \lim_{\Delta t \to 0} \frac{\Delta s}{\Delta t} = \frac{ds}{dt}$$

For the sake of simplicity, if we assume both initial space and time are equivalent to zero, the average velocity can be rewritten for a wave using the definitions of period and wavelength previously given, and is equal to

$$v = \frac{\lambda}{T}$$

Clearly the speed, that is the magnitude of the velocity relative to a single cycle is the same for the whole wave and for all its cycles.

The inverse of the period T defines another important parameter called *frequency, $f\,(=1/T)$*.

The SI unit for frequency is Hertz, *Hz* (Heinrich Hertz, German physicist, 1857–1894, Figure 1.2). The frequency represents the number of times that the cycle is repeated in the unit of time.

FIGURE 1.2   H. Hertz.

Curiosities

Looking at a car that passes us whose wheels are rotating clockwise, it sometimes seems that its wheels rotate counterclockwise. To understand why, we have to recognize that the image in our eye remains stationary for a few moments. This is verifiable by simply looking carefully at a light bulb and then

closing one's eyes: the image of the light will be seen again for a moment. This is because a picture remains "frozen" in our eye for a fraction of a second and our eye "resumes sight" only at this time. If we look at a detail of the wheel of the car that rotates in a clockwise direction (as the initial point in Figure 1.3), when we "see it again" after a certain time (now in the final position) we will have the sensation that the wheel has turned in the contrary direction.

By increasing the velocity of rotation of the wheel, there is a moment where it seems to cease rotating. This is called the *strobe* condition: the time the wheel takes to complete a turn is the same time our eyes require to be able to see the next picture.

FIGURE 1.3 A wheel that turns at a certain speed seems to spin in the opposite direction.

At the cinema, we have the sensation of seeing images as if they were moving normally, while we know that they are only a series of pictures (the *frames*) displayed one after the other.

Similarly, we see a lightbulb as if it always has the same brightness and is never off. In reality, the light bulb uses household electricity, which is not continuous but alternating according to a sinusoidal law (it starts at null, increases to reach a maximum value, then decreases in value to null again and so on, periodically) and, in reality, the light bulb is subjected to increases and decreases in brightness accordingly.

Given that the image remains in the eye for about 1/24 of a second, in order for us to see the light bulb always turned on, the *frequency* of the household electricity should have a value sufficiently greater than 24. In fact, it is equal to 60 Hz in the United States. and 50 Hz in Europe.

FIGURE 1.4
G. Galilei.

Galilei (Galileo Galilei, Italian physicist, mathematician, astronomer, and philosopher, 1564–1642, Figure 1.4) was the first to define and measure speed as the distance covered per unit of time.

The equation for velocity can also be written as

$$v = \frac{\lambda}{T} = \lambda f$$

whereby knowing the velocity v and the frequency f of a wave, it is possible to calculate the wavelength λ.

A real signal is necessarily limited in time and, therefore, every real signal presents frequency bounds. In electronic jargon, these bounds determine the *band*, a term that actually means *interval*, so that a *frequency interval* corresponds to a *band of frequency*, and every real signal has its band limitation.

Table 1.2 shows some typical bands of frequency for signals.

Examples

In the United States, the frequency of household electricity is $60\,Hz$, meaning the sinusoidal wave repeats itself 60 times in a second, which corresponds to a period of $0.1\overline{6}\,s$. In Europe the value is $50\,Hz$, so the corresponding period is $0.02\,s$.

When we listen to a European radio network, we probably do so in the *FM* (*Frequency Modulation*) band (range 87–108 MHz). By choosing a radio station in that band, we set the radio to a single frequency, for example, the frequency broadcast at $f = 100\,Mhz$. The signal that we receive at that frequency has a propagating speed close to that of light in vacuum, which is around $v \cong 300.000\,Km/s$. As a result, we have a wavelength λ equal to

$$\lambda = \frac{v}{f} \cong \frac{300.000\,km/s}{100\,MHz} = \frac{3{*}10^8\,m/s}{10^8\,Hz} = 3\,m$$

(as $1/Hz = s$). Therefore, the broadcast wave repeats every 3 meters and 100 million times per second.

One of the frequencies around which cellular phone signals are transmitted is $1800\,MHz$, and as result, the wavelength is

$$\lambda = \frac{v}{f} \cong \frac{300.000\,km/s}{1800\,MHz} = \frac{3{*}10^8\,m/s}{18{*}10^8\,Hz} \cong 16.6\,cm$$

Observation

The value of the frequency of a constant function is zero. In fact, we can imagine a constant as a limit of a periodic function with period $T = \infty$ (so $f = \frac{1}{T} = \frac{1}{\infty} = 0$).

TABLE 1.2
Typical Bands of Frequency of Some Signals

Type of signal	Frequency range
Seismic signals	1 Hz–200 Hz
Electrocardiogram	0.05 Hz–100 Hz
Audio signals	20 Hz–20 kHz
Video signals	50 Hz–4.2 MHz
Radio signals AM	540 kHz–1600 kHz
Radio signals FM	88 MHz–107 MHz
Video VHF	54 MHz–60 MHz
Video UHF	470 MHz–806 MHz
Portable cellular phone (GSM 900)	880 MHz–935 MHz
Portable cellular phone (GSM 1800)	1710 MHz–1880 MHz
Satellite video signals	3.7 GHz–4.2 GHz
Microwave communication	1 GHz–50 GHz

The ensemble of bands forms the *spectrum*, in which each band takes its own name as Figure 1.5 illustrates.

If we limit the spectrum to visible light, it is divided into frequencies of individual colors, as shown in Figure 1.6.

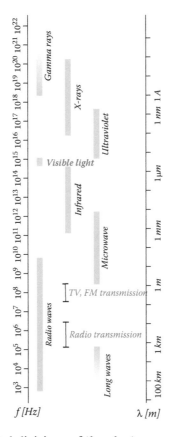

FIGURE 1.5 Subdivisions of the electromagnetic spectrum.

Curiosity

In addition to frequency, the term *band* can also be adopted to indicate the *intervals* of several other variables. As a result, we can talk about a *band structure* (those ranges of energy that an electron within the solid may have), a *band society* (the simplest form of human society generally consists of a small kin group), a *mathematic band* (a semi-group in which every element is idempotent, i.e., equal to its own square), and so on.

A periodic function is characterized by its *shape* (sinusoidal, square, saw-toothed, etc.), by its *amplitude*, namely the maximum value that it has within a cycle, and by the *phase* (measured in degrees, 0°–360°, or in radians, 0°–2π), namely its relative position at the origin of the Cartesian axes (Figure 1.7a through d).

The concept of phase can be understood as *absolute*, in the case that the signal wave is considered with respect to the Cartesian axes as a reference, or *relative*, when it is compared to another wave, as shown in Figure 1.8a and b.

According to the information an analog signal is asked to convey, it may vary in frequency (period), phase, or amplitude in response to changes in physical phenomena (light, sound, heat, position, etc.).

FIGURE 1.6 Spectrum of visible light.

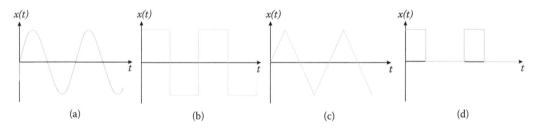

(a) (b) (c) (d)

FIGURE 1.7 Examples of different types of periodic waveforms:
(a) sinusoidal, (b) square, (c) triangular, and (d) train of pulses.

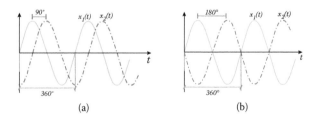

(a) (b)

FIGURE 1.8 (a) Function with difference of phase = 90°;
(b) function with difference of phase = 180°.

The dubber is asked to *synchronize* his/her voice with the lip movements of an actor on the screen. That is, the voice and lip movements must be in phase.

1.2.1 Sine Waves

The class of periodic waves belonging to the *sinusoidal* (Figure 1.9) is generally described by the equation

$$f(t) = F_M \sin(\omega t + \alpha)$$

where

→ F_M is the *maximum*, or *peak*, value that the sinusoidal wave assumes.
→ $\omega(= 2\pi f)$ is the *angular frequency* (also referred to by the terms angular speed, radial frequency, circular frequency, orbital frequency, and radian frequency), measured in radians/sec.
→ α is the *phase*.

Two sine curves that are out of phase with each other by 90° are called *orthogonal*, while those that phase with each other by 180° are in *phase opposition* and, as Figure 1.8b shows, are identical but have opposite signs (when one is positive the other is negative and vice versa).

Observation

Note that zero is obtained by adding two sine curves with the same frequency out of phase by 180°.

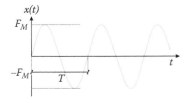

FIGURE 1.9 Sinusoidal waveform.

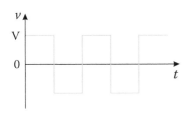

FIGURE 1.10 Square waveform.

1.2.2 Square Waves

Square waves are another example of periodic functions, since they regularly repeat two constant levels, alternating them and switching cleanly and instantaneously between them as in Figure 1.10.

The square wave can be made from an infinite sum of sine waves at different frequencies, starting from the so called *fundamental (angular) frequency* ω and adding "weighted" *harmonics* at odd-integer multiples of ω. The "weight" is represented by a multiplying coefficient of decreasing value as in Figure 1.11.

The infinite sum of sine waves at different frequencies is the *Fourier series* (details of which are given later in Section 1.7), which for the square wave can be written as

$$v(\omega t) = \frac{4}{\pi} \sum_{n=1}^{\infty} \frac{\sin[(2n-1)\omega t]}{2n-1}$$

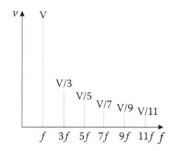

FIGURE 1.11 Coefficients as "weights" of the harmonics of the square wave.

Given multiplying coefficients of decreasing value (the denominator increases with n), the "weight" of the harmonics becomes less and less relevant.

1.2.3 Characterizations

1.2.3.1 Average Value

The *average value* of a generic periodic function $f(t)$ is typically indicated with a subscript, "f_{AVG}" (AVG: *average*), or with a top line, "$\bar{f}$," and is defined as

$$f_{AVG} = \bar{f} = \frac{1}{T} \int_{t_0}^{t_0+T} f(t)\,dt$$

(The average value calculated for a whole period is T. If instead of T in the formula, we use $T/2$, we will achieve the average value of a semi-period).

In the particular case of a sinusoidal function $f(t) = F_M \sin(\omega t)$, such value is null:

$$f_{AVG} = \frac{1}{2\pi} \int_{0}^{2\pi} F_M \sin(\omega t)\,dt = 0$$

In the case of the periodic function, $f_M(t) = |F_M \sin(\omega t)|$, which is graphically represented in Figure 1.12, the average value is

$$f_{M,AVG} = \frac{1}{2\pi} \int_{0}^{2\pi} |F_M \sin(\omega t)|\,dt = \frac{2}{\pi} F_M \cong 0.637 F_M$$

(apparently the same average that we would have achieved considering the sinusoidal function only for a semi-period). We will meet this particular function in Section 9.1.3.

Later in the book, the average value will be addressed also as the *DC component* for reasons discussed in Chapter 4, Section 4.5.

1.2.3.2 Root Mean Square Value

The *root mean square*, or *RMS*, value of a generic periodic function $f(t)$, typically indicated with a subscript "f_{RMS}," is defined as the root of the quadratic average, that is

$$f_{RMS} = \sqrt{\frac{1}{T} \int_{t_0}^{t_0+T} f^2(t)\,dt}$$

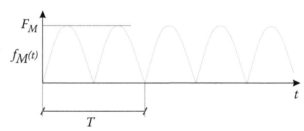

FIGURE 1.12 Representation of the absolute value of the sinusoidal function.

where the integration is performed over one cycle.

In the case of a sine wave, for example, the RMS value is

$$f_{RMS} = \sqrt{\frac{1}{T}\int_0^T F_M^2 \sin^2(\omega t)\, dt} = \sqrt{\frac{1}{T}\int_0^T F_M^2 \left(\frac{1-\cos(2\omega t)}{2}\right) dt} = \frac{1}{\sqrt{2}} F_M \cong 0.7 F_M$$

Therefore its square

$$\left(\left(\frac{1}{\sqrt{2}} F_M\right)^2 = \frac{1}{2} F_M^2\right)$$

is equal to half of the maximum value of the function $f^2(t) = F_M^2 \sin^2(\omega t)$.

1.2.3.3 Descriptive Values

In addition to *waveform, amplitude,* and *phase,* a periodic wave can be characterized with

→ *Maximum (or peak) value, F_M:* maximum amplitude
→ *Peak to peak value, $F_{P\text{-}P}$:* distance between max and min
→ *Crest factor, F_{CR}:* ratio between F_M and F_{RMS} (for a sinusoidal wave $F_{CR} = 1.414$)
→ *Average crest-factor, $\overline{F_{CR}}$:* ratio between F_M and F_{AVG}
→ *Form factor, F_F:* ratio between F_{RMS} and F_{AVG}

Examples of periodic waves and some of the related descriptive factors are given in Table 1.3.

1.3 FUNDAMENTAL TRIGONOMETRIC FORMULAE

Periodic physical quantities with temporal and spatial variation that can be expressed with the mathematical functions *sine* and *cosine* are discussed later in this chapter. For this, it is useful to review trigonometric formulas.

But such review does not have to be exhaustive, assuming the reader has some really basic knowledge of trigonometry.

1.3.1 Pythagorean Identity

The *Pythagorean identity* establishes a primary equation of trigonometry.

Given a unit circle, we define as *sine* and *cosine* functions, sin (α) and cos (α), respectively, the ratios

$$\frac{\overline{PH}}{\overline{OA}} \text{ and } \frac{\overline{OH}}{\overline{OA}}$$

But, given the hypothesis $\overline{OA} = 1$, this is simply equal to the lengths $\overline{PH}$ and $\overline{OH}$, as diagrammed in Figure 1.13.

Applying the Pythagorean theorem to the triangle OHP, we have

$$\overline{PH}^2 + \overline{OH}^2 + \overline{OP}^2 = 1$$

TABLE 1.3
Examples of Periodic Wave Functions and Values of Some Descriptive Factors

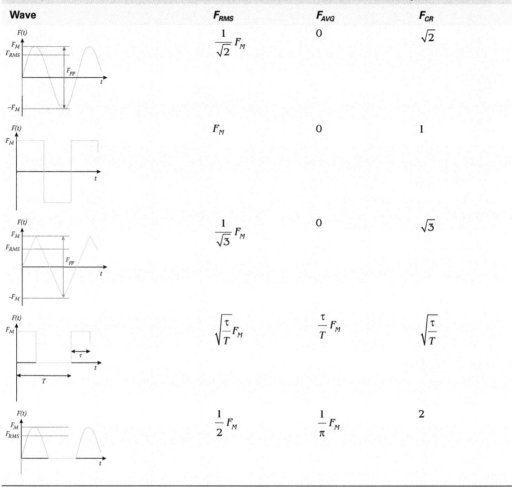

Wave	F_{RMS}	F_{AVG}	F_{CR}
	$\dfrac{1}{\sqrt{2}} F_M$	0	$\sqrt{2}$
	F_M	0	1
	$\dfrac{1}{\sqrt{3}} F_M$	0	$\sqrt{3}$
	$\sqrt{\dfrac{\tau}{T}} F_M$	$\dfrac{\tau}{T} F_M$	$\sqrt{\dfrac{\tau}{T}}$
	$\dfrac{1}{2} F_M$	$\dfrac{1}{\pi} F_M$	2

from which

$$\sin^2(\alpha) + \cos^2(\alpha) = 1$$

The equation is known as the *Pythagorean identity*.

1.3.2 Addition and Subtraction

Given two angles, α and β, the *sine* of their sum is *not* equal to the sum of their *sine*, meaning $\sin(\alpha + \beta) \neq \sin(\alpha) + \sin(\beta)$.

In fact, for example,

$$\sin\left(\frac{\pi}{2} + \frac{\pi}{2}\right) \neq \sin\left(\frac{\pi}{2}\right) + \sin\left(\frac{\pi}{2}\right)$$

since

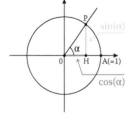

FIGURE 1.13
Representation of sine and cosine in the unit circle.

$$\sin\left(\frac{\pi}{2}+\frac{\pi}{2}\right)=\sin(\pi)=0$$

while

$$\sin\left(\frac{\pi}{2}\right)+\sin\left(\frac{\pi}{2}\right)=1+1=2$$

This example shows that the trigonometric functions *are not linear.*

The *sine* and *cosine* of the sum (or difference) of two angles have a more complex expression than a linear equation, and are equal to

$$\sin(\alpha\pm\beta)=\sin(\alpha)\cos(\beta)\pm\cos(\alpha)\sin(\beta)$$

$$\cos(\alpha\pm\beta)=\cos(\alpha)\cos(\beta)\mp\sin(\alpha)\sin(\beta)$$

1.3.3 Werner Formulas

The Werner formulas (named after Johann Werner, a German mathematician, 1468–1522) allow you to transform the product of trigonometric functions in terms of sums and differences.

From the first of the two equations shown in Section 1.3.2, by adding member–by-member the same equation, considered once positively and once negatively, you can obtain

$$\sin(\alpha)\cos(\beta)=\frac{1}{2}\left[\sin(\alpha+\beta)+\sin(\alpha-\beta)\right]$$

whereas by the second equation, using the same method, you can obtain

$$\sin(\alpha)\sin(\beta)=\frac{1}{2}\left[\cos(\alpha-\beta)-\cos(\alpha+\beta)\right]$$

$$\cos(\alpha)\cos(\beta)=\frac{1}{2}\left[\cos(\alpha-\beta)+\cos(\alpha+\beta)\right]$$

Note that from the arguments (α and β) of the initial functions, you arrive at the same arguments as the final functions, but added together algebraically ($\alpha+\beta$ and $\alpha-\beta$).

1.3.4 Duplication Formulas

Knowing the trigonometric functions related to a given angle (α), duplication of formulas allow us to calculate the trigonometric functions related to an angle doubled in value (2α).

From the last Werner formula, setting $\alpha=\beta$, we obtain

$$\cos(2\alpha)=\cos^2(\alpha)-\sin^2(\alpha)$$

If we combine this equation with the Pythagorean identity, it allows you to produce

$$\cos(2\alpha)=1-2\sin^2(\alpha)$$

$$\cos(2\alpha)=2\cos^2(\alpha)-1$$

1.3.5 Prosthaphaeresis Formulas

The inverses of the Werner formulas are named the *Prosthaphaeresis formulas* (also known as *Simpson's formulas*). The name originates from the union of the two Greek words *prosthesis*, meaning addition, and *aphaeresis*, meaning subtraction. Indeed the *Prosthaphaeresis formulas* allow you to transform additions and subtractions of the *sine* and *cosine* of two angles into products of the *sine* and *cosine* of half of the sum of the same angle.

Formula 1: $\sin(\alpha) + \sin(\beta) = 2\sin\left(\dfrac{\alpha+\beta}{2}\right)\cos\left(\dfrac{\alpha-\beta}{2}\right)$

Formula 2: $\sin(\alpha) - \sin(\beta) = 2\cos\left(\dfrac{\alpha+\beta}{2}\right)\sin\left(\dfrac{\alpha-\beta}{2}\right)$

Formula 3: $\cos(\alpha) + \cos(\beta) = 2\cos\left(\dfrac{\alpha+\beta}{2}\right)\cos\left(\dfrac{\alpha-\beta}{2}\right)$

Formula 4: $\cos(\alpha) - \cos(\beta) = -2\sin\left(\dfrac{\alpha+\beta}{2}\right)\sin\left(\dfrac{\alpha-\beta}{2}\right)$

These expressions are particularly important because they establish that by adding two sinusoidal signals with different frequencies, we can obtain a new signal equal to the product of a sine and a cosine of new frequencies that originated from the linear combination of the initial frequencies.

1.4 COMPLEX NUMBERS

A *complex number* is formed from two orthogonal components, represented on the Cartesian plane and defined in the ensemble of real numbers *R* and imaginary numbers *I*. The Cartesian notation is

$$z = x + jy$$

where *x* represents the real part ($x = Re\ (z)$), and *y* represents the coefficient of the imaginary part ($y = Im\ (z)$), with *j* the imaginary unit.

This number is represented in the complex Cartesian plane *C* by the point *P* with coordinates *x*, *y*, and by the vector that has a *magnitude* ρ (namely its length, that is the distance of the point *P* from the origin of the Cartesian axis) and a *phase* φ (or *argument*, namely the angle that the segment of magnitude ρ forms with the x-axis). With the description of the vector, we have automatic knowledge of the point *P* (Figure 1.14).

Obviously, we can obtain the value of the magnitude ρ with the Pythagorean theorem (Pythagoras of Samos, Greek philosopher and mathematician, 569–475 B.C.):

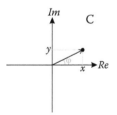

FIGURE 1.14
A representation of a complex number in the Cartesian plane.

$$\rho = |z| = \sqrt{x^2 + y^2}$$

We can obtain the *phase* φ by observing that $y = x * \tan(\varphi)$, from which

$$\varphi = \tan^{-1}\left(\frac{y}{x}\right)$$

Analyzing Figure 1.14 we can set

$$\begin{cases} x = \rho\cos(\varphi) \\ y = \rho\sin(\varphi) \end{cases}$$

whereby

$$z = x + jy = \rho\cos(\varphi) + j\rho\sin(\varphi)$$
$$\Rightarrow$$
$$z = \rho\big(\cos(\varphi) + j\sin(\varphi)\big)$$

which is named *trigonometric notation*. Lastly, considering Euler's formula (Leonard Euler, Swiss mathematician and physicist, 1707–1783), $e^{jx} = \cos(x) + j\sin(x)$, this becomes

$$z = \rho e^{j\varphi}$$

which is known is *polar notation*, where φ is expressed in radians.

1.5 PHASE VECTOR (OR PHASOR)

Complex numbers are really useful in the representation of sinusoidal waves. Therefore, we shall explore this relationship.

Given a generic co-sinusoidal function $f_c(t) = F_{Mc}\cos(\omega t + \varphi)$, it's possible to demonstrate that

$$f_c(t) = F_{Mc}\cos(\omega t + \varphi) = Re\left[F_{Mc}e^{j(\omega t + \varphi)}\right]$$

while considering a sinusoidal function $f_s(t) = F_{Ms}\sin(\omega t + \varphi)$

$$f_s(t) = F_{Ms}\sin(\omega t + \varphi) = Im\left[F_{Ms}e^{j(\omega t + \varphi)}\right]$$

Demonstration

By Euler's formula, $e^{jx} = \cos(x) + j\sin(x)$; making the real and the imaginary parts equal, you get

$$\begin{cases} \cos(x) = Re\left[e^{jx}\right] \\ \sin(x) = Im\left[e^{ix}\right] \end{cases}$$

In the first of the two equations we substitute

$$x = \omega t + \varphi$$

getting $\cos(\omega t + \varphi) = Re\left[e^{j(\omega t + \varphi)}\right]$, so multiplying both sides by F_M we have

$$F_M \cos(\omega t + \varphi) = F_M Re\left[e^{j(\omega t + \varphi)}\right]$$

F_M of the right-hand side can be inserted inside the real part.

The study of the vector $F_M e^{j(\omega t + \varphi)}$ provides the necessary information (through the real/imaginary part) about the trend over time on the periodic wave.

In conclusion a general periodic wave

$$v(t) = V_0 \cos(\omega t + \varphi)$$

can be considered the real part of a complex quantity such as

$$\vec{v}(t) = V_0 e^{j(\omega t + \varphi)} = V_0 e^{j\varphi} e^{j\omega t} = \overrightarrow{V_0} e^{j\omega t}$$

where $\overrightarrow{V_0} = V_0 e^{j\varphi}$ is a constant complex number, with magnitude V_0 and phase φ, while $e^{j\omega t}$ is a complex number with a unitary magnitude and a linearly variable phase with time. Therefore, $\vec{v}(t)$ is a vector with a constant magnitude V_0 that rotates in the complex plane with angular frequency ω and initial phase φ as shown in Figure 1.15.

Having said that, given the generic periodic co-sinusoidal function

$$f(t) = F_M \cos(\omega t + \varphi)$$

we can represent it by the complex number

$$\vec{F} = F_M e^{j\varphi}$$

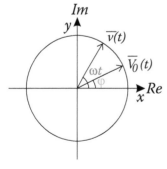

FIGURE 1.15 Vector $v(t)$ rotating in the complex plane.

known as a *phasor* (we will adopt it in Chapter 6 in Section 6.8).

It is important to say that knowing the phasor of the sine curve, we determine the amplitude, $F_M = |f(t)|$, and the phase, $\varphi = arg\left[f(t)\right]$, but *not* the frequency.

In summary, the phasor is a complex number equivalent to a sinusoidal function, but since its angular frequency is implicit, it can only be used in problems of known frequencies or iso-frequency (and it is demonstrable that this it is possible only for *linear* electrical circuits).

Considering a sinusoidal periodic function, we have seen how it is possible to extract its phasor.

The reverse is also possible. Given the complex number $\vec{F} = F_M e^{j\varphi}$ with a known angular frequency ω, it is possible to obtain the sine curve:

$$f(t) = Re\left[\vec{F}e^{j\omega t}\right] = Re\left[F_M e^{j(\omega t + \varphi)}\right] = Re\left[F_M \cos(\omega t + \varphi) + jF_M \sin(\omega t + \varphi)\right]$$

$$= F_M \cos(\omega t + \varphi)$$

The purpose of using phasors is to simplify the analysis of circuits endowed with sinusoidal inputs, assuming their *transient response* (i.e., different from equilibrium, or *steady state*), conditions is exhausted. Therefore, it is necessary that phasors satisfy the following requirements:

→ Addition of two sinusoidal functions should be equivalent to the addition of corresponding phasors.
→ The product of a sine curve with a real number has to be equivalent to the product of the corresponding phasor with the same real number.
→ A coherent rule has to exist to express the derivative of a phasor, in such a way that, by applying that rule with a phasor and the "inverse-transform," you will be able to re-obtain the derivative of the original sine curve: if $\vec{F}$ is the phasor of the function $f(t)$, then $j\omega\vec{F}$ is the phasor of the function

$$\frac{df(t)}{dt}$$

1.6 LAPLACE TRANSFORM

The *Laplace transform* (Pierre Simon, the Marquis of Laplace, French mathematician, physicist, and astronomer, 1749–1827, Figure 1.16) is a useful analytical tool to transform a function $x(t)$ with a real argument t (≥ 0) to a function $X(s)$ with a complex argument s. It is useful because the elaborate integral and differential equations in the time domain t can be reduced into more simple algebraic equations of the new complex variable's domain s.

The Laplace transform of a time-continuous function $x(t)$ is

$$X(s) = \int_0^\infty x(t)e^{-st}dt$$

The variable s that appears in the integral's exponent is complex, and therefore has real and imaginary parts:

$$s = \sigma + j\omega$$

Therefore, between the signal $x(t)$ and its associate transform $X(s)$ there is the "Laplace transform," L:

$$X(s) = L\left[x(t)\right]$$

In addition, it is possible to define the reverse operator L^{-1}:

$$x(t) = L^{-1}\big[X(s)\big]$$

becomes

$$x(t) = \frac{1}{2\pi j} \int_{\sigma-j\omega}^{\sigma+j\omega} X(s)e^{st}ds$$

From the characteristics of the Laplace transform, we can write for the property of differentiation

$$L\left[\frac{dx(t)}{dt}\right] = s\,X(s)$$

(when initial conditions are null) and, similarly, for the property of integration

$$L\left[\int x(t)\,dt\right] = \frac{1}{s}X(s)$$

These properties show us that derivatives and integrals in the domain of t become simple multiplications and divisions in the domain of s.

We will adopt the results of the Laplace transform in Chapter 7 where we will treat circuits in the frequency domain.

1.7 TAYLOR SERIES

The *Taylor series* (Brook Taylor, English mathematician, 1685–1731) allows one to approximate a function $x(t)$ about a point (t_0) in a polynomial form. The polynomial may be of any order (even infinite), but the greater the degree, the better the approximation results. Since the polynomials are simple functions to handle mathematically, the approximation of a function with a polynomial is an evident advantage.

However, note that the Taylor series cannot develop for just any function. In order to find such a series, some conditions have to be in place: the function $x(t)$ has to be infinitely differentiable (we can find the first derivative, second derivative, third derivative, and so on); and the function $x(t)$ has to be defined in a region near the value $t = t_0$.

Considering the most important developments of the Taylor series (those that usually have applications in electrical engineering and electronics problems), recall

$$\sin(x) = x - \frac{x^3}{3!} - \frac{x^5}{5!} - \frac{x^7}{7!} + \dots$$

$$\cos(x) = 1 - \frac{x^2}{2!} + \frac{x^4}{4!} - \frac{x^6}{6!} + \dots$$

$$e^x = 1 + x + \frac{x^2}{2!} + \frac{x^3}{3!} + \frac{x^4}{4!} + \dots$$

1.8 FOURIER SERIES AND INTEGRAL

In electronics, though not exclusively, the *Fourier series* (Jean Baptiste Joseph Fourier, French mathematician and physicist, 1768–1830, **Figure 1.17**) is of fundamental importance because it allows one to simplify the study of signals that have complex periodic waveforms. According to the principles of a Fourier series:

> Any periodic waveform $f(t)$, even if it is very complex, may be decomposed into the sum of a certain number (including infinity) of simple sine curves (with proper amplitude, frequency and phase displacement) and of a constant value equal to the average of the initial waveform.

Particularly, given a periodic waveform, in the sum of the sine curves with which it may be described, you can highlight

→ A pure Sine curve having the same frequency f (angular frequency ω) of the wave to decompose, known as a *harmonic*

→ A sine curve with double the frequency $2f$ (angular frequency 2ω) but a smaller amplitude than the fundamental harmonic ($F_2 < F_1$), named the *second harmonic*

→ A sine curve with triple the frequency $3f$ (angular frequency 3ω), but with an amplitude that is even smaller ($F_3 < F_2$), named the *third harmonic*

... and so on. Sine curves always have multiple frequencies with respect to their precedents and their amplitude diminishes more and more with the increasing of the frequency.

A possible constant component (namely with frequency zero) of value F_0 may be added to these sine curves, with value equal to the average of the signal to decompose. Therefore:

$$f(t) = F_0 + F_1 \sin(\omega t + \varphi_1) + F_2 \sin(2\omega t + \varphi_2)$$
$$+ F_3 \sin(3\omega t + \varphi_3) + \ldots$$

If we express the equation compactly, it can be written as

$$f(t) = F_0 + \sum_{n=1}^{\infty} F_n \sin(n\omega t + \varphi_n)$$

Figure 1.18 displays an example of the decomposition of the waveform of an odd function (namely symmetric with respect to the origin of the axis, such as $f(x) = -f(-x)$).

Another example of decomposition can be furnished for the function shown in Figure 1.12 (and later discussed in Chapter 9, Section 9.1.3), for which we can write

FIGURE 1.17 J. Fourier.

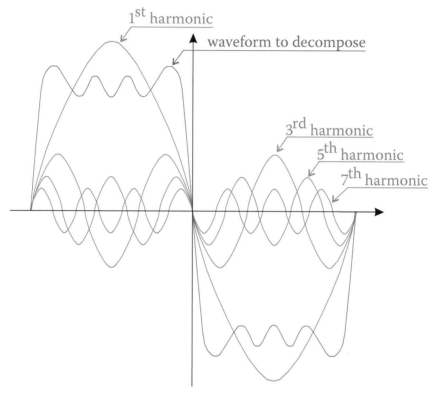

FIGURE 1.18 Decomposition of an odd periodic waveform. As a particular case, here the even harmonics of the series are null.

$$f(x) = \frac{2}{\pi} - \frac{4}{\pi}\left(\frac{\cos(2x)}{1.3} + \frac{\cos(4x)}{3.5} + \frac{\cos(6x)}{5.7} + \frac{\cos(8x)}{7.9} + \ldots\right)$$

which includes a constant component $(2/\pi)$ and cosine curves of multiple frequencies.

Since the amplitude decreases with an increase in frequency, with the same sequence of the harmonic, it generally happens that the decrease may become so significant that from a certain frequency value the terms of the development may become negligible.

The harmonics are represented by means of the *frequency spectrum*, which is a graphical representation indicating which harmonics are present and their amplitude (Figure 1.19).

In conclusion, thanks to the Fourier series, it is possible to write a periodic function as the sum of sinusoidal functions with proper frequencies, amplitudes, and phases. Therefore, it is possible to divide a periodic function (such as one that represents a type of sound into its fundamental components and recompose it by simply adding the individual components. Therefore, it is possible to reduce the study of a complex function to simpler ones.

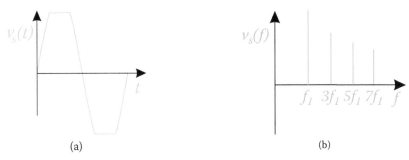

(a)   (b)

FIGURE 1.19   (a) A time-domain signal and (b) its frequency spectrum.

When the function to decompose is not periodic, the Fourier sum turns into an integral. Therefore, we will no longer discuss a series, but instead Fourier's integral, which can be considered a generalization of the series. The decomposition of a function with Fourier's integral is defined as the *Fourier transform*.

Observation

The signals that we handle are typically variable with continuity in time. Therefore, it appears appropriate to represent and mathematically analyze signals in the time domain. However, in practice, it is preferred to operate in the frequency domain, and it is possible to decompose the signal to its fundamental components (its amplitude, frequency, and phase), which are simpler to handle mathematically. Fourier provides us with the exact tool that we need to switch from the time domain to the frequency domain.

Curiosities

The Fourier series was introduced as a method to solve problems about the flow of heat through common materials.

As an application of the concept of Fourier series, audio samples can be composed as sums of waves of various frequencies.

1.9 MODULATION

The aforementioned Werner formulas (see Section 1.3.3) become particularly useful when it is necessary to translate a signal within frequencies. In fact, if we multiply a signal $m(t) = A_m \sin(2\pi f_m t)$ by a signal $c(t) = A_c \sin(2\pi f_c t)$, we obtain the resulting signal

$$r(t) = \frac{A_m A_c}{2}\left\{\cos\left[2\pi(f_c - f_m)t\right] - \cos\left[2\pi(f_c + f_m)t\right]\right\}$$

which has the characteristic of being "shifted" in frequency.

The original signal $m(t)$ is called *modulating*, the signal that multiplies $c(t)$ is called the *carrier*, and the obtained signal $r(t)$ is called *modulated*.

In electronics, the transmitted information is associated to the modulating signal. The carrier signal is the one that is modified from the modulating and is useful for shifting frequency. The modulated signal is the one actually broadcasted (by wire or by ether).

The whole process is called *modulation* and represents the technique of transmitting a signal (modulating) through another signal (carrier).

Curiosity

The tone-modulated Morse code (Samuel Finley Breese Morse, American inventor, 1791–1872) was invented for telegraphy and is now practically used only by a few amateur radio operators. It uses a binary (two-state) digital code that is similar to the widely used code nowadays adopted by modern computers.

1.10 KEY POINTS, JARGON, AND TERMS

→ The *International System of Units, SI*, is the modern metric system of measurement. It adopts *prefixes* to form multiples and submultiples of basic unity.

→ A *periodic waveform* repeats its exact same shape with space and/or time.

→ A *signal* is any wave to which it is possible to associate information.

→ *Band* is a synonym of *interval*.

→ A periodic wave can be basically described by its *amplitude* (or *maximum* or *peak*), *frequency*, *period*, and *phase*. Other important descriptive values are *average*, *root mean square*, *peak-to-peak*, *crest factor*, *average crest factor*, and *form factor*.

→ A *sine wave* is the simplest periodic wave to be described mathematically.

→ Fundamental trigonometric formulae for sinusoidal waveforms are the *Pythagorean identity*, the *Werner formulas*, the *duplication formulas*, and the *Prostaphaeresis formulas*. The *sine* of the sum of two angles is not equal to sum of their sines.

→ A complex number is any number z of the form $z = a + jb$, where a and b are real numbers and $j = \sqrt{-1}$ represents the unit imaginary number.

→ A *phasor* (or *phase vector*) is a vector, expressed with complex numbers, that represents a sinusoidally varying quantity.

→ The *Laplace transform* is a powerful tool that is very useful in electronics. It provides a method for representing and analyzing linear systems using algebraic methods. In fact, it converts integral and differential equations into algebraic ones.

→ A *Taylor series* is a useful tool to approximate a function, in a polynomial form, generally around a non-zero value.

→ The idea of a *Fourier series* is that you can write a periodic function as an infinite series of sine waves with proper frequencies, amplitudes, and phases.

When the function to decompose is not periodic, the sum turns into an *Fourier integral*.

→ Modulation is the process of adding information of a generally known *modulating* signal to another higher frequency signal named a *carrier*.

1.11 EXERCISES

EXERCISE 1
Find the average value of the function $v^2(\omega t) = V_M^2 \sin^2(\omega t)$ *here schematized in Figure 1.20.*

ANSWER

From the Werner formulas, we can write the useful equation $\sin^2(\omega t) = \dfrac{1 - \cos(2\omega t)}{2}$ to determine the average value:

$$\bar{f} = \frac{1}{\pi}\int_0^\pi V_M^2 \sin^2(\omega t)\, d(\omega t) = \frac{V_M^2}{\pi}\int_0^\pi \left[\frac{1 - \cos(2\omega t)}{2}\right] d(\omega t) = \frac{V_M^2}{\pi}\left[\omega t - \frac{\sin(2\omega t)}{2}\right]_0^\pi = \frac{V_M^2}{2}$$

This result is evident even with a visual inspection of Figure 1.20.

EXERCISE 2
Consider the function defined by

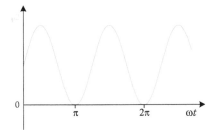

$$\begin{cases} v_g = V_M \sin(\omega t) \text{ for } 2n \le \omega t \le (2n+1)\pi \\ v_g = 0 \qquad\qquad\qquad elsewhere \end{cases}$$

(with n an integer), as in Figure 1.21.

Calculate

FIGURE 1.20 Sine-squared function.

- The average value (DC component) V_{AVG}
- The RMS value V_{RMS}

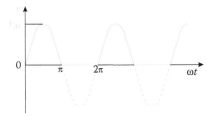

ANSWER

According to the definition of average value

FIGURE 1.21 Full-wave rectified sine function.

$$V_{AVG} = \frac{1}{\pi}\int_0^\pi v_g\, d(\omega t) = \frac{1}{\pi}\int_0^\pi V_M \sin(\omega t)\, d(\omega t) = \frac{V_M}{\pi}$$

and for the RMS value

$$V_{RMS} = \sqrt{\frac{1}{\pi} \int_0^\pi v_g^2 d(\omega t)} = \sqrt{\frac{1}{\pi} \int_0^\pi V_M^2 \sin^2(\omega t) d(\omega t)} = \frac{V_M}{2}$$

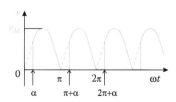

FIGURE 1.22 Stepwise sinusoidal function.

EXERCISE 3

Consider the periodic function schematized in Figure 1.22.

Calculate

· The average value (DC component) V_{AVG}

· The RMS value V_{RMS}

ANSWER

The average value is defined as

$$f_{AVG} = \overline{f} = \frac{1}{T} \int_{t_0}^{t_0+T} f(t) dt$$

which must be applied to the given function for which $T = \pi$ and which is not null only when $\alpha \le \omega t \le \pi$; as a consequence the limits of integration will be

$$V_{AVG} = \frac{1}{\pi} \int_\alpha^\pi v_M \sin(\omega t) d(\omega t) = \frac{v_M}{\pi} \left[-\cos(\omega t) \right]_\alpha^\pi = \frac{v_M}{\pi} (1 + \cos\alpha)$$

Observation

When $\alpha = 0$, $V_{AVG} = 2\frac{v_M}{\pi}$

Let's recall the definition of RMS value:

$$f_{RMS} = \sqrt{\frac{1}{T} \int_{t_0}^{t_0+T} f^2(t) dt}$$

Applying it to our problem, and recalling the *duplication formula* (Section 1.3.4)

$$\sin^2 x = \frac{1}{2}(1 - \cos 2x)$$

we have

$$V_{RMS} = \sqrt{\frac{1}{\pi} \int_\alpha^\pi v_M^2 \sin^2(\omega t) d(\omega t)} = \sqrt{\frac{v_M^2}{2\pi} \int_\alpha^\pi (1 - \cos 2\omega t) d(\omega t)}$$

$\Rightarrow$

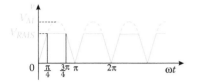

$$V_{RMS}^2 = \frac{v_M^2}{2\pi}\left[\omega t - \frac{1}{2}\sin 2\omega t\right]_\alpha^\pi = \frac{v_M^2}{2\pi}\left(\pi - \alpha + \frac{1}{2}\sin 2\alpha\right)$$

and finally

FIGURE 1.23 Sine wave limited for half of its period.

$$V_{eff} = v_M\sqrt{\frac{\pi - \alpha + \dfrac{1}{2}\sin 2\alpha}{2\pi}}$$

EXERCISE 4

Consider a sinusoidal function $v(t) = \left|V_M \sin(\omega t)\right|$, limited for half of its period at the value

$$V_{RMS} = \frac{V_M}{\sqrt{2}}$$

as in Figure 1.23.

Calculate

- The average value V_{AVG}
- The RMS value V_{RMS}
- The form factor F_F
- The crest factor F_C
- The average crest factor $\overline{F_C}$

ANSWER

Let us consider that the function can be written as a sine wave within the interval

$$0 - \frac{1}{4}\pi$$

as a constant of value

$$V_{RMS} = \frac{V_M}{\sqrt{2}}$$

within the limits

$$\frac{1}{4}\pi - \frac{3}{4}\pi$$

and again as a sine wave within the interval

$$\frac{3}{4}\pi - \pi$$

so the average value can be calculated by breaking the integral into three parts:

$$V_{AVG} = \frac{1}{\pi}\int_0^{\pi/4}\left|V_M \sin(\omega t)\right|d(\omega t) + \frac{1}{\pi}\int_{\pi/4}^{3/4\pi}\frac{V_M}{\sqrt{2}} + \frac{1}{\pi}\int_{3/4\pi}^{\pi}V_M \sin(\omega t)d(\omega t)$$

This equation can be simplified since the first and third addends furnish the same result, so it is convenient to just consider the first addend twice:

$$V_{AVG} = \frac{2}{\pi} \int_0^{\pi/4} V_M \sin(\omega t)\, d(\omega t) + \frac{1}{\pi} \int_{\pi/4}^{3/4\,\pi} \frac{V_M}{\sqrt{2}}$$

$$V_{AVG} = \frac{2}{\pi}\left[-\cos(\omega t) \right]_0^{\pi/4} + \frac{V_M}{\pi\sqrt{2}}\left(\frac{3}{4}\pi - \frac{1}{4}\pi \right) = \frac{2}{\pi}\left(-\frac{2}{\sqrt{2}}V_M + V_M \right) + \frac{V_M}{\pi\sqrt{2}}\left(\frac{3}{4}\pi - \frac{1}{4}\pi \right)$$

Finally:

$$V_{AVG} \cong 0.55 V_M$$

To calculate the RMS value, let's take into account the same considerations as before, obtaining

$$V_{RMS}^2 = \frac{2}{\pi} \int_0^{\pi/4} V_M^2 \sin^2(\omega t)\, d(\omega t) + \frac{1}{\pi} \int_{\pi/4}^{3/4\,\pi} \frac{V_M^2}{2} \cong 0,34 V_M^2$$

from which

$$V_{RMS} \cong 0.58 V_M$$

The form factor is defined as $F_F \underset{def}{=} \dfrac{V_{eff}}{V_{AVG}}$, so

$$F_F = \frac{0.58 V_M}{0.55 V_M} \cong 1.05$$

The crest factor is defined as

$$F_C \underset{def}{=} \frac{V_{MAX}}{V_{eff}}$$

so taking into account that the maximum value of the function is $V_{MAX} = V_M / \sqrt{2}$, we have

$$F_C = \frac{V_M / \sqrt{2}}{0.58 V_M} \cong 1.22$$

Finally, for the average crest factor $\overline{F_C} \underset{def}{=} \dfrac{V_{MAX}}{V_{eff}}$

$$\overline{F_C} = \frac{V_M / \sqrt{2}}{0.55 V_M} \cong 1.29$$

EXERCISE 5

A sawtooth signal $v(t)$ has the form

$$v(t) = 400t$$

for $0 < t < 10\,msec$, and then periodically replicates as in Figure 1.24.

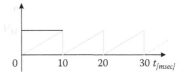

FIGURE 1.24   Sawtooth signal.

Calculate

- The RMS value V_{RMS}
- The average value V_{AVG}
- The form factor F_F

ANSWER

Let's consider the function at the instant $t = 10\,msec$. When it reaches its maximum value, the result is $v(t)\big|_{t=10msec} = 400 \times 10 * 10^{-3} = 4 = V_M$, so

$$V_{RMS}^2 = \frac{1}{10 * 10^{-3}} \int_0^{10 \cdot 10^{-3}} (400t)^2 \, dt = \frac{1}{10 * 10^{-3}} \left[\left(\frac{t^3}{3} \right) \right]_0^{10 \cdot 10^{-3}}$$

$$V_{RMS} \cong 2.3$$

As our sawtooth signal is linear, the average value is simply

$$V_{AVG} = 2$$

Finally for the form factor the result is

$$F_F = \frac{2.3}{2} = 1.15$$

EXERCISE 6

Given a periodic function

$$v = 200 \sin(\omega t) + 100 \sin(3\omega t) + 50 \sin(5\omega t)$$

Calculate its RMS value V_{RMS}.

ANSWER

The period of the function is $T = 2\pi$, so from the definition of RMS value we can write

$$V_{RMS} = \sqrt{\frac{1}{2\pi} \int_0^{2\pi} v^2 d(\omega t)} = \sqrt{\frac{1}{2\pi} \int_0^{2\pi} \left[200 \sin(\omega t) + 100 \sin(3\omega t) + 50 \sin(5\omega t) \right]^2 d(\omega t)}$$

$$= \sqrt{\frac{1}{2\pi} \left(\frac{200^2}{2} + \frac{100^2}{2} + \frac{50^2}{2} \right) 2\pi} \cong 162V$$

EXERCISE 7

Given the triangular wave in Figure 1.25, calculate its RMS value V_{RMS}.

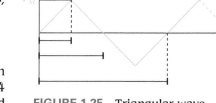

FIGURE 1.25 Triangular wave.

ANSWER

Because of the function symmetry, we can reduce the limits of integration to 0 and $T/4$ and then consider four times the obtained result:

$$V_{RMS} = \sqrt{\frac{4}{T} \int_{0}^{T/4} \left(\frac{4V_M t}{T} \right)^2 dt} = \sqrt{\frac{64V_M^2}{3T^3} \left(\frac{T}{4} \right)^3} = \frac{V_M}{\sqrt{3}}$$

2

Physical and Electrical Background

Electronics is defined as the scientific study of the behavior and design of electronic devices and circuits, based on some fundamental physical and electrical definitions.

2.1 FORCE, WORK, ENERGY, POWER

FIGURE 2.1
I. Newton.

The (mechanical) *force* $\vec{F}$ refers to any (physical) cause capable of modifying the condition of constant speed of a body (note that even a stationary body has a constant speed, equal to zero), or of deforming it.

The SI unit of force is the Newton (N) (Sir Isaac Newton, an English physicist, mathematician, astronomer, natural philosopher, alchemist, and theologian, 1642–1727, Figure 2.1).

A force $\vec{F}$ acting on a body of mass m causing its movement along a distance d, produces a *work W*:

$$W = \vec{F}d$$

The ability of a (physical) system to *work* on another (physical) system is known as *energy* (E), a scalar physical quantity.

FIGURE 2.2   J.P. Joule.

The SI unit of energy is the Joule (*J*) (James Prescott Joule, English physicist, 1818–1889, Figure 2.2).

It follows that for a mechanical work, with no other energy-transfer processes involved, the result is $\Delta E = W$. But the energy can be generally transferred from one system to the other in different forms, as *heat (Q), radiation, motion, sound,...* so the general form is

$$\Delta E = W + Q + E$$

where E represents other additional advected energy terms.

Curiosities

Kinetic energy is derived from the Greek term *kinema*, meaning "movement," from which the term *cinema* is also derived.

FIGURE 2.3
G.G. de Coriolis.

Coriolis (Gaspard-Gustave de Coriolis, French mathematician, mechanical engineer and scientist, 1792–1843, Figure 2.3) coined the term *work* to represent the transfer of energy by a force acting through a distance.

Leonardo's studies (Leonardo da Vinci, Italian polymath, 1452–1519, Figure 2.4) involved many fields such as aeronautics, anatomy, astronomy, botany, cartography, civil engineering, chemistry, geology, geometry, hydrodynamics, mathematics, mechanical engineering, optics, physics, pyrotechnics, and zoology. His inventions were practically perfect, but lacked one thing: *the energy*. If he had the sources of energy, the improvements in the quality of life for humans would have been absolutely exceptional.

FIGURE 2.4
L. da Vinci.

The SI unit of force, currently the *Newton*, was initially proposed to be named *vis*, the Latin word for force. This proposal was made by Italian physicists but was rejected by French colleagues since *vis* in their language indicates a certain masculine "feature." A similar situation for the French is the term *bit*, used in computer informatics. In fact, a vulgar French term to indicate that "feature" of men is *bite*. So in French texts we often find *chiffre binare* (*binary figures*) or *unité élementaire d'information* (*basic unit of information*) instead of *bits* (the binary figure), and *octet* rather than *bytes* (the character composed of eight bits).

Observations

Speaking of *energy*, it is extraordinary the implication that emerges from the Einstein equation (Albert Einstein, Nobel Prize-winning German physicist, 1879–1955):

$$E = mc^2$$

that is, energy and mass are equivalent! (*c*, the speed of light, is equal to ≅300,000 [*km*/sec]; nothing adds to the equation except that an enormous amount of energy can be drawn from a small mass—the bombs of Hiroshima and Nagasaki tell us something about the meaning of Einstein equation).

In a way, we can think about matter as "condensed" energy, while thinking of energy as "rarefied" matter.

Because of the match between energy and mass, the energy that an object has, as a result of its movement, is added to its mass. Step by step, this makes increases in speed more and more difficult, even if such difficulty is considerable only when approaching the speed of light. In fact, no object could ever be accelerated up to the speed of light, because in such a condition the mass will be infinity, and it will require an infinite amount of energy.

The letter *c* that indicates the *speed of light* comes from the word *celeritas*, which means *speed* in Latin.

The rate at which the energy *E* is transmitted or transformed or used is defined as the *power P*. So, an amount of work Δ*W* performed in an amount of time Δ*t* defines an *average power P*, as

$$P = \frac{\Delta W}{\Delta t}$$

while the instantaneous power P_i becomes

$$P_i = \lim_{\Delta t \to 0} \frac{\Delta W}{\Delta t} = \frac{dW}{dt}$$

For the sake of simplicity, assuming initial work and time equal to zero, the average power can be rewritten equal to

$$P = \frac{W}{t}$$

FIGURE 2.5
J. Watt.

The SI unit of power is the Watt (W) (James Watt, Scottish mathematician and engineer, 1736–1819, **Figure 2.5**).

The concept of power is primarily the *rate* (given the presence of the term *t*) with which it occurs in any transformation of energy. This implies that, for example, despite the enormous power in lightning strikes, energy released is not of great value, since the phenomenon is very short in time.

If the power of $1W$ in 1 *hour* does the work of $3600J$, this work is indicated by *Watt–hour* [*Wh*]. *KiloWatt-hour* (*kWh*) is the unit used to calculate what we pay for the electricity bill of a house.

FIGURE 2.6
A. Celsius.

2.2 HEAT AND TEMPERATURE

In 1824, Carnet (Nicolas-Léonard Sadi Carnet, French physicist, 1796–1832) showed how *heat Q* is a form of energy. In particular the heat "moves" between two bodies, or between two parts of the same body, which are at different thermal conditions. The heat is energy in transit: it always flows from points of higher temperature to those of lower temperature until it reaches thermal balance.

Temperature provides a way to measure the quantity of heat of a body and the three units of measurement that are commonly used to express it are the degree Celsius or Centigrade °*C*, the degree Fahrenheit °*F*, and the Kelvin *K*.

2.2.1 Degree Celsius or Centigrade (°C)

Celsius (Anders Celsius, Swedish physicist and astronomer, 1701–1744, **Figure 2.6**) arbitrarily established two values to reference temperature—corresponding to the temperature of melting ice (giving it a value of zero), and the temperature of boiling water at sea level (giving it a value of 100). Every grade of interval between zero and 100 is, therefore, a *degree Centigrade* or a *degree Celsius*.

2.2.2 Degree Fahrenheit (°F)

FIGURE 2.7   D.G. Fahrenheit.

Fahrenheit (Daniel Gabriel Fahrenheit, German physicist and engineer, 1686–1736, Figure 2.7) considered two values to reference temperature unlike those of the Celsius scale. In particular, he considered the lower reference of temperature being the temperature of the fusion of ice that occurs when it is in an ammonia solution (with ammonium chloride, NH_4Cl, a salt that is used to lower the melting point of ice on the road in winter to avoid the formation of ice), while the higher reference value of temperature was fixed as the average temperature of the healthy human body.

Divided into one hundred intervals between those two values, it appears that a degree Fahrenheit is equal to about 0.55 degrees Celsius. The temperature measured in Fahrenheit (°F) of boiling water is therefore approximately +212°F while zero Centigrade corresponds to +32°F. The range of 100 degrees Celsius is 180 degrees Fahrenheit, with the relationship shown here:

$$°F = °C + 1.8 + 32; \ °C = \frac{°F - 32}{1.8}$$

Curiosities

The CPU (*central processor unit*), which is the heart of a computer, has a normal operational temperature of around 104°F or 40°C. An increase of temperature to 144°F or 65°C can cause failure or even damage of the CPU.

Manufacturers of computer components and other electronic devices specify a maximum and a minimum operating temperature for their products.

2.2.3 Kelvin (K)

Kelvin (Lord William Thomson Kelvin, English physicist and mathematician, 1824–

FIGURE 2.8 W.T. Kelvin.

1907, Figure 2.8), considered the unit value of his temperature reference to correspond to the fraction 1/273.16 of the thermodynamic temperature of the triple point of water (points determined by the coexistence of the three states: ice, liquid, and vapor).

The value of that fraction comes from the fact that the volume of any gas (hydrogen, nitrogen, helium, oxygen, etc.) increases (decreases) exactly by 1/273.16 times its initial value when heated (cooled) from 0° to +1°C (from 0° to −1°C). Then, assuming an ideal gas, it could be cooled only to −273.16°C, the temperature at which its volume would be zero! This is why zero Kelvin is the absolute coldest temperature that is possible in the entire universe.

Between the degree Celsius and the Kelvin there is one conceptual difference. While the first is derived from the different indications between two specific passages of the state of water (liquid to solid: $0°C$ and liquid to gas: $100°C$), the second is associated with the absolute measure of the total energy of a system. Given that any system considered cannot have a negative value of energy (nor can a gas have a volume smaller than zero), it follows that the unit of measurement of Kelvin cannot assume negative values. The temperature of zero Kelvin is *absolute zero* and corresponds to no energy. A system with no energy is, however, only conceivable theoretically; therefore, zero Kelvin is only a value of abstraction that is not reachable in reality.

Curiosities

The work needed to remove heat from a gas increases as the temperature decreases, and an infinite amount of work would be needed to cool a matter to absolute zero. This means that "absolute zero" is a theoretical expression.

But, it is interesting to know how the concept of "zero" as a number is incredibly fairly recent. In fact it was born in India only in the ninth century AD.

So, it is no surprise that "zero" is still not named the same in many fields today.

Let's consider the "tennis score," for which the corresponding call for zero is "love." The reason seems to be that it is derived from the French expression "l'œuf" for egg, which looks like just the number zero.

But the term "love" can be traced to the seventeenth-century expression "play for love," meaning "to play without any wager," or "to play for nothing".

The failure of a team to score in a game, expecially for American ten-pin bowlers, is referred as a "goose egg," representing the numerical symbol "0."

In England, different sports traditionally use their own measure of "no score": "nil" in soccer, "nought" in cricket, "ow" in athletics timings. "Ow," or "double-O," is also adopted for a telephone number, and even James Bond's "007" serial number.

The expression for "zero" during the Second World War was "zilch," a slang referring to anyone whose name was not known, or the alliterative alternative "zip," currently widely adopted in computer science.

A very popular computer-animated film is titled *Finding Nemo*, a story of a fish. "Nemo" is a Latin word meaning "no one."

FIGURE 2.9
W. Rankine.

In the United States, the Rankine value is sometimes used (William John Macquorn Rankine, Scottish civil engineer, physicist and mathematician, 1820–1872, Figure 2.9); this is a thermodynamic absolute temperature scale. The symbol is $°R$ and the zero represents an absolute value, as with the Kelvin scale. $1[°R]$ is defined as equal to $1[°F]$ rather than the $1[°C]$.

We believe that in our universe there are some limits that cannot be passed. For instance, the maximum velocity is that of light, c = 299.792.458 [m/s], the lowest quantum of action is the Planck constant, $h = 6.62606957 * 10^{-34} [J * s]$, and the lowest temperature is absolute zero or zero Kelvin.

Conversely, it seems that a negative asbolute temperature has been reached. The paper that describes this strange phenomenon was published in *Science*, January 4, 2013, Vol. 339, no. 6115, pp. 52–55.

2.3 ELECTRIC CHARGE

It is a well known that matters, which have a mass, *attract* each other. We refer to their gravitational force, and the earth and moon are examples. Similarly, but at the atomic scale, there are bodies with a force between them, but that may be of *attraction* or of *repulsion*. This is named *electrostatic force* and exists in bodies with *electric charge*.

The *electric charge q* is an intrinsic physical property of some fundamental particles that form matter. Its SI unit is the Coulomb [C] (Charles Augustine Coulomb, French physicist, 1736–1806, Figure 2.10), and the value of a fundamental electrical charge is

$$q \cong 1.6 * 10^{-19} [C]$$

But, different from the gravitational force, electric charges can *attract* or *repulse* other charges by means of *electrostatic force*, so we classify them in two types, *positive* and *negative*, names that are purely conventional and suggested for the first time by Franklin (Benjamin Franklin, American politician, scientist, and inventor, 1706–1790).

FIGURE 2.10
C. Coulomb.

2.4 COULOMB FORCE

Coulomb's law states that, given two electrical charges, q_1 and q_2, the magnitude of the *electrostatic force* $\vec{F_C}$ between them is

$$|F_C| = k \frac{|q_1 q_2|}{r^2}$$

in which

$$k = 8.99 * 10^9 \left[\frac{Newton * m^2}{Coulomb} \right]$$

is a constant, and r is the distance that separates the charges. (Note the similarity with the expression of the gravitational force:

$$F_g = G \frac{m_1 m_2}{r^2}$$

in which G is the universal constant of gravitation, m_1 and m_2 the two masses, and r their distance).

The value of the *fundamental charge q*, one that has an electron or a proton, is $1.6 * 10^{-19}$ [C] and is conventionally positive, in the case of the proton, and negative, in the of case the electron.

For example, the force between an electron and a proton, distanced $r = 10^{-8}$ [m], is

$$F_C = 9 * 10^9 \frac{-1.6 * 10^{-19}}{\left(10^{-8}\right)^2} \cong -2.3 * 10^{-12} \, [\text{Newton}]$$

This force is negative, so the two charges will attract. If they were two charges of the same sign, the value of the force would have been positive and the charges would repel each other.

Observation

The electrostatic force is $4.17 * 10^{42}$ times stronger than the gravitational force, regardless of the distance between two charges.

2.5 ELECTRIC FIELD

Given an electric charge *q*, its noncontact electrostatic force of attraction/repulsion is generally represented with exiting/entering radial vectors (the direction conventionally depends on the type of charge, positive or negative) of the same charge, as represented in Figure 2.11a and b.

The electric field is stronger nearer the charge, and weakens as you move further away. For a positive charge, the arrows point outwards, and for a negative charge the arrows point inwards.

We can imagine the electric charge of any charged body producing a particular deformation of space around itself, as a physical entity. To each point of the space where such deformation occurs, it is associated with a vector of force. The ensemble of such vectors define the *electric field Ē* represented in Figure 2.11, mathematically equal to

$$\vec{E} = \frac{\vec{F_C}}{q}$$

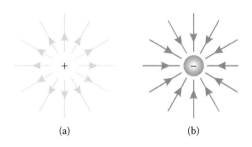

(a)   (b)

FIGURE 2.11 Vectors of the electric field: outgoing by (a) positive charges and ingoing by (b) negative charges.

Thus, the SI units of the electric field are Newtons per Coulomb [N/C]. If we have a "test charge" q at a point p in an electric field $\vec{E}$, it experiences a force $\vec{F_C} = q\vec{E}(p)$.

Two charged objects, with equal but opposite electric charges, that are separated by a distance, define an *electric dipole* (Figure 2.12).

As a consequence, an electric dipole is formed when the centers of positive and negative charges do not coincide.

The electric field between two isolated charges of equal sign is shown in Figure 2.13.

A very interesting case occurs when the electric field is between *sets of charges of opposite signs*, in a sort of "cylindrical configuration," as represented in Figure 2.14. Between them, a *uniform electric field* is formed, consisting of equally distanced parallel electric field lines.

In electronics, there are several possible applications of a uniform electric field, as we shall find out (Chapter 8, Figure 8.16).

Curiosities

It is sometimes useful to view the electric field as if it were an elastic cloth: a "source" charge ($+q$) deforms the cloth and a "test" charge ($-q$) follows the curvature of the sheet approaching the source (Figure 2.15a and b).

The concept of electric field was introduced by Faraday (Michael Faraday, English scientist, 1791–1867, Figure 2.16).

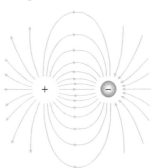

FIGURE 2.12   Electric field in an electric dipole.

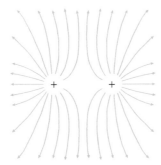

FIGURE 2.13   Electric field between two charges of equal value.

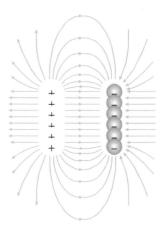

FIGURE 2.14   Electric field between two sets of charges of opposite value.

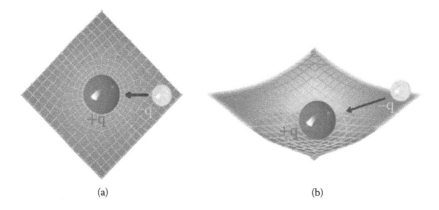

(a)   (b)

FIGURE 2.15   Spatial model of charge arrangement: (a) top and (b) lateral view.

2.6 ELECTRIC INDUCTION

Electrical charges within a body can be redistributed by the influence of other electrical charges existing in the surroundings within another body. The body subjected to redistribution, so to become *charged*, is called *induced*, the body where the conditioning charges are present, the *charging* one, is called the *inducer*, and the phenomenon is known as *electric induction*.

Figure 2.17 shows an already charged body (marked "*A*") near which we put a second one initially discharged (marked "*B*").

FIGURE 2.16   M. Faraday.

By induction on the second one, we have a redistribution of charges. One of the charges is turned toward the inductor body (negative in the figure), and the other is turned in the opposite direction. Putting the inductor body further away, the two opposite charges of the body "*B*" are remixed, neutralizing each other. On the other hand if we remove from body "*B*" the charge present on the side opposite

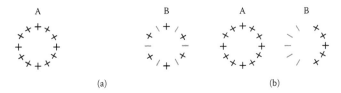

(a)   (b)

FIGURE 2.17 Bodies A and B are (a) initially distant and (b) near: opposite charges are attracted and on B negative charges are sorted in the direction of the positive ones of A.

the inductor (for example touching it with our finger), we still have the other charge on it, even after the inductor is taken away.

2.7 ELECTRIC CURRENT

Electric current is defined as the time rate at which electrical charges flow through a surface (e.g., the cross section of a wire). So, it is a measure of the amount of electrical charge Q transferred per unit time t, and its average value I is written as

$$I = \frac{\Delta Q}{\Delta t}$$

while the *instantaneous* current I_i is given by

$$I_i = \lim_{\Delta t \to 0} \frac{\Delta Q}{\Delta t} = \frac{dQ}{dt}$$

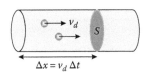

FIGURE 2.18 Mobile carriers crossing a surface.

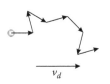

FIGURE 2.19 Flow of an electric charge in a material.

The charges that flow represent the *carriers* of the current.

With reference to Figure 2.18, n is the *volume density* of carriers and S the crossed surface in the x direction, and in volume Sx there will be nSx carriers. If q is the charge of each carrier, the charge in the volume is $\Delta Q = nqS\Delta x$.

But the flow of electrical charges is influenced by *scattering phenomena*, that is, the interactions that they suffer in the material (with other charges and/or with the lattice). Then, the flow deviates from a straight trajectory, as in Figure 2.19.

As a result, the carriers have a net velocity in one direction, known as (*axial*) *drift velocity* $\vec{v_d}$.

Since $\Delta x = \vec{v_d}\Delta t$, we can write $\Delta Q = nqS\vec{v_d}\Delta t$, from which

$$I = \frac{\Delta Q}{\Delta t} = nq\vec{v_d}S$$

The latter expression provides an operative equation. Since we can statistically evaluate the volume charge density n in a wire (it depends on the physical dimensions and the type of the wire), we know the drift velocity $\vec{v_d}$ (it depends

FIGURE 2.20 A.M. Ampere.

on the material of the wire), we can measure the cross section S, and the value of the fundamental charge is $q \cong 1.6*10^{-19}[C]$.

The SI unit of electrical current is the Ampere A (André-Marie Ampere, French physicist, mathematician, and philosopher, 1775–1836, Figure 2.20).

According to the conventional flow notation, the direction of *conventional current* is given by the direction in which positive charge moves. So, a flow of electrons is considered in the opposite direction.

Curiosities

Imagine two straight, parallel, and indefinitely long wires having negligible circular cross sections, placed in a vacuum, with 1 *meter* of distance between them. The value of $1A$ corresponds to the current that, flowing through them, determines $2*10^{-7}[N]$ force per meter of conductor.

When the first experiments were carried on electrical current, there was no idea which particles were the "carriers" of the current. So, conventionally, the current was defined to flow in the same direction as the positive charged carriers. Now we know that the electrons have negative charge, and they are the main carriers of the current.

Current density J is defined as a flow of electrical charges per unit area of cross section S. Therefore,

$$J = \frac{I}{S} = \frac{nq\overrightarrow{v_d}S}{S} = nq\overrightarrow{v_d}$$

In electrical wires, the current is due to a flow of negative charges (electrons), but there are systems in which the electrical current can also be due to positive charges. For example, in an electrolyte it is a flow of electrically charged atoms and/or molecules (ions), which can be both negative and positive.

Contrary to what one might think, the drift velocity $\overrightarrow{v_d}$ of electrons, flowing through a wire, is not very high. Let's consider a copper conductor with a cross section $S = 3 mm^2$ and a current $I = 10A$ flowing through it. We know that the volume density of carriers in it is $n=8.48*10^{28}$ [*electrons/m³*], so

$$v_d = \frac{I}{nqS} = \frac{10[A]}{\left(8.48*10^{28}\left[\frac{carriers}{m^3}\right]\right)\left(1.6*10^{-19}[C]\right)\left(3*10^{-6}[m^3]\right)} \cong 2.5*10^{-4}\left[\frac{m}{sec}\right]$$

Therefore, 1 *meter* is traveled in about 68 *minutes*! Common sense suggests that the current would have a much higher velocity than the calculated one.

So, why does your light bulb almost immediately light up the moment you turn on the electricity?

It is the electric field associated with the carriers, and not the single carrier, that propagates practically instantaneously.

Curiosity

Think of the carriers flowing in a wire as a river of warm putty. The electric field can "flow" fast because the wire is already filled with this putty. If you push on one end of a column of putty, the far end moves almost instantly. The field (and the energy related to it) "flows" fast, yet the velocity of each carrier is very slow.

Observation

The diameter of an electron is incredibly small and still is not completely known today. It is supposed to be less than one hundredth of a thousandth of a billionth of a millimeter, namely $10^{-17} m$. For an electron to travel 1 *meter* is roughly like one person covering a distance of 10^{17} m, which is hundreds of billions of billions of *km*!

Therefore, the velocity previoulsy calculated, even if it does appear to be low, is actually extraordinarily high for an electron.

The extremely low value of $q = 1.6 * 10^{-19} C$, implies that to obtain a current $I = nq\vec{v_d}S$ of the order of *Amperes* we must have an extremely high number of charges n.

We distinguish between *direct* and *alternating* current. In the case of *direct current*, named *DC*, the carriers flow in the same direction at all times. For *alternating current*, named *AC*, the direction of the current's flow is reversed, or alternated, on a regular basis. A mixed current can also occur, namely the sum of direct current and alternating current.

2.8 ELECTRIC VOLTAGE

The *voltage* (or *electric voltage* or *electric tension* or *electric potential*) V is a physical quantity closely linked to the energy of the electric field and to the work done to move the electrical charges in the electric field.

In particular, the *voltage difference* or *electrical potential difference* ΔV between two given points is equal to the amount of work ΔW that we must produce, against a static electric field, to move an unit-charge q between the same points:

$$\Delta V = \frac{\Delta W}{q}$$

So, voltage can be considered as the measure of specific potential energy per unit charge *between two points*, which is why it is also referred as voltage *drop*.

FIGURE 2.21 Birds don't get electrocuted when they sit on a power line because they do not experience a voltage "drop."

FIGURE 2.22 A. Volta.

In Figure 2.21, birds don't get electrocuted since they experience a certain voltage value but not its "drop."

The SI unit of voltage is the *Volt* $[V]$, thanks to Volta (Alessandro Giuseppe Antonio Anastasio Volta, Italian physicist and inventor, 1745–1827, **Figure 2.22**), corresponding to $[J/C]$.

Therefore, *voltage* is *electric potential energy per unit charge*, often simply referred to as *electric potential*, which must be distinguished from *electric potential energy* by underlining that the *potential* is a *per-unit-charge* quantity.

But the work ΔW corresponds to a force $\vec{F}$, which produces a displacement Δl ($\Delta W = \vec{F}\Delta l$), and the force is linked to the value of the electric field

$$\vec{E} \; (\vec{E} = \frac{\vec{F}}{q})$$

Then, a fundamental direct link follows between potential and electric field:

$$\Delta V = \vec{E}\Delta l$$

This equation, expressed in infinitesimal terms, becomes

$$dV = \vec{E}dl$$

If we generalize the three-dimensional case, we obtain

$$V = \int_0^L \vec{E}dl; \; E = -grad(V) = -\nabla V$$

with L the total distance covered.

FIGURE 2.23 Instrument to measure the voltage (1950s).

In electronics, the absolute value of the voltage does not matter, but the difference that it has at the head of a device or a circuit is important. For this reason, it is often indicated with the term *potential difference*, or *voltage drop*, or *voltage across* two points, ΔV. In Figure 2.23, an old-fashioned instrument useful to measure the voltage drop.

Given an electric field, every point of the field is characterized by a specific electric potential value. Therefore, it is convenient to use the quantity of *electric potential* to make quantitative comparisons between different electric fields or different points of the same field.

When the value of voltages between two points stays unchanged over a period of time, it is referred to as *direct voltage*. When the value of voltage between two points reverses periodically, namely the point with higher voltage assumes the value of the point with lower voltge and vice versa, we talk about *alternating voltage*. A further possibility is mixed voltage, namely the sum of direct voltage and alternating voltage.

Actually, even if DC refers to *direct current*, it is currently referred to as *direct voltage*, too. In the same way, AC is referred to as *alternating voltage*.
   The main (AC electrical) supply has a voltage with a frequency of $60\,Hz$ in the United States and $50\,Hz$ in Europe.

2.9 ELECTRON MOBILITY

The drift velocity $\overrightarrow{v_d}$ of an electron moving in a material depends on the physical characteristic of the material and on the value of the pulling electric field. To take into account both factors, we define the *electron mobility* or, more generally, the *charge* or *carrier mobility*

$$\mu \left[\frac{m^2}{V * s} \right]$$

as the average drift velocity of carriers per unit electric field strength

$$\mu = \frac{v_d}{E}$$

Practically, the carrier mobility is a measurement of the ease with which a particular type of charged particle moves through a material under the influence of an electric field.
   But, the previous equation states that one can make a carrier move as fast as we like just by increasing the electric field. Obviously this cannot be true. In fact, at more than the *saturated drift velocity* v_{sat} no charges can be stressed.
   So, the drift velocity can be empirically and more generally given as

$$v_d = \frac{\mu_0 E}{\sqrt[\beta]{1 + \left(\dfrac{\mu_0 E}{v_{sat}} \right)^{\beta}}}$$

(the parameter β depends on the material). Therefore $v_d|_{E \to 0} \cong \mu_0 E$ and $v_d|_{E \to \infty} \cong v_{sat}$.
   Finally, take into account that the mobility of a carrier in a given solid may vary with temperature.

2.10 ELECTRIC ENERGY AND POWER

The expression *electric energy* can be somewhat misleading, since it can refer to energy "stored" in an electric field, or to the (potential) energy of a charge in an electric field, or even the energy made available by the flow of electric charge in a wire.
   But, generally, *electric energy* is that associated with charges and transferred when the charges travel from place to place. So, and in accordance with the

aformentioned definition of electric voltage, the *electric energy* can be generally defined as potential energy that is assumed by electrical charges (equivalent to points, regions, or bodies) having a electric potential (voltage) different from that of a point of reference that we conventionally consider as potential zero. As a consequence, this is the energy made available by the flow of electric charge through a material.

In physics, *average power P* is the rate at which energy is transferred, used, or transformed. So, in terms of rate at which work is performed, it is written as

$$P = \frac{\Delta W}{\Delta t}$$

and *instantaneous power P_i* can be written as

$$P_i = \lim_{\Delta t \to 0} \frac{\Delta W}{\Delta t} = \frac{dW}{dt}$$

Accordingly, electric power is defined as the rate at which electrical energy is transferred (used or transformed) by an electric circuit, and its instantaneous value is related to the time needed for the potential V to transfer a charge q between two points:

$$P_i = \frac{dW_i}{dt} = \frac{d(qV_i)}{dt} = V_i \frac{dq}{dt} = V_i I_i$$

or, in the average occurrence

$$P = VI$$

Therefore electric power is a function of both voltage and current.

The SI unit of electric power is the Watt $[W]$ (James Watt, British engineer and inventor, 1736–1819).

When the voltage and the current alternate between positive and negative values, the electric power (as the product) is also variable over time. Thus, when we want a well-defined value, it makes sense to consider the *average* of the electric power. The portion of power flow, averaged over a complete cycle of the AC waveform, results in net transfer of energy in one direction, which is known as *real* (or *active*) power (the adjective meaning will be clarified later).

Assuming sinusoidal voltage and current waveforms, the average of their product over a period is

$$P = \frac{1}{2\pi} \int_0^{2\pi} V_M \sin(\omega t) I_M \sin(\omega t) \, dt$$

resulting in (it may be useful to apply Werner formulas, Chapter 1, Section 1.3.3)

$$P = \frac{V_M}{\sqrt{2}} \frac{I_M}{\sqrt{2}} = V_{RMS} I_{RMS}$$

which is equal to the product of the RMS value of voltage and current.

The relationship between *electric energy E* and *electric power P* in a given time *t* is

$$E = Pt$$

Thus, an equivalent definition of power can be the *rate of using energy.*

The SI unit of energy is Joule [*J*].

If the time is measured in hours, energy may be measurable in *watts-per-hour* (or *kilowatts-per-hour*, or *kWh*, to indicate its multiple), a unit of measurement preferred by electricity distribution companies.

The electricity meter provides a measurement of kilowatts-per-hour (*kWh*) consumed; therefore, the bill we pay is the energy (and not the power) used.

2.11 CONVENTIONAL NOTATIONS

To indicate an electric quantity (voltage, current, power, etc.), letter symbols should be used with consistency.

Uppercase letters (e.g. *E, P, V, I*, etc.) are used for DC (steady), average, RMS, maximum, minimum, and quantities with particularly slow variations that could be comparable to constants.

Lowercase letters (e.g., *v, i, p*, etc.) are used for instantaneous values that vary with time.

Distinguishing subscripts can be attached to uppercase or lowercase letters. For example, I_C indicates a constant current that flows in the material called "C." Maximum, minimum, and average are indicated by subscripts (e.g., $V_{MAX}, V_{min}, I_{MAX}, I_{min}$ ...).

Lowercase letters with uppercase subscripts indicate the instantaneous value of the total waveform, DC + AC, (e.g., i_C, P_R, etc.).

Lowercase letters with lowercase subscripts indicate the instantaneous value of the AC component only, (e.g., i_c, P_r, etc.).

A voltage, applied across two points, has two subscripts. For example, the notation v_{ab} indicates a variable potential applied across the points "a" and "b."

A value of an electric quantity can be mentioned as *nominal, rated*, or *real*. There is a bit of confusion in the utilization of these terms. An example may help: *Nominal* means electricity is 120V in North America and 230V in the European Union, but a variation within ±5% is allowed in the first and within ±10% in the second case. So the *real* value can be any delivered value in the range of 114V–126V or 207V–253V. Mains electricity is *rated* to deliver nominal voltage of 120V or 230V.

2.12 MAGNETIC FIELD

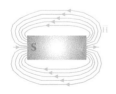

A *magnet* is generally considered to be a matter with the property, either natural or induced, of attracting iron or steel or nickel or cobalt, etc. Actually, in the region of space around a *magnet, magnetic forces* act on a theoretical unit magnetic pole in free space.

The ensamble of the magnetic forces form the *magnetic field H̄*. So, in addition to the existence of the aforementioned *electric fields* (see Section 2.5) in nature, there are also *magnetic fields* (Figure 2.24).

FIGURE 2.24 Magnet section and force lines of the magnetic field it produces.

The SI unit of magnetic field is *Ampere per meter [A/m]*.

In a magnet, we can identify two *poles* conventionally named *north*, from which come out the force lines, and *south* in which enter the force lines of the magnetic field. The designation of poles as north and south is arbitrary, just as electric charges are arbitrarily called positive and negative. The conventional names are due to the fact that if the magnet is left free to orientate itself in space, it always turns the end identified as north pole to the geographical North and the other to the geographical South.

Similarly to the *electric dipole* (Section 2.5) consisting of the combination of a positive and a negative electric charge, a bar magnet, consisting of a north and a south magnetic pole, is referred as a *magnetic dipole*.

It is important to note that, unlike electric fields, in magnetic fields the force lines are always closed. For this reason, there is no monopole; a magnet is always made of a *dipole* that has simultaneously (opposed) a north pole and a south pole.

2.12.1 Biot–Savart Law

It is fundamental to know that a magnetic field can be produced not only by a magnet, but also by moving electric charges (i.e., an electric current), or by an electric field variable in time, too.

Consider the magnetic field generated by electric currents. In a straight wire (considered, for simplicity, with infinite length) through which a current flows with intensity I, the magnetic field $\vec{H}$ in the surrounding space has force lines as shown in Figure 2.25 and its magnitude (intensity) in a point at distance d from the wire is equal to

$$H = \frac{1}{2\pi}\frac{I}{d}$$

From this, we deduce the reason why the SI unit of $\vec{H}$ is *Ampere/meter* ([A/m]).

This equation is known as the Biot–Savart law (Jean-Baptiste Biot, French physicist, mathematician and astronomer, 1774–1862; Félix Savart, French physicist, 1791–1841).

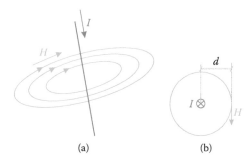

(a)   (b)

FIGURE 2.25 An electric wire through which a current I flows generates, in the surrounding area, a magnetic field H ideally placed on a plane perpendicular to the wire.

Therefore, the value of the magnetic field decreases inversely to the distance from the wire through which the current flows. The direction of the magnetic field is determined according to the *right-hand rule* (see Figure 2.26): if the thumb of the right hand points in the direction of the current then the fingers point in direction of $\bar{H}$.

The expression given for the Biot–Savart law derives from simplified hypothesis. Examination of more complicated current distributions invariably leads to lengthy, involved, and extremely unpleasant calculations. In any case, it is worth mentioning the occurrence of the magnetic field $\bar{H}$ created by electric current I in a circular loop as in Figure 2.27.

The magnetic field is more concentrated in the center of the loop than outside it.

Observation

In the Earth's core, there is a huge amount of metals, such as nickel and iron, which are ionized (i.e., charged) and in motion and therefore produce a meaningful magnetic field, which is fundamental for the existence of life. In fact, without this magnetic field, the atmosphere would be swept away by the wind of charged particles from the sun. This happened on Mars, where the magnetic shield is absent.

2.12.2 Magnetic Properties of Matter

We have seen that an electric current produces a magnetic field. But electric current is just due to the movement of charges, so electrons that rotate around a

FIGURE 2.26   Picture of the "right-hand rule."

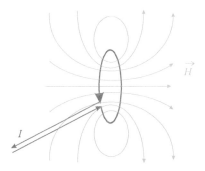

FIGURE 2.27   Magnetic field (green lines) produced by a loop current (in red).

FIGURE 2.28 Irregular arrangement of magnetic dipoles.

core produce their own magnetic field, clearly with very low intensity because of the extremely reduced number of electrons. The magnetic field produced by an atom defines what is called a *fundamental magnet*. But the motion of electrons of an atom is not aligned with the motion of an adjacent one, so the individual magnetic fields produced in the individual atoms have different and random directions, as shown in Figure 2.28.

Secondly, because of the incredibly high number of atoms inside a material, statistically, each single vector forming the magnetic field will annul each other. This is because for each magnetic field of an atom, there exists a field of another atom but in the opposite direction. The conclusion is that normally materials do not manifest their own magnetic field.

If we apply an external magnetic field to a test material, a phenomenon known as *Larmor precession* (Sir Joseph Larmor, Irish physicist, mathematician and politician, 1857–1942) can occur, namely the fundamental magnets are inclined to orient themselves according to the external magnetic field, such that the material may produce its own magnetic field.

Example

It is known that if a metal screwdriver (non-magnetic material) is rubbed with a magnet, then it will be able to attract screws. Therefore, clearly, it can be *magnetized*.

There are substances that, although submitted to an external magnetic field, are not magnetized anyway. After having defined the vector *magnetic induction*, we will understand the reason for this behavior.

2.13 MAGNETIC INDUCTION

The internal magnetic field within a material exposed to an $\vec{H}$ field is known as *magnetic induction*, $\vec{B}$ (sometimes also called the *magnetic field*, thus generating misunderstanding with $\vec{H}$):

$$\vec{B} = \mu\vec{H}$$

where μ, called *magnetic permeability*, can be written as

$$\mu = \mu_0\mu_r$$

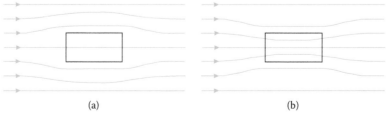

(a)   (b)

FIGURE 2.29 Different behaviors of various materials toward the magnetic field: (a) diamagnetic material; (b) paramagnetic material.

μ_0 being the *absolute magnetic permeability* of the empty space a universal constant equal to $4\pi^* 10^{-7} [Tm/A]$, and μ_r the *relative magnetic permeability*, its value depending on the material.

In SI units $\bar{B}$ is Tesla, $[T]$ (Nikola Tesla, Croatian-born Serbian-American inventor, 1856–1943).

The presence of a strong permeability means the material is made of something that can align strongly to an external magnetic field. To understand how the same vector $\bar{H}$ can produce different results in different materials, we consider two materials that are *diamagnetic* and *paramagnetic* (Figure 2.29).

Diamagnetic materials are such that the force lines of the magnetic field $\bar{H}$ tend to be rejected to the outside, and thus are sparse inside. On the contrary, for the *paramagnetic* materials the force lines of the field $\bar{H}$ tend to be concentrated in the material itself.

Examples of diamagnetic materials are the noble gases, nitrogen, hydrogen, graphite, gold, copper, silver, rock salt, and water (Table 2.1).

Paramagnetic materials include aluminum, magnesium, manganese, chrome, platinum, sodium, potassium, oxygen, and air (Table 2.2).

In addition to diamagnetic and paramagnetic materials, we also have the *ferromagnetic* ones. They have the same properties of paramagnetic materials but in addition, when at room temperature, they may be sources of magnetic fields themselves, if they were previously immersed in a magnetic field. Ferromagnetic substances lose their properties and become paramagnetic if they are subjected to a temperature equal or higher than the *Curie temperature* (Pierre Curie, French physicist, 1859–1906), which differs from one material to another. Iron, nickel, cobalt, and special alloys are ferromagnetic (Table 2.3).

In ferromagnetic materials, the propensity to "capture" the force lines of the magnetic field is particularly accentuated. This is used in order to create a *magnetic*

TABLE 2.1

Some Diamagnetic Substances and Their Relative Permeability

Substance	μ_r
Hydrogen	0.999998
Water	0.999991
Copper	0.999990
Silver	0.999980
Gold	0.999964

TABLE 2.2

Some Paramagnetic Substances and Their Relative Permeability

Substance	μ_r
Air	1.0000004
Oxygen	1.0000017
Aluminium	1.0000220
Platinum	1.00029
Manganese	1.0011

TABLE 2.3
Some Ferromagnetic Substances and Their Relative Min-Max Permeability

Substance	μ_{r-min}	μ_{r-max}
Iron	300	250,000
Nichel	300	2500
Iron-Nichel alloys	2500	250,000
Permalloy (77% Ni, 22% Fe)	8,000	100,000

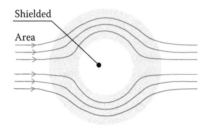

FIGURE 2.30 The scheme illustrates how it is possible to create a shielded area by exploiting the ferromagnetic material's properties.

shield that makes its internal space particularly insensitive to external magnetic fields (see Figure 2.30).

There are various applications of the magnetic shield. For example, in some instruments the shielding is used to avoid the earth's magnetic field or other false magnetic fields generated in the laboratory that would affect the measured values.

The previously defined parameter of magnetic permeability μ provides a "measure" of how much the material accentuates or attenuates the magnetic field $\vec{H}$ in the material itself. From that, it is easy to understand that

→ For diamagnetic material, $\mu_r < 1$.
→ For paramagnetic material, $\mu_r > 1$.
→ For ferromagnetic material, $\mu_r \gg 1$.

Diamagnetic materials, considering their low μ_r, are not able to be magnetized if subjected to an external magnetic field. Therefore, for these materials, we don't have the phenomenon of *Larmor precession* discussed in the previous section. Again, considering the value of μ_r, paramagnetic materials are inclined to be slightly magnetized, while ferromagnetic materials are those that are able to present an induction magnetic field with a relatively high value if they are subjected to an external magnetic field.

Curiosity

As we know, a lightning strike is an electrical discharge, namely a high value current in a very short time. If the lightning strikes a metallic material, a current will start up within it and consequently will start up magnetic force lines that enclose the material. But even if the generated current is very strong, the metal will not be magnetized since in order for it to be magnetized, the force lines of the field should have a direction parallel to the current, rather than perpendicular.

2.14 UNDULATORY PHENOMENA

Sounds reach us by means of waves, we perceive light as waves, and antennas emit special waves that radio and TV turn into sound and vision; we can affirm that it is through waves that we get almost all our information. Transmitted (or propagating) waves can be referred as undulatory phenomena.

2.14.1 Electromagnetic Waves

We have seen that electric fields and magnetic fields have a relationship (Section 2.12.1). When one of them exists, the other can exist too. Indeed, the charges have their own electric field all around them and, if moving, will generate a magnetic field. Therefore, in general it is easy to see how many physical phenomena are both electric and magnetic in nature; these phenomena are called *electromagnetic.*

Electromagnetic phenomena have periodicity and spread as waves, as a sphere in all directions into space. *Electromagnetic waves* are related to many different natural phenomena: radio and television signals, mobile telephone signals, X-rays, γ-rays, heat, and light are all electromagnetic waves; the only difference is in their wavelengths (Figure 1.5 in Chapter 1).

Electromagnetic waves, unlike the mechanical ones, do not need any medium to propagate within. For example, light waves travel to Earth from the sun through empty space.

Any propagation of electromagnetic waves that occurs through empty space will be at the speed of light $c \cong 300,000 \, km/sec$. If the propagation occurs in a medium, the wave speed will be, intuitively, lower.

Curiosity

It is possible that, if the mobile phone rings while we are listening to the radio, we will hear a "noise." This is because the current that activates the mobile phone produces emissions of electromagnetic waves that interfere with the radio signals.

2.14.1.1 Light

In summary, each particle with an electric charge has an electric field all around it, whose perturbation implies the generation of a magnetic field that when associated to the electric one defines the electromagnetic phenomenon. If this phenomenon has a particular wavelength, it is named *light.*

The visible light wavelengths are between $0.4 * 10^{-6} \, m$ and $0.8 * 10^{-6} \, m$ (400 nm–800 nm) and when it propagates in the atmosphere, it has a speed that decreases in a negligible way compared to its speed in empty space. It follows that its frequency range ($f = c/\lambda$, Chapter 1, Section 1.2) is approximately between $3.75 * 10^{14} \, Hz$ and $7.5 * 10^{14} \, Hz$, i.e., hundreds of *teraHertz* [*THz*]. This means that our eye is able to distinguish an oscillating phenomenon that is repeated hundreds of billions of billions of times per second!

Actually, electromagnetic phenomena are considered to have a double nature: *undulatory*, namely the propagation that happens through waves, and *corpuscular*, namely the propagation that happens through the motion of particles. This dualism (still not fully understood) is an essential basis for the modern *quantum theory of physics*.

Curiosities

Light speed is at its maximum in a vacuum, $c \cong 300,000 \, km/sec$, and decreases if it propagates in a medium. However, the speed reduction is negligible in the majority of cases. But in the laboratory, light has been exceptionally decelerated. It has been sent through a cooled gas of sodium with a temperature close to absolute zero and the speed measured was just half a meter per second!

Pratchett (Sir Terence David John Terry Pratchett, English writer, 1948–) observed of the speed of light:

> Light thinks it travels faster than anything but it is wrong.
> No matter how fast light travels, it finds the darkness has always
> got there first, and is waiting for it.

2.14.1.2 Electromagnetic Waves in a Medium

Electromagnetic waves that spread through a vacuum have the maximum possible speed in nature, namely the speed of light c. But if the propagation takes place in a nonconductive medium different from a vacuum, then the wave speed v depends on the parameters of *permittivity* (ε), and permeability (μ), linked to the medium itself, such that

$$v = \frac{1}{\sqrt{\varepsilon\mu}}$$

where $\varepsilon = \varepsilon_r\varepsilon_0$ and $\mu = \mu_r\mu_0$ are given from an absolute value equal to that of the vacuum (indicated with the subscript "$_0$") multiplied by a relative value due to the transmission medium (indicated with the subscript "r").

The meaning of the parameters ε and μ will be analyzed through the electric field (particularly in Section 4.6.1 of Chapter 4 about the *capacitor*) and magnetic field (particularly in Section 4.7.1 about the *inductor*).

With the appropriate substitutions, we can write

$$v = \frac{1}{\sqrt{\varepsilon\mu}} = \frac{1}{\sqrt{\varepsilon_0\varepsilon_r\mu_0\mu_r}} = \frac{1}{\sqrt{\varepsilon_0\mu_0}} \frac{1}{\sqrt{\varepsilon_r\mu_r}} = c \frac{1}{\sqrt{\varepsilon_r\mu_r}}$$

so we deduce that the speed in a medium other than a vacuum is always lower than the speed of light c since ε_r and μ_r are always equal to or greater than 1. The values of the parameters in a vacuum are, respectively,

$$\varepsilon_0 = \frac{1}{36\pi} 10^{-9} \left[\frac{F}{m} \right]; \; \mu_0 = 4\pi * 10^{-7} \left[\frac{T*m}{A} = \frac{H}{m} \right]$$

Example

What is the speed v of an electromagnetic wave in *teflon*? For *teflon*, we have $\varepsilon_r = 2.1$ and $\mu_r = 1$ (*permeability* is equal to that in a vacuum, characteristic of all non-ferrous and non-magnetic material). Then $v = 0.69c$, in other words the speed of an electromagnetic wave that spreads in the *teflon* is reduced approximately 30% compared to the speed in empty space.

2.14.2 Mechanical Waves

Mechanical waves are very different from the electromagnetic waves that don't need a medium in which to propagate. *Mechanical waves* are vibration of a physical medium, thus they need a medium to propagate. They manifest as a local oscillation of material that does not move far from its initial equilibrium position. So mechanical waves transport energy but not material. Think, for example, of a rope that vibrates: its movement also moves part of the air that surrounds the rope. The consequence is the local variation of air density and pressure. Usually these variations are very small compared to the equilibrium values, but it is precisely these variations that determine the beginning, in the air, of mechanical waves.

Figure 2.31 shows the rarefaction and compression of air molecules that is produced by the propagation of a mechanical wave. The term $P_a(t)$ indicates the value of the atmospheric pressure; where the pressure increases we have compression, but where the pressure decreases we have rarefaction of the air.

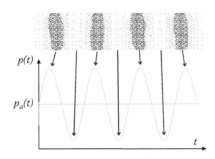

FIGURE 2.31 Air compression and rarefaction that define a mechanical wave.

Mechanical waves are often called *audio waves*, but this expression is inappropriate since audio waves are only a subset of mechanical waves. Audio waves are those within the frequency range audible to humans. Mechanical waves that have a frequency lower than the audible range are named *infrasonic* and those that are a higher frequency are named *ultrasonic*.

The propagation speed of a mechanical wave varies according to the material in which it propagates. For example, in air it has a speed of about $340\,m/s$ = $(1224\,km/h)$, in water about $1480\,m/s$, in soil about $3000\,m/s$, in steel around $5050\,m/s$, and so on.

Two waves that have same frequency but spread in different media will have different wavelengths, according to the equation $\lambda = v/f$. Therefore, for example, a wave with frequency $f = 100\,Hz$ in the air has wavelength $\lambda = 340/100 = 3.4\,m$ but in the water, the wavelength would clearly be larger since $\lambda = 1480/100 = 14.8\,m$.

Relatively, for the audio spectrum, the wave with frequency $f = 440\,Hz$ is the one we know in the musical field as the note "A." Assuming that in the air the audio wave propagation speed is about $v = 340\,m/s$, oscillating with frequency $f = 440\,Hz$, the wavelength of the "A" is

$$\lambda = \frac{v}{f} = \frac{340\,\dfrac{m}{s}}{440\,Hz} \cong 0.77\,m$$

it is not a coincidence that this is exactly the length of a guitar-string "pizzicato" to ring up the "A."

With a frequency double that of $440\,Hz$ ($2*440\,Hz = 880\,Hz$) we have the "A" in one higher octave, with a quadruple frequency ($2*2*440\,Hz = 1760\,Hz$) we have the "A" two octaves higher, etc. It is the same for the note "A" one octave lower ($440/2\,Hz = 220\,Hz$) or in two octaves lower ($440/4\,Hz = 110\,Hz$), etc. The octave that "A" is situated in is named the *basic octave* (see Figure 2.32).

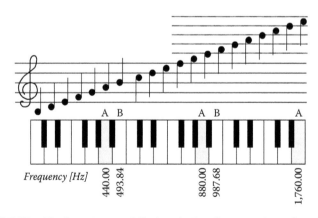

FIGURE 2.32 Music notes and their relative frequencies of oscillation.

Observations

It is possible to locate the source of a sound by noting the intensity (or amplitude or magnitude) difference of the sound heard, and/or by the phase difference (delay time) due to the differences in length of the route between the sound source and each one of our ears.

However, the intensity difference is perceptible only if the sound has a short wavelength. Otherwise, the sound may diffract and "bend" around the head. The result is having equal intensity in the two ears.

Also the phase difference is hardly distinguishable with intermediate audio frequencies, namely around $4000\,Hz$, which corresponds, for example, to the sound produced by a cricket (it is for this reason that it is difficult to locate crickets in a field).

Helmholtz (Herman von Helmholtz, German physiologist and physicist, 1821–1894) hypothesized that the inner ear (the cochlea) analyzes a sound depending on its frequency, and that different neurons are for different frequencies. Békésy (Georg von Békésy, Hungarian neurophysiologist, 1961 Nobel Prize winner in medicine, 1899–1972) confirmed the hypothesis by means of experimental works, and demonstrated the fundamental mechanism.

Curiosities

The jargon of musicians is certainly different from that of electricians or technicians in general, but we are referring to the same concepts even though they have different terms. For example,

- The *height* of a sound for musicians defines if a sound is grave or acute; a technician refers to this concept by using the term *frequency*.
- The *intensity* for musicians is the one that technicians call *amplitude* or *magnitude*.
- The *timbre* represents the audio "quality," or what makes a particular musical sound different from another. Each musical instrument has a characteristic timbre. A technician calls this the *waveform*.
- What musicians denominate as *intonation* of a sound is the frequency of the *fundamental harmonic* of the Fourier series for technicians.

The human ear is sensible to sounds in the range $20\,Hz$–$20\,kHz$. As well as for the aforementioned string of the guitar, this range depends on the dimension of the ear itself. It is not a coincidence that adults speak to little children in a sort of *falsetto*, which is approximately one octave just above the *modal voice*.

2.15 KEY POINTS, JARGON, AND TERMS

→ *Force* is capable of modifying the condition of constant speed of a body, or of deforming it.

→ A force does *work* if movement of a body is produced.

→ The ability of a system to to work on another system is known as *energy*.

→ *Power* is the rate of transfer of energy or the rate at which work is done.

→ *Heat* is a form of energy that is transferred from one system to another system by thermal interaction. *Temperature* provides a way to measure the quantity of heat.

→ The *electric charge* is a physical basic property of a unit of matter that causes it to experience a force when near other electrically charged matter.

→ Coulomb's law states that any two point charges exert a force on each other that is proportional to the product of their charges and inversely proportional to the square of the distance between them.

→ The *electric field* strength at a point is the force per unit charge exerted on a positive charge placed at that point.

→ An *electric dipole* is made of two charged objects, with equal but opposite electric charges, that are separated by a distance.

→ The drift velocity of electrons in a wire is proportional to the current that flows in the wire.

→ The electrical current is defined as the time rate at which electrical charges flow through a surface. To move electric charges, an electric force is necessary. The electric force acting on a charge causing its movement produces an electric work. The electric voltage is the amount of electric work necessary to move an charge between two points against a static electric field.

→ *Carriers* are the charges forming an electric current.

→ A DC current is non-varying electric current. The direction of the flow of positive and negative charges is one way, and does not change with time.

→ An AC current reverses its own direction periodically a specified number of times per second. The average value in a period is equal to zero. Even if the term AC is used to describe a current, it has become common to use the term to identify a source of alternating voltage.

→ A DC voltage is a voltage that does not change polarity across two points with time.

→ An AC voltage is a voltage continually changing between positive and negative values.

→ The generation or use of electric power $[W]$ over a period of time $[h]$ is often expressed in kilowatt-hours $[kWh]$.

→ A *magnet* is a body having the property of producing a magnetic field external to itself.

→ The *magnetic field* is the region around a magnet where magnetic forces act.

→ *Magnetic induction* represents how the magnetic field changes within a material.

→ The *permeability* of a matter is the measure of degree to which the magnetic lines of force can penetrate into the matter itself. Numerically, it is defined as the ratio of magnetic induction B to the magnetic field H.

→ According to *magnetism*, matter can be divided into *diamagnetic, paramagnetic*, and *ferromagnetic* types.

→ The absolute magnetic permeability μ_0 is a universal constant.

→ We get information through *waves*. *Electromagnetic* and *mechanical* waves are fundamental.

→ In the following table are the described quantities and their SI units:

Quantity [symbol]	Unit [symbol]
Force $\left[\vec{F}\right]$	Newton [N]
Energy [E]	Joule [J]
Work [W]	Joule [J]
Power [P]	Watt [W]
Temperature [T]	Degree Celsius or Centigrade [°C] Degree Fahrenheit [°F] Kelvin [K]
Charge [q]	Coulomb [C]
Electric field $\left[\vec{E}\right]$	Newtons per Coulomb $\left[\dfrac{N}{C}\right]$
Current [I]	Ampere [A]
Voltage [V]	Volt [V]

2.16 EXERCISES

EXERCISE 1

Two objects have the same diameter $d = 1\,mm$, and are separated by a distance of $s = 2\,mm$. The first object has an excess of $5*10^8$ electrons on it, the second object an excess of $4*10^{10}$ electrons. What is the electrostatic force that they exert on each other?

ANSWER

$$q_1 = 5*10^8 * \left(1.6*10^{-19}\,[C]\right) = 8*10^{-11}$$

$$q_2 = 4*10^{10} * \left(1.6*10^{-19}\,[C]\right) = 6.4*10^{-9}$$

$$|F_C| = k\frac{|q_1 q_2|}{r^2} = 8.99*10^9 \left[\frac{N*m^2}{C}\right] \frac{\|8*10^{-11}\,[C] * 6.4*10^{-9}\,[C]\|}{0.001\,[m]^2} \cong 0.46\,[mN]$$

EXERCISE 2

In a copper wire with a section of $S = 1.5 [mm^2]$ flows a current of $I = 10 [A]$. If the drift velocity of the electrons is $v_d = 5*10^{-4} [m/s]$, how many electrons are involved as carriers?

ANSWER

$$n = \frac{I}{q\overline{v_d}S} = \frac{10[A]}{1.6*10^{-19}[C]*5*10^{-4}\left[\frac{m}{s}\right]*1.5*10^{-6}[m^2]} = 8.33*10^{28} \left[\frac{electrons}{m^3}\right]$$

EXERCISE 3

A 15 Watt lightbulb is turned on for two hours. Determine how many joules of electric energy have been dissipated.

ANSWER

$$E = Pt = 15[W]*7200[s] = 108000 [J]$$

Please note that because the joule is such a small unit, quantities of energy are often given in *kJ*.

EXERCISE 4

Let's generate electricity with a stationary-bike power generator, which typically produces 100W in 1 hour. If you pedal for 2 hours a day, for 30 consecutive days, and the cost of electricity is 10 [cents/kWh], how much money do you save in that period?

ANSWER

The total produced energy is

$$E = 100[W]*(2*30)[h] = 6\,kWh$$

which corresponds to

$$Money = 6[kWh]*10\left[\frac{cents}{kWh}\right] = 60[cents] = 0.6[\$]$$

It does not appear to be so convenient.

EXERCISE 5

If your electric company charges 10 [cents/kWh] for electricity, how much would you pay to operate a 60[W] bulb for 24 hours?

ANSWER

$$Money = 10 \left[\frac{cents}{kWh} \right] {}^*0.06\,[kW]\,{}^*24\,[h] = 0.144\,[\$]$$

EXERCISE 6

Find the magnetic induction field $\vec{B}$ at the center of a circular current loop of r = 2 cm radius, carrying a current of 3A.

ANSWER

$$B = \mu_0 \frac{I}{2r} = 4\pi^*10^{-7} \left[\frac{T^*m}{A} \right] \frac{3\,[A]}{2^*0.02\,[m]} \cong 0.94\,\,[T]$$

3

Nature of Matter

Electrical and electronic engineering essentially originate from exploiting the motion of electrical charges. These constitute fundamental parts of matter, which makes it essential to know its basic elements, namely the atoms. In the past, many theories regarding the structure of an atom has been hypothesized; let's consider here the most relevant.

3.1 ATOMIC MODEL

We can start with the model of Democritus (Democritus, Greek philosopher, 460–370 BCE) who hypothesized that all matter was discontinuous, so in trying to divide it iteratively, there was a limit beyond which the matter itself could not be decomposed anymore. Therefore, he thought a tiny, fundamental basic unit existed and called it the *atom* (from the greek *atomos*: not divisible).

The theoretic-philosophical approach of Democritus was proved experimentally when Dalton (John Dalton, English chemist and physicist, 1766–1844, Figure 3.1) observed that in compounds, that is substances formed by the union of basic elements, the proportion between the masses of the individual elements is always constant. This result is possible only if the hypothesis of the existence of the atoms was true.

Therefore, matter shall be constituted by a basic component with a well-defined mass. Dalton hypothesized the atom as a unique sphere.

Afterwards, Thomson (Sir Joseph John Thomson, English scientist, 1856–1940, Figure 3.2), thanks to experiments with *cathode rays* (luminescent rays coming from the negative terminal, named the *cathode*, of a tube filled with gas to

FIGURE 3.1   J. Dalton.

FIGURE 3.2   J.J. Thomson.

FIGURE 3.3   E. Rutherford.

which was applied a difference of voltage) discovered that the atom had to be made of moving negatively charged particles, *electrons*, having a mass 2000 times smaller than the lightest known atom (hydrogen). But, being neutral matter, those negative charges had to be balanced by positive charges.

The atomic model proposed was again spherical, but not indivisible anymore. Rather it had its internal negative particles immersed in a "dust" of thin positively charged matter.

The successor of Thomson at the Cavendish laboratories of Cambridge University was Rutherford (Ernest Rutherford, New Zealand physicist, 1871–1937, Figure 3.3), who used the *alpha particles* (rays emitted by radioactive materials) as "bullets" to penetrate atoms of a thin sheet of gold.

If the model of Thomson was correct, as alpha particles are 7400 times heavier than electrons, they would have crossed the thin sheet of gold without any problems. However, certain particles were unexpectedly diverted. The correction to the Thomson model led to the new hypothesis of the Rutherford model: the atom is still a spherical structure, but with a relatively small core, the *nucleus*, containing all the positive charges and most of the mass (and therefore also constituted by particles without charge, *neutrons*), while the electrons are distributed in the rest of the atom and orbiting around the nucleus like planets around the sun (Figure 3.4).

The solar system model by Rutherford was also replaced, as it did not justify subsequent observations regarding the light that was emitted from atoms excited from electric discharge.

Bohr (Niels Henrik David Bohr, Danish physicist and mathematician, 1885–1962, Figure 3.5), a student of Rutherford, formulated the assumption that the electrons in atoms can move only along certain circular orbits, allowed each to correspond to a very specific energy level of electrons. The orbits are said to be *discrete*, that is, having a finite or countable, not continuous, set of values.

Accordingly, each orbit corresponds to a specific value of energy, so the energy is *discrete* as well.

The electrons can "transit" from one energy level to another (from one orbit to another), by the release or absorption of what is called *photon* of energy, a discrete value, equal to the "energy drop," that is, the difference between the initial and the final energy. A transition from a higher energy level to a lower level,

FIGURE 3.4 A schematization of the
atomic model proposed by Rutherford.

therefore, corresponds to the loss of energy with the emission of a photon, vice
versa with the absorption of a photon.

Observation

According to the theory of Einstein, light has the dual nature of both an electromagnetic wave and a cluster of "packages" of pure energy without a mass
associated with it. Each of these individual "packages" is called a *photon*, from
the Greek *photos*, meaning light.

A photon that "strikes" an electron involves the disappearance of the photon
and subsequent increased kinetic energy of the electrons so that, if the energy
it has acquired is sufficient, it can make a transition from one energy level to
the other.

In the final analysis, energy is transmitted from one point to another "carried"
by the photons.

3.1.1 Shells, Subshells, Orbitals

Today, the accepted atomic model is called *quantum-
mechanical*, according to which the orbits are replaced
by *electronic shells*. The shells (or *levels*) consist of *subshells* (or *sublevels*, even just one), and the subshells
consist of *orbitals* (even only one), where electrons exist
(each orbital can hold up to two electrons).

In matter, electrons do not "orbit" with the same meaning given to how the planets orbit around the Sun, but,
even having defined discrete energy levels, do not have
precise orbits around the nucleus, meaning that it is not
possible to simultaneously determine the position and

FIGURE 3.5 N. Bohr.

the speed of the electron. The electrons can only be described by regions within which there is a high probability of finding them.

Moreover, while the solar system has the form of a "plate," given that the planets have orbital trajectory on a plane, the shells expand in a volume.

 In nature, there are four fundamental forces:

- *Strong Nuclear Force*: This is the strongest of the four, and it acts in small distances, around 10^{-15} m and below.
- *Weak Nuclear Force*: This is 10^{-8} times weaker than the strong nuclear force, is responsible for radioactivity, and acts in distances around 10^{-17} m and below.
- *Electromagnetic Force*: This is 10^{-3} times weaker than the strong nuclear force, and acts between particles having electric charge at any distance.
- *Gravitational Force*: This is the weakest of the four, equal to 10^{-45} times that of the strong nuclear (so 10^{-42} times that of electromagnetic!), and acts for any distance.

The gravitational force is the only force that seems always to act exclusively in the form of attraction. The other forces may occur as attraction or as repulsion. That it is extremely weaker than the others is absolutely negligible, as we are dealing with elementary particles and atoms. But the fact that it acts at any distance and that it is only attractive implies that all its effects add to each other (for the other forces for which both attraction and repulsion are possible, the effects can diminish until all is nullified!). So when the number of particles is extremely large, such as those that are in a planet or in a star, the gravitational forces may dominate all the others. This is the reason why gravity determines the evolution of the universe.

Although the positive charge of the protons ensures there is repulsion between them, they remain together in the atomic nucleus. This is because their "glue" is the strong nuclear force, which is stronger than electromagnetic forces.

The *shells* are described by the *principal quantum number n*, a positive integer, and the maximum number of electrons that each shell can host is $2n^2$. So the first shell ($n = 1$) may "host" a maximum of two electrons, the second shell ($n = 2$) can host a maximum of eight, the third ($n = 3$) 18. The shells are sometimes referred to with letters instead of numbers. In particular K denotes the first shell ($n = 1$), L the second shell ($n = 2$), M the third shell ($n = 3$), N the fourth shell ($n = 4$), O the fifth shell ($n = 5$), P the sixth shell ($N = 6$), Q the seventh and final shell ($n = 7$).

The first shell ($n = 1$) is the lowest in energy $E_{n=1}$, the second shell ($n = 2$) with $E_{n=2}$ next, and so on: $E_{n=1} < E_{n=2} < E_{n=3} < E_{n=4} < E_{n=5} < E_{n=6} < E_{n=7}$ (Table 3.1).

Curiosity

It is visually useful to set up an analogy (with obvious limits) of shells with the seats in an amphitheatre: exactly in the same way a person chooses the seat on which to sit, an electron "chooses" a shell on which to stay; the seats in an amphitheatre have fewer seats in the rows that are closer to the center (the stage) and more seats in the rows most external. Similarly the more internal shells can "host" fewer electrons in comparison to more external shells; just as a person always tries to sit in the most internal row (right next to the stage), an electron prefers the most internal possible shell (next to the nucleus, if that orbital is not already fully booked). As in an amphitheatre, there are *no* seats between one row and another, so there is *no* possibility of sitting between one row and another, just like electrons may *not* find a place between one shell and another (Figure 3.6).

FIGURE 3.6   Aspendos Greek/Roman Amphitheatre in Antalya, Turkey.

Within each shell, there are *subshells* that are themselves composed of *orbitals*. We think of the subshells as divisions of a shell, like how a road is divided into lanes. In reality, things are more complicated than that, given that the sub-shells have forms that can be very different. The number of subshells is described by the quantum number l, the angular momentum, a positive integer that also includes zero. The number of subshell of a shell is equal to the value of the principal quantum number n. So the first shell ($n = 1$) has 1 subshell ($l = 0$), the second shell ($n = 2$) has 2 subshells ($l = 0$, $l = 1$), the third shell ($n = 3$) has 3 subshells ($l = 0$, $l = 1$, $l = 2$), and the fourth shell ($n = 4$) has 4 subshells ($l = 0$, $l = 1$, $l = 2$, $l = 3$).

TABLE 3.1

The Energy Level in Atoms Increases with the Principal Quantum Number n

Energy	Shell	Subshell
↑	$n = 4$	$4s, 4p, 4d, 4f$
	$n = 3$	$3s, 3p, 3d$
	$n = 2$	$2s, 2p$
	$n = 1$	$1s$

An old but still-used notation identifies the subshells with letters: s (from *shape*) identifies the subshell $l = 0$, p (from *principal*) identifies $l = 1$, d (from *diffuse*) identifies $l = 2$, and f (from *fundamental*) identifies $l = 3$. The notation of the subshell letters s, p, d, f is often used to describe the electronic configuration of the most external orbital of an atom, that is, the *valence orbital*.

All electrons in a subshell have the same energy that increases with l: $E_{l=0,s} < E_{l=1,p} < E_{l=2,d} < E_{l=3,f}$.

The *orbitals* can hold a total of two electrons each, and can be considered the electron paths inside a subshell. In detail, the orbital is a mathematical function that describes the physical region where electrons can be found around the nucleus.

3.1.2 Spectroscopic Notation

To know how many electrons an atom has and which shells and subshells are filled, you use *spectroscopic notation*. It consists of a number, which represents the shell, a letter, which represents the subshell, and a superscript, which represents the total number of electrons in the subshell.

Therefore, for example, the $1s^1$ notation indicates the atom of hydrogen, with its single electron placed in subshell s of the first shell.

The notation $1s^2$ states that electrons occupy the $n = 1$ energy level, electrons are in the s ($l = 0$) subshell, and two electrons are in the $1s$ subshell.

For lithium, which has three electrons, the notation is $1s^2 2s^1$, so the first shell (K, $n = 1$) has two electrons in its single subshell (s, $l = 0$), while the second shell (L, $n = 2$) has one electron in the first of two subshells (s, $l = 0$).

Table 3.2 shows the number of electrons and the relative shell(s)/subshell(s) for some atoms, grouped by row of the periodic table. The table is intentionally divided at the limit of the Silicon element, because it is about the latter that we will focus our attention in the following (see Section 3.8, and Chapter 8).

TABLE 3.2

Electrons in Shells and Subshells for Different Atoms

Shell		K(n = 1)	L (n = 2)		M (n = 3)	
subshell		1s	2s	2p	3s	3p
		(l = 0)	(l = 0)	(l = 1)	(l = 0)	(l = 1)
1	H	1				
2	He	2				
3	Li	2	1			

TABLE 3.2
Electrons in Shells and Subshells for Different Atoms (*Continued*)

	Shell	K($n = 1$)	L ($n = 2$)		M ($n = 3$)	
	subshell	1s	2s	2p	3s	3p
		($l = 0$)	($l = 0$)	($l = 1$)	($l = 0$)	($l = 1$)
4	Be	2	2			
5	B	2	2	1		
6	C	2	2	2		
7	N	2	2	3		
8	O	2	2	4		
9	F	2	2	5		
10	Ne	2	2	6		
11	Na	2	2	6	1	
12	Mg	2	2	6	2	
13	Al	2	2	6	2	1
14	Si	2	2	6	2	2

Curiosity

To get an idea of atomic dimensions, think, for example, of the hydrogen atom. It is the simplest in nature, since it is made of only one proton and one electron (one electron, one proton, and one neutron form *deuterium*, or *heavy hydrogen*). The radius of the atom in the fundamental state, that is, the average distance between the core and the orbiting electron, is approximately $5.29 * 10^{-11} m$. Consider then that the diameter of the whole hydrogen atom is on the order of $10^{-10} m$, while the diameter of the nucleus is around 10,000 times smaller (on the order of $10^{-14} m$), it follows that the very small volume occupied by the atom is essentially made up of vacuum! For the mass of the whole atom, we can pretty much consider only the nucleus, since the proton has a mass that is about 2000 times greater than that of the electron and slightly lower than that of the neutron.

3.1.3 Octet Rule

Let us now consider any of the shells, for example the one indicated with the letter M, that is, the third shell ($n = 3$). It has three subshells ($l = 0$, $l = 1$, $l = 2$) and, as we have seen, can host up to $2n^2 = 18$ electrons. The N shell can host even more, namely 32, the O shell can contain 50, the P shell can have up to 72, and the Q shell can have even up to 98. But, despite those numbers, the shells cannot be completely filled.

Let's now limit our attention to considering the atoms with fewer than 20 electrons in total. A chemical rule of thumb states that if they have fewer than eight electrons in their outer shell, that is, the valence shell, they tend to combine with others in such a way to obtain the same electronic configuration as a noble gas, that is, a condition of *saturation* with just eight electrons in the outer shell.

This behavior is known as the *octet rule*. An atom with eight electrons in the outermost shell is in its lowest state of energy, given that everything in nature just likes to occupy the lowest energy level possible. This implies that the atoms of noble gases do not bind with others because they already have eight electrons in their outermost shell, and therefore are in their condition of minimum energy. For the atoms of the other elements, this implies that they will naturally tend to combine with other atoms, forming molecules. Two or more atoms that complete their octet closely linked reduce their respective energy, which is discharged externally.

Consider, for example, a flame for welding: it initiates from hydrogen *H* and oxygen *O*; their combination happens with instantaneous release of large quantities of energy and the production of water, H_2O. Two hydrogen atoms combine with one oxygen atom and the molecule that is born has its own full octet, so the sum of their energy is less than that of the atom's constituents. Energy in excess is released and the molecule born is more stable than the three individual atoms of origin.

3.2 LATTICE

Among all types of materials available in nature—solids, liquids, and gases—the solids are by far the most used in electronics. They may be classified as *amorphous*, *polycrystalline*, and *crystalline*, with the distinction based on the characteristics of the distribution of the atom's components. In particular, *amorphous* materials have a disordered atomic distribution, *crystals* have a regular distribution, and the *polycrystals* are made up of "islands" of crystalline structures (thus they are crystalline only in part).

In this section, we shall put the emphasis on solids in their crystal form, because they are adopted more in electronics.

A crystal has the following properties:

→ It's a solid in which all the atoms are arranged in regular intervals (for each direction in the space *x*, *y*, *z*).
→ Its atoms are in equivalent positions.
→ Each atom "sees" the same boundary conditions.
→ The complete structure has translational symmetry.

In a crystal, identical structural units repeat periodically in space, and the periodicity is described by a (mathematical) *lattice*, consisting of an array of points at specific coordinates in space. The identical structural units, named *bases* or *unit cells*, are the atoms in some specific arrangement, unambiguously placed at every lattice point.

Each *unit cell* has

→ The same shape
→ The same content
→ The same volume

Lattices are classified according to seven basic symmetry groups: *triclinic, monoclinic, orthorhombic, tetragonal, rhombohedral, hexagonal,* and *cubic*.

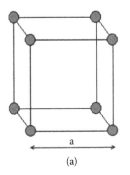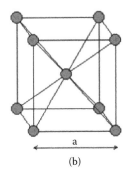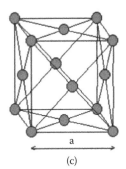

a   a   a

(a)   (b)   (c)

FIGURE 3.7 (a) Simple cubic cell, (b) body-centered cubic cell, (c) face-centered cubic cell.

Observation

A lattice is not a crystal, even though the two words are often used synonymously in colloquial language.

As examples, Figure 3.7 represents three typical cubic unit cells, *simple cubic (SC), body-centered cubic (BCC),* and *face-centered cubic (FCC).*

For the *Simple Cubic (SC)* cell, it has the equivalent of only 1 lattice point, given that each point at the vertexes contributes to the cell for 1/8 of its volume $(1/8 \times 8 = 1)$ (Figure 3.8a).

The *body-centered cubic (BCC)* cell has the equivalent of two lattice points (the sum for each point to the vertexes and the central point) (Figures 3.8a and b).

The *face-centered cubic (FCC)* cell has the equivalent of four lattice points, considering the points to the vertexes and the six points on the faces that contribute to the cell for 1/2 of their volume $((1/8)8 + (1/2)6 = 4)$ (Figures 3.8a, b, and c).

To describe the unit cells, the fundamental constants are the *lattice parameters* (or *lattice constants*), which are the constant distances, at specific temperature, in the three dimensions of the volume, namely a, b, and c, between unit cells. In the special case of cubic crystal structures, $a = b = c$, so it is simply referred to as a or a_0.

Examples of elements with BCC unit cells are chromium (*Cr*) with $a_{0(Cr)} = 0.289\,nm$, iron (*Fe*) with $a_{0(Fe)} = 0.287\,nm$, and sodium (*Na*) with $a_{0(Na)} = 0.429\,nm$.

Among elements with FCC unit cells are aluminium (*Al*) with $a_{0(Al)} = 0.405\,nm$, gold (*Au*) with $a_{0(Au)} = 0.408\,nm$, and copper (*Cu*) with $a_{0(Cu)} = 0.361\,nm$.

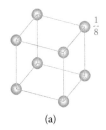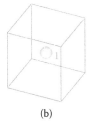

(a)   (b)   (c)

FIGURE 3.8 Contribution of each reticular point in different unitary cells. (a) Points at the vertexes, (b) central point, (c) points on the faces

3.3 WAVE–PARTICLE DUALITY

Electronics is based on knowledge of the physics of electrical charges and of their motion in vacuum and in matter (gas, solid, liquid), under different conditions (temperature, pressure, etc.). The goal of this knowledge is to develop and utilize devices capable to treat and modify physical quantities such as currents and voltages. Such devices are mostly based on solids that we distinguish into three classes: *conductors*, *insulators*, and *semiconductors*, according to their *electric conductivity*, that is, "how well" they conduct electricity, a property largely based on the arrangement of electrons in the outer shell of each atom involved.

To understand how electric conductivity works, let's consider the atomic model, and think about that *quantum* physics assumes, for matter and electromagnetic radiation, properties of both waves and particles simultaneously, a theory known as *wave–particle duality*. Accordingly, the electron can be studied both as a particle or as an electromagnetic wave.

Observation

Quantum physics was born in the beginning of the last century thanks to the Planck quantum principle (Max Planck, German physicist, 1858–1947) and the Heisenberg uncertainty principle (Werner Heisenberg, German physicist and Nobel Prize winner, 1901–1976).

In 1900, Planck assumed that light, X-rays, or other electromagnetic waves could not be issued with an arbitrary rhythm, but only in the form of *wave packages* named *quanta* that have a certain energy proportional to the wave frequency.

In 1926, Heisenberg hypothesized that when measuring the physical state of a quantum system there's a fundamental limit to the amount of precision that can be achieved. In fact, the more precisely the measure of the momentum of a particle the less precise the measurement of its position.

The Nobel prize winner de Broglie (Louis de Broglie, French mathematician and physicist, 1892–1987) hypothesized that each particle moves with momentum p ($p = mv$, m: mass, v: velocity) and is associated with a plane wave with wavelength equal to

$$\lambda = \frac{h}{p} = \frac{h}{mv}$$

where $h = 6.626 \times 10^{-34} \left[J/sec \right]$ is *Planck's constant*.

For Einstein's hypothesis, energy can be expressed through the relation

$$E = hf$$

Therefore, the frequency f and the wavelength λ ($= v/f$) have links with the energy E. On the other hand, Figure 3.9 shows that a rotating electron around the nucleus must necessarily associate an integer number of wavelengths since the sine curve must end without discontinuity. Thanks to this, and thanks to

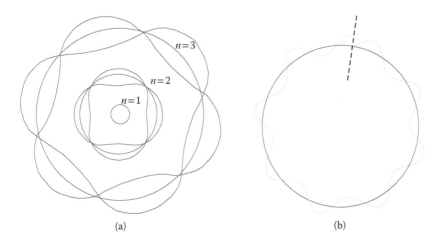

(a)   (b)

FIGURE 3.9   Atomic model with wave–particle duality: (a) "*n*"
represents the number of wavelengths, which can only be an
integer; (b) situation not possible with "*n*" not an integer.

the two previous equations, we can affirm that each fixed shell, with *principal quantum number n* and wavelength λ, corresponds to a fixed energy value *E*. Only an integer, and not a fraction, of wavelength can ensure stability to the revolutionary motion of the electron around the nucleus, and only *quantized* values of energy *E* will be possible, namely only discrete and not continuous values.

So, it is not possible for an electron to be at just any distance from the nucleus; it can be at only well-defined distances. Also, the shells must necessarily be separated and each shell is associated with a well-defined energy value. If an energetic surplus is provided to an electron, only two possibilities can occur: that surplus forces the electron to "jump" to a more external shell, or that surplus is not "absorbed" from the electron at all. This is not only valid for an electron to transit from a more internal shell (with lower energy) to a more external shell (with higher energy), but also to move from the last available external shell in the atom until it abandons the atom itself.

The opposite process may also happen: an electron will lose a certain amount of energy so that, if external to the atom, it goes back to bind with the atom or, if already part of the atom, falls to a more internal shell. The released energy may manifest itself as heat or light.

Unlike with macroscopic physics, in atomic physics the SI unit adopted for energy is the *electronvolt* (eV), which is the kinetic energy acquired from 1 electron when it is accelerated by an electric field associated to 1 *Volt* ($1\,eV = 1.602 \times 10^{-19}\,J$). The kinetic energy of a particle is proportional to the square of its speed and experimental measures show that an electron moving with a speed of 600 *km/sec* has energy equal to 1 *eV*, while it has a speed of 6000 *km/sec* if its energy is equal to 100 *eV*.

Observation

According to the Einstein equation, $E = mc^2$ (mentioned in Section 2.1), the energy equivalent to the mass of the proton, which at rest is 1.67×10^{-30} grams, is about 950 *MeV*! This is what we would obtain if we were able to convert the whole mass of the proton into energy.

3.4 ELECTROMAGNETIC RADIATION

From Section 2.14.1, it is clear how electromagnetic energy can spread through waves. But in general for the propagation of electromagnetic energy we use the term *radiation* (from Latin *radius*, namely *ray*; it is clear the association is to the rays that the sun uses to spread energy), considering that it can be associated with both a wave phenomenon and with particle motion (corpuscle with or without electrical charge). Electromagnetic radiation is generally classified according to its energy or according to the wavelength it has.

Such wave–particle duality allows us to explain a wide range of phenomena associated with electromagnetic radiation that could not be interpreted considering the undulatory nature or the particle nature alone. For example, diffraction and interference phenomena may be explained only if the radiation is represented as a wave, where vector (electric field and magnetic field) quantities oscillate in space and time, keeping orthogonal to the direction of propagation; instead, ionization and photoelectric effect phenomena require what we consider the radiation as a flow of corpuscular particles.

3.4.1 Photoelectric Effect

Experimentally it has been shown that when light "strikes" a material (typically a metal) electrons can be emitted. Such a phenomenon is named the *photoelectric effect*.

Curiosity

The *photoelectric effect* has different technological application—for example closing elevator sliding doors. An electromagnetic ray goes through the mirror of the door when it is open and hits a photoelectric cell that provides the electrons necessary to the electrical circuit that adjusts the closing door. When we cross the door, the ray is interrupted and, thus, so is the electric current. For this reason the closing of the door is inhibited.

With regard to the photoelectric effect, standard physics (which supports the undulatory nature of electromagnetic waves) may explain that

→ The electrons leave the material surface because they receive sufficient energy from electromagnetic radiation.
→ The number of electrons emitted increases with the intensity of the incident light.

But there are other phenomena associated with the photoelectric effect for which the classical theory fails:

→ The kinetic energy with which the electrons are emitted does not depend on the intensity of the radiation but rather on the frequency.
→ By increasing the intensity of the radiation, the photoelectric effect increases the number of electrons emitted but *not* their kinetic energy.

→ There is a frequency limit (different in different materials) where below it, there is no longer electron emission.

→ The delay between the electromagnetic wave striking the surface and the electron emission is lower than 10^{-9} sec.

So, we have to admit a corpuscular nature of the radiation, whereby the electrons of the material struck by radiation absorb a quantum of energy with value $E = hf$. If the energy with which the electron is bound to the material is W_0 (named *ionization energy*, i.e., the minimum energy required to remove, to infinity, an electron from the atom isolated in free space and in its ground electronic state), then the electron may be emitted when $hf \geq W_0$ and will have kinetic energy equal to $E_{kin} = hf - W_0$. This explains the linear dependence of the kinetic energy with the frequency of the incident radiation.

The kinetic energy equation also helps to experimentally obtain the value of Planck's constant, $h = 6.626 \times 10^{-34}$ J/sec.

Curiosity

The interpretation of the *photoelectric effect* is so important in modern physics that Einstein won the Nobel prize for it (not for his *theory of relativity*, as is the popular belief).

3.4.2 Production of Radiation

Electromagnetic radiation may be either natural or artificial. As already seen (Section 2.14.1) a current that flows in an electrical conductor is already a source of radiation (see the electromagnetic field associated) and the electric current can exist both from natural phenomena and from artificial causes.

But the electromagnetic radiation can be generated by several other natural and/or artificial phenomena such as

→ Oscillating circuits
→ Thermal radiation
→ Atomic transitions/laser
→ X-ray tube
→ Nuclear transitions/accelerators

Theoretically, any of these phenomena are able to generate electromagnetic radiation, just by being at a temperature that is higher than absolute zero, with intensity emitted and spectrum portion covered varying from body to body. We have materials for which the level of radiation produced can be significantly higher than others, so these materials can be conveniently adopted to generate artificial radiation.

3.4.3 Radiation and Ionization

Let's analyze the causes of radiations closely linked to the nature of the matter, namely what happens in the atom, at the level of the nucleus or the shell.

Electromagnetic radiation born in the nucleus is due to *radioactivity*, a term that means the spontaneous emission of particles and/or radiation. Such subatomic particles emitted by the nucleus are mainly three types: *alpha*, *beta*, and *gamma*:

→ The *alpha rays* are made of helium nucleus, that is, with two neutrons and two protons, and are thus positively charged.
→ The *beta rays* are made of electrons, and are thus negatively charged.
→ The *gamma rays* have no mass nor electric charge.

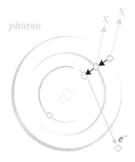

photon

FIGURE 3.10
Illustration of the photoelectric effect with X-ray emission.

In addition to these particles, we can also have spontaneous emissions of protons and neutrons.

The electromagnetic radiation at the shell level is due to interactions between photons and electrons resulting in the emission of *X-rays* by the *photoelectric* or *Compton effect* (Figure 3.10).

The photoelectric effect (that we analyzed in Section 3.4.1) exists when a photon with a certain energy value interacts with an electron belonging to a more internal shell of an atom (usually of the shell K). The electron increases its kinetic energy to a value equal to the difference between the incident photon energy and its binding energy. The result is a readjustment of the electron from the outer shell to the inner shell, considering that it aims to have the minimum energy value possible. The corresponding energy is released by *emissions of X-radiation*, particular to the type of material.

Curiosity

In 1895, during an experiment with vacuum tubes, Röntgen (Wilhelm Conrad Röntgen, German physicist, Nobel laureate, 1845–1923) noticed the emission of electromagnetic rays with an unknown nature. So he wrote about them in the magazine of *Wurzburg's Physical-Medical Society* in an article entitled "On a new type of ray. First communication." Röntgen called such radiation "X" to underline its unknown origin. But even though it is known today, the name remains. In 1901, Röntgen was awarded the first Nobel Prize for physics thanks to this discovery.

In a booklet dated 1917, titled *The Amateur Electricians*, written by a certain engineer named Fiorentino, published by the Library of People, on pp. 58–59 you can read:

> If you produce electric shocks in a dark room inside the Crookes tube and then generate X-rays, the fluorescent screen presented in face of the tube and with a distance of not more than one meter and a half, a bright square will appear. (...) From this, it is clear that, if we put in front of the screen a wooden box with keys, nails or other things inside, for example, we will see the projected shadow of the matters enclosed in the box, while this will give at most a light shadow (...). The most surprising effect is that one obtained by putting a part of the human body between the tube and the screen. With this process, you can clearly

see bones emerge, while the muscles and the sanguine arteries appear a little clearer and the flesh will just give a shadow of its contour. It is unnecessary to mention the enormous benefits of this application for medicine and surgery, from extracting bullets, to fractures, etc.

It is very curious and instructive to realize how the X-rays were utilized in that period as a clear sign of the events the beginning of last century.

The *Compton effect* takes place when a photon interacts with an electron that is weakly linked to the nucleus due to being on an external shell, yielding all or part of its energy to it, but still enough to abandon the atom (Figure 3.11).

An atom is *ionized*, and is named *ion*, when it has lost (acquired) one or more of its electrons, becoming positively (negatively) charged. The *ionization*, that is, the process that ionizes the atom, may have natural or artificial causes but is substantially due to external electromagnetic radiation, then photons that, with sufficient energy, interact with the electrons, which "tear off" from the atom.

The chemical behavior of the ion is obviously different from that of a neutral atom, and this alters the material, for example, a cell that the ion is a part of.

The sources of ionizing radiation are distinguished as *radioactive sources* and *radiogenic equipment*: the former emits radiation continuously because it is made of solid, liquid, or gaseous compounds that contain naturally radioactive atoms (*isotopes*), while the latter are instruments (X-ray tubes, accelerators, etc.) that produce radiation only when they are switched on.

Observation

Ionizing radiation is also used in the medical field for *radiotherapeutic treatments* and *diagnostic* purposes (X-ray radiography, computerized axial tomography—CAT, positron emission tomography—PET).

Non-ionizing radiations are those that do not have enough energy to produce ionization. Therefore, they cannot, for example, produce genetic mutations of DNA, but are able to induce currents that can still produce harmful effects for humans.

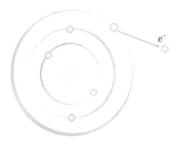

FIGURE 3.11   Graphic description of the Compton effect releasing an electron.

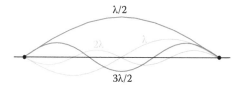

FIGURE 3.12 Some oscillations of a vibrant rope fixed at its ends.

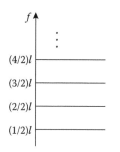

FIGURE 3.13 Discretization of oscillation frequencies.

3.5 ENERGY BAND THEORY

The *energy band theory* provides an interesting explanation of the behavior of electrons in solids. To do an analysis, at least qualitatively, the analogy of the vibrant rope phenomenon is applicable. Given an elastic rope with length l, if we imagine it fixed at its ends, we can make it vibrate only with frequencies f that satisfy the condition $l = n(\lambda / 2)$, where n is an integer, seen in Figure 3.12.

Considering

$$f = \frac{v}{\lambda} = v\left(\frac{n}{2l}\right)$$

(where $n = 1, 2, 3...$) means that, in a fixed transmission medium, f is proportional to the value

$$\frac{n}{2l}\left(f \propto \frac{n}{2l}\right)$$

Therefore, the rope essentially can vibrate only with certain frequencies and not with all frequencies. We can note the phenomenon named *frequency discretization*, as represented by Figure 3.13.

The discretization phenomenon of frequencies exists also, similarly, for a mass connected to a spring fixed to a wall (Figure 3.14).

But the interesting phenomenon is when we have two oscillators with the same frequency of oscillation. It is possible to show each frequency of the isolated oscillators break into two distinct frequencies for the system given from the oscillating couple, as schematically shown in Figure 3.15 and 3.16a through c.

 Curiosity

It is useful to think of racing cars as an analogy to the phenomenon of splitting two frequencies from a single one when there is an interaction between the oscillators. Consider two twin cars on twin tracks. The energy required for the first car to cover the distance of the first track, other things being unchanged, is exactly the same energy required for the second car on the second track. The value of this energy depends on cunning aerodynamics that acts to increase grip on a racetrack. Airflow over a car's body and wings "pushes" the car on to the ground, enabling it to take turns at higher speeds than usual. But, when there are two cars running in series, one following the other, things get more complicated. The lead car will have the benefit of undisturbed "clean" air to run in, but leave a trail of turbulent "dirty" air in its wake. This is a problem for the car when it gets to a corner, since the wings and spoilers do not provide a good aerodynamic grip, missing enough air flow over them. The rear car takes advantage from following in this "dirty" air when going straight ahead, exploiting the lead object's slipstream, and getting a few extra miles an hour for free. As a result, the lead car will require more energy and the rear one will need less energy to run the same track than the energy needed going round a track on its own.

Therefore, the value of the energy "splits" into two (near) values when the cars run "while influencing" each other.

The conclusion is similar for a system of three coupled (as shown in Figure 3.17) and it is possible to combine it into a system of n coupled oscillators.

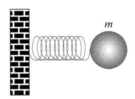

FIGURE 3.14   Mass connected to a spring fixed to a wall.

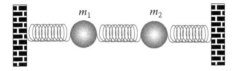

FIGURE 3.15   A system of two bonded masses.

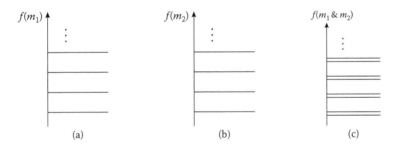

(a)   (b)   (c)

FIGURE 3.16   Each frequency of a single oscillator breaks down into a pair of frequencies for a couple of oscillators.

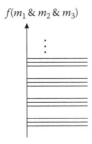

FIGURE 3.17   Split frequencies for a system of three coupled oscillators.

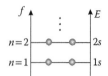

FIGURE 3.18   Energy levels for the $1s$ and $2s$ scenario in an atom.

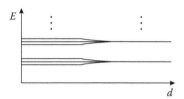

FIGURE 3.19   Progress of the energy levels of a system of three atoms by varying the distance.

This matter may be discussed in the atomic field, considering the electrons as particles revolving around the nucleus with discrete frequencies.

Therefore, consider two twin atoms placed, initially, at a great distance apart. For simplicity, we consider the electrons just in shell s ($1s$ and $2s$), whose energy levels are placed as in Figure 3.18, for both atoms. Clearly for the more external levels, they will compete with a higher energy, so we will also be able to report the conditions to an axis of energy E.

Bringing the first isolated pair of atoms near, we may have an interaction that will be higher as the distance is lower. When the distance enables the two atoms to bind together with a covalent bond, they will realize a system similar to the one described for the oscillators. The energy levels separate and a pair of energy values will be created for each of the beginning energy levels.

With the atoms far apart (i.e., with a such distance that it is not possible to warn interaction) there is a single energy level $1s$ for both, but when the two atoms are at a distance close enough so they can interact (making a bond), the energy level $1s$ splits into two new levels. The same applies to level $2s$ from which two separate energy levels are generated.

When the number of atoms that can interact is three, level $1s$ of each can be broken down into three new common levels. Figure 3.19 symbolizes the phenomenon indicating the specificity among both the values of the energy E and the distance d.

So, iterating, N interacting atoms will correspond to N energy levels $1s$.

Given a crystal with atomic density on the order of $10^{23}\,[atoms\,/\,cm^3]$, the number of atoms interacting is so high that it is impossible to consider the discrete number of new energy level, but it will be considered reasonably *continuous*. So you have those that are defined *bands* (namely *intervals*) of energy. Each individual initial energy level now resolves in a band and the bands may be far from each other or partially overlying (Figure 3.20).

If the solid has L energy levels, knowing that each level admits at maximum two electrons, we know that in each band there may be at most $2L$ electrons. The band will be defined as *full* in the case that it has been occupied by all the $2L$ electrons, *half-full* if the electrons that occupy it are not equal to the maximum, and *empty* if it is not occupied by any electron. The way the bands are occupied, and the distance that separates them, defines the difference between a *conductor*, an *insulator*, and a *semiconductor*. To investigate the properties of such materials, it is useful to adopt a simplified graphic representation of the energy bands.

Referring to Figure 3.21, the energy bands formed inside a solid can be schematized, for simplicity, using ranges along the energy axes. For our purpose, between all bands of energy admitted for an atom within a material, we will consider only the band with the higher energy values. It is called the *valence band* because it is

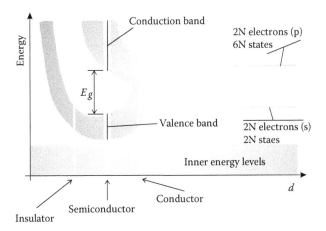

FIGURE 3.20   Energy bands in a crystal.

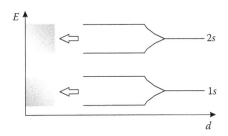

FIGURE 3.21   Simplified
representation of the energy bands.

occupied by valence electrons, that is, those of the outer shells of the atom, from which covalent bonds can form between the atoms. It is precisely the valence electrons that interest us for the purpose of electrical conduction, since these electrons may be released from the atom they belong to with a lower energy expense compared to all the other electrons that form the atom. The transfer of a valence electron from an atom occurs if we provide a high enough energy value to the electron to transit from the valence band to the *conduction band*, the term with which we denote the range of energies proper of an electron released from the atomic bond but still belonging to the material.

Relating to Figure 3.21, we define as an *energy gap* (or *bandgap*) E_g, the minimum energetic leap necessary to transform a valence electron, bonded to the atom, to a *carrier* (definition in Section 2.6), an electron released from the atom but still belonging to the material, to which it is possible to associate a current. As the energy gap (also known as the *prohibited band of energy*) is defined, it is between the maximum value of the valence band E_v, and the minimum value of the conduction band E_c.

It is the value of $E_g \left(= E_c - E_v\right)$ that properly categorizes the materials as *insulators, conductors,* and *semiconductors* (Figure 3.22).

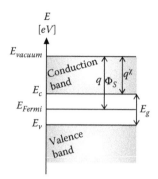

FIGURE 3.22
Band diagram.

- For conductors, the energy-gap value is negligible, $E_g \approx 0.01\,eV$, or zero.
- For semiconductors, typical values are in the range $0.5 < E_g < 2\,eV$.
- For insulators, typical values are $E_g > 2.5\,eV$.

Observation

A semiconductor is sometimes referred as a small bandgap insulator. In the same manner, a semi-insulator can be considered a large bandgap conductor.

Other references include

→ E_{vacuum}: *level of vacuum*, is the energy limit beyond which the electron is no longer bonded to its atom nor the material. Its excess energy has a kinetic nature.

→ E_F: *Fermi level or Fermi energy* (Enrico Fermi, Italian physicist, Nobel laureate, 1901–1954) is the energy value that, in terms purely of operating at absolute zero temperature, is placed in the middle between the valence and conduction bands of a material.

→ $q\Phi_s$: *work function*, the amount of energy required to pass from the Fermi level to the energy level of a vacuum.

→ q^x: *electronic affinity*, the amount of energy required to pass from the minimum conduction energy to the energy level of a vacuum.

3.6 CONDUCTORS

Conductors are materials that allow the flow of charged particles. Candidates to be good conductors are elements with few valence electrons and with a small energy gap, that is, the metals. Classic examples are copper, aluminum, silver, and gold.

It is important to stress that the thermal energy at room temperature is just sufficient to make the valence electron transit to the conduction band. Statistically we have one electron for each atom to become a carrier. So, Cu with FCC unit cells, the equivalent of four lattice points, and lattice parameter $a_{0(Cu)} = 3.62\,A$, at room temperature furnishes

$$\frac{4}{\left(3.62\times10^{-8}\right)^3}\cong8.4\times10^{22}\left[\frac{atoms}{cm^3}\right]$$

as carriers in the valence band (Figure 3.23).

When an electric voltage is applied across a conductor wire, the free electrons move in the direction imposed by the related electric field, so creating a flow of charges, that is, an electric current.

To understand where the free electrons are in a conductor, it is convenient to refer to *Gauss's law* (Johann Carl Friedrich Gauss, German mathematician and scientist, 1777–1855, Figure 3.24), also known as *Gauss's flux theorem*, which states that the electric field within a spherical surface uniformly charged is null (Figure 3.25).

Considering that the electric field is zero within a conductor, it must vary from zero only along the surface, so if the conductor is charged, the net charge only distributes on the surface.

The force lines of the electric field are orthogonal to the surface, since it is an equipotential surface, otherwise the conduction charges on the surface would not be in equilibrium.

3.7 INSULATORS

Insulator materials are those for which the necessary energy to transfer the valence electrons to the conduction band is much higher than conductor materials. As most of the electrons stay bonded to the respective atoms, the sufficient number of carriers to obtain a reasonable electric current is not reached. In practice, the insulator allows just a very small amount of current through, practically negligible for most applications, called *leakage current*.

In the case of insulators, the energy gap E_g has an energetic range of some eV, and the higher its energetic range, the better the insulator it will be (Figure 3.26).

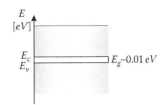

FIGURE 3.23 Energy band with really narrow gap.

FIGURE 3.24   J. Gauss.

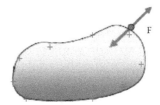

FIGURE 3.25 Electrical charges arranged only on the surface of a conductor.

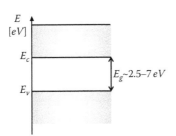

FIGURE 3.26   Energetic band with wide gap.

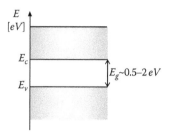

E
[eV]

E_c

$E_g \sim 0.5\text{–}2\,eV$

E_v

FIGURE 3.27   Typical E_g range for a semiconductor.

3.8 SEMICONDUCTORS

Most electronic devices exploit the junction properties (interface) between semiconductor materials characterized by different properties, or between metal and semiconductor.

Therefore, the study of such devices requires the knowledge of the chemical, physical, thermal, and electrical properties of the semiconductor.

The *semiconductor* has an energetic gap that is not too wide (Figure 3.27), but not negligible either (typically does not exceed $E_g \cong 1.5\,eV$).

Curiosities

Faraday realized and made the first documented observation of a *semiconductor effect* in 1833.

The term "semiconducting" was used for the first time by Volta in 1782.

3.8.1 Intrinsic and Extrinsic Semiconductors

A classification scheme has semiconductors divided into *intrinsic* and *extrinsic* types.

Intrinsic semiconductors are chemically pure, containing nothing but semiconductive material.

Extrinsic semiconductors have other types of atoms, called *impurities*, added to them and the addiction process is known as *doping*.

The intrinsic semiconductors do not have a sufficient number of electrons in the conduction band to guarantee a reasonable value of electric current. To increase this number, we need to furnish sufficient energy to the electrons to promote them from the valence band. A way to do this is to raise the temperature of the material, since heat is a type of energy; this method is obviously impracticable. This is why we refer to the doping process (details in Chapter 8), which makes the semiconductor *extrinsic*.

Curiosity

In 1940, Ohl (Russell Shoemaker Ohl, American engineer, 1898–1987) discovered that a doped semiconductor presents new very interesting electric properties.

Later, in 1946, Ohl patented the modern solar cell.

3.8.2 Commonly Used Semiconductors

A semiconductor can be composed of a single element or a compound of elements from two, three, or more different groups of the periodic table. The compound can be a binary (two elements), ternary (three elements), quaternary (four elements), or even quinary (five elements) alloy.

Examples of single element semiconductors are *carbon (C), silicon (Si), germanium (Ge),* and *tin (Sn),* but the electronics industry generally utilize *Si* or *Ge,* and of these two, silicon is much more widely used. They belong to Group IV of the periodic table; this means they have four valence electrons in their outer shell, thus they can give or accept electrons equally well to complete their *octet* (see Section 3.1.3).

Compound semiconductors are generally synthesized using elements from Groups II through VI, e.g., from Group III and V (III–V compounds) or II and VI (II–VI compounds). Examples of binary compounds are *silicon carbide (SiC), gallium nitride (GaN), indium phosphide (InP), gallium phosphide (GaP), indium antimonide (InSb),* and *gallium arsenide (GaAs).* Ternary compounds include *aluminium gallium arsenide (AlGaAs), indium gallium arsenide (InGaAs),* and *indium gallium phosphide (InGaP).*

Among all the compound semiconductors, the one currently in widest use is GaAs. One of the fundamental parameters to consider for semiconductors is the value of their *energy gap* E_g. Normally, we consider its value at the room temperature, but E_g varies with the temperature *T*. Examples include the following:

$$E_{g(Ge)}(T) = 0.785 - 3.7 \times 10^{-4}T$$

$$E_{g(Si)}(T) = 1.17 - \frac{4.73 \times 10^{-4}T^2}{T + 636}$$

$$E_{g(GaAs)}(T) = 1.52 - \frac{5.4 \times 10^{-4}T^2}{T + 204}$$

$$E_{g(InP)}(T) = 1.421 - \frac{4.9 \times 10^{-4}T^2}{T + 327}$$

These are not exact but experimental formulas. In fact, for instance, some labs report the following for *Si*: $E_{g(Si)} = 1.21 - 3.60 \times 10^{-4}T$. According to the equations we can calculate the energy gap values for different semiconductors at different temperatures (Table 3.3).

TABLE 3.3

E_g Values for Some Semiconductors as a Function of *T*

Energy gap [eV]

T [K] Semic.	0	100	200	300	400
Ge	0.785	0.748	0.711	0.674	0.637
Si	1.170	1.164	1.147	1.125	1.097
GaAs	1.520	1.502	1.466	1.424	1.377
InP	1.421	1.409	1.384	1.351	1.313

3.8.2.1 Germanium

The very first electronic device was made of germanium, but devices soon transitioned from germanium to silicon. With respect to *Si*, *Ge* has substantially higher mobility of carriers, thus it is interesting for higher-speed devices but, because of its lower energy gap, the (instrinsic) *Ge* (Figure 3.28) suffers from higher current fluctuations. Currently, very few devices are made with *Ge* and it is mostly used combined with *Si*, forming *silicon germanium* (*SiGe*).

3.8.2.2 Silicon

Silicon is the second most common element, 27.7%, in the earth's crust by mass (Figure 3.29). It is produced by heating *silicon dioxide* (SiO_2) with carbon to temperatures approaching 2200°C. SiO_2 commonly takes the form of ordinary sand, but also exists as quartz, rock crystal, amethyst, agate, flint, mica, jasper, opal, and dust. So, with respect to *Ge*, *Si* is less expensive due to its greater abundance, and it is also a more "stable" and strong material, in terms of "fluctuations" of electric current, thanks to its higher bandgap.

 Si in its neutral state has 14 protons and 14 electrons.

 The electronic configuration of *Si* reports two electrons in the *K*-shell ($n = 1$), eight electrons in the *L*-shell ($n = 2$), and four electrons in the *M*-shell ($n = 3$), which is half filled. The four electrons that orbit the nucleus in the outermost, or *valence*, *M*-shell energy level can hypothetically be given to, accepted from, or shared with

FIGURE 3.28   Instrinsic germanium.

FIGURE 3.29   Silicon dioxide compound (SiO_2)
(top), instrinsic silicon (Si) (bottom).

other atoms. But, generally (though not exclusively), Si does not give off its electrons easily and readily forms covalent bonds to complete its M-shell.

The crystal of silicon has a *diamond cubic crystal structure*, as seen in Figure 3.30, with a *lattice parameter*, at $300\,K$, of approximately of $a_0 = 5.43\,Å$.

The number of equivalent atoms for a cell is given considering the contribution of the vertexes $(1/8)\,8 = 1$, the contribution of surfaces $(1/2)\,6 = 3$, and the contribution of interior atoms $4*1 = 4$; in total there are eight equivalent atoms.

This number becomes important in the calculation of the volumetric density of atoms, namely in knowing how many atoms to a cubic centimeter a crystal of silicon has

$$\frac{8}{(5.43\times10^{-8})^3} = 5\times10^{22}\left[\frac{atoms}{cm^3}\right]$$

a number which will be useful in Section 8.2.2.

Curiosity

Silicon does expand when it freezes, similar to water and differently from a great number of other substances.

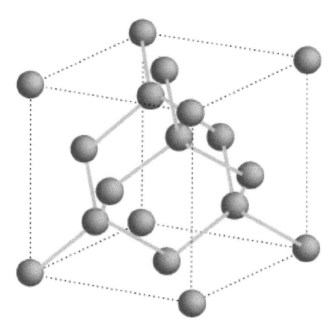

FIGURE 3.30   Lattice of silicon cell.

How many of these $5*10^{22}$ atoms for cm^3 release one electron as a carrier at room and other temperatures? The volumetric density of free electrons, that is, promoted from the valence to the conduction band, is called *intrinsic carrier concentration (per unit volume)* n_i $[cm^{-3}]$, and depends on both material and temperature, according to the equation

$$n_i^2 = BT^3 e^{\left(-\frac{E_g}{kT}\right)} \left[cm^{-6}\right]$$

where

- → E_g: Energy gap $[eV]$
- → k: Boltzmann constant ($8.62 \times 10^{-5}\,[eV/K]$) (Ludwig Edward Boltzmann, Austrian physicist, 1844–1906)
- → T absolute temperature $[K]$
- → B: a value function of the material and the temperature (for silicon $B_{Si}\big|_{T=300K} = 1.08 \times 10^{-31}\,[K^{-3}cm^{-6}]$)

According to the previous equation and as indicated in the following table, the free electrons at room temperature in the silicon are on the order of $10^9\,[electrons\,/\,cm^3]$, a number that is absolutely negligible compared to the $5*10^{22}\,[atoms\,/\,cm^3]$ existing atoms. Furthermore, that number decreases down to almost zero when the temperature is at $-100°C$ (Table 3.4).

Manufacturing procedures to obtain instrinsic silicon to be used in electronics include crystal growth and the wafer-slicing process. Figure 3.31a through c represents pure monocrystalline silicon, which is sliced into very thin wafers, and processed to obtain eletronic devices and circuits.

TABLE 3.4
Intrinsic Carrier Concentration per Unit Volume for Silicon at Different Temperatures

T [°C]	−50	0	20	50	100	150	200
T [K]	223	273	293	323	373	423	473
n_i [cm^{-3}]	$2*10^6$	$9*10^8$	$4*10^9$	$6*10^{10}$	$1*10^{12}$	$2*10^{13}$	$1*10^{14}$

Some fundamental properties of silicon at room temperature are

→ Lattice constant $a_{0(Si)} = 5.43 \left[\AA \right]$

→ Bandgap energy $E_{g(Si)} = 1.12 \left[eV \right]$

→ Intrinsic carrier concentration $n_{i(Si)} = 1.0 \times 10^{10} \left[cm^{-3} \right]$

→ Electron mobility $\mu_{n(Si)} = 1500 \left[cm^2/V^*s \right]$

→ Relative dielectric constant $\varepsilon_{r(Si)} = 1.19$

3.8.2.3 Gallium Arsenide

GaAs is a binary semiconductor alloy with the advantage of some electronic properties that are better than those of Si. A higher carrier mobility allows higher operational speed and low power consumption. A *direct* energy gap (that is, the momentum ($p = mv$) of carriers, m being the mass and v the velocity, is the same in both the conduction band and the valence band) allows the fabrication of devices capable of emitting light in the infrared and red frequencies, as used in CD and DVD systems.

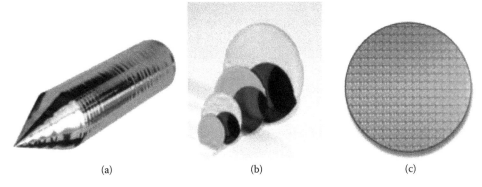

(a) (b) (c)

FIGURE 3.31 (a) Monocrystal silicon ingot; (b) silicon wafer from silicon ingots of different diameters; (c) several devices made on the silicon wafer surface.

GaAs-based structures are less susceptible to radiation and hot-carrier damage, which results in increased reliability.

However, *GaAs* fabrication technology is not as mature as that for *Si*, which has been driven by the enormous commercial resources in so many fields of application in consumer electronic industries.

3.8.2.4 Organic Semiconductors

Some electronic circuits are made of devices based on semiconductor polymer-based materials.

These organic semiconductors are mainly used for displays, RFID (radio-frequency identification), solar cells, and smart textiles.

With respect to their inorganic counterparts, advantages come from easy-to-implement fabrication steps and the possibility of making devices that are transparent, flexible, cheap, and even biodegradable (being made from carbon). Disadvantages come from fragility, shorter lifetimes, dependency on stable environment conditions, and the conduction of current not being particularly good, because of low electric mobility μ_{os}, limited to

$$\mu_{os} \approx 1-10 \left[\frac{cm^2}{V \times s} \right]$$

two or three order of magnitude lower than that in conventional inorganic semiconductors. Hence, applications are limited to those devices where requirements on mobility of charges are not too high.

In order to have carriers for the electric current in an inorganic semiconductor, we know that electrons must be promoted from the valence to the conduction band. Also, in organic material we must have electron excess free to move in the whole solid. But for organic semiconductors, we must replace the concept of valence and conduction bands with other similar ones named, respectively, the HOMO (highest occupied molecular orbital) and LUMO (lowest unoccupied molecular orbital) of the molecule. HOMO and LUMO levels may take the shape of an energy band, but the bands are rather narrow. What we called the *energy gap* for inorganic semiconductors here is known as the *HOMO-LUMO gap*.

3.8.2.5 Others

Other types of semiconductors have been investigated and utilized. Among them *IGZO* and *doped synthetic diamond* are interesting to know.

IGZO (*indium gallium zinc oxide*) has been recently used in LCD screens. It allows faster devices as compared to amorphous silicon, and smaller pixels, for high screen resolutions (higher than HDTV) and reaction screen speed. The first smartphone with a low-power *IGZO* display was the 4.9-inch Sharp Aquos Phone Zeta SH-02E, in 2012.

Pure *diamond* is a super-tough electrical insulator, but with doping it becomes a semiconductor with a very high electron mobility, three times more than silicon. Because diamond is also the best thermal conductor on Earth, it would be possible in principle to prepare synthetic diamonds to make electronic devices that handle high-power signals but do not require power-hungry cooling systems. The first hint for the exploration of diamond as a semiconductor material was noted in 1952, when natural diamond crystals were found to show a semiconductor type of conductivity. But the realistic assessment of the progress for active diamond electronics is that the research and development in this area did decrease significantly after 1995, and no practical electronic devices have been successfully developed.

3.9 ELECTRICAL RESISTANCE AND JOULE HEATING

The difficulty with which the carriers move in the material determine what is known as the *resistance R* of the material to the flow of current. Electrical resistance for carriers is similar to mechanical friction for objects or substances in contact with each other.

To derive the physical meaning of the resistance, let's consider here the averaged expression (for simplicity with null starting values) of the current:

$$I = nqv_dS$$

For the mobility, we have

$$\mu = \frac{v_d}{E}$$

and for the voltage, we have

$$V = El$$

Mixing the previous equations we obtain

$$I = nqv_dS = nq\mu ES = nq\mu\frac{V}{l}S$$

The resistance represents the proportional link between voltage and current, so

$$R = \frac{1}{nq\mu}\frac{l}{S}$$

(additional details will be given in Section 4.4.7).

The SI unit for resistance R is the Ohm (Ω) (Georg Simon Ohm, German physicist, 1787–1854, Figure 3.32).

Resistance causes some of the electrical energy to turn into heat, so some electrical energy is lost. This conversion is known as the *resistive heating effect* (or *Joule heating*, or *ohmic heating*), first studied by Joule.

3.10 TEMPERATURE COEFFICIENT

The value of the electric current depends strictly on the *number* of carriers available for conduction. This is the number of electrons that acquire enough energy to transit from the valence to the conduction band. The energy for the transitions is available without any effort, since it is provided by the heat available at room temperature. But it is

FIGURE 3.32
G.S. Ohm.

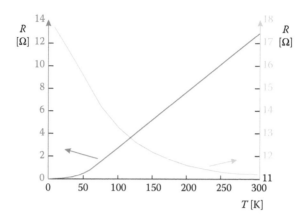

FIGURE 3.33   Typical variation of resistance
according to the temperature for a conductor
(in red) and a semiconductor (in blue).

not necessarily true that an increase in temperature to higher values with respect to the room temperature corresponds to an increase of current.

In fact, conductors and semiconductors perform in the opposite manner.

In conductors, all valence electrons are in the conduction band just at room temperature, so an increase of temperature does not increase the number of carriers, but corresponds to more energy for electrons already in the conduction band. This excess of energy means an increase in *scattering phenomena* (see Section 2.6) and, therefore, a reduction of the electron mobility, and hence of the value of the current.

In semiconductors, just a small percentage of valence electrons are in the conduction band at room temperature, so an increase of temperature provides the energy for increasing the number of electrons that can become carriers; hence the current increases.

Conductors are said to have a *positive coefficient of temperature* (the higher the temperature, the more the resistance), and semiconductors have a *negative coefficient of temperature* (the higher the temperature, the less the resistance), as Figure 3.33 represents.

For the insulator, things are almost unchanged because a reasonable increase of room temperature, within a certain limit, is not enough to break the atomic bond and bring a sufficient number of electrons in the conduction band. So the resistance of the material to the current flow (which is already considerable) remains essentially unchanged.

3.11 KEY POINTS, JARGON, AND TERMS

→ An atom is the smallest particle of an element with all the properties of the element. It consists of a nucleus, made up of protons and neutrons, and an electron cloud made up of electrons that "move" in "shells" around the nucleus.

→ The *orbital* is a 3D space around a nucleus in which electrons are most likely to be found.

→ The *shells* represent the arrangement of electrons in energy levels around an atomic nucleus. Each shell is composed of one or more *subshells* made of a set of electron orbitals.

→ The *octect rule* is a chemical bonding theory, which assumes that atoms with fewer then 20 electrons exhibit a tendency for their valence shells to have a full complement of eight electrons. So atoms will lose, gain, or share electrons to achieve the electron configuration of the nearest noble gas having eight valence electrons.

→ Electronics is based on knowledge of the physics of electrical charges and of their motion in vacuum and in matter, under different conditions.

→ Solids can be classified as *amorphous, polycrystalline,* and *crystalline.*

→ The wave–particle duality theory states that matter and electromagnetic radiation have properties of both waves and particles simultaneously.

→ An atom is called *ionized,* and is named an *ion,* when it has lost (acquired) one or more of its electrons, becoming positively (negatively) charged.

→ The way the valence and conduction bands are occupied and the distance between them, named the *energy gap,* defines the difference between a *conductor,* an *insulator,* and a *semiconductor.*

→ A commonly used solid semiconductor material is silicon.

→ In its pure state, a semiconductor is called *intrinsic.* When atoms of other elements are added, a process called *doping,* the semiconductor is called *extrinsic.*

→ Semiconductors can be doped with other atoms to add or subtract electrons.

→ Conductors have a positive coefficient of temperature, semiconductors a negative one.

→ The opposition offered by a material to the flow of a current defines the *electrical resistance,* which causes a conversion of electrical energy into heat (and radiation), named *Joule heating.*

.

4

Two-Terminal Components

4.1 DEFINITIONS

An *electrical component* (or *device*) is an element that allows a flow of charges with a specific voltage–current relationship. It is provided with *terminals*, two or more, that connect leads or metallic pads, as access and exit points for the electric current, across which the voltage is applied. The simplest electrical component has only two terminals, so only one current and one voltage is available. The voltage–current relationship is known as a *transfer function* (or *constitutive equation*). Two terminals of the same component are also known as *one-port*, so the two-terminal component is a *one-port device* (Figure 4.1a).

The symbol of a two-terminal component is given in Figure 4.2 where the schematization of the applied voltage and the current flowing is highlighted.

When a component is made of a single device, it is said to be *elementary*, but a two-terminal component can also be made by a *set* of electrical devices. Examples of elementary components are *resistors*, *capacitors*, and *inductors*, defined later.

In the case of three terminals, we talk about a *three-terminal* electrical component (Figure 4.1b), when the terminals are generally *N*, the component is referred as *N*-terminals (or an *N*-pole/*N*-terminal device). The terminals are also referred to as *pins*.

An electrical component can be *passive* or *active*.

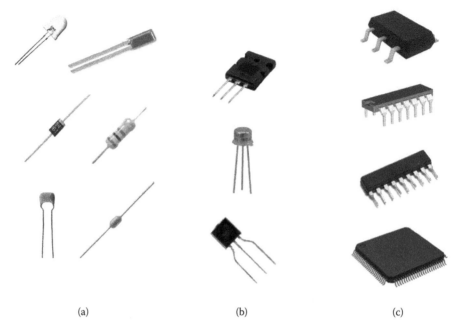

(a)   (b)   (c)

FIGURE 4.1   Examples of (a) two-terminal,
(b) three-terminal, and (c) N-terminal components.

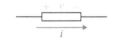

FIGURE 4.2   Rrepresentation of a generic two-terminal electric component.

It is *active* if it contains at least one elementary active component. An *active component* furnishes energy to other components. The active components are typically voltage and current sources or generators.

The *passive component* contains no source that could add energy to the signal, but provides an energy transformation or dissipation.

Observation

There are some misleading definitions of *active component*. In fact, they include semiconductor devices, which are unable to furnish energy alone. These semiconductor devices, such as the transistors (treated in Chapter 8, or tunnel diodes) are sometimes referred as *active* because these are able *to amplify* a signal, i.e. to increase its strength. But this amplification is possible only thanks to synergetic effects with external power supplied by other components. So, recalling a different definition, *passive* refers to transform or to dissipate the energy and *active* to drive it.

A component is said to be *linear* when its voltage versus current relationship can be described by a *linear* function. Please remember that a function $y = f(x)$ is *linear* if and only if $f(a_1x_1 + a_2x_2) = a_1y_1 + a_2y_2$ for any two inputs x_1 and x_2, with a_1 and a_2 constant values.

When electrical components (whether two-terminal or more generally, N-terminal) are connected to allow the flow of current, their set defines an *electrical circuit*.

To solve a circuit means to mathematically determine the values of the current that flows through, and the voltage that is across, each of the components.

An AC/DC circuit is a combination of circuit elements plus one or more AC/DC generator(s) or source(s), which provide(s) the alternating/constant current and/or voltage.

FIGURE 4.3 Conventional direction for the current in a passive component B.

4.2 CONVENTIONAL NOTATIONS

Conventionally, the following notation is assigned: for passive components, the current flows from higher to lower voltage; therefore, the tip of the arrow that represents the current is on the side of the "−" sign of the voltage (Figure 4.3). This is because the current naturally flows from higher to lower voltage. To make it flow in the opposite way, we need *active* components for which, therefore, the notation states the current flows coming from the side with lower voltage (Figure 4.4).

FIGURE 4.4 Conventional direction for the current in an active component G.

4.3 TOPOLOGY OF INTERCONNECTIONS

The interconnections among the circuit components is known as the *topology* of a circuit. For two-terminal components, we can distinguish between the *series, parallel, bridge, star* (or *Y*), and *triangle* (or *Δ*) topologies.

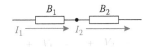

FIGURE 4.5 Series topology for two-terminal devices.

4.3.1 Series Topology (Voltage Divider)

One possibility is to connect two two-terminal components in *series*, as depicted in Figure 4.5.

In the series topology, the two components have one terminal in common. Because of that bond, the current flowing is the same in both the components, while the voltage at the heads of the series is the sum of the individual voltages:

$$I = I_1 = I_2; V = V_1 + V_2$$

One fundamental application of the series topology is the *voltage divider*, schematized in Figure 4.6. It represents how the total voltage V_T across the series is fractioned between the two components in V_1 and V_2.

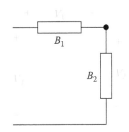

FIGURE 4.6 Voltage divider.

4.3.2 Parallel Topology (Current Divider)

In a *parallel* topology, the two components have both terminals in common (Figure 4.7).

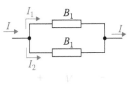

FIGURE 4.7 Parallel connection of two bipolars.

For such a topology, the flowing current is different in each component, while the voltage is the same for both, and is equal to the total voltage:

$$I = I_1 + I_2; \quad V = V_1 = V_2$$

One fundamental application of the parallel topology is the *current divider,* schematized in Figure 4.8. It represents how the total current I across the parallel is fractioned between the two components in I_1 and I_2.

Observation

Two components connected by themselves in a single loop circuit are in parallel and in series at the same time.

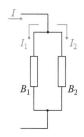

FIGURE 4.8   Current divider.

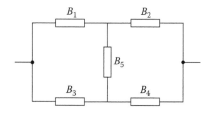

FIGURE 4.9   Bridge topology.

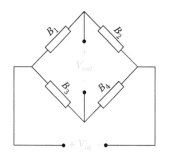

FIGURE 4.10   Bridge topology without the central B5 device.

4.3.3 Bridge Topology

Figure 4.9 represents the *bridge* (or *H-bridge*) topology, so called since the component B_5 forms a sort of "bridge" between the pair B_3, B_4 and the pair B_1, B_2.

This configuration is often used to withdraw the output voltage (V_{out}) at two opposite terminals of the bridge when a voltage (V_{in}) is applied at the other two terminals, as reported in Figure 4.10. (Please note that Figure 4.10 represents the same topology of Figure 4.9; for our purposes, the connecting wires can be drawn in any format we please).

The component B_5 is omitted in some configurations.

4.3.4 Star and Triangle Topologies

The *star* and *triangle* topologies take their names from the geometric figures that emerge from the arrangement of the components, which are said to form a *star* (or *Y*, or *T* if it is seen "upside down") topology when the currents flowing through all of them converge at a common point, as in Figure 4.11a. Therefore, all the components have only one terminal in common (but, unlike the series topology, here there must be at least tree components).

Components are said to form a triangle (or *delta,* Δ) topology when the series is closed on itself, as in Figure 4.11b.

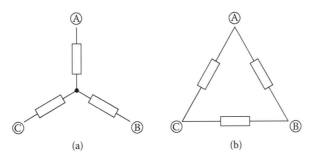

(a)   (b)

FIGURE 4.11   (a) Star and (b) triangle topologies.

4.4 RESISTORS

4.4.1 Resistance, Resistivity, Conductance, and Conductivity

If electrons move in a vacuum, they continue with the same speed without ever stopping, because without friction or other obstructive forces, they remain in their state of motion or rest (*first Newtonian law*).

Otherwise, the electrons that move in a material meet forces that oppose their motion because of interactions with the lattice. The intensity with which these opponent forces act is somehow "measured" with a fundamental parameter: the *electrical resistance R* (aforementioned in Section 3.9 of Chapter 3). It is a greatly exploited property in electric and electronic circuits, and it is a function both of the physical nature of the material, through the parameter *resistivity* (or *specific resistance*) ρ, and of its geometry expressed by its length *l*, and its section *S*, for a wire-form material (Figure 4.12).

As a consequence, we have the function

$$R = \rho \frac{l}{S} \ [\Omega]$$

FIGURE 4.12   Wire-form material with section *S* and length *l*.

The inverse of resistance is defined as *conductance* $G(= 1/R)$, the unit of which was named the Siemens [S] (Ernst Werner von Siemens, German engineer and industrialist, 1816–1892, Figure 4.13).

Sometimes the conductance is called the *mho*, which is just ohm written backwards.

From the resistance equation, we can easily deduce the SI unit for the resistivity ρ, equal to Ωm. Some typical values of ρ for common conductors are reported in Table 4.1.

The inverse of resistivity is named *conductivity*, or *specific conductance* σ (= 1/ρ). Figure 4.14 shows the range of variability of the conductivity (expressed with $\Omega^{-1}cm^{-1}$) for insulators, semiconductors, and conductors and as we can see it is a very large range ($10^{-24} < \sigma < 10^7$).

FIGURE 4.13   E.W. Siemens.

TABLE 4.1
Resistivity of Some Common Conductors and Percentage Variation per Centigrade

Materials	Composition	ρ (@20°C) [*10⁻⁶Ωm]	Δρ/°C [Δ%/°C]
Aluminum		0.028	0.36
Silver		0.016	0.41
Costantan	Cu + 45%Ni + 1%Mn	0.50	0.002
Iron		0.097	0.39
Gold		0.024	0.40
Manganin	Cu + 12%Ni + 2%Ni	0.42	0.001
Mercury		0.984	0.07
Nickel-Chrome	80%Ni + 20%Cr	1.07	0.0011
Platinum		0.1	0.50
Copper		0.018	0.41
Tungsten		0.05	0.43

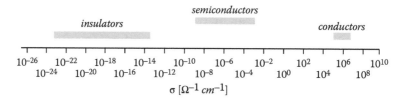

FIGURE 4.14 Conductivity variation range for insulators, semiconductors, and conductors.

Following is a summary of the parameters introduced here:

→ *R*: resistance [*Ohm*, Ω]
→ ρ: resistivity, or specific resistance [*Ohm × meter*, Ω × m]
→ *G*: conductance [*Siemens, S*]
→ σ: conductivity, or specific conductance $\left[\dfrac{1}{Ohm \times meter}, \Omega^{-1} \times m^{-1}, or \dfrac{S}{m}\right]$

Observation

From what we have said until now, it is reasonable to think that resistance is expressed by a real number, but it may happen to be expressed with a complex number, too. In that case, we will no longer talk about resistance *R*, but more generally about *impedence Z*. This important concept will be taken up and extended in Chapter 7, relating to the "frequency domain."

Resistance is the intrinsic property of a fundamental device: the *resistor*. Thanks to its resistance, the resistor is able to limit or regulate the flow of electrical current in an electrical circuit.

The term *resistor* refers to a device, while the term *resistance* is its feature. However, it often happens that in common usage, we talk improperly about resistance meaning the device itself.

Resistors are an example of components and are passive devices, so they *dissipate* and do not generate electrical power. Their symbol is shown in Figure 4.15.

Figure 4.16 shows some commercially available resistors. Please note the colors stamped on the body of the resistors. They represent codes that indicate the resistance value, as detailed in Section 4.4.3.

4.4.2 Classification

The resistors can be classified as in Table 4.2.

The first distinction is about the nominal value of the resistor, whether it is fixed or variable. Then between the variable resistors, we distinguish those whose nominal value is changeable in a mechanical way (manually or with a tool) or because of other external causes (temperature, light, pressure, etc.).

The structure with which the resistors are produced generally respects the classification shown in Table 4.3.

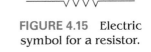

FIGURE 4.15   Electric symbol for a resistor.

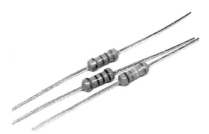

FIGURE 4.16   Commercially available resistors.

TABLE 4.2
Classification of Resistors

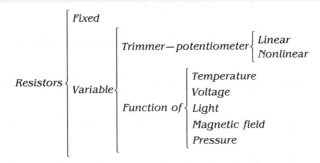

$$Resistors \begin{cases} Fixed \\ \\ Variable \begin{cases} Trimmer-potentiometer \begin{cases} Linear \\ Nonlinear \end{cases} \\ \\ Function\ of \begin{cases} Temperature \\ Voltage \\ Light \\ Magnetic\ field \\ Pressure \end{cases} \end{cases} \end{cases}$$

TABLE 4.3
Classification of the Structure of Resistors

$$Structure \begin{cases} Foil \\ \\ Grid \begin{cases} Carbon \\ Metal \begin{cases} Thick\ film \\ Thin\ film \end{cases} \end{cases} \end{cases}$$

Curiosity

Not all resistors are used for all purposes. For instance, we can mention resistors for medical applications, when they are requested to support high voltages. These resistors have critical features such as *linearity*, expressed by voltage coefficient (*VCR*) and temperature coefficient (*TCR*), and *long-term stability* under voltage stress.

4.4.3 Coding Scheme

The standard resistors are commonly made of a cylinder of a material (typically a carbon composition) with a known resistivity, covered with insulating material on which is stamped the Ohm nominal value of resistance R_N, according to a code that uses a color-coding scheme. The real value R of a single resistor is in the range $R = R_N \pm \Delta R_N$, namely the nominal value, algebraically added to the tolerance, that may be expressed in absolute value ΔR_N or, more commonly, in relative value percentage $\Delta R_N / R_N \times 100 = \Delta R_N \%$.

The color-coding scheme typically consists of four colors to which are associated values, as in Table 4.4.

The first color represents the most significant digit of the value of R_N, the second color the second significant digit, the third color corresponds to the multiplier, that is the number of zeros to be added to the first two digits, and lastly, the fourth colored band corresponds to the value of the tolerance $\Delta R_N \%$ (Figure 4.17).

For example, a resistor with the bands colored yellow/violet/green/gold has a resistance value of $4.7\,M\Omega$ (yellow: 4, violet: 7, green: $\times 10^5$), with a tolerance of $\pm 5\%$ (gold: 5).

Sometimes you can find resistors with a fifth color that stands for a so-called *quality factor*, namely the probability of failure after 1000 hours of operation. Other times, five color bands have the nominal value indicated by the first three colors, the fourth band indicates the multiplier value, and the fifth band indicates the tolerance.

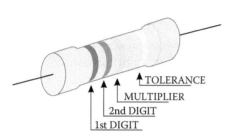

FIGURE 4.17 Color-coding scheme printed on a resistor.

In commerce, resistors with any value of resistance do not exist, of course. The commercial values that we can find are only those that follow specific standards

TABLE 4.4

Color Codes

Color	Resistance value [Ohm]			Tolerance
	1st digit	2nd digit	3rd digit	
None	-	-	-	± 20%
Silver	-	-	10^{-2}	± 10%
Gold	-	-	10^{-1}	± 5%
Black	0	0	10^0	-
Brown	1	1	10^1	± 1%
Red	2	2	10^2	± 2%
Orange	3	3	10^3	-
Yellow	4	4	10^4	-
Green	5	5	10^5	± 0.5%
Blue	6	6	10^6	-
Violet	7	7	10^7	-
Grey	8	8	10^8	-
White	9	9	10^9	-

and that give a *normalized value* and *tolerance*. *Normalized value* (also known as *preferred value*) is the one from which all the other values are obtainable, by multiplying or dividing it by 10 or multiples of 10. The series on the market are identified by the letter "E" followed by a number. In particular, the series named *E6* is related to resistors with tolerance of ±20%, for the series *E12* the tolerance is ±10%, for *E24* the tolerance is ±5%, for *E48* the tolerance is ±2%, for *E96* the tolerance is ±1%, and lastly for *E192* the tolerance is ±0.5%. For standardized values of some for the mentioned series we can refer to Table 4.5.

TABLE 4.5
Normalized Values of Some Series of Resistors

Series	E96	E48	E24	E12
Normalized value	100	100	100	100
	102			
	105	105		
	107			
	110	110	110	
	113			
	115	115		
	118			
			120	120
	121	121		
	124			
	127	127		
	130		130	
	133	133		
	137			
	140	140		
	143			
	147	147		
	150		150	150
	154	154		
	158			
			160	
	162	162		
	165			
	169	169		
	174			
	178	178		
			180	180
	182			
	187	187		
	191			
	196	196		
	200		200	
	205	205		

(Continued)

TABLE 4.5
Normalized Values of Some Series of Resistors (*Continued*)

Series	E96	E48	E24	E12
	210			
	215	215		
			220	220
	221			
	226	226		
	232			
	237	237		
			240	
	243			
	249	249		
	255			
	261	261		
	267			
			270	270
	274	274		
	280			
	287	287		
	294			
			300	
	301	301		
	309			
	316	316		
	324			
			330	330
	332	332		
	340			
	348	348		
	357			
			360	
	365	365		
	374			
	383	383		
			390	390
	392			
	402	402		
	412			
	422	422		
			430	
	432			
	442	442		
	453			
	464	464		
			470	470
	485			
	487	487		
	499			

TABLE 4.5
Normalized Values of Some Series of Resistors (*Continued*)

Series	E96	E48	E24	E12
			510	
	511	511		
	523			
	536	536		
	549			
			560	560
	562	562		
	576			
	590	590		
	604			
	619	619		
			620	
	634			
	649	649		
	665			
			680	680
	681	681		
	698			
	715	715		
	732			
	750	750	750	
	768			
	787	787		
	806			
			820	820
	825	825		
	845			
	866	866		
	887			
	909	909		
			910	
	931			
	953	953		
	976			

4.4.4 Power Resistors

When choosing a resistor, it is wrong to consider only its "ohmic" value. The maximum power that it can dissipate without damage must also be considered. For the standard resistors, typical power values that can be dissipated are $1/8\,W$, $1/4\,W$, and $1/2\,W$.

This is generally defined as *nominal power*—the value of power that the resistor can dissipate, when at room temperature of 25°C, without permanently altering its structure. A *power resistor* is able to dissipate nominal powers equal to $1\,W$, $2\,W$ and more. Commonly, it has a parallelpiped shape (rather than cylindrical) and a ceramic body. The resistive material is mainly made of a nickel-chromium wire.

For power resistors, the ohmic value is sometimes stamped as an alphanumeric code rather than the aformentioned color-coding scheme, with a letter *R* (indicating that this is a *R*esistance) placed instead of the symbol Ω. This code is preceded by the value of the power that the resistor is able to dissipate, with the letter *W*, which stands for Watts (Figure 4.18).

FIGURE 4.18   Power resistor.

4.4.5 Trimmer and Potentiometer

Trimmers (or *presets*) and *potentiometers* (colloquially known as *pots*) are particular resistors (or *rheostats*) whose nominal resistance value can be varied with external action. They are provided with a sliding contact within the body of the resistor, which is a third (central) terminal, in addition to the two (external) classic ones. The sliding contact can be moved as we desire, acting from the outside. Typically, we move that contact with a screw, in the case of a trimmer, and with a handle or a slide in the case of the potentiometers.

Considering the central terminal and one of the two external ones, the ohmic value of the resistor will change depending on the position that the sliding contact assumes (Figure 4.19a through c).

The trimmers and the potentiometers may be linear or nonlinear in type. For the case of linear, the resistance varies with direct proportionality to the sliding contact movements. In the case of nonlinearity, the principal functional features are shown in Figure 4.20b through d, where we see the resistance percentage value *R* graphed versus the sliding contact point "*d*" (Figure 4.20).

4.4.6 SMD Resistors

To increase the possibility of circuital integration and the density of components on the surface unit, we have commercial resistors named *SMD* (*Surface Mounted Device*) that have an interesting feature of being particularly small (Figures 4.21 and 4.22).

For this type of resistor, the identification is given by a code with three numbers: the first two numbers identify the nominal value, while the third is the number of zeros to add to this value. Tacitly the Ohm is the unit of measurement. So, for example, the code "334" represents a resistor with 330,000 Ω (i.e., the number 33 followed by 4 zeros), while the code "120" represents a resistance of 12 Ω (i.e., the number 12 followed by no zeros).

For *SMD precision resistors*, the identification code is four digits, the first three representing the first three numbers of the resistance value, and the fourth representing the power of ten. So, for instance, "1001" stands for $100 \times 10^1 = 1000\,\Omega = 1k\Omega$.

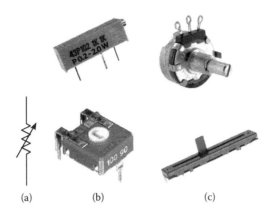

(a)   (b)   (c)

FIGURE 4.19 (a) Electric symbol for a variable resistor, and examples of commercial (b) trimmers and (c) potentiometers.

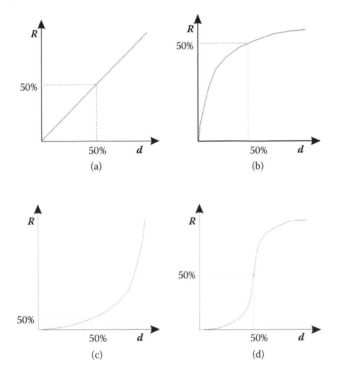

FIGURE 4.20 Some possible variations of R for the trimmers and potentiometers.

4.4.7 Ohm's Law

Ohm's law expresses the link between the voltage V across a resistor of resistance R with current I flowing into it:

$$V = RI$$

(please refer to the discussion in Section 3.9 of Chapter 3).

FIGURE 4.21   SMD resistors.

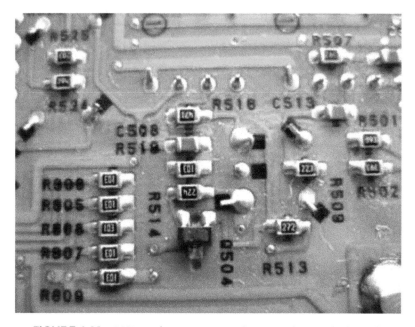

FIGURE 4.22   SMD resistors mounted on an electronic board.

Examples

 Let's consider a classic conductor in the form of the electrical cable, made of copper, common in our homes. Since the electrical cable has an approximate cylindrical shape, it may be considered a standard resistor, even if this is not technically correct because it is not made to limit or regulate the flow of current, as is required from real resistors. In fact, the ideal electrical cable should have a value of zero resistivity but instead (even if it is really low), it is approximately $\rho = 1.8 \times 10^{-8} \, \Omega m$ (as shown in Table 4.1). For electrical wiring used to carry electricity in homes (building wiring), the electrical cables for electric-lights circuitry usually have a section of $S = 1.5 \, mm^2$. So if we use, for example, a conductor $l = 100 \, m$ long, with a current of $10 A$, then the voltage across the cable's ends is

$$V = RI = \rho \frac{l}{S} I = 1.8 \times 10^{-8} \; \Omega m \frac{2 \times 100 \; m}{1.5 \times 10^{-6} \; m} \times 10 \, A = 24 \, V$$

(the number 2 that multiplies the length of the conductor considers that the electric circuit is made of starting and return cables). This means that if at the beginning of the conductor there are $110 \, V$ at the end of the circuit we will have $110 - 24 = 86 \, V$, which may be a problem for devices designed to be powered with $110 \, V$. The problem is solved, for example, by choosing a conductor with a bigger section S.

FIGURE 4.23 Lightning crossing the atmosphere.

As another example let's apply Ohm's law to the case of lightning that crosses the atmosphere (this application is valid only approximately, since we are assuming the atmospheric medium has a constant resistance, which is only partly real) (Figure 4.23).

In this case, the typical voltage value is $V = 20 \times 10^9 \, V$, and considering the medium mainly made of water droplets for which $R = 2.6 \times 10^5 \, \Omega$, we will have a current equal to

$$I = \frac{20 \times 10^9}{2.6 \times 10^5} \cong 7.7 \times 10^4 \, A$$

Since, on average, the human body can tolerate currents on the order of $30 \, mA$ it is convenient to not challenge thunderstorms.

 Curiosity

Only for ease of memorization, the so-called "magic triangle" can be useful in the representation of Ohm's law:

To know V it is sufficient to cover it with a finger and read RI; to know R it is sufficient to cover it and read V/I; to know I it is sufficient to cover it and read V/R.

Observation

 About linearity:

Ohm's law may be represented on a V-I Cartesian diagram (Figure 4.24).

The voltage-current characteristic is linear with R. In fact Ohm's law is expressed with the equation of a straight line expressed by the *slope–intercept form*, $y = mx + q$, without the y-intercept (the q term), and for which the slope m is equal to $1/R$. With the example of Figure 4.24, we can see how a higher resistance ($R_1 > R_2$) matches a lower slope of the straight line. When equal voltage is applied across R_1 a lower current will flow compared to R_2.

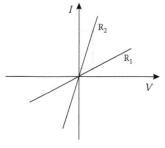

FIGURE 4.24 Graphic representation of the V-I characteristic for generic resistances.

4.4.8 Joule's Law

Let's apply the electric power definition, $P = VI$, to the case of a resistance, and since we know Ohm's law, $V = RI$, we get

$$P = VI = RI^2 = \frac{V^2}{R}$$

so from the value of R and the voltage V across it, or from the current I flowing through it, it is possible to identify the value of the dissipated power.

Example

We can apply the equation of electric power to the case of a common lamp, 100 W, fed with the house voltage of 110 V. Its filament resistance is $R = V^2/P = 121\,\Omega$, and the current flowing is $I = P/V \cong 0.91\,A$.

The power we are talking about is completely dissipated in the resistor. In fact, when the current flows in it, part of the potential electrical energy owned by the electrical charges is transformed into heat, and the electric voltage decreases accordingly. This effect is named *Joule's law*, also known as the *Joule effect*. The heat increases the temperature of the resistor proportionally to the current squared, so when the currents become too high, we risk damage to the resistor and/or damage to the insulation that coats it.

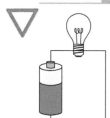

Example

Having a light bulb that is 110 V, 75 W, we can calculate the current that flows when it is turned on and the resistance it gives to the passage of current.

From the electrical power definition we extract $I = P/V = 75/110 \cong 0.68\,A$, while from Ohm's law we extract $R = V/I = 110/0.68 \cong 161.7\,\Omega$.

4.4.9 Real Power

If the supply voltage is not a continuous function but sinusoidal, namely half of the time positive and half of the time negative, the current will also be sinusoidal, but the power that is the product of the two will be positive all the time. This means that the direction of energy flow does not reverse, even if the voltage and the current both reverse. It can be compared to a person who goes back and forth a few steps on a "moving walkway," for whom the net motion is "one way."

In fact, a sinusoidal voltage

$$v(t) = V_M \sin (\omega t)$$

(V_M is the peak amplitude) applied across a resistor R, determines a current in phase with the voltage, as

$$i(t) = I_M \sin (\omega t)$$

(I_M is the peak amplitude), so the instantaneous electric power, as the product, is

$$p(t) = v(t)i(t) = V_M \sin (\omega t)\, I_M \sin (\omega t) = V_M I_M \sin^2(\omega t)$$

(see Figure 4.25).

But recalling that $\sin^2 (\omega t) = \dfrac{1 - \cos (2\omega t)}{2}$ we could rewrite power as

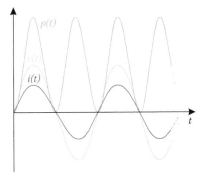

FIGURE 4.25 $v(t)$, $i(t)$, and their product $p(t)$ when voltage and current are in phase.

$$p(t) = V_M I_M \frac{1 - \cos (2\omega t)}{2} = \frac{1}{2} V_M I_M - \frac{1}{2} V_M I_M \cos (2\omega t)$$

which is the algebraic sum of a constant term

$$\left(\frac{1}{2} V_M I_M \right)$$

and a cosinusoidal one

$$\left(-\frac{1}{2} V_M I_M \cos (2\omega t) \right)$$

of double frequency ($2\omega t$) with respect to the voltage and current.

Recalling that the RMS value of a sinusoidal function is equal to the maximum value of the function divided by $\sqrt{2}$ (namely $F_{RMS} = \dfrac{F_M}{\sqrt{2}}$), we could write the previous expression as

$$p(t) = \frac{1}{2} V_M I_M - \frac{1}{2} V_M I_M \cos (2\omega t) = \frac{V_M}{\sqrt{2}} \frac{I_M}{\sqrt{2}} - \frac{V_M}{\sqrt{2}} \frac{I_M}{\sqrt{2}} \cos (2\omega t)$$

$$= V_{RMS} I_{RMS} - V_{RMS} I_{RMS} \cos (2\omega t)$$

This expression of $p(t)$ represents the value of the *instantaneous power*, being at time t. So, the power consists of a DC ($V_{RMS} I_{RMS}$) part and an AC ($V_{RMS} I_{RMS} \cos (2\omega t)$) part with double frequency with respect to both voltage and current. But what about the mean power delivered P? It is the average, named *real power*, and consists of the total energy converted in one cycle divided by the period T of the cycle:

$$P \underset{\text{def}}{=} \frac{1}{T} \int_{t=0}^{T} p(t)\, dt = \frac{1}{2\pi} \int_{0}^{2\pi} v(t)i(t)\, dt = \frac{1}{2\pi} \int_{0}^{2\pi} \frac{V_M^2}{R} \sin^2 (\omega t)\, dt$$

$$= \frac{1}{2\pi} \frac{V_M^2}{R} \int_{0}^{2\pi} \frac{1}{2} (1 - \cos (2\omega t))\, dt$$

Now the sinusoidal term averages to zero over any number of complete cycles, so we simply obtain

$$P = \frac{V_{RMS}^2}{R} = RI_{RMS}^2 = V_{RMS}I_{RMS}$$

4.4.10 Physical Meaning of the RMS Value

Let's consider a time-varying AC signal that delivers to a resistor, in a time t, a certain amount of real power P. What would be the value of a constant DC signal capable of delivering to the same resistor, in the same time t, exactly the same amount of power P? This constant DC signal corresponds to the RMS value of the AC signal.

So, the RMS (or *effective*) value of a signal is equal to the DC value that provides the same average power to a resistor.

Let's detail the reason for this by considering a signal of current $i(t)$ (Figure 4.26).

Figure 4.26 represents the functions $i(t) = I_M \sin (\omega t)$ and $i^2(t)$. As we can see, the sinusoidal function squared assumes only positive values and double frequency, with a maximum value I_M^2 and an average value $I_M^2/2$.

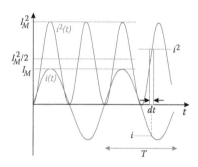

FIGURE 4.26 $i(t)$ and $i^2(t)$.

We assume that the sinusoidal current $i(t)$ flows in a resistor with resistance R. In the interval of time dt (in reality, it is infinitesimal, but represented as finite in Figure 4.26 for graphical reasons) we can consider the current as constant, equal to i^2, and the energy delivered $dE = Ri^2dt$. The quantity i^2dt corresponds to the rectangle with infinitesimal area dA. In a whole period T there will be an infinite number of these rectangles, so the delivered energy will be related to the sum of all these, $dE = R(dA_1 + dA_2 + dA_3 + ...)$. The sum of the infinite terms dA corresponds to the area A under the curve i^2dt. Such area A, for obvious graphic reasons, is also equal to $A = I_M^2/2\,T$, therefore we can write

$$dE = R\left(dA_1 + dA_2 + dA_3 + ...\right) = R\frac{I_M^2}{2}T$$

But, by definition, the *RMS* value of the function $i(t)$ is $I_{RMS} = \frac{I_M}{\sqrt{2}}$, so we can say

$$dE = R\frac{I_M^2}{2}T = R\frac{I_M}{\sqrt{2}}\frac{I_M}{\sqrt{2}}T = RI_{RMS}^2T$$

or else the power delivered by the resistor R in the time T is due to the RMS value of the current squared. Finally, we note that exactly the same result is obtainable with a DC current of value I_{RMS}.

In summary, for a periodic signal, the RMS value of an AC signal is exactly the value of a corresponding DC signal that would have resulted in the same average power dissipation.

The mains (AC electrical) supply is rated by its *RMS* values. So, the sinusoidal voltage has a peak amplitude of $110 \times \sqrt{2} \cong 155[V]$ in the United States and $220 \times \sqrt{2} \cong 311.1[V]$ in Europe.

4.4.11 Resistors in Series

FIGURE 4.27 Resistors in series.

We adopt the generic definition given for the series topology (Section 4.3.1) for the case of resistors.

Two resistors in series have one terminal in common, as in Figure 4.27.

When in series and inserted in a circuit, their current is the same, so indicating with I_{R1} the current flowing in the resistor R_1 and with I_{R2} the current flowing in the resistor R_2, we will have

$$I = I_{R1} = I_{R2}$$

Instead, the total voltage V will be the sum of the voltage V_{R1} at the terminals of R_1 and the voltage V_{R2} at the terminals of R_2, therefore

$$V = V_{R1} + V_{R2}$$

The described situation is schematized in the following figures, where resistors can be considered lightbulbs (Figure 4.28).

For both the resistors, you can write the equations due to Ohm's law:

$$V_{R1} = R_1 I_{R1}; \; V_{R2} = R_2 I_{R2}$$

Combining the last two equations with the previous ones we will have

$$V = V_{R1} + V_{R1} = R_1 I_{R1} + R_2 I_{R2} = R_1 I + R_2 I = (R_1 + R_2)I$$

We obtain, then, that the total resistance is the sum of resistances:

$$V = R_{TOT} I; \; R_{TOT} = R_1 + R_2$$

However, this result obtained with only two resistors is also valid for any number of them, so in the general case of N series resistors we have

$$R_{TOT} = \sum_{n=1}^{N} R_n$$

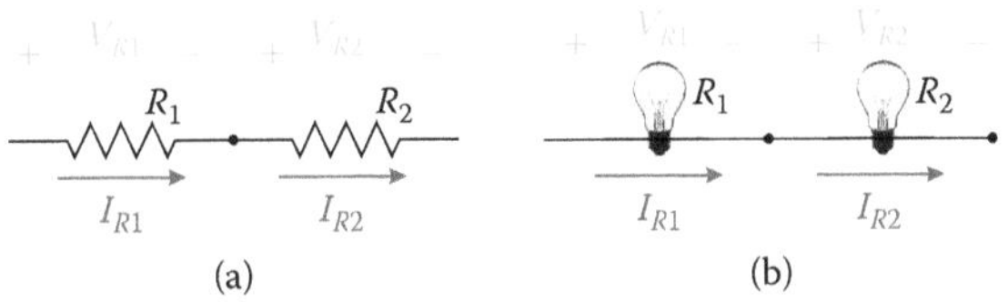

(a) (b)

FIGURE 4.28 Resistors in series, schematized by (a) symbols and (b) lightbulbs.

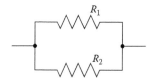

FIGURE 4.29   Resistors in parallel.

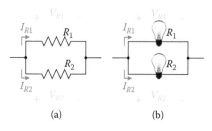

(a)   (b)

FIGURE 4.30   Resistors in parallel, schematized by a (a) symbol and (b) lightbulbs.

4.4.12 Resistors in Parallel

We adopt the generic definition given for the parallel topology (Section 4.3.2) for the case of resistors.

Two resistors in parallel have two terminals in common, as in Figure 4.29.

In the following, and really on many occasions, we'll adopt the notation $R_1 || R_2$ to represent the two resistors R_1 and R_2 in a parallel topology.

When the resistors are in parallel, they have the same voltage across them, so if we take V_{R1} as the voltage across the resistor R_1 and V_{R2} as the voltage across the resistor R_2, we will have (Figure 4.30)

$$V = V_{R1} = V_{R2}$$

The total current I will be the sum of the current I_{R1} that flows in R_1 and the current I_{R2} that flows in R_2, therefore

$$I = I_{R1} + I_{R2}$$

For both the resistors we can write the equations due to Ohm's law:

$$V_{R1} = R_1 I_{R1}; \; V_{R2} = R_2 I_{R2}$$

Combining the last two equations with the previous ones, we will have

$$I = I_{R1} + I_{R2} = \frac{V_{R1}}{R_1} + \frac{V_{R2}}{R_2} = \frac{V}{R_1} + \frac{V}{R_2} = V\left(\frac{1}{R_1} + \frac{1}{R_2}\right) \Rightarrow V = \left(\frac{1}{\dfrac{1}{R_1} + \dfrac{1}{R_2}}\right) I$$

We then obtain the total resistance as

$$R_{TOT} = \left(\frac{1}{\dfrac{1}{R_1} + \dfrac{1}{R_2}}\right) \Rightarrow \frac{1}{R_{TOT}} = \frac{1}{R_1} + \frac{1}{R_2}$$

namely the reciprocal of the total resistance is equal to the sum of the reciprocals of the resistances in parallel.

However, this result obtained with only two resistors is also valid for any number of them, so in the general case of N parallel resistors we have

$$\frac{1}{R_{TOT}} = \sum_{n=1}^{N} \frac{1}{R_n}$$

In general, by connecting more resistors in series we obtain increases in the total resistance, while with the parallel topology the total resistance decreases. In particular for series resistors, the total resistance is higher than the largest of the resistances, while for the parallel resistors, the total resistance is lower than the smallest of the resistances. This is reasonable if we think that the series connection corresponds, practically, to an increase in the length of the conductor, keeping the same section (and the value of R is proportional to the length of the conductor l), while the parallel connection has the same length but increases the total section through which the total current can be divided (and the value of R is inversely proportional to the section of the conductor S).

4.4.13 Resistor Bridge

The bridge topology, we already described in the case of generic components (Section 4.3.3), now is specialized for resistors. In this case, the configuration is named the *Wheatsone bridge* (Charles Wheatstone, English physicist, 1802–1875, Figure 4.31).

A *balanced bridge* refers to the case in which no current flows in the resistor that acts as a bridge, otherwise the configuration is called *unbalanced* (Figure 4.32).

FIGURE 4.31   C. Wheatstone.

4.4.14 Resistors in Star and Triangle Connections

We already defined the triangle and star connection between generic components in Section 4.3.4. Now we will specialize such definitions to resistors.

Resistors in a star (known also as *wye* or *Y* or *T*) connection, can be schematized according to the equivalent representations in Figure 4.33.

Cases (a) and (b) in Figure 4.33 are obviously identical since, for us, it is not important how the components are positioned, but only how they are connected.

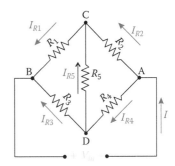

FIGURE 4.32   Wheatstone bridge.

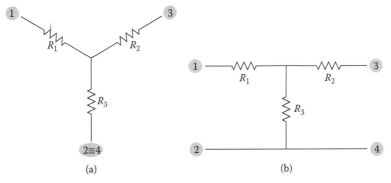

(a)   (b)

FIGURE 4.33   Resistors in a star connection: (a) Y and (b) Π scheme.

Also the connection wire between points 2 and 4 in Figure 4.33b is not really necessary, but simply useful to show how a voltage can be applied as input to the left of the topology (across points 1 and 2) and how a voltage can be obtained as output (across points 3 and 4) to the right of the same scheme.

Despite the two schemes being identical, the one in Figure 4.33a is used more in electrotechnics while the other in Figure 4.33b is more adopted in electronics (for reasons that will become clear later) and is also named the *T scheme*, clearly relating to its shape.

As reported for connections schematized in Figure 4.33, we can also assume equivalent topologies for the triangle (or Δ) connections, which are drawn in Figure 4.34, where the (b) case is also known as Π scheme.

It is possible to determine appropriate values for the resistors in order to move from a star scheme to a triangle one and vice versa, i.e., the two schemes are equivalent with the right resistance values. For example, let's now obtain the values of the resistors of the star configuration starting with those in the triangle configuration. We denote by

→ $R_{ij}(Y)$ the equivalent resistance between the points i and j ($i = 1, 2, 3; j = 1, 2, 3$) of the star connection

→ $R_{hk}(\Delta)$ the equivalent resistance between the points h and k ($h = 1, 2, 3; k = 1, 2, 3$) of the triangle connection

Then

$$R_{12}(Y) = R_1 + R_3; R_{12}(\Delta) = R_B \mid\mid (R_C + R_A) = \frac{R_B(R_A + R_C)}{R_A + R_B + R_C}$$

$$\Rightarrow R_1 + R_3 = \frac{R_B(R_A + R_C)}{R_A + R_B + R_C} \tag{eq. I}$$

$$R_{13}(Y) = R_1 + R_2; R_{13}(\Delta) = R_C \mid\mid (R_B + R_A) = \frac{R_C(R_A + R_B)}{R_A + R_B + R_C}$$

$$\Rightarrow R_1 + R_2 = \frac{R_C(R_A + R_B)}{R_A + R_B + R_C} \tag{eq. II}$$

$$R_{23}(Y) = R_3 + R_2; R_{23}(\Delta) = R_A \mid\mid (R_B + R_C) = \frac{R_A(R_B + R_C)}{R_A + R_B + R_C}$$

$$\Rightarrow R_2 + R_3 = \frac{R_A(R_B + R_C)}{R_A + R_B + R_C} \tag{eq. III}$$

By subtracting eq. III from eq. I, we obtain an equation for $R_1 - R_2$ that, added to the eq. II, furnishes

$$R_1 = \frac{R_B R_C}{R_A + R_B + R_C}$$

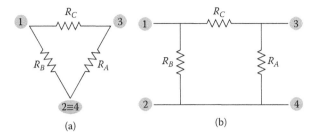

FIGURE 4.34   Resistors in a triangle connection: (a) classic and (b) ∏ scheme.

By combining the previous equations in a different way we can obtain the values for R_2 and R_3 too.

4.4.15 Voltage Divider

The circuit schematized here, in two versions but completely equivalent (Figure 4.35), realizes a voltage divider with resistors. So, with V the voltage across the series, if we consider the voltage only across the resistor R_2 we can write

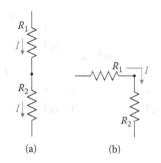

FIGURE 4.35   A voltage divider made of resistors in series.

$$V_{R2} = \frac{R_2}{R_1 + R_2} V$$

which represents a very useful and adopted equation for the resistor voltage divider.

It is easily demonstrable because it equals the two currents $I = V_{R2}/R_2$ and $I = V/(R_1 + R_2)$.

We can interpret the obtained result by observing how the voltage V is divided between the two resistors. The value of each portion of voltage (V_{R1} and V_{R2}) at the ends of one resistance is given by the total voltage V proportionally reduced. Then

$$V_{R1} = \frac{R_1}{R_1 + R_2} V; \; V_{R2} = \frac{R_2}{R_1 + R_2} V$$

If we apply that result to a more generic case of N series resistances (Figure 4.36), we have for the ith resistance (where i can be any number between 1 and N)

$$V_{Ri} = \frac{R_i}{\sum_{n=1}^{N} R_n} V$$

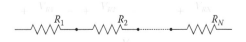

FIGURE 4.36   Voltage distribution among a series of resistances.

FIGURE 4.37 A current divider made of two resistors in parallel.

4.4.16 Current Divider

Figure 4.37 represents a resistor current divider.

It splits the source current I between the two resistors R_1 and R_2, respectively. So, the current that flows in the resistor R_1 is equal to

$$I_{R1} = \frac{V}{R_1} = \frac{(R_1||R_2)I}{R_1} = \left(\frac{R_1 R_2}{R_1 + R_2}\right)\frac{I}{R_1} = \frac{R_2}{R_1 + R_2}I$$

$$\left(with\ R_1||R_2 = \frac{R_1 R_2}{R_1 + R_2}\right)$$

The same can be considered for I_{R2}, so we can write the final equations as

$$I_{R1} = \frac{R_2}{R_1 + R_2}I;\ I_{R2} = \frac{R_1}{R_1 + R_2}I$$

We can express the equations according to the conductance G, instead of resistance R, obtaining

$$I_{R1} = \frac{G_1}{G_1 + G_2}I;\ I_{R2} = \frac{G_2}{G_1 + G_2}I$$

Generalizing for N parallel resistors, we have for the ith resistor (where i can be any number between 1 and N)

$$I_{Ri} = \frac{G_i}{\sum_{n=1}^{N} G_n}I$$

4.5 ELECTRICAL SOURCES

An *electrical source supply* or *electrical source* or, simply, a *source*, is a device that supplies electrical power. We can distinguish two main sources: the voltage and the current, each of them subdivided into *direct* (DC) or *alternating* (AC) sources. The symbols for AC and DC are shown in Figure 4.38,

(a)   (b)

FIGURE 4.38   Circuit symbol for (a) alternating current/voltage and (b) direct current/voltage.

4.5.1 Ideal Voltage Source

An *ideal voltage source* is defined as a two-terminal device capable of supplying and maintaining the same voltage $v_s(t)$ across its terminals for any current flowing through it (requested by the external circuit).

Therefore, as defined, the output voltage is calculated according to the following equation:

$$v_s(t) = f_v(t)$$

where $f_v(t)$ refers to an arbitrary function that does not have any dependence on other electrical parameters. Considering such independence, the source we are talking about is also said to be an *independent voltage source*.

The independent voltage source is typically used as a model for many real electrical sources such as batteries, electric accumulators, alternators, dynamos, and so on.

In the case that the function has a constant value, $v_s(t) = cost$, we will talk generally about a *constant voltage source*, a typical example being the electric battery. Figure 4.39 reports two possible symbols and the V-I graph.

More generally, the function $v_s(t)$ can take a periodic course in time (typically sinusoidal). In this case, we will talk about a *variable voltage source* (typically *sinusoidal voltage source*), of which the possible symbols are shown in Figure 4.40.

Nominal voltage value is defined to be the value of voltage provided by the source, a value typically stamped on the body of the device.

The term *biasing* means the deliberate application of a DC voltage across a component, directly or indirectly by means of a DC source, an operation fundamental in many of the electronic circuits designed, detailed, and discussed in the following.

Ideal voltage sources cannot be connected in parallel.

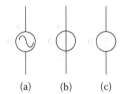

FIGURE 4.39 (a), (b) Voltage source symbols, (c) I-V characteristic of an ideal voltage source.

FIGURE 4.40 Some commonly adopted symbols for the variable voltage source.

Curiosity

The first practical method of obtaining an electric voltage, as a reliable source of electrical current, is credited to Alessandro Volta of Italy. In 1800, he built the voltaic pile, the first battery in history, a device that produces electricity from a chemical reaction.

4.5.2 Ideal Current Source

An *ideal current source* is defined as a two-terminal device capable of producing and maintaining a defined current $i_s(t)$ for any voltage at its terminals.

Therefore, as defined, we describe a constitutive equation by

$$i_s(t) = f_i(t)$$

where $f_i(t)$ refers to an arbitrary function that does not have any dependence on other electrical parameters. Considering such independence, the source we are talking about is also said to be an *independent current source*.

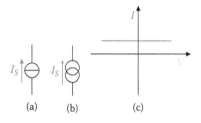

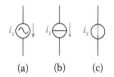

FIGURE 4.41 (a), (b) Current source symbols, (c) I-V characteristic of an ideal current source.

FIGURE 4.42 Some commonly adopted symbols for the variable current source.

The independent source of current is typically used as a model for various electronic circuits containing real active elements.

When the function has a constant value, $i_s(t) = cost$, we will talk generically about a *constant current source*. Two of its commonly adopted symbols and and the V-I graph is shown in Figure 4.41.

More generally, the function $i_s(t)$ can take a periodic course in time (typically sinusoidal). In this case, we will talk about a *variable current source* (typically *sinusoidal current source*), of which the possible symbols are shown in Figure 4.42.

Nominal current value is defined to be the value of current provided by the source, a value typically printed on the body of the device.

Ideal current sources cannot be connected in series.

4.5.3 Non-Ideal Voltage and Current Sources

The model given from ideal voltage and current sources, although used often, does not exactly match those that are the real ones. In fact, ideal sources are mathematical abstractions, and their model can also lead to certain incongruities. For example, an ideal voltage source can provide infinite current with non-zero voltage difference, and an ideal current source can provide infinite voltage difference with non-zero current, so either device is theoretically capable of delivering infinite power! We know that for two devices in a parallel topology, the voltage across them is the same, but then what happens if those two devices are two ideal voltage sources, each of them providing a voltage of different value? We will have the paradox that the voltage should be the same, but each of the sources can impose a different one.

An analogous paradox is for two current sources in series, each of them providing a different current value. What will be the real amount of current flowing through them?

These incongruities find solutions when considering the independent sources as a real model that takes into account their unwanted but unavoidable internal electrical resistance (as defined in Section 3.9 of Chapter 3). Since these devices are made of materials with non-zero resistive value, their internal resistance cannot be null. So, this internal resistance will be such to distinguish the ideal model from the real one.

For the voltage source this means that it may not always provide a voltage of constant value, at any time and for every load. Therefore, the voltage it supplies could be lower (for it to be higher would be an absurdity) than the nominal one. Therefore, a quantity (of course the smallest possible) of the voltage can be subtracted from the nominal value. Such behavior can be schematized with a series resistance, the reason for the difference between the nominal voltage and the effective one applied to the load (Figure 4.43).

Therefore, a real voltage source V_s can be schematized with an ideal one V_{si} with a series resistor R_s, named *internal resistance* because it is not accessible from the outside, being part of the source itself.

The value of the internal resistance typically changes with the utilization of the real source. The typical example of a constant voltage source is the *battery*, and it is well known that it will be exhausted with usage, that is it provides a value of voltage that steadily decreases. This behavior can be schematized with a value of internal resistance R_s that, step by step, becomes higher, and the greater this value is, the lower the voltage V_s we can obtain from the device.

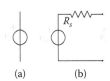

FIGURE 4.43 (a) Ideal and (b) real (ideal + resistance) voltage source model.

The lower value of the internal resistance, the more a real voltage source can be approximated to an ideal one.

This is also true for the real current source. In particular, to create a scheme that can provide a real value of current i_s lower than the ideal one i_{si}, it is necessary that we put its internal resistance R_p in parallel with the ideal source. In fact, it is with the parallel topology, and not with the series one, that we can obtain an output with a reduced current. So part of the current supplied by the ideal source will flow on the parallel resistance rather than on the load.

The higher the value of the internal resistance, the more a real current source can be approximated to an ideal one (Figure 4.44).

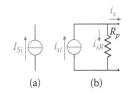

FIGURE 4.44 (a) Ideal and (b) real (ideal + resistance) current source model.

4.6 CAPACITORS

4.6.1 Capacitance

In electronics it is fundamental to have the possibility to "store" electric charge. The measure of the capacity of storing electric charge Q for a given potential difference ΔV is named *capacitance* C:

$$C = \frac{Q}{\Delta V}$$

The SI unit for capacitance is coulombs per volt, or Farad [F]. This is a large unit, so a typical capacitance is in the picofarad [pF] to millifarad [mF] range.

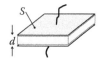

FIGURE 4.45 A parallel plate capacitor. The conducting plates are blue, and the dielectric is yellow.

The *capacitor* (sometimes simply referred to as *cap*) is the electronic two-terminal device for storing charge. It consists of two conducting *plates*, flat and parallel in their simplest form, separated by an insulating material called the *dielectric*, as seen in Figure 4.45.

The dielectric keeps the opposite sign charges separated, avoiding their recombination and the nullifying of the electric field between them. The property of the dielectric is the *dielectric constant* ε, which measures the "ability" to store electrical energy in an electric field or, from a different point of view, the resistance encountered when forming an electric field in it. In fact, as the capacitor is biased (for example by means of a battery), positive charges are collected on one plate and negative ones on the other, and an electric field is developed across the dielectric. In an ideal capacitor, the electric field should be made of straight lines, forming a uniform field, contained entirely between the plates. Because the plates are of finite size, the electric field lines bend more and more approaching the edges, a phenomenon know as the *edge effect* (Figure 4.46).

The non-uniform fields near the edge are known as *fringing fields*. In any case, for the most part in applications the edge effect is ignored.

The capacitance C of a flat plate capacitor is directly proportional to the surface areas of the plates S, and to the dielectric constant ε of the insulating material separating the plates, and is inversely proportional to their distance d:

$$C = \varepsilon \frac{S}{d}$$

with $\varepsilon = \varepsilon_0 \varepsilon_r$, $\varepsilon_0 = 8.854 \times 10^{-12}$ [F/m] the dielectric value of vacuum, and ε_r the dielectric constant relative (also called *relative permittivity*) to the adopted insulator.

Observation

In the formula $C = \varepsilon \, (S/d) = \varepsilon_0 \varepsilon_r \, (S/d)$ the contribution of ε_0, of the order of magnitude of 10^{-12} [F/m], makes it really hard for standard capacitors to have a capacitance of 1 Farad. So, the usual values are in the range of pF, nF, μF.

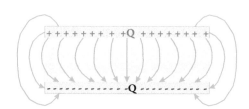

FIGURE 4.46 Lines of the electric field (green) within the plates (blue) of the capacitor.

A capacitor in an electrical circuit stores a quantity of charge Q equal in magnitude for both plates, but of opposite sign. Meanwhile in the dielectric, we have the polarization phenomenon: the molecules arrange to form a chain of oriented electrical dipoles that allow the passage of the effect of the current (and not the current itself) within the capacitor (Figure 4.47).

More complex plates can be cylindrical (Figure 4.48a) or spherical (Figure 4.48b).

For the first, the capacitance C is usually stated per unit length l:

$$\frac{C}{l} = \frac{2\pi\varepsilon}{\ln\left(\dfrac{b}{a}\right)}$$

where a is the radius of the inner and b of the outer conductors, respectively.

For the second the capacitance C is expressed by

$$C = \frac{4\pi\varepsilon}{\left(\dfrac{1}{a} - \dfrac{1}{b}\right)}$$

where a is the radius of the inner and b of the outer conductors, respectively.

Figure 4.49 shows the circuital symbols of a capacitor, in (a) common capacitors and in (b) capacitors that have a definite polarity, i.e., for which the bias cannot be never reversed.

The voltage to current ratio v/i of a capacitor is not "stable" as for the resistor, for which it is a constant equal to the resistance R. In fact, the current flowing into the capacitor is the rate of change of the charge across the capacitor plates:

$$i = \frac{dQ}{dt} = \frac{d}{dt}(Cv) = C\frac{dv}{dt}$$

hence

$$i(t) = C\frac{dv(t)}{dt}$$

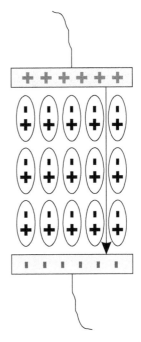

FIGURE 4.47 Orientation of the dielectric molecules subjected to the electric field between the plates.

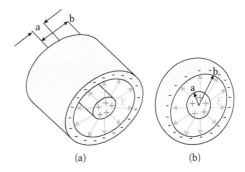

(a)   (b)

FIGURE 4.48 (a) Cylindrical and (b) spherical capacitor.

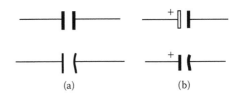

(a)   (b)

FIGURE 4.49 Symbols to represent capacitors in circuits: (a) standard and (b) fixed polarity capacitors.

so the current is not proportional to the voltage but to its variations. There is a higher current for higher variation of the voltage, even if the nominal value of the latter is not so high.

By reversing such an expression, we have the value of the voltage $v(t)$ starting from the current $i(t)$:

$$v(t) = \frac{1}{C} \int i(t)\, dt + v_0$$

Variations of the current do not produce *simultaneous* variations of the voltage, therefore the capacitor introduces a certain inertia in the circuit, i.e., a propensity to maintain the status which it had before the appearance of the variation. So the opposition to change is no longer referred to as *resistance* (as for the resistor case), a real number, but *reactance X* and, particularly, *capacitive reactance X_c*, an imaginary number (details in Section 7.2.1 in Chapter 7).

The possibility of a capacitor to store electric energy is so important that it determines the behavior of a circuit with frequency, the energy capacity in a high-power system, and other fundamental aspects.

Observations

Leyden jars are considered history's first capacitors. Basically, the Leyden jar is a bottle made of a dielectric glass, partially filled with water, with two layers of lining metal foils, one inside and the other outside of the glass (Figure 4.50).

The Leyden jar was invented in 1745 by Musschenbroek (Pieter Van Musschenbroek, Dutch scientist, 1692–1761), at Leiden University in the Netherlands, and independently by Kleist (Ewald Jürgen Georg von Kleist, German jurist, Lutheran cleric, and physicist, 1700–1748).

FIGURE 4.50 The Leyden jar. A metal rod allows charges to flow to the inner metal foil from the outside.

Later, Franklin (Benjamin Franklin, American statesman and inventor, 1706–1790) worked with the Leyden jar and found that a flat piece of glass worked as well as the jar model, prompting him to develop the flat capacitor.

In 1926, Sprague (Robert C. Sprague, American engineer and entrepreneur, 1900–1991, pioneer in radio and television electronics) opened the way for the spiral-wound capacitor, made of a rolled pair of thin conducting foils separated by a paper insulating sheet.

A fundamental type of capacitor is the *electrolytic* capacitor. The modern version was patented by Lilienfeld (Julius Edgar Lilienfeld, Austro-Hungarian physicist, 1882–1963) in 1926.

In recent years, *super capacitors* have been developed. They are electric double-layer capacitors (EDLC), with an energy density hundreds of times greater than conventional capacitors. These new capacitors have the advantage of extremely reduced separation, of the order of nanometers, between the plates. In addition, the materials of the plates allow them to fit a much larger surface area into a given volume.

If we try to measure the resistance of a capacitor, it starts from a low value and, after a while, it results in an infinite value.

This is because the instrument provides a DC current and, measuring the voltage drop, determines the value of the resistance. As a consequence, the current charges the capacitor which, when completely charged, does not allow any more current to flow. According to Ohm's law, no current means infinite resistance.

Curiosity

Capacitors are occasionally referred to as *condensers*, which is actually an older term used by Volta in 1782. This was because electricity was thought to be a sort of "fluid" or "matter" that could be "condensed." But, to avoid confusion with other type of condensers, the name was changed to capacitors after 1926. Nobody knows who actually coined the new term.

4.6.2 Types of Capacitors

The main distinction between the different types of capacitors concerns the type of dielectric. Therefore, we distinguish *ceramic* capacitors, with a ceramic dielectric, *polyester* capacitors, where the dielectric is a film of plastic material, *electrolytic* capacitors, with dielectric sheets of porous insulator soaked with liquid electrolyte, and so on (Figures 4.51 and 4.52).

The electrolytic capacitors (Figure 4.51d) have the advantage of being able to achieve a greater capacitance, but the disadvantage of being biased, so they can only be inserted in circuit schemes that always keep the same polarity across them; otherwise they will be damaged.

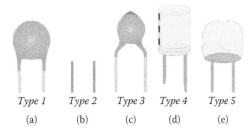

Type 1	Type 2	Type 3	Type 4	Type 5
(a)	(b)	(c)	(d)	(e)

FIGURE 4.51 Schematic representation of different types of capacitors.

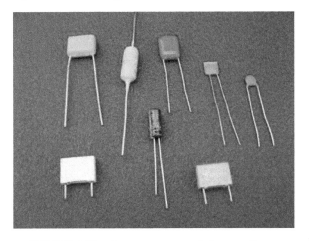

FIGURE 4.52 Some commercial capacitors.

4.6.3 Capacitor Markings

4.6.3.1 Ceramic Capacitors

The capacitance values, printed on the bodies of ceramic capacitors, consist of a code that may be different depending on the ranges of values considered:

$1\,pF$–$8.2\,pF$:

The following encodings are possible, depending on the manufacturer:

→ The simple numeric value is given; it is intended to be picoFarad.
→ The p, which stands for pico(Farad), appears in place of the dot in the printed number.

The values generally found in commerce are indicated in the following table. In the first row is the value, in the second one, the first code, and in the third one, the second code:

1 pF	1.2 pF	1.5 pF	1.8 pF	2.2 pF	2.7 pF	3.3 pF	3.9 pF	4.7 pF	5.6 pF	6.8 pF	8.2 pF
1	1.2	1.5	1.8	2.2	2.7	3.3	3.9	4.7	5.6	6.8	8.2
1p0	1p2	1p5	1p8	2p2	2p7	3p3	3p9	4p7	5p6	6p8	8p2

$10\,pF$–$82\,pF$:

The nominal value is reported, omitting the unit of measure, which is intended to be pF. The value is indicated in the table in the first row, in the second is the code:

10 pF	12 pF	15 pF	18 pF	22 pF	27 pF	33 pF	39 pF	47 pF	56 pF	68 pF	82 pF
10	12	15	18	22	27	33	39	47	56	68	82

$100\,pF$–$1000\,pF$:

The following encodings are possible, depending on the manufacturer:

→ Three numbers show the capacitance in pF: the first two are the first and second significant digits and the third is a multiplier code (the numbers of zeros to write after the first two digits).
→ The n, which stands for nano(Farad), appears in place of the dot in the printed number.
→ The printed number starts with a dot and the omitted unit is intended to be μF.

For the following table, the first row reports the value, the second row shows the first code, and the third row shows the second code:

100 pF	120 pF	150 pF	180 pF	220 pF	270 pF	330 pF	390 pF	470 pF	560 pF	680 pF	820 pF	1000 pF
101	121	151	181	221	271	331	391	471	561	681	821	102
n10	n12	n15	n18	n22	n27	n33	n39	n47	n56	n68	n82	1n

4.6.3.2 Polyester Capacitors

Polyester capacitors, generally with a parallelepiped shape, in addition to using the same marking we have already discussed for the ceramic capacitors, have higher values of capacitance; they also use the letter u indicating the μF. So for the letter n, the u replaces the dot.

An additional letter indicates the tolerance value and a further number indicates the working voltage. The working voltage is intended to be the amount of voltage the capacitor can withstand long term (for the normal life of the capacitor).

The following encodings are possible, depending on the manufacturer:

→ A simple numeric value is given, intended to be microFarad.
→ The u, which stands for micro(Farad), appears in place of the dot in the printed number

The values typically found on the market are

1 nF	1.2 nF	1.5 nF	1.8 nF	2.2 nF	2.7 nF	3.3 nF	3.9 nF	4.7 nF	5.6 nF	6.8 nF	8.2 nF
102	122	152	182	222	272	332	392	472	562	682	822
1n	1n2	1n5	1n8	2n2	2n7	3n3	3n9	4n7	5n6	6n8	8n2
0.001	0.0012	0.0012	0.0018	0.0022	0.0027	0.0033	0.0039	0.0047	0.0056	0.0068	0.0082

10 nF	12 nF	15 nF	18 nF	22 nF	27 nF	33 nF	39 nF	47 nF	56 nF	68 nF	82 nF
103	123	153	183	223	273	333	393	473	563	683	823
10n	12n	15n	18n	22n	27n	33n	39n	47n	56n	68n	82n
0.01	0.012	0.012	0.018	0.022	0.027	0.033	0.039	0.047	0.056	0.068	0.082
u01	u012	u015	u018	u022	u027	u033	u039	u047	u056	u068	u082

100 nF	120 nF	150 nF	180 nF	220 nF	270 nF	330 nF	390 nF	470 nF	560 nF	680 nF	820 nF	1 μF
104	124	154	184	224	274	334	394	474	564	684	824	105
100n	120n	150n	180n	220n	270n	330n	390n	470n	560n	680n	820n	1
0.1	0.12	0.12	0.18	0.22	0.27	0.33	0.39	0.47	0.56	0.68	0.82	1
u1	u12	u15	u18	u22	u27	u33	u39	u47	u56	u68	u82	1u

To complete the capacitor markings,

→ A letter indicates the tolerance. The most frequently used are M, to indicate a 20% tolerance on the value of capacitance, the letter K, for a 10% tolerance, and the letter J, for a 5% tolerance. To supplement:

B: ±0.10%	C: ±0.25%	D: ±0.5%	E: ±0.5%
F: ±1%	G: ±2%	H: ±3%	J: ±5%
K: ±10%	M: ±20%	N: ±0.05%	P: +100%, −0%
Z: +80%, −20%			

→ A number indicates the working voltage (in volts).

Just for example, the code $33J63$ indicates a capacitance of $330\,nF$, tolerance 5%, and a working voltage of $63\,V$.

4.6.3.3 Electrolytic Capacitors

Electrolytic capacitors are mostly cylindrical and have only one possible bias, identifiable from the longer positive terminal compared to the negative one, and from the negative sign printed on them (see Figure 4.51d).

Also, the nominal value of the capacitance is plainly printed on the body of the capacitor (no code adopted), as well as the value of the maximum working voltage that we cannot exceed before damaging the device. Typical values of maximum voltages for the electrolytic capacitors available on the market are $10\,V$, $16\,V$, $20\,V$, $25\,V$, $35\,V$, $63\,V$, $100\,V$, $250\,V$, and $400\,V$.

4.6.3.4 Color-Coding Scheme

For some capacitors with plastic dielectrics, the markings are given by a color code in the same way we have already discussed for resistors. In particular the identification is done according to Table 4.6.

4.6.3.5 SM Capacitors

As already seen for resistors, in order to increase the possibility of circuital integration and the density of components on the surface unit, we can find some capacitors on the market named SM (*Surface Mount*) that are particularly small.

TABLE 4.6

Color-Coding Scheme for Some Capacitors

Color	1st digit	2nd digit	3rd digit	Tolerance
Black	-	0	10^0	± 20%
Brown	1	1	10^1	± 1%
Red	2	2	10^2	± 2%
Orange	3	3	10^3	-
Yellow	4	4	10^4	-
Green	5	5	10^5	± 5%
Blue	6	6	-	-
Violet	7	7	-	-
Gray	8	8	-	-
White	9	9	-	± 10%

To identify such capacitors, some manufacturers use a code of three alphanu-meric characters. The first character indicates the manufacturer, the second one represents a number, made of two digits (A = 10, B = 12, C = 15, D = 18, etc.), which stands for the nominal value (in pF), and the third character is the multiplier number.

But quite often, the markings are not reported at all. Not even the size helps to understand the nominal value. A $22\,nF$ capacitor could be smaller in dimension than a $1\,nF$ capacitor because, for instance, of the different dielectric or working voltage. In those cases the only way to identify the value of the capacity is to measure it.

4.6.4 Charge and Discharge of a Capacitor

A DC current doesn't flow through a capacitor, because of the presence of its diele-tric (an insulator). However, if we connect a lightbulb to a battery with a capacitor in series (Figure 4.53), we can see that the lightbulb will turn on for an instant and then off again. This happens because in the first moment in the circuit (bulb included) a current flows (Figures 4.54a and 4.55a). After this moment, the current stops and the light bulb turns off. The capacitor was *charged* because charges were stored on its plates. If the charged capacitor is now connected to a resistor (the bulb in our case), the charge flows out of the capacitor (the bulb turns on for a short time), and the capacitor is *dis-charged* (Figures 4.54b and 4.55b).

The equation of the capacitor's behav-ior in a circuit seen previously,

$$i(t) = C\frac{dv(t)}{dt}$$

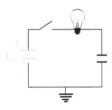

FIGURE 4.53 Charging circuit of a capacitor with a lightbulb in series.

does not take into account the *transient* time (charge and discharge instants in the previous example), but only the so-called *steady state* situation, i.e., the one following an initial state of "adjustment." According to such a formula, when apply-ing a constant voltage (the one of a bat-tery) to the terminals of a capacitor, no current flows (given that the derivative of a constant quantity is null). Therefore the for-mula does not justify very short-term, tran-sitory situations, known as *transient*, for which we would note that very short light-ing of the bulb. The formula refers then to the situation known as *steady state*.

If, at the terminals of a capacitor, we apply an alternating voltage instead of a continuous one, we get the phenomena of capacitor charge and discharge in every sinusoidal semiperiod, so the bulb would be continuously subjected to the lighting

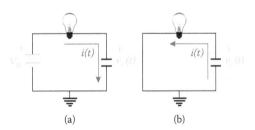

(a)   (b)

FIGURE 4.54 (a) Charge and (b) discharge of the capacitor

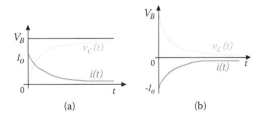

(a)   (b)

FIGURE 4.55 Voltage and current behavior in capacitor: (a) charge and (b) discharge.

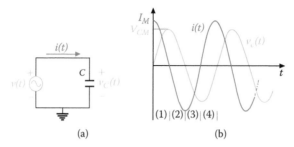

(a)   (b)

FIGURE 4.56   (a) A voltage source
charges a capacitor, which (b) shifts the
phase between voltage and current

phenomena mentioned before. If the frequency of bulb lighting corresponds to that of house voltage (60 Hz in the United States, 50 Hz in Europe) the bulb will always be lit (given the phenomenon of image persistence on our eye's retina that we have already discussed in the "Curiosities" note in Section 1.2 of Chapter 1).

Let us examine this phenomenon in detail (Figure 4.56):

→ During the first quarter, period (1) in the figure, a charge interval, voltage increases positively from zero to a maximum value V_{CM}. The capacitor must accordingly be charged by a positive current, which starts from its maximum value I_M, and gradually decreases until it is reduced to zero at the exact moment when the capacitor reaches its maximum charge state.

→ In the second quarter, period (2) since we have a discharge interval, the voltage across the capacitor decreases from V_{CM} to zero. Accordingly, the capacitor will have to discharge itself with a negative current having opposite direction with respect to the previous one.

→ In the third quarter, period (3), the charge is opposite in sign with respect to the quarter, so the voltage across the capacitor increases from zero to $-V_{CM}$, and the capacitor will receive a charge current varying from $-I_M$ to zero.

→ In the last quarter, period (4), since we have a discharge interval, the absolute value of the plates' voltage decreases from $|-V_{CM}|$ to zero, and the current varies analogously to what happened in interval (3) but opposite in sign.

It is important to note that the voltage across the capacitor must be equal to the sinusoidal one of the source, that is it must be $v(t) = v_C(t)$ and that the current, both during charge and discharge intervals, cannot vary exponentially, being bound by the sinusoidal voltage present at the terminals of the capacitor. Therefore, the current $i(t)$ too, will be sinusoidal, even if 90° phase shifted with respect to the voltage.

As happens for the resistance that opposes the flow of the current, similarly the capacitance opposes any changes to the voltage, but at a rate equal to that of change of the electrical charge on the plates.

4.6.5 Capacitors in Series

We have seen that for one capacitor the relation that expresses its stored charge is $Q = Cv$. Therefore, if we consider two capacitors in a circuit, the relation for the first one will be $Q_1 = C_1v_1$ and for the second $Q_2 = C_2v_2$. If those two capacitors are in series, as represented in Figure 4.57a and b, the charging current i flowing through them is the same for both, $i = i_1 + i_2$, as it only has one path to follow. As a consequence, both capacitors store the same amount of charge $Q = Q_1 = Q_2$, regardless of their capacitance value. In fact, the charge stored by a plate of one capacitor must have come from the plate of the other adjacent capacitor.

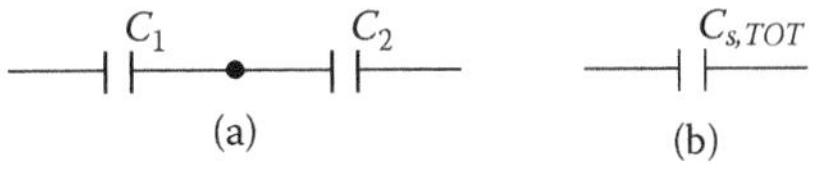

FIGURE 4.57 (a) Capacitors in series; (b) a unique capacitor equivalent to the two in series.

But, in any given series of components, the total voltage v is the sum of the partial voltages, therefore

$$v = v_1 + v_2 = \frac{Q}{C_1} + \frac{Q}{C_2} = Q\left(\frac{1}{C_1} + \frac{1}{C_2}\right)$$

then

$$\frac{1}{C_{s,TOT}} = \frac{1}{C_1} + \frac{1}{C_2}$$

with $C_{s,TOT}$ the capacitance value, which can be replaced equivalently to both capacitors in series.

As a general rule, in the case of N capacitors in series we can write

$$\frac{1}{C_{s,TOT}} = \sum_{n=1}^{N} \frac{1}{C_n}$$

Observations

The topology of capacitors in series is generally avoided in power circuits. This is because if one capacitor is slightly leaky, it will gradually transfer its voltage to the others, possibly exceeding their voltage rating and, consequently, damaging them.

If a series topology for capacitors is adopted, it is convenient to adopt capacitors with equal, or similar, capacitance values. This is because for very different capacitors in series, the voltage drop is higher for the smaller ones and, again, their voltage rating can be exceeded.

4.6.6 Capacitors in Parallel

Let's consider two capacitors that can store a charge of Q_1 and Q_2, respectively. We know that $Q_1 = C_1 V_1$ and $Q_2 = C_2 V_2$.

If we arrange them in parallel, as in the example in Figure 4.58, the total charge stored is $Q = Q_1 + Q_2$, so $Q = C_1 V_1 + C_2 V_2$.

But, for the parallel topology the voltage V_1 across the first capacitor is the same as voltage V_2 across the second capacitor, so $V = V_1 = V_2$, and $Q = C_1 V + C_2 V = (C_1 + C_2)V$ or, for the equivalent capacitor

$$Q = (C_1 + C_2)V = C_{p,TOT}V$$

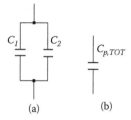

FIGURE 4.58　(a) Capacitors in parallel; (b) a unique
capacitor equivalent to the two in parallel.

This result can be generalized for a set of N capacitances in parallel:

$$C_{p,TOT} = \sum_{n=1}^{N} C_n$$

So, a parallel circuit is a way to increase the total storage of electric charge.

Observation

We can see how the laws found for the equivalence of capacitors in series and parallel are exactly inverted with regard to the laws found in the case of series and parallel resistances.

4.6.7 Energy Stored in Capacitors

We want to charge an initially uncharged parallel plate capacitor. To this aim, we must transfer the electric charges Q to the plates. The work done in transferring an infinitesimal amount of charge dq is

$$dW = Vdq = \frac{q}{C}dq$$

The overall work is computed by the integration that yelds

$$W = \int_0^Q \frac{q}{C}dq = \frac{1}{C}\frac{Q^2}{2}$$

The work we have done corresponds to the energy stored in the capacitor of capacitance $C = Q/V$, hence

$$W = \frac{1}{C}\frac{Q^2}{2} = \frac{1}{C}\frac{(CV)^2}{2} = \frac{1}{2}CV^2$$

This amount of energy is stored in the electric field generated between the plates, and the calculated expression can be generalized for a capacitor with any geometrical structure other than the parallel plate structure.

Let's now consider two capacitors with capacitance C_1 and C_2, respectively.

→ If connected in series the equivalent capacitor will have a capacitance $C_{s,TOT} = C_1 + C_2$.
→ If connected in parallel the equivalent capacitor will have a capacitance $C_{p,TOT} = \dfrac{C_1 C_2}{C_1 + C_2}$.

For the sake of simplicity, let's refer to capacitors with equal capacitance $C_1 = C_2$, then

$$C_{s,TOT} = 2C$$

$$C_{p,TOT} = \frac{C}{2}$$

Two identical capacitors have two identical voltage ratings V, so considering the total energy E_s stored by them in the series configuration, we have

$$E_s = \frac{1}{2} C_{s,TOT} V^2 = \frac{1}{2}(2C)V^2 = CV^2$$

and the total energy E_p for the parallel configuration is

$$E_p = \frac{1}{2} C_{p,TOT} V^2 = \frac{1}{2}\left(\frac{C}{2}\right)(2V)^2 = CV^2$$

It follows that $E_s = E_p$, as is obvious since it is unreasonable to change the total energy storage simply by reconnecting the same capacitors in different topologies.

Due to what we have observed for the series arrangment, and the latter conclusion, it follows that, when it is possible, it is better to adopt parallel capacitors rather than series ones, because they are safer and more reliable, but maintain the same stored energy.

4.6.8 Reactive Power

If we apply a sinusoidal voltage $v(t) = V_M \sin(\omega t)$ across a capacitor, since it results in

$$i(t) = C \frac{dv(t)}{dt}$$

the flow of current will be

$$i(t) = \omega C V_M \cos(\omega t) = \omega C V_M \sin\left(\omega t - \frac{\pi}{2}\right)$$

Therefore, the peak value of the current is $I_M = \omega C V_M$, with the factor ωC simply a constant of proportionality, and the voltage and current are phase shifted by $\pi / 2$. In particular, current leads voltage by a quarter phase or 90°.

Solving for $v(t) / i(t)$ gives

$$\frac{v(t)}{i(t)} = X_C = \frac{1}{\omega C}$$

a ratio having the units of Ohm; therefore, we can expect it behaves like a resistance. But we know the resistance as a property of the resistor, a *real* number, which determines power dissipation. Now things are different, X_C is named *capacitive reactance*, which behaves differently from the resistance. Details about this argument will be treated in greater detail later (Section 7.2.1 of Chapter 7), Now we want to focus on the power.

We know that by multiplying voltage and current we get the power p but this means that p has null mean value, as we can see in Figure 4.59.

Mathematically this is justified as follows:

$$p(t) = v(t)i(t) = V_M \sin(\omega t) \omega C V_M \sin(\omega t) = V_M I_M \sin(\omega t) \cos(\omega t)$$

But remembering that $\sin(2\omega t) = 2\sin(\omega t)\cos(\omega t)$, the following results:

$$p(t) = \frac{1}{2} V_M I_M \sin(2\omega t)$$

Therefore, the resulting power will be of double angular frequency compared to voltage and current, with equal parts above and below the horizontal axis.

Besides, remembering the *RMS* value for a sinusoidal signal ($F_{RMS} = F_M / \sqrt{2}$), we can write

$$p(t) = \frac{V_M}{\sqrt{2}} \frac{I_M}{\sqrt{2}} \sin(2\omega t) = V_{RMS} I_{RMS} \sin(2\omega t)$$

Therefore, in the case of a capacitor, the net value of power averaged over time has a null mean value. But a null mean power cannot characterize the circuit from an energetic point of view and the definition of power so far considered (real power, Section 4.4.8) cannot be valid here. Even the instruments for measuring power assign a zero value, as if there was no current flowing in the circuit. This incongruity has created the definition of so-called *reactive power*.

The fact that reactive power is alternated (positive and negative) means that it is bi-directional, that is, it is transferred from the source to the capacitor (positive)

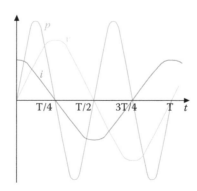

FIGURE 4.59 Voltage, current, and power for the capacitor.

and then returns from the capacitor to the source (negative). Therefore the capacitor does not dissipate energy but it can store it and then restore it.

4.7 INDUCTORS

4.7.1 Inductance

The term *solenoid* refers to a coil, that is, a series of loops wound into a tightly packed helix. A solenoid adopted in a circuit is called an *inductor,* which is a passive device capable of generating a magnetic field of significant value when current flows through it.

Just like resistors and capacitors, inductors are another example of two-terminal devices. An inductor is essentially a solenoid of conducting material, typically a copper wire covered by an insulator. The coil generally has a cylindric form and is wrapped around a *core.*

The inductor (Figure 4.61) is also known by the term *coil* or *reactor* or *inductance,* but, actually, to use correct terminology, *inductance* is the specific feature of an inductor; it is represented by the letter L and measured in Henry $[H]$ (Joseph Henry, Scottish-American scientist, 1797–1878, Figure 4.60).

Just as for resistors, whose resistance depends both on physical and geometrical parameters ($R = \rho(l/S)$), and capacitors, whose capacity depends both on physical and geometrical parameters ($C = \varepsilon(S/d)$, once again the inductance of inductors depends both on physical and geometrical parameters.

Regarding geometry, inductance increases proportionally with regard to both the number N and the section $S\ [m^2]$ of the coils, while it decreases with the core length $l\ [m]$ (Figure 4.62).

Regarding physics, the inductance depends on the magnetic permeability $\mu = \mu_0\mu_r\ [N/A^2]$ of the material around which the coiling takes place. This material can also be absent (coilings "in the air," $\mu = \mu_0 = 4\pi * 10^{-7}\ [N/A^2]$) but the inductance value increases significantly if we use a material with a high magnetic permeability.

For the inductance of an inductor formed by spires coiling around a core of cylindrical material we get

$$L = \mu N^2 \frac{S}{l}$$

FIGURE 4.60   J. Henry.

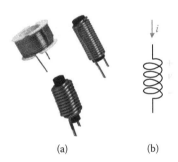

(a)   (b)

FIGURE 4.61   (a) Inductors and (b) their circuit representation and conventional symbology for applied voltage and current direction.

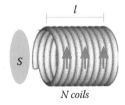

N coils

FIGURE 4.62   Scheme of an inductor.

Curiosity

The inductance value is increased considerably by the coils. A simple straight conductor, which is not coiled, has its own impedance value, which is equal to

$$L = al\left(\ln\frac{4l}{d} - 1\right)$$

in which a is a constant equal to $200*10^{-9}$, l the length, and d the conductor diameter.

So, for instance, a wire $1m$ in length and $1.5\,mm^2$ in section (thus $1.38\,mm$ in diameter) has around $1.4\,\mu H$ in inductance.

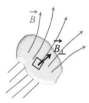

FIGURE 4.63 The schematization of magnetic flux through a surface.

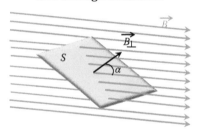

FIGURE 4.64 Flat surface immersed in an uniform magnetic field.

FIGURE 4.65 W. Weber.

4.7.2 Flux

For electrical applications it is fundamental to know the influence of a magnetic field on the surroundings. In particular, we have to measure the so called *magnetic flux* Φ_B, which is the number of lines of force of the magnetic field $\vec{B}$ passing through a surface S (closed or open), placed in a magnetic field (Figure 4.63):

$$\Phi_B = \int \vec{B}\,d\vec{S}$$

The surface S can be thought of as the one delimiting one turn of insulated wire, and in this case we talk about *flux linkage* (with the wire). To get the maximum flux linkage, it is preferred to adopt more turns of insulated wire, that is, an inductor.

If the field $\vec{B}$ is *uniform* over the turns, or the turns do lie in one plane, then the equation simplifies (Figure 4.64):

$$\Phi_B = SB\bot = BS\cos(\alpha)$$

and if the loop has N turns, it becomes

$$\Phi_B = NSB\bot = NBS\cos(\alpha)$$

The SI unit of magnetic flux Φ_B is Tm^2 or Weber $[Wb]$ (Wilhelm Eduard Weber, German physician, 1804–1891, Figure 4.65).

4.7.3 Electromagnetic Induction

Electromagnetic induction is the process of using magnetic fields to produce voltage, and in a closed loop circuit, a current.

To understand this process, Faraday experimentally demonstred how a magnetic field interacts with an electric circuit. He discovered that any change (no matter how it is produced) in the magnetic environment of a conducting coil of wire causes a voltage, named *emf* (*electro-motive force*), to be induced in the coil.

Faraday's law of electromagnetic induction (or Faraday's law of induction, sometimes referred to as the generator effect) states that the emf induced in a loop of conducting material made of a coil of N turns is equal the the rate of change of magnetic flux $\Delta\Phi_B$ through it:

$$emf = -N\frac{\Delta\Phi_B}{\Delta t}$$

But, since Lenz's law (Heinrich Friedrich Emil Lenz, Russian physicist of German ethnicity, 1804–1865) states that any induced current in a coil will result in a magnetic flux that is opposite to the original changing flux, the negative sign in the previous equation is necessary.

In any case, Faraday's law is basically used to find the *magnitude* of the induced voltage, since its sign is only useful to determine the *direction* of the current, which can be found using Ohm's law provided we know the conductor's resistance. Therefore we can affirm that, if a coil of N turns experiences a change in magnetic flux, then the instantaneous magnitude of the induced voltage v is given by

$$v = N\frac{d\Phi_B}{dt}$$

The size of the induced voltage depends on how quickly the flux through the coil is changing. Let's recall that $\Phi_B = BS\cos(\alpha)$, then

$$v = N\frac{d(BS\cos(\alpha))}{dt} = N\left\{S\cos\alpha\frac{dB}{dt} + B\cos\alpha\frac{dS}{dt} + BS\frac{d\cos\alpha}{dt}\right\}$$

so the instantaneous magnitude voltage v can be induced by changes of the field (dB/dt), by changes of loop area (dS/dt), and by changes of the orientation of the loop relative to the field

$$\left(\frac{d(\cos\alpha)}{dt}\right)$$

An example of application of Faraday's law is seen in the dynamic microphone. The mechanical waves of the voice make the diaphragm of the microphone move, and the coils move accordingly. The magnetic bar is stationary so that a voltage v is induced across the coils. The voltage's dynamic is the same as that of the voice, so that the mechanical energy is transformed into electric energy (Figure 4.66).

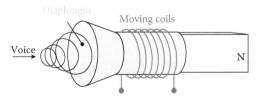

FIGURE 4.66 Dynamic microphone.

4.7.4 Self-Inductance

According to the aforementioned definitions and laws

- → A current flowing in a wire or, better, in a coil, produces a magnetic field around it (Biot–Savart law, Section 2.12.1 in Chapter 2) and, thus, a magnetic flux.
- → But, even intuitively, any changes in the current will cause changes in the magnetic flux.
- → The changes in the magnetic flux around the coil cause an *emf* across it (Faraday's law), which opposes the changing flux (Lenz's law).

For the coil forming a closed circuit, the *emf* produces a current. Consequently when this current changes, an *emf* is induced in that coil, which tends to oppose the change in the current.

This property of opposing the field-producing current is called the *self-inductance (L)*, (more correctly the *self-induction coefficient* of the circuit, but simply known as *inductance*) of the coil. It can be expressed as

$$v(t) = L\frac{di(t)}{dt}$$

Thus, we can affirm that inductance L is the relation between the voltage and time rate variation of the current.

By reversing this expression, we have the value of the current $i(t)$ starting from the voltage $v(t)$:

$$i(t) = \frac{1}{L}\int v(t)\,dt + i_0$$

The parameter L tends to be of higher value in coil-form circuits and coiled ferromagnetic cores. This is why here we will deal with the inductor, given that its form is ideal to cause a high inductance value. According to the last equations, we can affirm that the current in an inductor cannot change instantaneously, that is, inductors tend to resist any change in current flow.

Observations

 Please note how the previous formula is analogous to that of the capacitor as long as we substitute $v \leftrightarrow i$ and $L \leftrightarrow C$.

Actually, the voltage v and the *emf* define the same idea, as they use the same measurement unit $[V]$. But it is convenient to mention the *emf* when we treat magnetic flux, just to underline that there is not a "real source" of voltage in that case.

There seems to be an incongruity between the latter expression of the voltage v and the *emf* of the previous section, because of the "minus" sign, which reflects the implications of Lenz's law. But either polarity is reasonably usable, simply taking into account a phase shift. We prefer to adopt the notation with the *positive* sign.

The inductance assumes greater importance at high frequency, as there the current is changing rapidly so that the magnetic flux changes rapidly accordingly, leading to high voltage that must be taken into account.

4.7.5 Types of Inductors

The inductance L depends on the geometry and on the materials used to construct the inductor, but not on the current or the voltage supplied to it. The inductors differ basically on the types of magnetic cores.

We can distinguish cores depending on their shapes and materials. The most common shapes are *toroidal* or *"e"* ones. The most common materials are *ceramic, ferrite, Kool Mµ®* (mix of iron, silicon, and aluminum), *MPP (molypermalloy), powdered,* and *laminated.*

4.7.5.1 Solenoid

The magnetic field in a solenoid, with N turns, length l, and cross-section S, is assumed to be constant with value

$$B = \mu \frac{N}{l} i$$

So the magnetic flux can be written as

$$\Phi_B = \mu \frac{NS}{l} i$$

(Figure 4.67).

But, from Faraday's law we have

$$v = N \frac{d\Phi_B}{dt}$$

and from the definition of induction

$$v(t) = L \frac{di(t)}{dt}$$

therefore $N\Phi_B = Li$.

Combining the two results, we have

$$L = \mu N^2 \frac{S}{l}$$

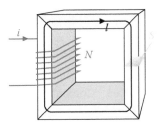

FIGURE 4.67 Inductor in a solenoid shape with N turns around a ferromagnetic core of section S.

Whenever electrons flow through an inductor, the magnetic field that will develop around that inductor can have considerable strength. So, it is believed that some illusionists use such magnetic fields to create the illusion of levitation (Figure 4.68).

FIGURE 4.68   A child levitating in front of witnesses. Illustration in the book *Saducismus Triumphatus* by Joseph Glanvill, 1681, London.

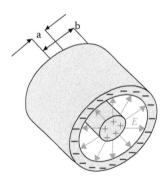

FIGURE 4.69   Coaxial inductor.

4.7.5.2 Coaxial

Consider a coaxial inductor, as represented in Figure 4.69.

We have an inner conducting cylinder of radius a, an outer conducting cylinder of radius $b > a$, and their length l. The magnetic flux loop, which is along the inner cylinder and closes along the outer one, results in

$$\Phi_B = \int \vec{B} d\vec{S} = \int_a^b \frac{\mu l i}{2\pi r} dr = \frac{\mu l i}{2\pi} \ln\left(\frac{b}{a}\right)$$

Therefore, the inductance L is

$$L = \frac{\mu l}{2\pi} \ln\left(\frac{b}{a}\right)$$

4.7.6 Inductors in Series

We have seen that for one inductor the relation that expresses its inductance is $v = L(di/dt)$. Therefore, if we consider two inductors in a circuit the relation for the

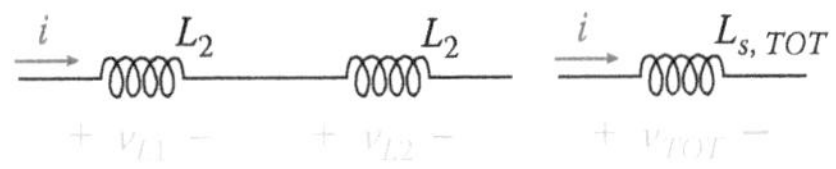

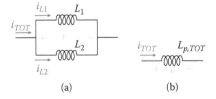

(a)　　　　　　　　　　(b)

FIGURE 4.70 (a) Two inductors in series, and (b) its equivalent representation.

first one will be $v_1 = L_1(di_{L1}/dt)$ and for the second $v_2 = L_2(di_{L2}/dt)$. If those two inductors are in series, their current will be the same, $i = i_{L1} = i_{L2}$, as represented in Figure 4.70.

The N inductors in series have the same current flowing through them, but the total voltage v_{TOT} will be the sum of each voltage v_n, with $n = 1,2...N$, so

$$v = L_1\frac{di}{dt} + L_2\frac{di}{dt} + ... + L_N\frac{di}{dt} = (L_1 + L_2 + ... + L_N)\frac{di}{dt}$$

In conclusion, for N inductors in series the total inductance $L_{s,TOT}$ is equal to the sum of the single inductances L_n, with $n = 1,2..N$:

$$L_{s,TOT} = \sum_{n=1}^{N} L_n$$

4.7.7 Inductors in Parallel

If we consider a parallel connection of two inductors, with the same voltage v across them, but differing in current as in the following representation, we have

$$i_{L1} = \frac{1}{L_1}\int_{t_0}^{t} v\,dt \text{ and } i_{L2} = \frac{1}{L_2}\int_{t_0}^{t} v\,dt$$

(Figure 4.71).
It follows that

$$i_{TOT} = i_{L1} + i_{L2} = \frac{1}{L_1}\int_{t_0}^{t} v\,dt + \frac{1}{L_2}\int_{t_0}^{t} v\,dt = \left(\frac{1}{L_1} + \frac{1}{L_2}\right)\int_{t_0}^{t} v\,dt$$

This is evident for N inductors in parallel too, hence for N inductors in parallel the inverse of the total inductance $1/L_{p,TOT}$ is equal to the sum of the inverse of the single inductances L_n, with $n = 1,2..N$:

$$\frac{1}{L_{p,TOT}} = \sum_{i=n}^{N} \frac{1}{L_n}$$

(a)　　　　　　　　　　(b)

FIGURE 4.71 (a) Two inductors in parallel and (b) its equivalent representation.

For cases of both series and parallel topology, the inductors are situated in each other's magnetic fields, the adopted approach is invalid due to so-called *mutual inductance*, which is the ratio of the *emf* in one inductor to the corresponding change of current in another neighboring inductor.

4.7.8 Energy Stored in Inductors

Inductors are capable of storing energy in a magnetic field. To evaluate how this can be possible, we can start considering an inductor of inductance L connected to a voltage supply v. The supply can vary so to increase the current flowing through the inductor from zero to some final value I. A varying current corresponds to an electro-motive force *emf* as

$$emf = -L\frac{di}{dt}$$

which acts to oppose the increase in the current.

A work W is done by the voltage source v against the *emf* to cause the current in the inductor. The instantaneous value of the work dW is

$$dW = Pdt = -emf\ idt = L\frac{di}{dt}\ idt = Lidi$$

and the total work W done in establishing the final current I is

$$W = \int_0^I Lidi$$

Therefore

$$W = \frac{1}{2}LI^2$$

4.7.9 Reactive Power

If we apply a sinusoidal voltage $v(t) = V_M \sin(\omega t)$ across an inductor, it results in

$$i(t) = \frac{1}{L}\int v(t)dt + i_0$$

and thus the flowing current will be

$$i(t) = \frac{V_M}{\omega L}\sin\left(\omega t - \frac{\pi}{2}\right)$$

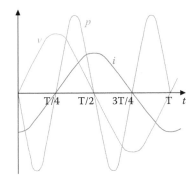

FIGURE 4.72 Voltage, current, and power at the terminals of an inductor.

(supposing zero as the constant term in addition to the integral). Therefore, the peak value of the current is $I_M = V_M/\omega L$, and voltage and current are phase shifted by $\pi/2$. In particular, the current lags behind the voltage by an angle of 90° (Figure 4.72).

Solving for $v(t)/i(t)$ gives

$$\frac{v(t)}{i(t)} = X_L = \omega L$$

X_L is named inductive reactance.

The instant power p is

$$p(t) = v(t)i(t) = V_M \sin(\omega t) I_M \sin\left(\omega t - \frac{\pi}{2}\right)$$

Considering Werner's formula $\sin(\alpha)\sin(\beta) = 1/2\left[\cos(\alpha - \beta) - \cos(\alpha + \beta)\right]$ (Section 1.2.6 in Chapter 1), we can rearrange the previous equation as

$$p(t) = \frac{1}{2}V_M I_M \left[\cos\left(\frac{\pi}{2}\right) - \cos\left(2\omega t - \frac{\pi}{2}\right)\right] = \frac{1}{2}V_M I_M \sin(2\omega t)$$

which is termed *reactive power.*

4.8 CAPACITOR–INDUCTOR DUALITY

In mathematical terms, the ideal capacitor can be considered as an inverse of the ideal inductor and vice-versa. In fact, the voltage–current equations of the two devices can be transformed into one another by exchanging voltage with current, and capacitance with inductance.

Let's recall their equations:

$$i(t) = C\frac{dv(t)}{dt} \text{ and } v(t) = L\frac{di(t)}{dt}$$

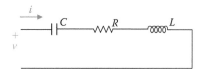

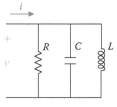

FIGURE 4.73 RLC
series circuit.

We can transform one into the other exchanging $v \leftrightarrow i$ and $C \leftrightarrow L$.

The same "duality" exists between resistance R and conductance G, so we can "translate" the equation of the circuit in series in Figure 4.73, which is

$$v = Ri + L\frac{di}{dt} + \frac{1}{C}\int i\,dt$$

to the one of the circuit in parallel in Figure 4.74, which result in

FIGURE 4.74 RLC
parallel circuit.

$$i = Gv + C\frac{dv}{dt} + \frac{1}{L}\int v\,dt$$

4.9 COMPLEX POWER

We talk about real power in the case of a resistor and about reactive power in the case of a capacitor. In general, in a circuit there are both resistors and capacitors. The former do not introduce a phase shift of voltage with current, the latter do so by 90°. Thus, in general, phase shift in a circuit, when both of such devices are present, is between 0° and 90°. Therefore, in order to be able to treat power we need to know the angle of phase shift between the sinusoidal signals of input and output.

Seeing as we are bound by the angle of phase shift between the input signal (for example a voltage) and the output signal (for example a current), let's consider the general case for which the input can be given by

$$v(t) = V_M \cos(\omega t + \varphi_V)$$

and the output by

$$i(t) = I_M \cos(\omega t + \varphi_I)$$

Observation

If $\varphi_V = 90°$ and $\varphi_I = 0°$ we come across a sinusoidal input and cosinusoidal output, the case of the capacitor, while with $\varphi_V = 0°$ and $\varphi_I = 0°$ we come across the case of the resistor, so the two signals considered are a generic case.

In such an occurrence the power is expressed by

$$p(t) = v(t)i(t) = V\cos(\omega t + \varphi_V)I\cos(\omega t + \varphi_I)$$

and for the sake of simplicity we can adjust the origin of time so that $\varphi_i = 0$:

$$p(t) = v(t)i(t) = V \cos(\omega t + \varphi_V - \varphi_I)I \cos(\omega t)$$

Now, recalling Werner's formulae for trigonometry

$$(\cos(A)\cos(B) = \frac{1}{2}[\cos(A - B) + \cos(A + B)])$$

the equation can be rewritten as

$$p(t) = \frac{1}{2}V_M I_M \cos(\varphi_V - \varphi_I) + \frac{1}{2}V_M I_M \cos(2\omega t + \varphi_V - \varphi_I)$$

The term $\varphi_V - \varphi_I$ is defined as *phase shift* while the constant term of the equation, i.e.,

$$\frac{1}{2}V_M I_M \cos(\varphi_V - \varphi_I)$$

which does not depend on ω, is exactly the *real power*. This can be both positive (in which case we talk about *dissipated power*) and negative (in which case we talk about *supplied power*).

Recalling that (Section 1.2.5 in Chapter 1) $\cos(\alpha \pm \beta) = \cos(\alpha)\cos(\beta) \mp \sin(\alpha)\sin(\beta)$, we can rewrite the expression of $p(t)$ as

$$p(t) = \frac{1}{2}V_M I_M \cos(\varphi_V - \varphi_I) + \frac{1}{2}V_M I_M \cos(\varphi_V - \varphi_I)\cos(2\omega t) - \frac{1}{2}V_M I_M \sin(\varphi_V - \varphi_I)\sin(2\omega t)$$

The term in red

$$(\frac{1}{2}V_M I_M \cos(\varphi_V - \varphi_I))$$

corresponds to the *real power*, the term in green

$$(\frac{1}{2}V_M I_M \sin(\varphi_V - \varphi_I))$$

corresponds to the *reative power*.

In the special case of a purely resistive component, the phase shift $\varphi_V - \varphi_I = 0$, so real power

$$P = \frac{1}{2}V_M I_M$$

(Joule effect), and reactive power $Q = 0$.

In the special case of a purely capacitive component, the phase shift $\varphi_V - \varphi_I = -90°$, so real power $P = 0$ (this is because energy is stored in the electric field of the capacitor but later returned to the circuit), and reactive power

$$Q = -\frac{1}{2}V_M I_M$$

In the special case of a purely inductive component, the phase shift $\varphi_V - \varphi_I = +90°$, so real power $P = 0$ (this is because energy is stored in the magnetic field of the inductor but later returned to the circuit), and reactive power

$$Q = +\frac{1}{2}V_M I_M$$

Please note that

→ The *real power P* is bound to the resistive part of the circuit (resistances do not produce a phase shift between voltage and current) by which it is completely converted to heat (*Joule effect*). Real power is fundamentally the average power. It is expressed in *watts* $[W]$.

→ The *reactive power Q* is bound to the so-called *reactances* (capacitors and inductors), which restore it to the power supply every half period (deduced from the fact that the power is positive for a half-period and negative in the following half-period). It is expressed in volt-ampere reactive $[VAR]$.

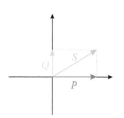

Real and *reactive* power can be visualized in a complex Cartesian coordinate system, with the real power P on the x-axis (the real numbers) and the reactive power Q on the y-axis (the imaginary numbers). The vector sum of the two powers furnishes the *apparent* or *complex power S*, expressed in volt-ampere $[VA]$. So real and reactive powers are the cathetuses and the complex power the hypotenuse of the electric power triangle.

The ratio between real power and apparent power in a circuit is called the *power factor*, which is equal to the absolute value of the cosine of the angle phase φ between the current and voltage: $|\cos\varphi|$ (Figure 4.75).

For the sake of completeness, we have to mention also the *apparent power* A_p, defined as the magnitude of complex power $A_p = |S|$, measured in volt-ampere $[VA]$.

FIGURE 4.75 Real power *P* (red), reactive power *Q* (blue), and complex power *S* (green) represented in the complex plane.

Observation

Differently from what it is commonly believed, the power used by a household appliance cannot be calculated simply by multiplying the voltage and current to that load. This is true for resistive loads only, such as incandescent lightbulbs or heating devices. In fact, other loads such as TVs, computers, washers, refrigerators, fans, etc., introduce a phase shift between voltage and current, resulting in a complex power rather than a simply real one.

The electric companies charge for real power.

The apparent power is fundamental for calculating the power factor, which is a key merit figure for the measure of the energy efficiency of electric consumption.

4.10 SUMMARY OF CONSTITUTIVE RELATIONS

Summing up what we saw for resistors, capacitors, and inductors we have formulae that indicate their characteristic value (resistance, capacitance, and inductance respectively) and formulae that indicate their behavior inside a circuit (i.e., that current flows through if we apply a voltage at their terminals or vice-versa what voltage we get at their terminals if we flow through a current). The results are shown in Table 4.7.

4.11 KEY POINTS, JARGON, AND TERMS

→ An *electrical component* allows a flow of charges. Voltage can be applied across its *terminals*.

→ An *active* component can furnish energy to other components. The *passive* component provides an energy transformation or dissipation.

→ An *electrical circuit* is defined as a set of electrical components connected in a way to allow the flow of current. To *solve* a circuit means to mathematically determine the values of the current that flows and the voltage that exists across each of the components.

→ The convention defines current to flow from higher to lower voltage for passive components, the contrary for active components.

→ The ± orientation of a voltage drop is called the *polarity* or *bias*. To *bias* is the deliberate application of a DC voltage between two points.

→ The *topology* of the circuit is the form of interconnections among components. For two-terminal components we can distinguish the *series, parallel, bridge, star,* and *triangle* topologies.

→ The law for an ideal wire is $v = 0$. This is because we assume that circuit wires have no resistance. Actually real wires always have voltage across them if there is current flowing through them, but we consider the voltage value to be negligible (since their resistance is supposed to be negligible).

→ Circuit components can absorb or release power. Conventionally, (+)power is dissipated by the component, (−)power is delivered from the component.

→ Associated reference directions refers to defining the current through a passive circuit component as *positive* when entering the terminal associated with the "+" reference for voltage.

TABLE 4.7

Summary of Laws for Resistors, Capacitors, Inductors

Device	Parameter	Behavior
Resistor	$R = \rho \dfrac{l}{S}$	$v = R;\ \left(i = \dfrac{v}{R} \right)$
Capacitor	$C = \varepsilon \dfrac{S}{d}$	$i(t) = C \dfrac{dv(t)}{dt};\ \left(v(t) = \dfrac{1}{C}\int i(t)\,dt \right)$
Inductor	$L = \mu \dfrac{S}{l} N^2$	$v(t) = L \dfrac{di(t)}{dt};\ \left(i(t) = \dfrac{1}{L}\int v(t)\,dt \right)$

→ The *resistor* is an electrical component used to control or limit current in an electric circuit by providing *resistance*. For it we distinguish *resistance* R [Ohm, Ω]; *resistivity* ρ [$Ohm^{*}meter, \Omega^{*}m$]; *conductance* G [*Siemens, S*]; and conductivity

$$\sigma \left[\frac{1}{Ohm^{*}meter}, \; \Omega^{-1*}m^{-1} \right]$$

→ A color-coding scheme identifies the value of the resistance.

→ *Ohm's law* states that the current flowing in a conductor is directly proportional to the voltage across the conductor itself. The constant of proportionality is the resistance of the conductor.

→ *Joule's law* states that when an electric current passes through a resistor, heat energy is produced.

→ *Real power* is basically average power. It is bound to the resistive part of the circuit.

→ The RMS (or *effective*) value of a signal is equal to the DC value that provides the same average power to a resistor.

→ Resistors in series form a *voltage divider*. The equivalent resistance of *resistors in series* is the sum of each resistance. Resistors in parallel form a *current divider*. The inverse of the equivalent resistance of *resistors in parallel* is the algebraic sum of the inverses of the individual resistances.

→ The bridge topology with resistors is named the *Wheatstone bridge*.

→ The law for an ideal voltage source is $v = V$, regardless of the current. Ideal voltage sources guarantee the voltage across two terminals at the specified potential.

→ The law for an ideal current source is $i = I$, regardless of the voltage. Ideal current sources guarantee the specified current flowing between two terminals.

→ Resistance refers to the property of opposing the flow of the current for a fixed voltage; capacitance refers to the property of opposing any changes to the voltage at a rate equal to the rate of change of the electrical charge on the plates; inductance refers to the property of opposing the producing current of a magnetic field.

→ The *real power P* is the average power associated with resistances. The *reactive power Q* is useful to measure the energy given by the power supply, during the first half-cycle of a sinusoid, to the inductors and capacitors, and returned by them to the supply in the second half-cycle. The vector sum of the real and the reactive powers furnishes the *apparent* or *complex power S*.

→ The measure of the capacity of storing electric charge for a given potential difference is named *capacitance*. The *capacitor* is the electronic two-terminal device for storing charge. In the electric field generated between the capacitor's plates is stored energy.

→ For capacitors, variations of the current don't produce simultaneous variations of the voltage. This asynchronous opposition to change is referred as capacitive reactance, an imaginary number.

→ For capacitors in series, the total capacitance is less than any one of the series capacitors' individual capacitances. For capacitors in parallel, the total capacitance is the sum of the individual capacitors' capacitances.

→ A current in a coil produces a magnetic field such as with a bar magnet.

→ *Magnetic flux* is the number of lines of force of the magnetic field passing through a surface.

→ *Electromagnetic induction* is the process of using magnetic fields to produce voltage, and in a closed loop circuit, a current.

→ *Self-inductance* is the property of opposing the field-producing current.

→ The coil's response depends on signal frequencies; the higher the frequency the less easily the signal flows.

→ For inductors in parallel, the total inductance is less than any one of the parallel inductors' individual inductance. For inductors in series, the total inductance is the sum of the individual inductors' inductances.

→ *Reactive power* describes the background energy movement in an AC circuit arising from the production of electric and magnetic fields. It is alternated, and thus bi-directional, i.e., it is transferred from the source to the capacitor and/or inductor and then returns from the capacitor and/or inductor to the source. The reactive power is bound to the so-called reactances.

→ The vector sum of real and reactive powers furnishes the *apparent* or *complex power*.

4.12 EXERCISES

EXERCISE 1

Indicate the nominal value R_N of a resistor with four color bands as in Figure 4.76.

FIGURE 4.76 Resistor with bands respectively of brown, black, green, and brown

ANSWER

Brown: 1 (first digit), black: 0 (second digit), green: 5 (multiplier), brown: ±1% (tolerance). So $R_N = 1 M\Omega$ with a tolerance of ±1%.

EXERCISE 2

Indicate the nominal value R_N of a resistor with five color bands as in Figure 4.77.

FIGURE 4.77 Resistor with bands respectively of orange, orange, white, black, and brown.

ANSWER

Orange: 3 (first digit), orange: 3 (second digit), white: 9 (third digit), black: 1 (multiplier), brown: 1% (tolerance). So $R_N = 339\Omega$ with a tolerance of 1%.

EXERCISE 3

An electric iron has a resistance of 15Ω, and takes a current of $7A$, when switched on for $1 min$. Calculate the heat produced.

ANSWER

The heat is a form of energy, so

$$E = Pt = VIt = RI^2t = 15*7^2*60 = 44.1 \left[kJ\right]$$

EXERCISE 4

Demonstrate that the power dissipated by a resistor of resistance R, across which there is a sinusoidal voltage $v(wt) = V_M \sin(wt)$, is obtained by the RMS values of current I_{RMS} and voltage V_{RMS} multiplied together.

ANSWER

The instantaneous power is given by

$$p = vi = \frac{v^2}{R} = Ri^2$$

and in a cycle the real power corresponds to

$$P = \frac{1}{T}\int_0^T pdt = \frac{1}{T}\int_0^T Ri^2 dt = \frac{1}{T}\int_0^T R[I_M \sin(wt)]^2 \, dt = \frac{RI_M^2}{2T}\int_0^T 2\sin^2(wt)dt$$

$$= \frac{RI_M^2}{2T}\int_0^T [1 - \cos(2wt)] \, dt$$

with $\cos(2x) = 1 - 2\sin^2(x)$. Solving, we obtain

$$P = R\frac{I_M^2}{2}$$

but the RMS value of a sinusoidal waveform is $\frac{1}{\sqrt{2}}$ times its peak value, so

$$P = R\frac{I_{RMS}}{\sqrt{2}}\frac{I_{RMS}}{\sqrt{2}} = RI_{RMS}^2 = \frac{V_{RMS}^2}{R} = V_{RMS}I_{RMS}$$

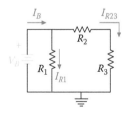

FIGURE 4.78 Example of a current divider.

EXERCISE 5

Let's consider the circuit, forming a current divider, shown in Figure 4.78, for which: $V_B = 20\,V$; $R_1 = 100\,\Omega$; $R_2 = 150\,\Omega$; $R_3 = 50\,\Omega$.
Determine the currents that flow in the three resistances.

ANSWER

The battery provides the following values of voltage and current:

$$I_B = I_{R1} + I_{R23}; V_B = R_1 I_1 = (R_2 + R_3)I_{R23}$$

From these equations, we can obtain the values of each current:

$$I_{R1} = \frac{R_2 + R_3}{R_1} I_{R23} = \frac{R_2 + R_3}{R_1}(I_B - I_{R1})$$

$$\Rightarrow I_{R1} = \frac{R_2 + R_3}{R_1 + R_2 + R_3} I_B$$

$$I_{R23} = \frac{R_1}{R_2 + R_3} I_{R1} = \frac{R_1}{R_2 + R_3}(I_B - I_{R23})$$

$$\Rightarrow I_{R23} = \frac{R_1}{R_1 + R_2 + R_3} I_B$$

The total current I_B provided by the battery is calculated by first obtaining the resistance equivalent to the three resistances of the circuit:

$$R_{eq} = R_1 \mid\mid (R_2 + R_3)$$

Then $I_B = \dfrac{V_B}{R_{eq}}$.

Finally, by replacing the numerical values we obtain

$$I_B = \frac{3}{10}[A]; I_{R1} = \frac{2}{10}[A]; I_{R23} = \frac{1}{10}[A]$$

with the same current value flowing in R_2 and R_3.

EXERCISE 6

Define the topology of the connection among components shown in Figure 4.79.
    How are R_1, R_2, R_3 connected?

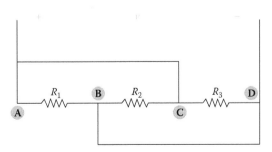

FIGURE 4.79   Topology to be recognized.

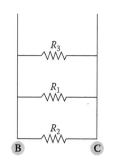

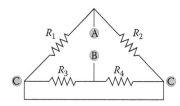

FIGURE 4.80 Different schematization for the topology to be recognized.

FIGURE 4.81 Determine the equivalent resistance of the schematized topology.

ANSWER

This exercise is very useful to understand how it can be fundamental to draw the circuit topology. In fact, Figure 4.79 does not allow us to easily recognize that the resistors form a parallel connection. This is because

R_1 is between (A) and (B);

R_2 is between (C) and (B);

R_3 is between (C) and (D);

But, clearly (A) and (C) represent the same point (only a wire without resistance between them), and the same for (B) and (D). So, the three resistors are only between (B) and (C), as represented in the topology shown in Figure 4.80.

EXERCISE 7

Given the circuit as shown in Figure 4.81 with $R_1 = 1200\,\Omega$; $R_2 = 1700\,\Omega$; $R_3 = 2200\,\Omega$; $R_4 = 3200\,\Omega$, calculate the equivalent resistance between the points (A) and (B).

ANSWER

Rearranging, we can draw Figure 4.82.

Now, it is clear that R_1 and R_2 form a parallel connection, as well as R_3 and R_4.

Therefore

$$R_{12} = R_1 \mid\mid R_2 \cong 703\,\Omega$$
$$R_{34} = R_3 \mid\mid R_4 \cong 1304\,\Omega$$

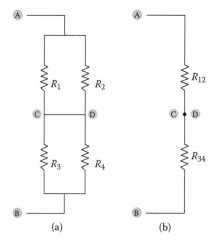

(a)   (b)

FIGURE 4.82 Simplifications of the previous figure.

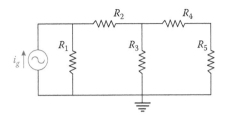

FIGURE 4.83 A simple electric circuit made up of a current source and resistors.

The equivalent resistors R_{12} and R_{34} clearly form a series topology, so we can write

$$R_{AB} = R_{1234} = R_{12} + R_{34} \cong 2007\,\Omega$$

EXERCISE 8

In a circuit arranged as in Figure 4.83 the current source $i_g = 5\,mA$ drives five resistors of values

$R_1 = 12\,k\Omega; R_2 = 4.7\,k\Omega; R_3 = 10\,k\Omega; R_4 = 6\,k\Omega; R_5 = 1\,k\Omega$ *in a mix of series and parallel connections.*

Calculate the voltage v_g across the current source, the electric power P_g supplied by the source, and the power dissipated in the resistor R_5.

ANSWER

To solve the problem, it is convenient to rearrange as in Figure 4.84.
From Figure 4.84b we obtain R_{45} as

$$R_{45} = R_4 + R_5$$

Numerically $R_{45} = 7\,k\Omega$.
From Figure 4.84c, we obtain R_{345}:

$$R_{345} = R_3 \,||\, R_{45} = \frac{R_3 R_{45}}{R_3 + R_{45}}$$

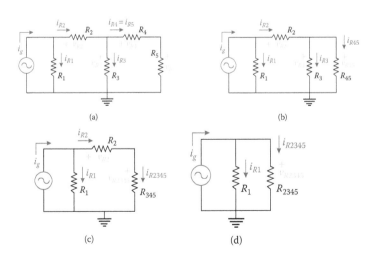

FIGURE 4.84 (a) We point out the current flowing and simplify as (b) $R45$ is $R4$ in series with R5, (c) $R345$ is $R3$ in parallel with $R45$, and (d) $R2345$ is $R2$ in series with $R345$.

Numerically $R_{345} \cong 4.1 k\Omega$.

From Figure 4.84d, we obtain R_{2345}:

$$R_{2345} = R_2 + R_{345}$$

$$R_{12345} = R_1 \,||\, R_{2345} = \frac{R_1 R_{2345}}{R_1 + R_{2345}}$$

Numerically $R_{345} \cong 4.1 k\Omega$.

So, the voltage and power supplied by the source are

$$v_g = R_{12345} i_g$$

$$P_g = v_g i_g$$

Numerically $v_g \cong 25.4 V$ and $p_g \cong 0.13 W$.

Now, the currents flowing in resistor R_1 resistor R_{2345} are, respectively,

$$i_{R1} = \frac{v_g}{R_1}$$

$$i_{R2345} = i_g - i_{R1}$$

Numerically $i_{R1} \cong 2.1 mA$ and $i_{R2345} \cong 2.9 mA$.

Easily, the following is obtained:

$$v_{R3} = v_g - R_2 i_{R2345}$$

$$i_{R45} = \frac{v_{R3}}{R_{45}}$$

$$v_{R5} = R_5 i_{R45}$$

Numerically, $v_{R3} \cong 11.87 V$; $i_{R45} \cong 1.7 mA$; and $v_{R3} \cong 11.7 V$.

Finally, the power dissipated by R_5 is

$$P_{R5} = v_{R5} i_{R45}$$

Numerically, $p_{R5} \cong 2.9 mW$.

EXERCISE 9

A circuit is constructed with five resistors and a current source as shown in Figure 4.85.

The values for the resistors are $R_1 = 1 k\Omega$; $R_2 = 2 k\Omega$; $R_3 = 1 k\Omega$; $R_4 = 500\,\Omega$; and $R_5 = 1.5 k\Omega$.

The source current is $i_g = 5A$.

What is i_{R3}, the current that flows through resistor R_3?

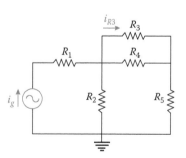

FIGURE 4.85 A circuit made up of five resistors and a current source.

ANSWER

The topology can be simplified, as in the topologies shown in Figure 4.86, with $R_{34} = R_3 \,||\, R_4$ and $R_{345} = R_{34} + R_5$.

The current i_{R345} can be found with the rule of the current divider between the resistors R_2 and R_{345}:

$$i_{R345} = \frac{R_2}{R_2 + R_{345}} i_g$$

Referring now to the circuit of Figure 4.85, and reapplying the current divider rule, we can write

$$i_{R3} = \frac{R_4}{R_3 + R_4} i_{R345} = \frac{R_4}{R_3 + R_4} \frac{R_2}{R_2 + R_{345}} i_g$$

The numerical values are

$$R_{34} = R_3 \,||\, R_4 \cong 333\,\Omega;\; R_{345} \cong 1333\,\Omega;\; i_{R345} = 3\,A;\; i_{R3} = 1\,A$$

EXERCISE 10

Star and triangle topologies can be equivalent. Referring to Figure 4.87a and b, this is true when the value of the resistance between two points in a star connection is the same as that of the triangle's counterparts. Therefore, for instance, a resistance value between the Ⓐ *and* Ⓑ *points of the star topology must be equal to that between the* Ⓐ *and* Ⓑ *points of the triangle.*

Determine the relationships between the resistances of the two topologies.

ANSWER

From triangle to star topology:
Let's start by imposing the equality between the Ⓒ and Ⓑ points:

$$R_y + R_z = R_3 \,||\, (R_1 + R_2)$$

$$\frac{1}{R_y + R_z} = \frac{1}{R_3} + \frac{1}{R_1 + R_2}$$

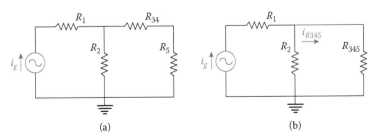

(a)   (b)

FIGURE 4.86   Simplified topologies with (a) parallel and (b) series resistors.

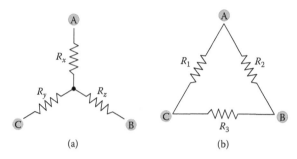

FIGURE 4.87   Resistors in (a) star and (b) triangle topologies.

Now between the Ⓒ and Ⓐ points:

$$\frac{1}{R_y + R_x} = \frac{1}{R_1} + \frac{1}{R_2 + R_3}$$

And between the Ⓐ and Ⓑ points:

$$\frac{1}{R_x + R_z} = \frac{1}{R_2} + \frac{1}{R_1 + R_3}$$

So

$$\begin{cases} R_y + R_z = \dfrac{R_3\left(R_1 + R_2\right)}{R_1 + R_2 + R_3} \\[2mm] R_y + R_x = \dfrac{R_1\left(R_2 + R_3\right)}{R_1 + R_2 + R_3} \\[2mm] R_x + R_z = \dfrac{R_2\left(R_1 + R_3\right)}{R_1 + R_2 + R_3} \end{cases}$$

Subtracting the third equation from the first one, we obtain

$$R_x - R_y = \frac{R_1 R_2 - R_2 R_3}{R_1 + R_2 + R_3}$$

and adding the second equation,

$$R_x = \frac{R_1 R_2}{R_1 + R_2 + R_3}$$

Similarly, we can obtain

$$R_y = \frac{R_1 R_3}{R_1 + R_2 + R_3}$$

$$R_z = \frac{R_2 R_3}{R_1 + R_2 + R_3}$$

From star to triangle topology:

Let's start from the latest equations, and note that

$$R_x R_y = \frac{R_1 R_2}{R_1 + R_2 + R_3} \frac{R_1 R_3}{R_1 + R_2 + R_3} = \frac{R_1^2 R_2 R_3}{\left(R_1 + R_2 + R_3 \right)^2}$$

$$R_x R_z = \frac{R_1 R_2^2 R_3}{\left(R_1 + R_2 + R_3 \right)^2}$$

$$R_y R_z = \frac{R_1 R_2 R_3^2}{\left(R_1 + R_2 + R_3 \right)^2}$$

Now, summing,

$$R_x R_y + R_x R_z + R_y R_z = \frac{\left(R_1 + R_2 + R_3 \right) R_1 R_2 R_3}{\left(R_1 + R_2 + R_3 \right)^2} = \frac{R_1 R_2 R_3}{R_1 + R_2 + R_3}$$

and dividing by R_z

$$\frac{R_x R_y + R_x R_z + R_y R_z}{R_z} = \frac{\dfrac{R_1 R_2 R_3}{R_1 + R_2 + R_3}}{\dfrac{R_2 R_3}{R_1 + R_2 + R_3}} = R_1$$

Similarly

$$\frac{R_x R_y + R_x R_z + R_y R_z}{R_y} = R_2$$

$$\frac{R_x R_y + R_x R_z + R_y R_z}{R_x} = R_3$$

EXERCISE 11

Given the circuit in Figure 4.88, with $R_1 = 2200\,\Omega$; $R_2 = 3200\,\Omega$; $R_3 = 4700\,\Omega$; $R_4 = 1200\,\Omega$; $R_5 = 3000\,\Omega$, calculate the equivalent resistance.

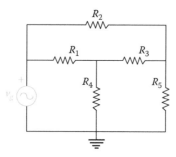

FIGURE 4.88   A voltage source with five resistors with a particular topology.

ANSWER

We can solve by noting that R_1, R_2, R_3 are arranged with a triangle connection, which can be rearranged in star topology, with the equivalent resistors R_x, R_y, R_z of values (Figure 4.89)

$$R_x = \frac{R_1 R_3}{R_1 + R_2 + R_3} \cong 1024\,\Omega$$

$$R_y = \frac{R_1 R_2}{R_1 + R_2 + R_3} \cong 697\,\Omega$$

$$R_z = \frac{R_2 R_3}{R_1 + R_2 + R_3} \cong 1489\,\Omega$$

From analysis of Figure 4.89b, it is easy to obtain

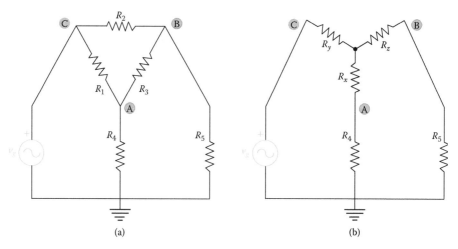

(a)   (b)

FIGURE 4.89   Equivalent topology of the given circuit,
with (a) triangle and (b) star connections.

$$R_{x4} = R_x + R_4 \cong 2224\,\Omega$$

$$R_{z5} = R_z + R_5 \cong 4489\,\Omega$$

Consequently,

$$R_g = R_y + R_{x4} \,||\, R_{z5} \cong 2184\,\Omega$$

EXERCISE 12

Let's consider Figure 4.90, where $R_1 = 10\,\Omega$; $R_2 = 20\,\Omega$; $R_3 = 30\,\Omega$; $R_4 = 40\,\Omega$; $R_5 = 50\,\Omega$; and $V_B = 15\,V$.

Calculate the values of the total current I and those of all the currents flowing in each single resistor I_{R1}, I_{R2}, I_{R3}, I_{R4}, and I_{R5}.

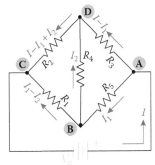

ANSWER

Looking at the figure we can write

$$(V_B - A - B - C)\,\text{loop}: V_B = R_5 I_1 + R_1 (I_1 - I_2)$$

$$(V_B - A - D - C)\,\text{loop}: V_B = R_3 (I - I_1) + R_2 (I - I_1 + I_2)$$

FIGURE 4.90 Example of Wheatstone bridge.

$$(A - D - B)\,\text{loop}: 0 = R_5 I_1 + R_4 I_2 - R_3 (I - I_1)$$

from which we obtain the system

$$\begin{cases} (R_1 + R_5) I_1 - R_1 I_2 = V_B \\ (-R_2 - R_3) I_1 + R_2 I_2 + (R_2 + R_3) I = V_B \\ (R_5 + R_3) I_1 + R_4 I_2 - R_3 I = 0 \end{cases}$$

numerically equal to

$$\begin{bmatrix} 60 & -10 & 0 \\ -50 & 20 & 50 \\ 80 & 40 & -30 \end{bmatrix} \begin{bmatrix} I_1 \\ I_2 \\ I \end{bmatrix} = \begin{bmatrix} 15 \\ 15 \\ 0 \end{bmatrix}$$

where the solution is

$$\begin{bmatrix} I_1 \\ I_2 \\ I \end{bmatrix} \cong \begin{bmatrix} 0.24 \\ -0.06 \\ 0.56 \end{bmatrix}$$

It is useful to solve the matrix system with a *MATLAB®* routine:

```
% MATLAB routine
M = (60 −10 0; −50 20 50;80 40 −30);
V = (15
15
0);
I = linsolve(M,V);)
```

From this, it is possible to identify all the current values:

$$I_{R1} = I_1 - I_2 \cong 0.29\,A$$
$$I_{R2} = I - I_1 + I_2 \cong 0.27\,A$$
$$I_{R3} = I - I_1 \cong 0.32\,A$$
$$I_{R4} = I_2 \cong -0.06\,A$$

(the latter value is not zero, so the example is for an *unbalanced bridge*)

$$I_{R5} = I_1 \cong 0.56\,A$$

The meaning of the negative value of I_{R4} is that the current flows in the opposite direction with respect to that of the figure. Applying Ohm's law to each resistance, we determine all the voltage values too:

$$V_{R1} = R_1 I_{R1} \cong 2.9\,V$$
$$V_{R2} = R_2 I_{R2} \cong 5.3\,V$$
$$V_{R3} = R_3 I_{R3} \cong 9.7\,V$$
$$V_{R4} = R_4 I_{R4} \cong -2.3\,V$$
$$V_{R5} = R_5 I_{R5} \cong 12\,V$$

EXERCISE 13

A discharged capacitor of $C = 80\,\mu F$, in series with a resistor of $R = 500\,\Omega$, is connected to a +45V battery.
  What is its final charge?
  What is its charge after $t = 0.1\,s$?

ANSWER

The final amount of charge is

$$q = CV = 80 \times 10^{-6} * 45 = 3.6\,mC$$

The time constant τ is

$$\tau = RC = 500*80 \times 10^{-6} = 40\,msec$$

$t = 0.1\,s$ is equal to 2.5τ, so

$$q\big|_{t=0.1sec} = 3.24\,mC$$

EXERCISE 14

Determine the value of the capacitive reactance X_c of a $1\,\mu F$ capacitor at a frequency of $50\,Hz$.

ANSWER

The angular frequency is defined as $\omega = 2\pi f$ and from it we can determine the capacitive reactance X_C:

$$X_C = \frac{1}{\omega C}$$

Numerically: $\omega \cong 314{,}2\left[\dfrac{rad}{sec}\right]$ $X_C \cong 3{,}182\,k\Omega$.

EXERCISE 15

In a purely inductive AC circuit, the supplied voltage is $v_L = 10te^{-5t}\ [V]$. If the current is initially zero, for $t_0 = 0$, which will be its value at $t_1 = 0.2\,s$?

ANSWER

According to the inductor's equation

$$i_L = \frac{1}{L}\int_0^{0.2} v(t)\,dt = \frac{1}{L}\int_0^{0.2} 10te^{-5t}\,dt = \frac{1}{L}\left[\frac{-e^{-5t}}{25}(1+5t)\right]_0^{0.2} \cong 1.05A$$

EXERCISE 16

An inductor has an inductance of $L = 10\,mH$. If the flux through it changes at a rate of

$$\frac{\Delta\Phi_B}{\Delta t} = 0.005\left[\frac{Wb}{s}\right]$$

determine the rate of change of the current $\Delta I/\Delta t$ through the inductor.

ANSWER

The voltage v induced across the inductor can be written both as $v = \dfrac{d\Phi_B}{dt}$ or $v = L(di/dt)$. Equating and solving for di/dt gives

$$\frac{di}{dt} = \frac{1}{L}\frac{d\Phi_B}{dt} = \frac{1}{10*10^{-3}}0.005 = 0.05 \left[\frac{A}{s}\right]$$

EXERCISE 17

A hollow air-cored inductor coil consists of $N = 400$ turns of conductive wire. When a DC current $I = 5[A]$ flows in it, a magnetic flux $\Phi_B = 25[mWb]$ is produced.
   Calculate the self-inductance L of the coil.

ANSWER

$$L = N\frac{\Phi_B}{I} = 400\frac{25*10^{-3}}{5} = 2[H]$$

EXERCISE 18

In New York City, the local Earth magnetic field has a horizontal component of $B_- = 1.5*10^{-5}[T]$ that points toward geographic north, i.e., parallel to the ground, and a vertical component of $B_\perp = 5.0*10^{-5}[T]$ that points downward, i.e., perpendicular to the ground. If a house has a floor of a surface $S = 20m*15m$, what is the magnetic flux in it?

ANSWER

The horizontal component B_- of the magnetic field does not contribute to the flux since the field is parallel to the floor. Therefore

$$\Phi_B = BS\cos(\alpha) = B_\perp S = (5.0*10^{-5})(20*15) = 0.015\ [Wb]$$

Two-Port Networks

5.1 DEFINITIONS

Let us consider the scheme shown in Figure 5.1.

It is known as a *two-port network* (or *four-terminal network* or *quadripole* or, simply, *2-port*), a sort of *expansion* with respect to the two-terminal component previously detailed in Chapter 4. A 2-port is simply a network with four terminals arranged into pairs called *ports*, which serve respectively as input and output for voltages (v_i and v_o) and currents (i_i and i_o). Similarly to the aforementioned two-terminal component, the two-port network can be made of a single device or by a *set* of electrical devices, but also by part of a circuit or even by a complete circuit. It is sometimes useful to consider a complete circuit, or a part of it, as made of an ensemble of two or more two-port networks, so that the problem of solving the complete circuit can be broken into a set of manageable sub-problems, to solve them separately and then to link the sub-problem solutions together.

Examples of basic two-port networks are *transformers* (Section 5.2), while other elementary quadripoles include *matching networks* (Section 6.8 in Chapter 6), *filters* (Section 7.3 in Chapter 7), and even *transmission lines* and *coaxial cable* between cities.

The two-port network can be considered a *functional block*, a kind of "black box" to which an input voltage and current, v_i and i_i, is supplied and from which an output voltage and current, v_o and i_o, is obtained. Thus, it is called functional block because it links output and input values functionally:

$$\left(v_o, i_o\right) = f\left(v_i, i_i\right)$$

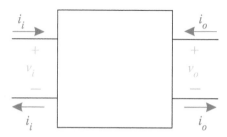

FIGURE 5.1 Schematic representation of a generic electric two-port network.

The knowledge of these two-port parameters (voltages and currents) enables us to treat the two-port as a "black box" that can be embedded within a large complete circuit.

A 2-port is called *active* if the output power is bigger than the input power (due to the presence of sources or generators), and *passive* if the output power is smaller than the input power (it could be that the voltage is bigger in the output or that the current is higher in the output, but the two cannot happen at the same time because the resulting power must be smaller). But, we will consider as general conditions for the 2-port that

→ No energy is stored within it.
→ No independent sources or generators are inside it (eventually, only dependent ones, according to the definition in Section 5.3).
→ The same current must enter and leave the same port (as schematized in Figure 5.1).

Actually, the two-port network is mostly referred to as a complete circuit used to process a signal. The signal is furnished at the input port by another network or by a signal source, then the two-port processes the signal, and the processed signal is fed into a load connected at the output port (Figure 5.2).

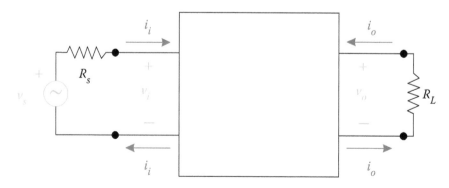

FIGURE 5.2 A voltage source furnishes a signal to a 2-port, which elaborates it and feeds a resistance load.

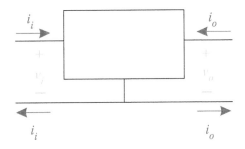

FIGURE 5.3 Schematization of a three-terminal network.

The analysis of two-port networks was pioneered in the 1920s by Breisig (Franz Breisig, German mathematician, 1868–1934).

In engineering, a *black box* is any device or system that can be viewed solely in terms of its input, output, and transfer characteristics. It means that you can even ignore its real internal workings, and the inner components can be eventually unavailable for inspection.

In a two-port network, if one terminal is in common with the input and the output port, then the network is said to be a *three-terminal network* (Figure 5.3).

In any case, it is commonly preferred to consider a three-terminal network as a simple instance of a two-port network.

5.2 TRANSFORMERS

A *transformer* is an example of a two-port network formed by a single device.

It is based on Faraday's law (Section 4.7.3 in Chapter 4) and on the ferromagnetic properties (Section 2.13 in Chapter 2) of an iron core. The transformer in its simplest form is made by a loop of steel laminations, a good conductor for magnetic fields, with two inductively coupled coils as in Figure 5.4.

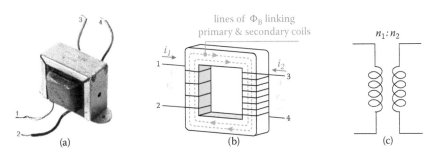

FIGURE 5.4 (a) The transformer, (b) its scheme, and (c) its circuit symbol.

A changing voltage v_1 across the first coil, called the *primary* coil, causes a current i_1 to flow (for closed circuit), which results in a magnetic flux Φ_B. The core carries the flux to the second coil, called the *secondary* coil, where the changing flux induces a current i_2 (for closed circuit) and, therefore, will result in a voltage v_2 on its terminals.

The transformer is so adopted to "transform," namely efficiently raise/lower AC voltages/currents from input to output port, since i_2 can differ from i_1 and v_2 can differ from v_1. In fact, if n_1 is the number of turns in the primary and n_2 that of the secondary, we get the following equation:

$$\frac{v_1}{v_2} = \frac{n_1}{n_2}$$

Defining n as the turns ratio, $n = n_2 / n_1$, we get

$$v_2 = nv_1$$

If there are fewer turns in the secondary winding n_2 than in the primary one n_1, that is, $n < 1$, the secondary voltage will be lower than the primary $v_2 < v_1$, and the transformer is known as a *step down transformer*. Otherwise, if $n > 1$, the transformer is known as a *step up transformer*.

If we assume that the transformer is an ideal element, the input power P_1 at the input port will have to be equal to the output power P_2 at the output port, so $P_1 = v_1 i_1 = P_2 = v_2 i_2$. Therefore, if the output voltage is higher/lower than the input one, the opposite will be true for the current; it must be lower/higher, so

$$i_2 = \frac{i_1}{n}$$

Comparing electrical engineering to classical mechanics, we can think of the transformer, with its coils, as equivalent to a pure rolling contact between two cylinders or disks—one the driver and the other the driven, as Figure 5.5 depicts. The *couple* (a system of forces with a resultant moment but no resultant force, producing a torque) and the *rotational speed* are analogues to the voltage and the current: the disk with the bigger diameter has a greater torque and a smaller rotational speed as compared to the disk with the smaller diameter, which has a smaller torque and a greater rotational speed; in the same way the impedance of a transformer with a bigger number of coils has higher voltage and lower current as compared to that with a smaller number of coils, which has lower voltage and higher current.

FIGURE 5.5   Two disks with rolling contacts.

Again, for comparison, the mechanical power is equal to

$$P = torque * rotational\ rotational\ speed$$

and it is maintained constant for the two disks.

Curiosities

The transformer was co-invented by **Bláthy** (Ottó Titusz Bláthy, Hungarian electrical engineer, 1860–1939), **Déri** (Miksa Déri, Hungarian electrical engineer, 1854–1938), **and Zipernowsky** (Károly Zipernowsky, Hungarian electrical engineer, 1853–1942).

The name "transformer" was conied by Bláthy.

In 1886 **Stanley** (William Stanley Jr., American physicist, 1858–1916) **built** the first commercially used transformer.

5.3 DEPENDENT SOURCES

A dependent source is another example of a two-port network.

Electrical *dependent sources* are similar to the sources we treated so far (*regular*, or *independent*, Section 4.5 in Chapter 4) except that the voltage or current now depends on other voltages or currents in the circuit. This behavior corresponds very closely to the way a number of interesting and useful electronic devices behave.

A dependent source is analytically represented by a *constraint* equation, which relates the dependent source's variable to the voltage or current that the source depends on in the circuit. This kind of source cannot be a power generator by itself, but generates power and supplies it to the circuit only if there are other independent voltage or current sources in the circuit.

A voltage source that depends on a voltage is referred to as a Voltage Controlled Voltage Source (VCVS); when it depends on a current it is referred to as a Current Controlled Voltage Source (CCVS).

Similarly, a current source that depends on a voltage is referred to as a Voltage Controlled Current Source (VCCS); when it depends on a current it is referred to as a Current Controlled Current Source (CCCS).

Whichever the type of dependent source, if the relation of dependence is of first order then the source is defined as *linearly dependent*.

In dependent sources, we distinguish an input port, from where we can control, and an output port, where we place the source. Therefore, dependent sources are another example of a two-port network (see Figure 5.6), at the output of which we have an electrical quantity supplied by the dependent source, proportional to an electrical quantity of the input port. Proportionality is given by a real constant, which constitutes the control parameter.

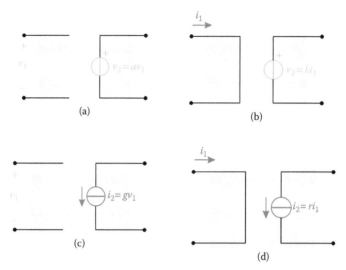

FIGURE 5.6   Four possible types of dependent sources:
(a) VCVS, (b) CCVS, (c) VCCS, (d) CCCS.

In summary, we have

→ Voltage Controlled Voltage Source (VCVS), $v_2 = av_1$
→ Current Controlled Voltage Source (CCVS), $v_2 = ki_1$
→ Voltage Controlled Current Source (VCCS), $i_2 = gv_1$
→ Current Controlled Current Source (CCCS), $i_2 = ri_1$

The control parameters of the four types of dependent sources are real constants respectively:

1) a *[dimensionless]*
2) k *[Ω]*
3) g *[Ω⁻¹]*
4) r *[dimensionless]*

There are different applications for dependent sources, which can provide a convenient means of converting between voltage and current, or vice-versa, changing resistance or impedance, converting one resistance in one conductance, or vice-versa, modelling the behavior of some devices (Section 11.3 in Chapter 11) or parts of circuits, and regulating voltage or current, etc.

5.4 MODELS OF TWO-PORT NETWORKS

Passive and linear circuits, or part of them, can be represented by a two-port network model. We shall describe here a mathematical representation assuming networks with passive elements (having characteristic values that do not change with boundary conditions), with only dependent source elements (no independent ones), and with the same current that must enter and leave a port (port condition).

5.4.1 Classification

For a two-port network, we have a total of four variables, that is, the electrical quantities voltages/currents v_1, i_1, v_2, and i_2 at the input and output ports, respectively. Two of these quantities can be considered independent variables and the other two dependent ones, which results in the resolution of a system of two equations in two unknown quantities. There are six possible combinations of four variables taken two by two, that is,

$$\binom{4}{2}$$

and the classification relies on the chosen combination. All possibilities are highlighted in Table 5.1.

The coefficients of these six matrixes are the *matrix parameters* and, obviously, form inverse relations two by two: $[Z] \leftrightarrow [Y]$, $[H] \leftrightarrow [G]$, $[A] \leftrightarrow [B]$. The matrices $[G]$ and $[B]$ are only occasionally adopted.

The names of the six matrices are derived as follows:

→ For $[Z]$ and $[Y]$ from the dimensional analysis of their elements, so $Z \to$ *impedances*, $Y \to$ *admittances*
→ For $[H]$ and $[G]$ from the fact that the elements are dimensionally mixed (we can find resistances, or conductances or dimensionless terms), so $[H] \to$ *hybrid* and $[G] \to$ *reverse hybrid*

TABLE 5.1

Mathematical Models for the Representation of a Two-Port Network

Var.	Analytical Notation	Matrix Notation	Name	Matrix, Parameters
i_1 i_2	$\begin{cases} v_1 = z_{11}i_1 + z_{12}i_2 \\ v_2 = z_{21}i_1 + z_{22}i_2 \end{cases}$	$\begin{bmatrix} v_1 \\ v_2 \end{bmatrix} = \begin{bmatrix} z_{11} & z_{12} \\ z_{21} & z_{22} \end{bmatrix} \begin{bmatrix} i_1 \\ i_2 \end{bmatrix}$	Current controlled	(Z), impedance
v_1 v_2	$\begin{cases} i_1 = y_{11}i_1 + y_{12}v_2 \\ i_2 = y_{21}i_1 + y_{22}v_2 \end{cases}$	$\begin{bmatrix} i_1 \\ i_2 \end{bmatrix} = \begin{bmatrix} y_{11} & y_{12} \\ y_{21} & y_{22} \end{bmatrix} \begin{bmatrix} v_1 \\ v_2 \end{bmatrix}$	Voltage controlled	(Y), admittance
i_1 v_2	$\begin{cases} v_1 = h_{11}i_1 + h_{12}v_2 \\ i_2 = h_{21}i_1 + h_{22}v_2 \end{cases}$	$\begin{bmatrix} v_1 \\ i_2 \end{bmatrix} = \begin{bmatrix} h_{11} & h_{12} \\ h_{21} & h_{22} \end{bmatrix} \begin{bmatrix} i_1 \\ v_2 \end{bmatrix}$	Hybrid 1	(H), hybrid
v_1 i_2	$\begin{cases} i_1 = gv_1 + g_{12}i_2 \\ v_2 = g_{21}v_1 + g_{22}i_2 \end{cases}$	$\begin{bmatrix} i_1 \\ v_2 \end{bmatrix} = \begin{bmatrix} g_{11} & g_{12} \\ g_{21} & g_{22} \end{bmatrix} \begin{bmatrix} v_1 \\ i_2 \end{bmatrix}$	Hybrid 2	(G), reverse hybrid
v_2 i_2	$\begin{cases} v_1 = a_{11}v_2 - a_{12}i_2 \\ i_1 = a_{21}v_2 - a_{22}i_2 \end{cases}$	$\begin{bmatrix} v_1 \\ i_1 \end{bmatrix} = \begin{bmatrix} a_{11} & a_{12} \\ a_{21} & a_{22} \end{bmatrix} \begin{bmatrix} v_2 \\ -i_2 \end{bmatrix}$	Transmission 1	(A), transmission
v_1 i_1	$\begin{cases} v_2 = bv_1 + b_{12}i_1 \\ i_2 = b_{21}v_1 + b_{22}i_1 \end{cases}$	$\begin{bmatrix} v_2 \\ i_2 \end{bmatrix} = \begin{bmatrix} b_{11} & b_{12} \\ b_{21} & b_{22} \end{bmatrix} \begin{bmatrix} v_1 \\ i_1 \end{bmatrix}$	Transmission 2	(B), reverse transmission

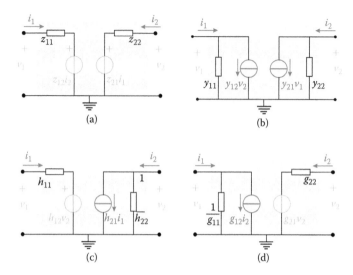

FIGURE 5.7 Equivalent circuits for the matrices. (a): (Z), (b): (Y), (c): (H), (d): (G)

→ For $[A]$ and $[B]$ the variables of one port are placed in relation to those of the other; the two-port network behaves as a transmission line, so $[A] \rightarrow transmission$ and $[B] \rightarrow reverse\ transmission$.

Given that all of these expressions are set in relation to the same variables, it is always mathematically possible to cross from one coefficient matrix to another.

Observation

The hybrid parameters are so called because they have mixed dimensions: h_{11} and h_{22} [*unitless*], h_{12} [Ω], and h_{21} [Ω^{-1}] .

5.4.2 Equivalent Circuits

Each of the mathematical equations in Table 5.1 expresses a sum of voltages or a sum of currents. So they are equivalent to circuits made either through a series (the sum of voltages) or a parallel (the sum of currents). These circuits are expressed as two-port networks, including dependent sources, as represented in Figure 5.7.

5.4.3 Examples of Conversion between Network Parameters

The same two-port network can be described in terms of different matrixes but equivalent. Let's consider some examples of conversions.

5.4.3.1 From [H] to [Z]
Given the circuit in Figure 5.8.

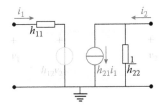

FIGURE 5.8 Two-port network described by (H) parameters.

Let us express it first in terms of matrix $[Z]$ and then in terms of matrix $[H]$.

From the definition of $[Z]$, we write

$$\begin{cases} v_1 = z_{11}i_1 + z_{12}i_2 \\ v_2 = z_{21}i_1 + z_{22}i_2 \end{cases}$$

So, we can express the matrix coefficient z_{11} as the ratio v_1 / i_1 as $i_2 = 0$, that is, without current flowing in the output port:

$$z_{11} = \frac{v_1}{i_1}\bigg|_{i_2=0}$$

In the same way, the matrix coefficient z_{12} is given by the ratio v_1 / i_2 imposing

$$i_1 = 0: \ z_{12} = \frac{v_1}{i_2}\bigg|_{i_1=0}$$

The parameters z_{21} and z_{22} can be similarly expressed.
Let's look at this in detail:

$$z_{11} = \frac{v_1}{i_1}\bigg|_{i_2=0} \qquad \begin{cases} v_1 = h_{11}i_1 + h_{12}v_2 \\ 0 = h_{21}i_1 + h_{22}v_2 \end{cases} \Rightarrow v_1 = h_{11}i_1 + h_{12}\left(-\frac{h_{21}}{h_{22}}i_1\right)$$

$$\Rightarrow z_{11} = \frac{h_{11}i_1 + h_{12}\left(-\dfrac{h_{21}}{h_{22}}i_1\right)}{i_1}$$

$$\Rightarrow z_{11} = \frac{h_{11}h_{22} - h_{12}h_{21}}{h_{22}} = \frac{\Delta h}{h_{22}}$$

with $\Delta h = h_{11}h_{22} - h_{12}h_{21}$

$$z_{12} = \frac{v_1}{i_2}\bigg|_{i_1=0} \qquad \begin{cases} v_1 = 0 + h_{12}v_2 \\ 0 = 0 + h_{22}v_2 \end{cases} \Rightarrow z_{12} = \frac{h_{12}}{h_{22}}$$

$$z_{21} = \frac{v_2}{i_1}\bigg|_{i_2=0} \qquad \begin{cases} v_1 = h_{11}i_1 + h_{12}v_2 \\ 0 = h_{21}i_1 + h_{22}v_2 \end{cases}$$

From the second equation: $z_{21} = -\dfrac{h_{21}}{h_{22}}$

$$Z_{22} = \frac{v_2}{i_2}\bigg|_{i_1=0} \qquad \begin{cases} v_1 = 0 + h_{12}v_2 \\ i_2 = 0 + h_{22}v_2 \end{cases}$$

From the second equation: $z_{22} = \dfrac{1}{h_{22}}$

Summarizing,

$$\begin{bmatrix} z_{11} & z_{12} \\ z_{21} & z_{22} \end{bmatrix} = \begin{bmatrix} \dfrac{h_{11}h_{22} - h_{12}h_{21}}{h_{22}} & \dfrac{h_{12}}{h_{22}} \\ -\dfrac{h_{21}}{h_{22}} & \dfrac{1}{h_{22}} \end{bmatrix}$$

5.4.3.2 From [Z] to [H]

We want to express the two-port network schematized in Figure 5.9 and represented in terms of (Z) parameters, by means of the (H) matrix.

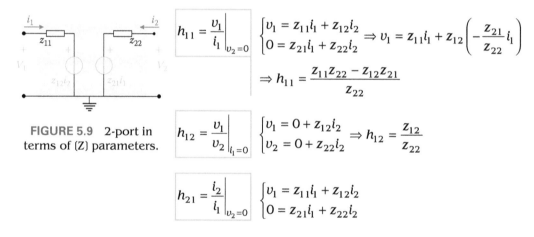

FIGURE 5.9   2-port in terms of (Z) parameters.

$$h_{11} = \frac{v_1}{i_1}\bigg|_{v_2=0} \qquad \begin{cases} v_1 = z_{11}i_1 + z_{12}i_2 \\ 0 = z_{21}i_1 + z_{22}i_2 \end{cases} \Rightarrow v_1 = z_{11}i_1 + z_{12}\left(-\frac{z_{21}}{z_{22}}i_1\right)$$

$$\Rightarrow h_{11} = \frac{z_{11}z_{22} - z_{12}z_{21}}{z_{22}}$$

$$h_{12} = \frac{v_1}{v_2}\bigg|_{i_1=0} \qquad \begin{cases} v_1 = 0 + z_{12}i_2 \\ v_2 = 0 + z_{22}i_2 \end{cases} \Rightarrow h_{12} = \frac{z_{12}}{z_{22}}$$

$$h_{21} = \frac{i_2}{i_1}\bigg|_{v_2=0} \qquad \begin{cases} v_1 = z_{11}i_1 + z_{12}i_2 \\ 0 = z_{21}i_1 + z_{22}i_2 \end{cases}$$

From the second equation: $h_{21} = -\dfrac{z_{21}}{z_{22}}$

$$h_{22} = \frac{i_2}{v_2}\bigg|_{i_1=0} \qquad \begin{cases} v_1 = 0 + z_{12}i_2 \\ v_2 = 0 + z_{22}i_2 \end{cases}$$

From the second equation: $h_{22} = \dfrac{1}{z_{22}}$

Summarizing,

$$\begin{bmatrix} h_{11} & h_{12} \\ h_{21} & h_{22} \end{bmatrix} = \begin{bmatrix} \dfrac{z_{11}z_{22} - z_{12}z_{21}}{z_{22}} & \dfrac{z_{12}}{z_{22}} \\ -\dfrac{z_{21}}{z_{22}} & \dfrac{1}{z_{22}} \end{bmatrix}$$

5.4.4 Conversion Table

It is useful to summarize with a table the possible conversions among the most commonly adopted matrices (Table 5.2).

TABLE 5.2
Conversions between Two-Port Parameters

To:	From: [Z]	[Y]	[H]	[A]
[Z]	$\begin{bmatrix} z_{11} & z_{12} \\ z_{21} & z_{22} \end{bmatrix}$	$\begin{bmatrix} \dfrac{y_{22}}{\Delta y} & -\dfrac{y_{12}}{\Delta y} \\ -\dfrac{y_{21}}{\Delta y} & \dfrac{y_{11}}{\Delta y} \end{bmatrix}$	$\begin{bmatrix} \dfrac{\Delta h}{h_{22}} & \dfrac{h_{12}}{h_{22}} \\ -\dfrac{h_{21}}{h_{22}} & \dfrac{1}{h_{22}} \end{bmatrix}$	$\begin{bmatrix} \dfrac{a_{11}}{a_{21}} & \dfrac{\Delta a}{a_{21}} \\ \dfrac{1}{a_{21}} & \dfrac{a_{22}}{a_{21}} \end{bmatrix}$
[Y]	$\begin{bmatrix} \dfrac{z_{22}}{\Delta z} & -\dfrac{z_{12}}{\Delta y} \\ -\dfrac{z_{21}}{\Delta y} & \dfrac{z_{11}}{\Delta y} \end{bmatrix}$	$\begin{bmatrix} y_{11} & y_{12} \\ y_{21} & y_{22} \end{bmatrix}$	$\begin{bmatrix} \dfrac{1}{h_{11}} & -\dfrac{h_{12}}{h_{11}} \\ \dfrac{h_{21}}{h_{11}} & \dfrac{\Delta h}{h_{11}} \end{bmatrix}$	$\begin{bmatrix} \dfrac{a_{22}}{a_{12}} & -\dfrac{\Delta a}{a_{12}} \\ -\dfrac{1}{a_{12}} & \dfrac{a_{11}}{a_{12}} \end{bmatrix}$
[H]	$\begin{bmatrix} \dfrac{\Delta z}{z_{22}} & \dfrac{z_{12}}{z_{22}} \\ -\dfrac{z_{21}}{z_{22}} & \dfrac{1}{z_{22}} \end{bmatrix}$	$\begin{bmatrix} \dfrac{1}{y_{11}} & -\dfrac{y_{12}}{y_{11}} \\ \dfrac{y_{21}}{y_{11}} & \dfrac{\Delta y}{y_{11}} \end{bmatrix}$	$\begin{bmatrix} h_{11} & h_{12} \\ h_{21} & h_{22} \end{bmatrix}$	$\begin{bmatrix} \dfrac{a_{12}}{a_{22}} & \dfrac{\Delta a}{a_{22}} \\ -\dfrac{1}{a_{22}} & \dfrac{a_{21}}{a_{22}} \end{bmatrix}$
[A]	$\begin{bmatrix} \dfrac{z_{11}}{z_{21}} & \dfrac{\Delta z}{z_{21}} \\ \dfrac{1}{z_{21}} & \dfrac{z_{22}}{z_{21}} \end{bmatrix}$	$\begin{bmatrix} -\dfrac{y_{22}}{y_{21}} & -\dfrac{1}{y_{21}} \\ -\dfrac{\Delta y}{y_{21}} & -\dfrac{y_{11}}{y_{21}} \end{bmatrix}$	$\begin{bmatrix} -\dfrac{\Delta h}{h_{21}} & -\dfrac{h_{11}}{h_{21}} \\ -\dfrac{h_{22}}{h_{21}} & -\dfrac{1}{h_{21}} \end{bmatrix}$	$\begin{bmatrix} a_{11} & a_{12} \\ a_{21} & a_{22} \end{bmatrix}$

5.4.5 Examples

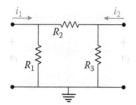

FIGURE 5.10 pi (Π) network.

We want to determine the parameters of the matrix $[Z]$ of the network represented in Figure 5.10.

From the definition of the matrix we have

$$\begin{bmatrix} v_1 \\ v_2 \end{bmatrix} = \begin{bmatrix} z_{11} & z_{12} \\ z_{21} & z_{22} \end{bmatrix} \begin{bmatrix} i_1 \\ i_2 \end{bmatrix}$$

therefore the single parameters can be deduced as

$$z_{11} = \frac{v_1}{i_1}\Big|_{i_2=0} \; ; z_{12} = \frac{v_1}{i_2}\Big|_{i_1=0} \; ; z_{21} = \frac{v_2}{i_1}\Big|_{i_2=0} \; ; z_{22} = \frac{v_2}{i_2}\Big|_{i_1=0}$$

$$z_{11} = \frac{v_1}{i_1}\Big|_{i_2=0}$$

As i_2 must be null, current i_1 flows through the parallel between R_1 and $R_2 + R_3$, so $v_1 = \left[R_1 \,||\, (R_2 + R_3)\right]i_1$ and consequently

$$z_{11} = R_1 \,||\, (R_2 + R_3) = \frac{R_1(R_2 + R_3)}{R_1 + R_2 + R_3}$$

$$z_{12} = \frac{v_1}{i_2}\Big|_{i_1=0}$$

As i_1 must be null, the current i_2 is divided between R_3 (i_{R3}) and the series $R_2 + R_1$ (i_{R21}). As already deduced in the example of the current divider (see Section 4.4.16 in Chapter 4), we have

$$i_{R21} = \frac{R_3}{R_1 + R_2 + R_3} i_2$$

Consequently

$$v_1 = R_1 i_{R_{21}} = \frac{R_1 R_3}{R_1 + R_2 + R_3} i_2$$

and then

$$z_{12} = \frac{R_1 R_3}{R_1 + R_2 + R_3}$$

$$z_{21} = \frac{v_2}{i_1}\Big|_{i_2=0}$$

As i_2 must be null, the current i_1 is divided between R_1 (i_{R1}) and the series $R_2 + R_3$ (i_{R23}), so

$$i_{R23} = \frac{R_1}{R_1 + R_2 + R_3} i_1$$

But

$$v_2 = R_3 i_{R3} = R_3 i_{R23} = \frac{R_1 R_3}{R_1 + R_2 + R_3} i_1$$

as a consequence

$$z_{21} = \frac{R_1 R_3}{R_1 + R_2 + R_3}$$

$$z_{22} = \left. \frac{v_2}{i_2} \right|_{i_1=0}$$

$$z_{22} = R_3 \mid\mid (R_1 + R_2) = \frac{R_3 (R_1 + R_2)}{R_1 + R_2 + R_3}$$

Summarizing,

$$[Z] = \begin{bmatrix} \dfrac{R_1 (R_2 + R_3)}{R_1 + R_2 + R_3} & \dfrac{R_1 R_3}{R_1 + R_2 + R_3} \\ \dfrac{R_1 R_3}{R_1 + R_2 + R_3} & \dfrac{R_3 (R_1 + R_2)}{R_1 + R_2 + R_3} \end{bmatrix}$$

From the same two-port network of the previous figure we now want to deduce the parameters of matrix $[H]$.

From the definition of the matrix we have

$$\begin{bmatrix} v_1 \\ i_2 \end{bmatrix} = \begin{bmatrix} h_{11} & h_{12} \\ h_{21} & h_{22} \end{bmatrix} \begin{bmatrix} i_1 \\ v_2 \end{bmatrix}$$

therefore the single parameters can be deduced as

$$h_{11} = \left. \frac{v_1}{i_1} \right|_{v_2=0} \quad ; h_{12} = \left. \frac{v_1}{v_2} \right|_{i_1=0} \quad ; h_{21} = \left. \frac{i_2}{i_1} \right|_{v_2=0} \quad ; h_{22} = \left. \frac{i_2}{v_2} \right|_{i_1=0}$$

As v_2 must be null, the exit port must result "closed" (a simple wire connects the two terminals), so the resistances R_1 and R_2 are found in parallel:

$$h_{11} = R_1 \mid\mid R_2 = \frac{R_1 R_2}{R_1 + R_2}$$

$$h_{12} = \left. \frac{v_1}{v_2} \right|_{i_1=0}$$

Voltage divider $v_1 = \dfrac{R_1}{R_1 + R_2} v_2$, so

$$h_{12} = \frac{R_1}{R_1 + R_2}$$

$$h_{21} = \left. \frac{i_2}{i_1} \right|_{v_2=0}$$

In this case resistance R_3 is shorted. At the exit node (the one where terminals R_2 and R_3 are joint) we have $i_2 + i_{R2} = 0$, but

$$i_{R2} = \frac{R_1}{R_1 + R_2} i_1$$

hence

$$i_2 = -\frac{R_1}{R_1 + R_2} i_1$$

and finally

$$h_{21} = -\frac{R_1}{R_1 + R_2}$$

$$h_{22} = \left. \frac{i_2}{v_2} \right|_{i_1=0}$$

Current divider

$$i_{R3} = \frac{R_1 + R_2}{R_1 + R_2 + R_3} i_2, \text{ and } v_2 = R_3 i_{R3} = \frac{R_3 (R_1 + R_2)}{R_1 + R_2 + R_3} i_2,$$

from which we get

$$v_2 = R_3 i_{R3} = \frac{R_3 (R_1 + R_2)}{R_1 + R_2 + R_3} i_2$$

Observation

Note how current divider equations were useful in the case of current relations, whereas voltage divider equations were useful in the case of voltage relations.

Summarizing,

$$[H] = \begin{bmatrix} \dfrac{R_1 R_2}{R_1 + R_2} & \dfrac{R_1}{R_1 + R_2} \\[3mm] -\dfrac{R_1}{R_1 + R_2} & \dfrac{R_1 + R_2 + R_3}{R_3 (R_1 + R_2)} \end{bmatrix}$$

5.5 INTERCONNECTIONS OF TWO-PORT NETWORKS

Two-port networks can be interconnected in different configurations, such as *series, parallel, cascade, series–parallel,* and *parallel–series* connections. For each configuration a certain matrix may be more useful than another to describe the network, so calculations can be simplified.

5.5.1 Series Connection

The scheme in Figure 5.11 shows a series connection of two-port networks. Note that the connection is called *series* because the voltages are summed.

In this case, a representation in terms of the impedance matrix makes calculations easier, therefore

$$v_1 = v_1' + v_1''\,; v_2 = v_2' + v_2''$$

$$i_1 = i_1' = i_1''\,; i_2 = i_2' = i_2''$$

$$\begin{bmatrix} v_1' \\ v_2' \end{bmatrix} = \begin{bmatrix} z_{11}' & z_{12}' \\ z_{21}' & z_{22}' \end{bmatrix}\begin{bmatrix} i_1' \\ i_2' \end{bmatrix}; \begin{bmatrix} v_1'' \\ v_2'' \end{bmatrix} = \begin{bmatrix} z_{11}'' & z_{12}'' \\ z_{21}'' & z_{22}'' \end{bmatrix}\begin{bmatrix} i_1'' \\ i_2'' \end{bmatrix}$$

$$\begin{bmatrix} v_1 \\ v_2 \end{bmatrix} = \begin{bmatrix} v_1' \\ v_2' \end{bmatrix} + \begin{bmatrix} v_1'' \\ v_2'' \end{bmatrix} = \begin{bmatrix} z_{11}' & z_{12}' \\ z_{21}' & z_{22}' \end{bmatrix}\begin{bmatrix} i_1' \\ i_2' \end{bmatrix} + \begin{bmatrix} z_{11}'' & z_{12}'' \\ z_{21}'' & z_{22}'' \end{bmatrix}\begin{bmatrix} i_1'' \\ i_2'' \end{bmatrix}$$

$$\Rightarrow$$

$$\begin{bmatrix} z_{11} & z_{12} \\ z_{21} & z_{22} \end{bmatrix} = \begin{bmatrix} z_{11}' & z_{12}' \\ z_{21}' & z_{22}' \end{bmatrix} + \begin{bmatrix} z_{11}'' & z_{12}'' \\ z_{21}'' & z_{22}'' \end{bmatrix}$$

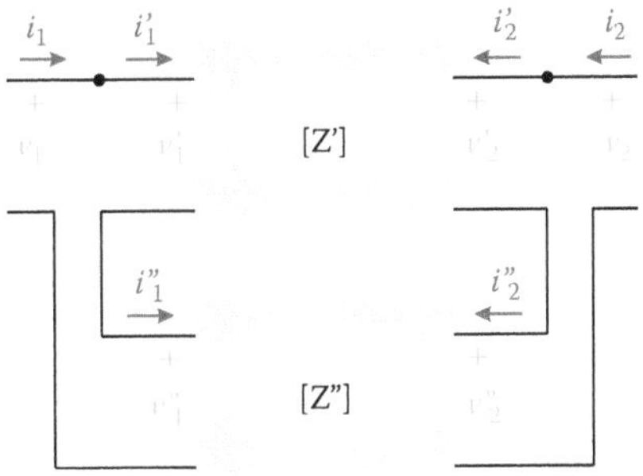

FIGURE 5.11 Series connection of two two-port networks described by [Z] matrixes.

So, the overall Z-parameter matrix for series connected two-port networks is simply the sum of Z-parameter matrices of each individual two-port network.

5.5.2 Parallel Connection

The scheme in Figure 5.12 shows a parallel (shunt) connection of two-port networks.

Note that the connection is called *parallel* because the currents are summed. In this case, representation in terms of the admittances matrix makes our calculations easier, therefore

$$v_1 = v_1' = v_1''\,; v_2 = v_2' = v_2''$$

$$i_1 = i_1' + i_1''\,; i_2 = i_2' + i_2''$$

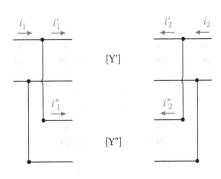

FIGURE 5.12 Parallel connection of two two-port networks described by [Y] matrixes.

$$\begin{bmatrix} i'_1 \\ i'_2 \end{bmatrix} = \begin{bmatrix} y'_{11} & y'_{12} \\ y'_{21} & y'_{22} \end{bmatrix}\begin{bmatrix} v'_1 \\ v'_2 \end{bmatrix}; \begin{bmatrix} i''_1 \\ i''_2 \end{bmatrix} = \begin{bmatrix} y''_{11} & y''_{12} \\ y''_{21} & y''_{22} \end{bmatrix}\begin{bmatrix} v''_1 \\ v''_2 \end{bmatrix}$$

$$\begin{bmatrix} i_1 \\ i_2 \end{bmatrix} = \begin{bmatrix} i'_1 \\ i'_2 \end{bmatrix} + \begin{bmatrix} i''_1 \\ i''_2 \end{bmatrix} = \begin{bmatrix} y'_{11} & y'_{12} \\ y'_{21} & y'_{22} \end{bmatrix}\begin{bmatrix} v'_1 \\ v'_2 \end{bmatrix} + \begin{bmatrix} y''_{11} & y''_{12} \\ y''_{21} & y''_{22} \end{bmatrix}\begin{bmatrix} v''_1 \\ v''_2 \end{bmatrix}$$

$$\Rightarrow$$

$$\begin{bmatrix} y_{11} & y_{12} \\ y_{21} & y_{22} \end{bmatrix} = \begin{bmatrix} y'_{11} & y'_{12} \\ y'_{21} & y'_{22} \end{bmatrix} + \begin{bmatrix} y''_{11} & y''_{12} \\ y''_{21} & y''_{22} \end{bmatrix}$$

So, the overall Y-parameter matrix for parallel connected two-port networks is simply the sum of Y-parameter matrices of each individual two-port network.

5.5.3 Series–Parallel and Parallel–Series Connections

A series–parallel connection consists of a series topology at the input port and a parallel topology at the output port, and the opposite for the parallel–series connection.

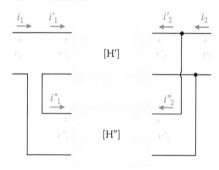

FIGURE 5.13 Series–parallel connection of a two-port network.

It can be easily shown for the series–parallel connection that the best choice of two-port parameter is the H-parameter, since the overall H-parameter matrix for series–parallel connected two-port networks is simply the sum of H-parameter matrices of each individual two-port network (Figure 5.13).

$$v_1 = v'_1 + v''_1 ; v_2 = v'_2 = v''_2$$

$$i_1 = i'_1 = i''_1 ; i_2 = i'_2 + i''_2$$

$$\begin{bmatrix} v'_1 \\ i'_2 \end{bmatrix} = \begin{bmatrix} h'_{11} & h'_{12} \\ h'_{21} & h'_{22} \end{bmatrix}\begin{bmatrix} i'_1 \\ v'_2 \end{bmatrix}; \begin{bmatrix} v''_1 \\ i''_2 \end{bmatrix} = \begin{bmatrix} h''_{11} & h''_{12} \\ h''_{21} & h''_{22} \end{bmatrix}\begin{bmatrix} i''_1 \\ v''_2 \end{bmatrix}$$

$$\begin{bmatrix} v_1 \\ i_2 \end{bmatrix} = \begin{bmatrix} v'_1 \\ i'_2 \end{bmatrix} + \begin{bmatrix} v''_1 \\ i''_2 \end{bmatrix} = \begin{bmatrix} h'_{11} & h'_{12} \\ h'_{21} & h'_{22} \end{bmatrix}\begin{bmatrix} i'_1 \\ v'_2 \end{bmatrix} + \begin{bmatrix} h''_{11} & h''_{12} \\ h''_{21} & h''_{22} \end{bmatrix}\begin{bmatrix} i''_1 \\ v''_2 \end{bmatrix}$$

$$\Rightarrow$$

$$\begin{bmatrix} h_{11} & h_{12} \\ h_{21} & h_{22} \end{bmatrix} = \begin{bmatrix} h'_{11} & h'_{12} \\ h'_{21} & h'_{22} \end{bmatrix} + \begin{bmatrix} h''_{11} & h''_{12} \\ h''_{21} & h''_{22} \end{bmatrix}$$

In the same manner when two-port networks are connected in a parallel–series configuration, the convenient choice of two-port parameter is the G-parameter, since the overall G-parameter matrix results from the sum of G-parameter matrices of each individual two-port network (Figure 5.14).

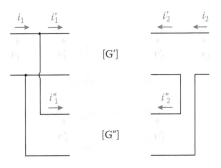

$$v_1 = v_1' = v_1''; v_2 = v_2' + v_2''$$

$$i_1 = i_1' + i_1''; i_2 = i_2' = i_2''$$

FIGURE 5.14 Parallel–series connection of a two-port network.

$$\begin{bmatrix} i_1' \\ v_2' \end{bmatrix} = \begin{bmatrix} g_{11}' & g_{12}' \\ g_{21}' & g_{22}' \end{bmatrix}\begin{bmatrix} v_1' \\ i_2' \end{bmatrix}; \begin{bmatrix} i_1'' \\ v_2'' \end{bmatrix} = \begin{bmatrix} g_{11}'' & g_{12}'' \\ g_{21}'' & g_{22}'' \end{bmatrix}\begin{bmatrix} v_1'' \\ i_2'' \end{bmatrix}$$

$$\Rightarrow$$

$$\begin{bmatrix} g_{11} & g_{12} \\ g_{21} & g_{22} \end{bmatrix} = \begin{bmatrix} g_{11}' & g_{12}' \\ g_{21}' & g_{22}' \end{bmatrix} + \begin{bmatrix} g_{11}'' & g_{12}'' \\ g_{21}'' & g_{22}'' \end{bmatrix}$$

5.5.4 Cascade Connection

The scheme seen in Figure 5.15 shows a cascade connection of two-port networks.

Note that the connection is called *cascade* insofar as the exit port variables of the first network correspond to the entry port variables of the second. In this case a representation in terms of a matrix of parameters of transmission makes our calculations easier, therefore we obtain

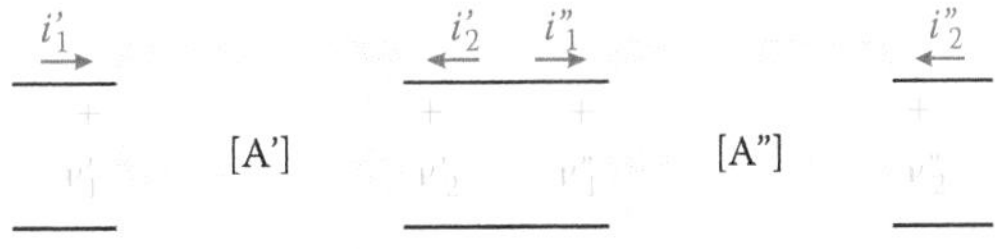

FIGURE 5.15 Cascade connection of two two-port networks described by (A) matrices.

$$v_1'' = v_2'; i_1'' = -i_2'$$

$$\begin{bmatrix} v_1' \\ i_2' \end{bmatrix} = \begin{bmatrix} a_{11}' & a_{12}' \\ a_{21}' & a_{22}' \end{bmatrix}\begin{bmatrix} v_2' \\ i_2' \end{bmatrix}; \begin{bmatrix} v_1'' \\ i_2'' \end{bmatrix} = \begin{bmatrix} a_{11}'' & a_{12}'' \\ a_{21}'' & a_{22}'' \end{bmatrix}\begin{bmatrix} v_2'' \\ i_2'' \end{bmatrix}$$

$$\begin{bmatrix} v_1' \\ i_2' \end{bmatrix} = \begin{bmatrix} a_{11}' & a_{12}' \\ a_{21}' & a_{22}' \end{bmatrix}\begin{bmatrix} v_1'' \\ i_1'' \end{bmatrix} = \begin{bmatrix} a_{11}' & a_{12}' \\ a_{21}' & a_{22}' \end{bmatrix}\begin{bmatrix} a_{11}'' & a_{12}'' \\ a_{21}'' & a_{22}'' \end{bmatrix}\begin{bmatrix} v_2'' \\ -i_2'' \end{bmatrix} = \begin{bmatrix} a_{11} & a_{12} \\ a_{21} & a_{22} \end{bmatrix}\begin{bmatrix} v_2'' \\ -i_2'' \end{bmatrix}$$

$$\Rightarrow$$

$$\begin{bmatrix} a_{11} & a_{12} \\ a_{21} & a_{22} \end{bmatrix} = \begin{bmatrix} a_{11}' & a_{12}' \\ a_{21}' & a_{22}' \end{bmatrix}\begin{bmatrix} a_{11}'' & a_{12}'' \\ a_{21}'' & a_{22}'' \end{bmatrix}$$

So, the overall A-parameter matrix for cascade-connected two-port networks is simply the product of A-parameter matrices of each individual two-port network.

5.6 KEY POINTS, JARGON, AND TERMS

→ A 2-port is simply a network with four terminals arranged into pairs called *ports*, which serve respectively as input and output for voltages (v_i and v_o) and currents (i_i and i_o).

→ A transformer is a two-port network device, to raise/lower AC voltages/currents from input to output port.

→ Electrical *dependent sources* are similar to *independent sources*, except that their voltage or current depends on other voltages or currents in the circuit.

→ There are four types of dependent sources: VCVS, CCVS, VCCS, CCCS.

→ Passive and linear circuits can be represented by a two-port network model.

→ A two-port network has four terminal variables: v_1, i_1, v_2, i_2. Since only two of them are independent, there are six possible sets, represented by the following matrices: $[Z]$, $[Y]$, $[H]$, $[G]$, $[A]$, $[B]$.

→ Two-port networks can be interconnected in five ways: in *series*, in *parallel*, in *series–parallel*, in *parallel–series*, and in *cascade*.

→ For 2-ports in series, it is convenient to adopt the $[Z]$ matrix. For 2-ports in parallel, it is convenient to adopt the $[Y]$ matrix. For 2-ports in series–parallel, it is convenient to adopt the $[H]$ matrix. For 2-ports in parallel–series, it is convenient to adopt the $[G]$ matrix. For 2-ports in cascade, it is convenient to adopt the $[A]$ matrix.

5.7 EXERCISES

EXERCISE 1

Referring to Figure 5.16, which is the equivalent resistance $R_{eq} = v_1 / i_1$ "seen" at the input port of the transformer?

ANSWER

From the condition $P_1 = P_2$, that is, $v_1 i_1 = v_2 i_2$, and since $R_{eq} = v_1 / i_1$ and $R = v_2 / i_2$, it is easy to obtain

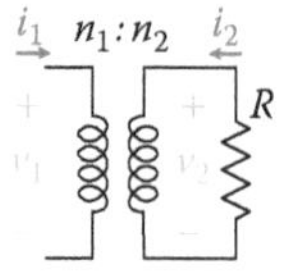

$$R_{eq} = \left(\frac{n_1}{n_2}\right)^2 R$$

FIGURE 5.16
Transformer with a load resistance.

So, $R_{eq} > R$ when $n_1 > n_2$.

EXERCISE 2

Given the simple two-port network made of two resistors shown in Figure 5.17, calculate its hybrid matrix (H).

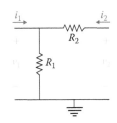

ANSWER

Starting from the definitions of the hybrid parameters, it is easy to write

FIGURE 5.17
Simple two-port network.

$$h_{11} \underset{\text{def}}{=} \left. \frac{v_1}{i_1} \right|_{v_2=0} = R_1 \,||\, R_2$$

$$h_{12} \underset{\text{def}}{=} \left. \frac{v_1}{v_2} \right|_{i_1=0} = \frac{R_1}{R_1 + R_2}$$

$$h_{21} \underset{\text{def}}{=} \left. \frac{i_2}{i_1} \right|_{v_2=0} = -\frac{R_1}{R_1 + R_2}$$

$$h_{22} \underset{\text{def}}{=} \left. \frac{i_2}{v_2} \right|_{i_1=0} = \frac{1}{R_1 + R_2}$$

Summarizing,

$$[H] = \begin{bmatrix} \dfrac{R_1 R_2}{R_1 + R_2} & \dfrac{R_1}{R_1 + R_2} \\ -\dfrac{R_1}{R_1 + R_2} & \dfrac{1}{R_1 + R_2} \end{bmatrix}$$

EXERCISE 3

$[Y]$ *to* $[Z]$ *conversion. Determine the* $[Z]$ *matrix in terms of* $[Y]$ *matrix elements.*

ANSWER

Let's recall the $[Y]$ matrix expression

$$\begin{bmatrix} i_1 \\ i_2 \end{bmatrix} = \begin{bmatrix} y_{11} & y_{12} \\ y_{21} & y_{22} \end{bmatrix} \begin{bmatrix} v_1 \\ v_2 \end{bmatrix}$$

and apply Cramer's rule:

$$v_1 = \frac{\begin{vmatrix} i_1 & y_{12} \\ i_2 & y_{22} \end{vmatrix}}{\begin{vmatrix} y_{11} & y_{12} \\ y_{21} & y_{22} \end{vmatrix}} = \frac{y_{22} i_1 - y_{12} i_2}{\Delta y} = \frac{y_{22}}{\Delta y} i_1 - \frac{y_{12}}{\Delta y} i_2$$

where $\Delta y = y_{11} y_{22} - y_{12} y_{21}$.

Similarly,

$$v_2 = \frac{\begin{vmatrix} y_{11} & i_1 \\ y_{21} & i_2 \end{vmatrix}}{\begin{vmatrix} y_{11} & y_{12} \\ y_{21} & y_{22} \end{vmatrix}} = \frac{y_{11}i_2 - y_{21}i_1}{\Delta y} = -\frac{y_{21}}{\Delta y}i_1 + \frac{y_{11}}{\Delta y}i_2$$

Now, from the expression

$$\begin{bmatrix} v_1 \\ v_2 \end{bmatrix} = \begin{bmatrix} z_{11} & z_{12} \\ z_{21} & z_{22} \end{bmatrix}\begin{bmatrix} i_1 \\ i_2 \end{bmatrix}$$

we can define every $[Z]$ element, in particular

$$z_{11} \underset{def}{=} \left.\frac{v_1}{i_1}\right|_{i2=0}$$

which thanks to the obtained expression for v_1 furnishes

$$z_{11} = -\frac{y_{22}}{\Delta y}$$

In the same way

$$z_{12} \underset{def}{=} \left.\frac{v_1}{i_2}\right|_{i1=0}$$

together with the expression of v_1 furnishes

$$z_{12} = -\frac{y_{12}}{\Delta y}$$

Again,

$$z_{21} \underset{def}{=} \left.\frac{v_2}{i_1}\right|_{i2=0}$$

together with v_2 determines

$$z_{21} = -\frac{y_{21}}{\Delta y}$$

Finally,

$$z_{22} \underset{def}{=} \left.\frac{v_2}{i_2}\right|_{i1=0}$$

with the obtained v_2 results in $z_{22} = \frac{y_{11}}{\Delta y}$

Summarizing,

$$[Z] = \begin{bmatrix} \dfrac{y_{22}}{\Delta z} & -\dfrac{y_{12}}{\Delta z} \\[3mm] -\dfrac{y_{21}}{\Delta z} & \dfrac{y_{11}}{\Delta z} \end{bmatrix}$$

EXERCISE 4

$[H]$ to $[Z]$ conversion. Determine the $[Z]$ matrix in terms of $[H]$ matrix elements.

ANSWER

Let's recall the $[H]$ matrix expression

$$\begin{bmatrix} v_1 \\ i_2 \end{bmatrix} = \begin{bmatrix} h_{11} & h_{12} \\ h_{21} & h_{22} \end{bmatrix} \begin{bmatrix} i_1 \\ v_2 \end{bmatrix}$$

and rewrite the second equation as

$$v_2 = -\frac{h_{21}}{h_{22}} i_1 + \frac{1}{h_{22}} i_2 \underset{\text{def}}{=} z_{21} i_1 + z_{22} i_2$$

while from the first equation we can write

$$v_1 = h_{11} i_1 + h_{12} v_2 = h_{11} i_1 + h_{12} \left(-\frac{h_{21}}{h_{22}} i_1 + \frac{1}{h_{22}} i_2 \right) = \frac{\Delta h}{h_{22}} i_1 + \frac{h_{12}}{h_{22}} i_2 \underset{\text{def}}{=} z_{11} i_1 + z_{12} i_2$$

with $\Delta h = h_{11} h_{22} - h_{12} h_{21}$.
Finally,

$$[Z] = \begin{bmatrix} \dfrac{\Delta h}{h_{22}} & \dfrac{h_{12}}{h_{22}} \\[3mm] -\dfrac{h_{21}}{h_{22}} & \dfrac{1}{h_{22}} \end{bmatrix}$$

EXERCISE 5

$[Z]$ to $[Y]$ conversion. Determine the $[Y]$ matrix in terms of $[Z]$ matrix elements.

ANSWER

Let's start from the definition of $[Z]$ and $[Y]$ matrices:

$$\begin{bmatrix} v_1 \\ v_2 \end{bmatrix} = \begin{bmatrix} z_{11} & z_{12} \\ z_{21} & z_{22} \end{bmatrix} \begin{bmatrix} i_1 \\ i_2 \end{bmatrix}$$

$$\begin{bmatrix} i_1 \\ i_2 \end{bmatrix} = \begin{bmatrix} y_{11} & y_{12} \\ y_{21} & y_{22} \end{bmatrix} \begin{bmatrix} v_1 \\ v_2 \end{bmatrix}$$

Now, replacing the first column array of the $[Z]$ matrix into the $[Y]$ matrix, we have

$$\begin{bmatrix} i_1 \\ i_2 \end{bmatrix} = \begin{bmatrix} y_{11} & y_{12} \\ y_{21} & y_{22} \end{bmatrix} \begin{bmatrix} z_{11} & z_{12} \\ z_{21} & z_{22} \end{bmatrix} \begin{bmatrix} i_1 \\ i_2 \end{bmatrix}$$

So, it is evident that

$$\begin{bmatrix} y_{11} & y_{12} \\ y_{21} & y_{22} \end{bmatrix} \begin{bmatrix} z_{11} & z_{12} \\ z_{21} & z_{22} \end{bmatrix} = \begin{bmatrix} 1 & 0 \\ 0 & 1 \end{bmatrix}$$

which corresponds to the equation set

$$\begin{cases} y_{11}z_{11} + y_{12}z_{21} = 1 \\ y_{11}z_{12} + y_{12}z_{22} = 0 \\ y_{21}z_{11} + y_{22}z_{21} = 0 \\ y_{21}z_{12} + y_{22}z_{22} = 1 \end{cases}$$

from which

$$y_{21} = -y_{22} \frac{z_{21}}{z_{11}}$$

$$y_{12} = -y_{11} \frac{z_{12}}{z_{22}}$$

$$y_{11} \left(\frac{z_{11}z_{22} - z_{12}z_{21}}{z_{22}} \right) = 1 \quad \Rightarrow \quad y_{11} = \frac{z_{22}}{\Delta z}$$

$$y_{22} \left(\frac{z_{11}z_{22} - z_{12}z_{21}}{z_{11}} \right) = 1 \quad \Rightarrow \quad y_{22} = \frac{z_{11}}{\Delta z}$$

We can substitute the latter two equations y_{11} and y_{22} into the previous ones y_{21} and y_{12}, obtaining

$$y_{21} = -\frac{z_{11}}{\Delta z} \frac{z_{21}}{z_{11}} = -\frac{z_{21}}{\Delta z}$$

$$y_{12} = -\frac{z_{22}}{\Delta z}\frac{z_{12}}{z_{22}} = -\frac{z_{12}}{\Delta z}$$

Summarizing,

$$[Y] = \begin{bmatrix} \dfrac{z_{22}}{\Delta z} & -\dfrac{z_{12}}{\Delta z} \\ -\dfrac{z_{21}}{\Delta z} & \dfrac{z_{11}}{\Delta z} \end{bmatrix}$$

EXERCISE 6

$[H]$ to $[Y]$ conversion. Determine the $[Y]$ matrix in terms of $[H]$ matrix elements.

ANSWER

Let's recall the $[H]$ parameters written as $\begin{cases} v_1 = h_{11}i_1 + h_{12}v_2 \\ i_2 = h_{21}i_1 + h_{22}v_2 \end{cases}$, and rewrite the first equation as

$$i_1 = \frac{1}{h_{11}}v_1 - \frac{h_{12}}{h_{11}}v_2 \underset{\mathrm{def}}{=} y_{11}v_1 + y_{12}v_2$$

This result allows us to write

$$i_2 = h_{21}\left[\frac{1}{h_{11}}v_1 - \frac{h_{12}}{h_{11}}v_2\right] + h_{22}v_2 = \frac{h_{21}}{h_{11}}v_1 + \frac{\Delta h}{h_{11}}v_2 \underset{\mathrm{def}}{=} y_{21}v_1 + y_{22}v_2$$

with $\Delta h = h_{11}h_{22} - h_{12}h_{21}$.
   It follows that

$$[Y] = \begin{bmatrix} \dfrac{1}{h_{11}} & -\dfrac{h_{12}}{h_{11}} \\ \dfrac{h_{21}}{h_{11}} & \dfrac{\Delta h}{h_{11}} \end{bmatrix}$$

EXERCISE 7

$[Z]$ to $[H]$ conversion. Determine the $[H]$ matrix in terms of $[Z]$ matrix elements.

ANSWER

Let's recall the $[Z]$ parameters written as $\begin{cases} v_1 = z_{11}i_1 + z_{12}i_2 \\ v_2 = z_{21}i_1 + z_{22}i_2 \end{cases}$, and rewrite the second equation as

$$i_2 = -\frac{z_{21}}{z_{22}} i_1 + \frac{1}{z_{22}} v_2 \underset{\text{def}}{=} h_{21} i_1 + h_{22} v_2$$

which rearranged into the first equation gives

$$v_1 = z_{11} i_1 + z_{12} i_2 = z_{11} i_1 + z_{12} \left[-\frac{z_{21}}{z_{22}} i_1 + \frac{1}{z_{22}} v_2 \right] = \frac{\Delta z}{z_{22}} i_1 + \frac{z_{12}}{z_{22}} v_2 \underset{\text{def}}{=} h_{11} i_1 + h_{12} v_2$$

with $\Delta z = z_{11} z_{22} - z_{12} z_{21}$.

It follows that

$$[H] = \begin{bmatrix} \dfrac{\Delta z}{z_{22}} & \dfrac{z_{12}}{z_{22}} \\[2ex] -\dfrac{z_{21}}{z_{22}} & \dfrac{1}{z_{22}} \end{bmatrix}$$

EXERCISE 8

The scheme in Figure 5.18 represents a so called T-network (please refer to the star topology discussed in Section 4.3.4 of Chapter 4).
Calculate its $[Z]$ matrix.

ANSWER

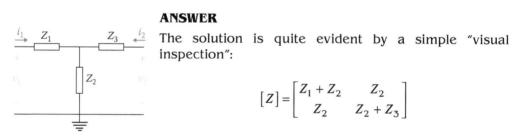

The solution is quite evident by a simple "visual inspection":

$$[Z] = \begin{bmatrix} Z_1 + Z_2 & Z_2 \\ Z_2 & Z_2 + Z_3 \end{bmatrix}$$

FIGURE 5.18
T-network.

EXERCISE 9

Calculate the four parameters h_{11}, h_{12}, h_{21}, h_{22}, of the hybrid matrix $[H]$ of the two-port T-network represented in Figure 5.18.

ANSWER

The hybrid parameters are obtained strarting from their definitions. In particular, we have

$$h_{11} \underset{\text{def}}{=} \left. \frac{v_1}{i_1} \right|_{v2=0}$$

$$h_{21} \underset{\text{def}}{=} \left. \frac{i_2}{i_1} \right|_{v2=0}$$

To calculate them, we impose $v_2 = 0$, as represented in Figure 5.19.

We write $\quad\quad\quad\quad\quad\quad\quad\quad\quad\quad\quad\quad\quad\quad\quad\quad\quad\quad\quad\quad$ h_{11}

$$h_{11} = Z_1 + Z_2 \,||\, Z_3$$

According to the current divider equation $-i_2 = i_{Z2} = \dfrac{Z_2}{Z_2 + Z_3} i_1$, so $\quad$ h_{21}

$$h_{21} = -\frac{Z_2}{Z_2 + Z_3}$$

By definition $\quad\quad\quad\quad\quad\quad\quad\quad\quad\quad\quad\quad\quad\quad\quad\quad\quad\quad\quad\quad\quad$ h_{12}

$$h_{12} \underset{\text{def}}{=} \left.\frac{v_1}{v_2}\right|_{i1=0}$$

therefore we have to consider zero the current i_1 at the input port and, as a consequence, the voltage drop across Z_1 is null, so v_1 is across Z_2. According to the voltage divider equation

$$v_1 = v_{Z3} = \frac{Z_2}{Z_2 + Z_3} v_2$$

therefore

$$h_{12} = \frac{Z_2}{Z_2 + Z_3}$$

$$h_{12} = \frac{1}{Z_2 + Z_3} \quad\quad\quad\quad\quad\quad\quad\quad\quad h_{22}$$

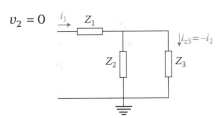

FIGURE 5.19 T-network with $v_2 = 0$.

Summarizing

$$[H] = \begin{bmatrix} Z_1 + Z_2 \mid\mid Z_3 & \dfrac{Z_2}{Z_2 + Z_3} \\[3mm] -\dfrac{Z_2}{Z_2 + Z_3} & \dfrac{1}{Z_2 + Z_3} \end{bmatrix}$$

EXERCISE 10

Calculate the elements of the admittance matrix $[Y]$ relative to the T-network represented in Figure 5.18.

ANSWER

Let's recall the definitions of the y_{11} and y_{21} parameters

$$y_{11} \underset{\text{def}}{=\!=} \left. \frac{i_1}{v_1} \right|_{v2=0}$$

$$y_{21} \underset{\text{def}}{=\!=} \left. \frac{i_2}{v_1} \right|_{v2=0}$$

and take into account the equations

$$\begin{cases} i_1 = y_{11}v_1 + y_{12}v_2 \\ i_2 = y_{21}v_1 + y_{22}v_2 \end{cases}$$

To calculate the unknown parameters we refer to the scheme of Figure 5.19, with $v_2 = 0$.

It is easy to obtain $\qquad\qquad\qquad y_{11}$

$$y_{11} = \frac{1}{Z_1 + Z_2 \mid\mid Z_3} = \frac{1}{Z_1 + \dfrac{Z_2 Z_3}{Z_2 + Z_3}} = \frac{Z_2 + Z_3}{Z_1 Z_2 + Z_1 Z_3 + Z_2 Z_3}$$

Referring to Figure 5.18, the voltage $v_x = -Z_3 i_2 \Rightarrow \qquad y_{21}$

$$i_2 = -\frac{v_x}{Z_3}$$

and (voltage divider's rule)

$$v_x = \frac{Z_2 \,||\, Z_3}{Z_2 \,||\, Z_3 + Z_1} v_1$$

so

$$\left. \frac{i_2}{v_1} \right|_{v2=0} = \frac{-\dfrac{v_x}{Z_3}}{v_1} = \frac{-\dfrac{1}{Z_3} \dfrac{Z_2 \,||\, Z_3}{Z_2 \,||\, Z_3 + Z_1} v_1}{v_1} = -\frac{1}{Z_3} \frac{Z_2 \,||\, Z_3}{Z_2 \,||\, Z_3 + Z_1} =$$

$$= \frac{-\dfrac{1}{Z_3} \dfrac{Z_2 Z_3}{Z_2 + Z_3}}{\dfrac{Z_2 Z_3}{Z_2 + Z_3} + Z_1} = -\frac{Z_2}{Z_2 + Z_3} \frac{Z_2 + Z_3}{Z_1 Z_2 + Z_1 Z_3 + Z_2 Z_3}$$

$$y_{21} = -\frac{Z_2}{Z_1 Z_2 + Z_1 Z_3 + Z_2 Z_3}$$

From the definitions of y_{12} and y_{22}

$$y_{12} \stackrel{\text{def}}{=\!=} \left. \frac{i_1}{v_2} \right|_{v1=0}$$

$$y_{22} \stackrel{\text{def}}{=\!=} \left. \frac{i_2}{v_2} \right|_{v1=0}$$

We can evaluate them by referring to the circuit in Figure 5.20, with $v_1 = 0$.

Similar to what we discussed about y_{12} the parameter Z_1 now, for symmetry reasons, swap Z_1 with Z_3:

$$y_{12} = -\frac{Z_2}{Z_1 Z_2 + Z_1 Z_3 + Z_2 Z_3}$$

With the same considerations y_{22} previously mentioned

$$y_{22} = \frac{Z_1 + Z_2}{Z_1 Z_2 + Z_1 Z_3 + Z_2 Z_3}$$

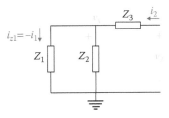

FIGURE 5.20 T-network with $v_1 = 0$.

Summarizing,

$$[Y] = \begin{bmatrix} \dfrac{Z_2 + Z_3}{Z_1 Z_2 + Z_1 Z_3 + Z_2 Z_3} & -\dfrac{Z_2}{Z_1 Z_2 + Z_1 Z_3 + Z_2 Z_3} \\ -\dfrac{Z_2}{Z_1 Z_2 + Z_1 Z_3 + Z_2 Z_3} & \dfrac{Z_1 + Z_2}{Z_1 Z_2 + Z_1 Z_3 + Z_2 Z_3} \end{bmatrix}$$

EXERCISE 11

Calculate the elements of the transmission matrix $[T]$ relative to the T-network represented in Figure 5.18.

ANSWER

The *T-network* can be considered composed of three two-port sub-networks in cascade, as in Figure 5.21: therefore a series impedance Z_1 cascades to the parallel impedance Z_2 and to the series impedance Z_3. The transmission matrices of the sub-networks are respectively equal to

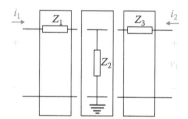

$$[T_1] = \begin{bmatrix} 1 & Z_1 \\ 0 & 1 \end{bmatrix}$$

$$[T_2] = \begin{bmatrix} 1 & 0 \\ Y_2 & 1 \end{bmatrix}$$

FIGURE 5.21 T-network as three two-port networks in cascade.

(with $Y_2 = \dfrac{1}{Z_2}$)

$$[T_3] = \begin{bmatrix} 1 & Z_3 \\ 0 & 1 \end{bmatrix}$$

It follows that the overall transmission matrix is their product:

$$\begin{bmatrix} v_1 \\ i_1 \end{bmatrix} = \begin{bmatrix} 1 & Z_1 \\ 0 & 1 \end{bmatrix} \begin{bmatrix} 1 & 0 \\ Y_2 & 1 \end{bmatrix} \begin{bmatrix} 1 & Z_3 \\ 0 & 1 \end{bmatrix} \begin{bmatrix} v_2 \\ i_2 \end{bmatrix} = \begin{bmatrix} 1 + Z_1 Y_2 & Z_1 \\ Y_2 & 1 \end{bmatrix} \begin{bmatrix} 1 & Z_3 \\ 0 & 1 \end{bmatrix} \begin{bmatrix} v_2 \\ i_2 \end{bmatrix}$$

$$= \begin{bmatrix} 1 + Z_1 Y_2 & (1 + Z_1 Y_2) Z_3 + Z_1 \\ Y_2 & Z_3 Y_2 + 1 \end{bmatrix} \begin{bmatrix} v_2 \\ i_2 \end{bmatrix}$$

Finally we can write

$$[T] = \begin{bmatrix} 1 + Z_1 Y_2 & Z_1 + Z_3 + Z_1 Z_3 Y_2 \\ Y_2 & 1 + Z_3 Y_2 \end{bmatrix}$$

EXERCISE 12

Calculate the $[Z]$ matrix of the two-port shown in Figure 5.22.

ANSWER

It is easy to solve considering that this network is similar to that of Figure 5.17; we simply use $Z_2' = Z_2 \,||\, Z_3$, and $Z_3' = Z_4 + Z_5$, as here schematized in Figure 5.23.

EXERCISE 13

The two-port network shown in Figure 5.24, here highlighted in a colored box, can be represented by the $[Y]$ matrix. At the network input is a voltage source v_s with its internal impedance Z_s, and at the network output a load impedance Z_L.

Calculate the impedance Z_i "seen" at the input port, the impedance Z_o "seen" at the output port, the current ratio

$$A_i \underset{def}{=\!=} \frac{i_2}{i_1}$$

and the voltage ratio

$$y_{22} \underset{def}{=\!=} \left. \frac{i_2}{v_2} \right|_{v1=0} .$$

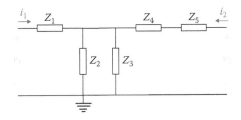

| FIGURE 5.22 Two-port network with five impedances in T configuration. | FIGURE 5.23 T-network considering series and parallel topologies. |

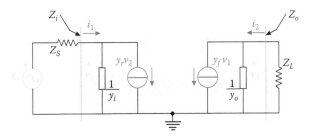

FIGURE 5.24 Two-port network, represented by an (Y) matrix, powered by a voltage source that feeds a load impedance.

ANSWER

The analytical notation of the admittance matrix follows

$$\begin{cases} i_1 = y_i v_1 + y_r v_2 & \text{(eq.I)} \\ i_2 = y_f v_1 + y_o v_2 & \text{(eq.II)} \end{cases}$$

At the output of the network

$$v_2 = -y_f v_1 \left(\frac{1}{y_o} \,||\, Z_L \right)$$

which, inserted into *eq.I* gives

$$i_1 = y_i v_1 - y_r y_f v_1 \left(\frac{1}{y_o} \,||\, Z_L \right) \Rightarrow i_1 = v_1 \left(y_i - \frac{y_f y_r Z_L}{1 + y_o Z_L} \right)$$

so

$$Z_i \underset{\text{def}}{=} \frac{v_1}{i_1} = \frac{1}{y_i - \dfrac{y_f y_r Z_L}{1 + y_o Z_L}}$$

Similarly

$$Z_o \underset{\text{def}}{=} \frac{v_2}{i_2} = \frac{1}{y_o - \dfrac{y_f y_r Z_g}{1 + y_i Z_g}}$$

which is evident due to the two-port symmetry.
For the current ratio

$$A_i \underset{\text{def}}{=} \frac{i_2}{i_1} = -\frac{y_f / Z_L}{y_i y_o - y_f y_r + y_i / Z_L} = -\frac{y_f / Z_L}{\Delta y + y_i / Z_L}$$

where $\Delta y = y_i y_o - y_f y_r$.
For the voltage ratio, we obtain

$$A_v \underset{\text{def}}{=} \frac{v_2}{v_1} = -\frac{y_f}{y_o + 1 / Z_L}$$

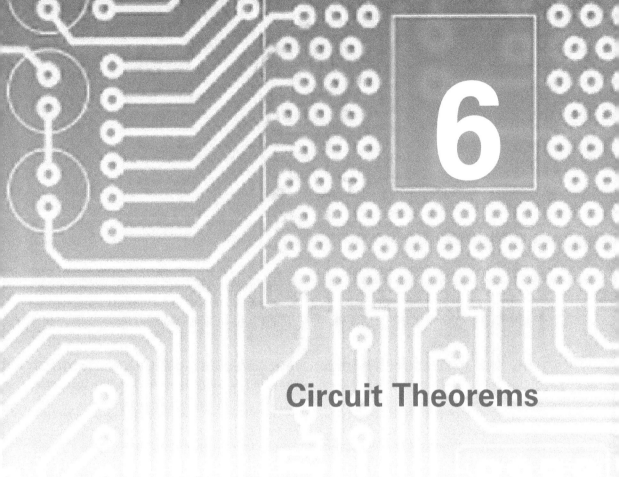

Circuit Theorems

6.1 DEFINITIONS

6.1.1 Electric Circuit and Its Elements

An *electric circuit* (or simply *circuit*) is a system of conductors and devices (sources, resistors, capacitors, inductors, transformers, etc.) interconnected to each other in a way to allow the flow of electrical current. The aim is to elaborate and/or transfer electric signals or electric power from one point to another.

Observation

There are many examples of electric circuits in nature.

The brain can be considered the interplay of varying voltages and currents. It contains a network of organized and interconnected nerve electrical cells, each with a particular purpose and each electrically connected with all of the other cells.

In the myocyte, the heart muscle cell, electric activation takes place from the inflow of sodium ions across the cell membrane. There is a specific electrical circuit pathway through the heart: sinoatrial (SA) node, atrioventricular (AV) node, to left and right bundle branches, and finally to the Purkinje fibers, which make the heart contract.

Colonies of bacteria use *nanowires* to transport electrons. These nanowires are constructed out of protein molecules that self-assemble themselves into long filaments and conduct current between two electrodes. In such a way the bacteria communicate and share energy.

One of the fundamental hypotheses for the circuits we will study is to consider the conductors that connect the elements as ideal, therefore equipotential, that is, without any voltage drop at their terminals. As a consequence we can say that energy (and therefore voltage) variations can only take place inside the devices and not inside the connecting conductors (Figure 6.1). When this hypothesis is satisfied we have *lumped-element* circuits.

One consequence of this hypothesis is that what is important in circuits is only the way in which the devices are interconnected (through terminals) and not their arrangement. The way the circuit devices are connected is called circuit *topology*. To illustrate this, we present two circuits that have the same topology, though they have different arrangements (Figure 6.2).

The two circuits shown have identical functions as far as the circuit is concerned but, for obvious reasons of clarity, we prefer to adopt the second topology.

Current flows in a circuit only if there are no interruptions in the circuit itself, which is known as a *closed circuit* or *closed loop circuit*. Vice-versa when there is an interruption in a circuit there can be no current flow ($i = 0$) and we have an *open circuit*.

A *short circuit* (or simply *short* or *s/c*) occurs when there is a very low-resistance connection (towards zero), usually accidental, between two points of an electrical circuit at different voltages. Such an event is typically followed by an instantaneous increase of the electrical current.

The *open circuit*, defined previously, is the exact opposite of the *short circuit*, that is, a state of non-connection between two points of a circuit. The current cannot therefore flow and the resistance between the two points of the circuit can be considered infinite (Figure 6.3).

A *linear circuit* is made only of *linear components* (according to the definition in Section 4.1 in Chapter 4), so that the output is, within a given dynamic range, linearly proportional to the input.

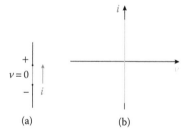

(a) (b)

FIGURE 6.1 Zero voltage drop at the terminals of any conductor section in the case of lumped-element circuits: (a) conductor segment chosen; (b) current–voltage characteristic (or I–V graph). Any current can flow where v = 0.

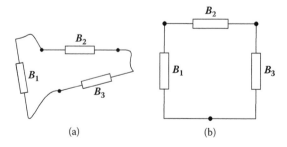

(a) (b)

FIGURE 6.2 Different topologies of the same circuit.

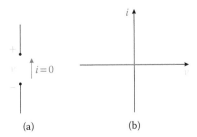

(a)   (b)

FIGURE 6.3 Null current in an open circuit. Between the opening points, as in (a), there can be any voltage value (determined by the rest of the circuit), as in (b).

Observation

Every component belonging to a network is characterized by a parameter. For example, the resistor by its resistance, the capacitor by its capacitance, and the inductor by its inductance. If the parameter does not depend on the voltage across the component or on the current flowing into it, then the component is considered to be *linear*.

According to this, the resistor, the capacitor, and the inductor are all linear elements.

6.1.2 Equivalent Networks

An electric network, or a part of it, or a component, can sometimes be more easily analyzed by replacing it with its *equivalent*. An *equivalent network* or *component* is made by a theoretical electric circuit with exactly the same electrical characteristics of the network/circuit or component it is intended to replace. This *equivalence* can be limited, too. For example, if we are interested in the working function of a circuit within a frequency band, we can admit an equivalence just limited within that band and not elsewhere.

Given a circuit, it has to be analyzed by *network analysis*, that is, the process of determining the voltages across and the currents through every component in the circuit. To this aim it can be convenient to replace the actual components with others with the same effect. The easiest example is to replace two resistors in series with the equivalent one ($R = R_1 + R_2$), or replace a Δ configuration with the Y equivalent. It is a bit more complex to replace a whole circuit or a part of it, or a component, with its equivalent. To reduce computational complexity, a single component can simply be replaced in terms of its mathematical functionality.

6.1.3 Node, Branch, Mesh

Some other essential definitions:

→ *Node*: A point at which two or more components are connected together with shorts (point "A" in Figure 6.4).
→ *Branch*: A part of a circuit between two nodes (as "A" and "B" in Figure 6.4b). It can result with two or more components in series.
→ *Mesh*: Any closed circuit (therefore any loop that starts and ends at the same node; three meshes in the example in Figure 6.4c).

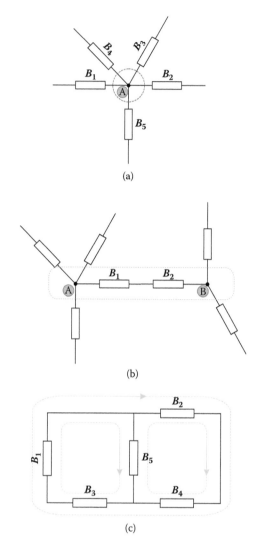

(a)

(b)

(c)

FIGURE 6.4 Example of (a) node "A," (b) branch between nodes "A"-"B," (c) mesh (three in blue).

In the example of Figure 6.5, we have four nodes (indicated by the letters A, B, C, D) and seven meshes (indicated by the relevant paths: R_1, R_2, R_3; R_2, R_4, V_B; R_3, R_5, R_4; R_2, R_3, R_5, V_B; R_3, R_4, V_B, R_1; R_2, R_1, R_5, R_4; V_B, R_1, R_5).

6.1.4 Ground and Floating Ground

In electrical and electronic circuits, several values of voltage differences are present at the same time at different points. Seeing as it is useful to relate all the voltages to a unique

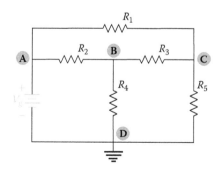

FIGURE 6.5 Example of a circuit with 4 nodes and 7 meshes.

reference value, we can adopt a particular point in the circuit as a common reference. *Ground* (or *earth*) can be the reference point in an electrical circuit from which other voltages are measured. The ground is generally a zero voltage value, otherwise it is called a *floating ground*.

The *ground* is usually idealized as an infinite source or sink for charge, which can absorb an unlimited amount of current without changing its zero voltage value. But this can be true only if it is connected to the real earth. In electronic equipment such as portable devices, the *ground* is usually a large conductor attached to one side of the power supply, which is attached to the chassis of the devices. So we call it a *chassis ground* (Figure 6.6).

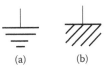

(a) (b)

FIGURE 6.6   Circuit symbols for (a) earth ground and (b) chassis ground.

6.1.5 Decibel

For the study of a generic electrical circuit it is useful to introduce the *decibel* (dB). It is a scale that derives from the expression of *gain* (or *loss*, i.e., *negative gain*) that in the past was used in telecommunication systems.

Such a (negative) gain was measured in *Bel* (Alexander Graham Bell, Scottish-American scientist and inventor, 1847–1922, Figure 6.7) and was used in the company labs of *Bell Telephone* to quantify the reduction of the audio level occurring in 1 mile of standard telephone cables. But the *Bel* unit was too high for everyday use, so the *decibel* was introduced, that is, one tenth of a Bel: decibel = 0.1 Bel.

FIGURE 6.7   A.G. Bell.

The *decibel* (or *dB*) represents the ratio between two levels of power, or *power gain*, equal to

$$dB = 10\log\left(\frac{P_2}{P_1}\right)$$

in which P_2 is output power of the circuit and P_1 is input power.

As defined, the *dB* is a *dimensionless* parameter (given that it is the ratio of two quantities having the same unit of measure).

From the ratio, we obtain the logarithm essentially to take advantage of the mathematical properties of the logarithm itself:

$$\log(P_1 P_2) = \log(P_1) + \log(P_2)$$

$$\log\left(\frac{P_1}{P_2}\right) = \log(P_1) - \log(P_2)$$

$$\log(P^n) = n\log(P)$$
$$\log(1) = 0$$

So in this way

→ The gains (or the losses by attenuation) can be summed instead of multiplied.
→ Negative decibels represent attenuations, positive decibels represent gains.
→ The measure scale is "compressed" (for example the enormous interval $1\,\mu W$–$100\,W$ is equal to "only" $80\,dB$).

According to the definition, expressing the gain G in dB, G_{dB}, we get the following:

→ If $P_2 = P_1 \Rightarrow G_{dB} = 0\,dB$.
→ If $P_2 = 2P_1 \Rightarrow G_{dB} = 3\,dB$.
→ If $P_2 = (1/2)P_1 \Rightarrow G_{dB} = -3\,dB$.
→ If $P_2 = 1,000,000P_1 \Rightarrow G_{dB} = 60\,dB$.

Therefore $0\,dB$ means that the circuit has produced neither gains nor losses on the signal; $3\,dB$ means that in the output the circuit has doubled its input power (active network); $-3\,dB$ means that in the output the circuit has half its input power (passive network); and $60\,dB$ represents a network that, in the output, produces a gain of one million times its input power. So, as we can see, the dB reduces the values to be considered, given its property of "compression," as we can also note in Figure 6.8.

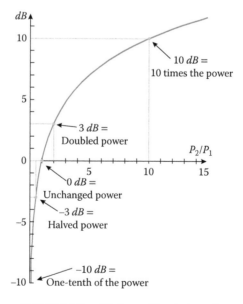

FIGURE 6.8 dB, logarithmic trend.

Curiosities

The human ear behaves in a highly nonlinear way, so much so that to hear a sound that is double the intensity of another, it needs to be 10 times more powerful. In this case, too, the logarithmic scale is preferable.

Capabilities of the human ear:

- The smallest detectable step-change in amplitude for a pure tone is around 0.3 dB and, more practically, even 1 dB.
- The smallest detectable frequency variation of a pure tone is around 0.2% in the central audio band.

Alexander Bell is still credited with the invention of the telephone. In reality, Bell realized the first practical telephone that was commercialized. The very first idea of the telephone, as a voice communication apparatus, is credited to Meucci (Antonio Santi Giuseppe Meucci, Italian-born, naturalized American inventor, 1808–1889, Figure 6.9).

FIGURE 6.9
A. Meucci.

We can express decibels regarding a specific point of reference such as μW, mW, W:

$$dB_{\mu W} = dB_{\mu} = 10 \log\left(\frac{P}{1 * 10^{-6}}\right)$$

$$dB_{mW} = dB_{m} = 10 \log\left(\frac{P}{1 * 10^{-3}}\right)$$

$$dB_{W} = 10 \log(P)$$

(As we note above, in the subscript the letter W can be omitted.)

Of the three units the one used the most is dB_m and, knowing the impedance value, we can express it through voltages.

6.2 KIRCHHOFF'S LAWS

Kirchhoff's circuit laws (Gustav Robert Kirchhoff, German physician and mathematician, 1824–1887, Figure 6.10) are fundamental for the analysis of any network no matter how complex. They derive from the principles of conservation of charge and energy in circuits.

Kirchhoff's voltage law (KVL) and Kirchhoff's current law (KCL) show the relationships between voltages and currents in circuits.

FIGURE 6.10   G. Kirchhoff

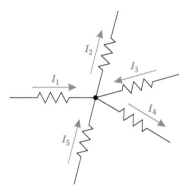

FIGURE 6.11 Example of entering and exiting currents in a node.

6.2.1 Kirchhoff's Current Law (KCL)

Kirchoff's current law (KCL) concerns nodes of any electrical network and states that the flow of carriers satisfies the so-called condition of continuity, hence the statement that *"for a given node in a circuit, the algebraic sum of the currents entering equals the sum of the currents leaving."*

This law expressed in mathematical terms gives the equation

$$\sum_{n=1}^{N} I_n = 0$$

in which N indicates the total number of currents in the node of interest. The algebraic sum states that the currents entering the node must be opposite in sign from the currents exiting it. To illustrate this, let us consider the case in Figure 6.11.

The sum of currents entering the node is equal to that exiting the same node, so $I_1 + I_3 + I_5 = I_2 + I_4$, which equals $I_1 + I_3 + I_5 - I_2 - I_4 = 0$; therefore the algebraic sum (i.e., sum that takes into account positive or negative signs of currents) is equal to zero.

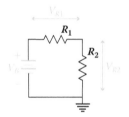

FIGURE 6.12 Example of a mesh for the application of Kirchhoff's voltage law.

6.2.2 Kirchhoff's Voltage Law (KVL)

Kirchhoff's voltage law (KVL) concerns meshes of any electrical network and states that *"around any closed loop in a circuit, the algebraic sum of the voltages across all elements is zero."*

This law, expressed in mathematical terms, gives the equation

$$\sum_{n=1}^{N} V_n = 0$$

in which N is the total number of voltages related to the examined mesh.

As an example let us consider the case in Figure 6.12.

From the analysis of the voltages in the mesh, we have a battery voltage that is distributed across the two resistances, so $V_B = V_{R1} + V_{R1}$, an equation which can be written as $V_B - V_{R1} - V_{R1} = 0$; therefore, there is an algebraic sum equal to zero.

Example

In the circuit shown in Figure 6.13,

KVL for mesh "ABD": $V_S = R_2 \left(I_2 - I_1 \right) + R_4 \left(I_2 - I_3 \right)$
KVL for mesh "ABC": $0 = R_1 \left(I_1 \right) + R_3 \left(I_3 - I_1 \right) + R_2 \left(I_2 - I_1 \right)$
KVL for mesh "BCD": $0 = R_4 \left(I_2 - I_3 \right) + R_3 \left(I_3 - I_1 \right) + R_5 \left(I_3 \right)$

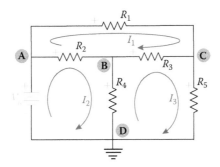

FIGURE 6.13 Network with four nodes and seven loops.

$$\Rightarrow \begin{bmatrix} V_S \\ 0 \\ 0 \end{bmatrix} = \begin{bmatrix} -R_2 & R_2 + R_4 & -R_4 \\ R_1 - R_3 - R_2 & R_2 & R_3 \\ -R_3 & R_4 & -R_4 + R_5 \end{bmatrix} \begin{bmatrix} I_1 \\ I_2 \\ I_3 \end{bmatrix}$$

6.3 THÉVENIN'S THEOREM

In the following sections, some theorems will be defined so that a large complex network can be simplified to reduce its mathematical analysis (Figure 6.14).

We start here with a form of *equivalent network*, as defined in Section 6.1.2 of Chapter 6, that comes from the Thévenin equivalent, detailed by Thévenin's theorem (Léon Charles Thévenin, French engineer, 1857–1926, Figure 6.15); this form is useful in many practical cases as it simplifies things significantly.

In particular Thévenin's theorem states that "any linear network, no matter how complex, made of a combination of independent (current and/or voltage) sources

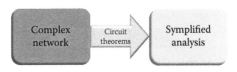

FIGURE 6.14 The circuit theorems allows us to simplify the analysis of a complex network.

FIGURE 6.15 L.C. Thévenin.

and resistors, from the viewpoint of any pair of terminals 'A' and 'B,' is equivalent to a single voltage source v_{TH} and a single series resistor R_{TH}" (Figure 6.16).

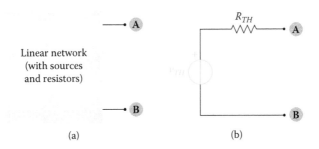

Linear network
(with sources
and resistors)

(a)   (b)

FIGURE 6.16 (a) Active linear network, (b) Thévenin's equivalent circuit.

The values of Thévenin's equivalent voltage source v_{TH}, and of Thévenin's equivalent series resistance R_{TH}, are calculated as follows:

→ R_{TH} is the resistance "seen" at the port A-B of the circuit and "summarizies" all its passive components. We need to replace the internal resistance of voltage and current sources. For the ideal voltage source this means replacing it with a short circuit, whereas for the ideal current source, it means replacing it with an open circuit. On the other hand, shorting or opening the circuit must *not* be done with dependent sources, as they present (or can present) non-null internal equivalent resistances, or have a characteristic that can be a function of a particular resistance inside the circuit. The resistance can then be calculated at the port "A-B" with the formulae for the equivalences of the topology of interconnections. The load impedance, if any, at the "A-B" port must be removed.
→ v_{TH} is the open-circuit voltage across "A"-"B" nodes, with no output load, so the load impedance, if any, must be removed.

Please be aware that circuits can be linear only within a certain frequency range, thus the Thévenin equivalent is valid only within this range.

Example

Consider the circuit shown in Figure 6.17 and determine Thévenin's equivalent to terminals A and B.

Calculation of Thévenin's voltage in terminals A-B (voltage divider):

$$V_{TH} = V_{AB} = \frac{R_2}{R_1 + R_2} V$$

From terminals A-B, shorting V, we obtain an R_{TH} equal to

$$R_{TH} = \frac{R_1 R_2}{R_1 + R_2}$$

If we want to know the current that would flow through a hypothetical load resistance R_L connected to terminals A-B, we will have

$$i_C = \frac{V_{TH}}{R_{TH} + R_L}$$

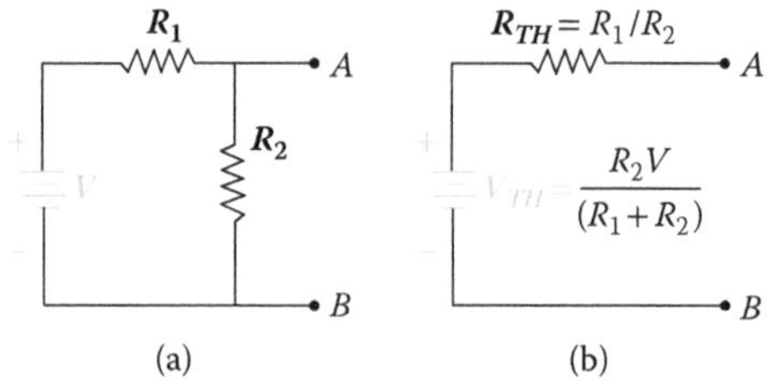

FIGURE 6.17   Example of Thévenin's equivalence.

6.4 NORTON'S THEOREM

Norton's theorem (Edward Lawry Norton, American electronics engineer, 1898–1983) states that "any linear network, no matter how complex, made of a combination of independent (current and/or voltage) sources and resistors, from the viewpoint of any pair of terminals 'A' and 'B,' is equivalent to a single current source i_N with a single parallel resistor R_N" (Figure 6.18).

→ Norton's resistance R_N is obtained in an analogous way to that of Thévenin's.
→ The current value of the source i_N is equal to the one that can be measured or deduced when the terminals "A" and "B" are shorted.

The Norton↔Thévenin equivalence is immediate, keeping in mind that $R_{TH} = R_N = R_{out}$. In fact, suppose we have at our disposal a Norton's equivalent circuit of a certain network, as illustrated in Figure 6.19.

To verify the equivalence, we need to connect a resistance R_L to both circuits and calculate the current value i_{RL}:

$$i_{RL} = i_N \frac{R_N R_L}{R_N + R_L} \frac{1}{R_L} \; ; \; i_{RL} = \frac{V_{TH}}{R_N + R_L} = \frac{i_N R_N}{R_N + R_L}$$

both of which, obviously, are equal.

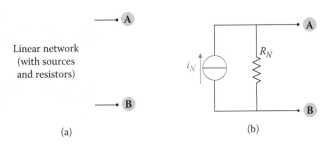

FIGURE 6.18 (a) Active linear network, (b) Norton's equivalent circuit

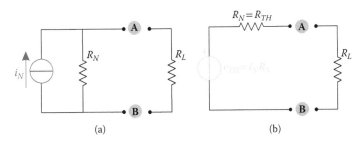

FIGURE 6.19 Verification schemes of Thévenin–Norton equivalence.

6.5 SUPERPOSITION THEOREM

Within the context of linear circuit analysis, both for transient or steady-state conditions, the *superposition theorem* is of fundametal importance. It is really useful for calculating the voltages and currents in a circuit with two or more (voltage and/or current) sources.

It asserts that *"if in a linear network more than one voltage and/or current source operates at the same time, the voltage between any two nodes in the network (or the current in any branch) is the algebraic sum of voltages (or currents) owing to each of the independent sources acting one at a time."*

Observation

To apply the superposition theorem to a network, it is fundamental to respect the linearity requirement.

Take into consideration, for instance, that power in an resistor is a nonlinear function, varying either with the square of voltage across it or with the square of current flowing through it. So, the superposition theorem cannot be applied to determine the power associated with a resistor.

Example

Let us consider the following circuit for which

$$V_{B1} = 9\,V; V_{B2} = 4.5\,V; R_1 = R_2 = R_3 = 3\,\Omega$$

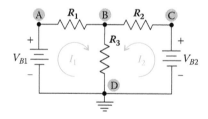

We calculate the current that flows through R_3.
The KVL equations for two loops of this circuit are

$$\begin{cases} V_{B1} = R_1 I_1 + R_3\left(I_1 + I_2\right) \\ V_{B2} = R_2 I_2 + R_3\left(I_1 + I_2\right) \end{cases} \Rightarrow \begin{cases} 9 = 6I_1 + 3I_2 \\ 4.5 = 3I_1 + 6I_2 \end{cases}$$

The following is the solution for the equations: $I_1 = 1.5\,A$, $I_2 = 0\,A$.
The same solutions can be found applying the superposition theorem. In fact, considering the two cases of only one source at a time, we have

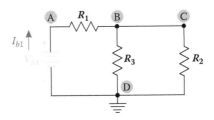

$$V_{B1} = \left(R_1 + R_2 \,|\,|\, R_3\right) I_{B1}$$

$$V_{R_1}' = R_1 I_{B1} = 6\,V \Rightarrow V_{R_2}' = V_{R_3}' = V_{B1} - V_{R_1} = 3\,V$$

Therefore $I_{R3}' = V_{R3}'/R_3 = 3/3 = 1\,A$. While for the second source

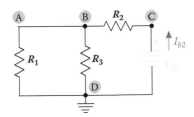

$$V_{B2} = \left(R_2 + R_1 \,|\,|\, R_3\right) I_{B2} \Rightarrow I_{B2} = \frac{4.5}{3+1.5} = 1\,A$$

$$V_{R_2}'' = R_2 I_{B2} = 3\,V \Rightarrow V_{R_1}'' = V_{R_3}'' = V_{B2} - V_{R_2} = 1.5\,V$$

Therefore $I_{R3}'' = V_{R3}''/R_3 = 1.5/3 = 0.5\,A$.

In conclusion $I_{R3TOT} = I_{R3}' + I_{R3}'' = 1.5\,A$.

6.6 MILLER'S THEOREM

In addition to the way to form *equivalent circuits* furnished by the aforementioned theorems of Thévenin and Norton, *Miller's theorem* (John Milton Miller, American physician and electrical engineer, 1882–1962) refers to the possibility of obtaining an equivalent circuit to partially reduce the complexity of performing a network analysis.

In particular, Miller's theorem asserts that "in any given linear two-port network having the output signal proportional to the input one, a shunt impedance Z directly connected between input and output ports can be replaced by two equivalent impedances Z_A and Z_B placed in the only input and output ports of the network respectively."

Let us consider a linear two-port network with an impedance Z that connects an input terminal at node "A" with an output one at node "B," as shown in Figure 6.20.

Miller's theorem states the equivalence between the scheme in the previous figure and the one represented in Figure 6.21.

The equivalence can be shown when input voltage V_A and output voltage V_B of the network are directly proportional, that is, the following relation is valid:

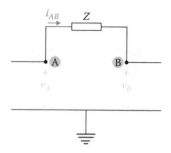

FIGURE 6.20 Impedance Z between the input and output ports of a two-port network.

$$v_B = K v_A$$

with K called the *Miller factor* (which can also be a complex value).

The two schematizations are equivalent if

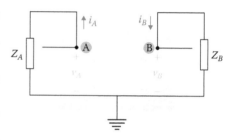

→ The voltage V_A across node "A" and the ground does not change.

→ The voltage V_B across the node "B" and the ground does not change.

→ The current i_{AB} flowing out of the node "A" and entering the impedance Z is the same as the current i_A flowing out of the node "A" and entering the impedance Z_A.

FIGURE 6.21 Equivalent circuit of the two-port network according to Miller's theorem.

→ The current i_{AB} flowing into the node "B" and leaving the impedance Z is the same and the current i_B flowing out of the node "B" and entering the impedance Z_B

From Figure 6.20, we get

$$i_{AB} = \frac{v_A - v_B}{Z} = v_A \frac{1 - K}{Z} = v_B \frac{\frac{1}{K} - 1}{Z}$$

and from Figure 6.21

$$i_A = \frac{v_A}{Z_A} \text{ and } i_B = \frac{v_B}{Z_B}$$

But, for equivalence reasons, we have to impose $i_A \overset{!}{=} i_{AB} \overset{!}{=} i_B$, so the aftermath is

$$Z_A = Z \frac{1}{1 - K}$$

$$Z_B = Z \frac{K}{K - 1}$$

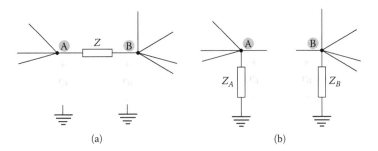

(a)   (b)

FIGURE 6.22   (a) An impedance bridging node "A" and "B"
is converted into (b) two grounded impedances.

This theorem can be suitably adopted not just by thinking of the impedance Z as a bridge between input and output ports of a two-port network, but also as a link between two given nodes inside a general network, provided with a Miller factor. The link between nodes can be just one impedance or a branch, made of two or more impedances.

In Figure 6.22, we see that a floating impedance Z can be converted to two grounded impedances Z_A and Z_B, again with the conditions

$$Z_A = Z \frac{1}{1-K}$$

and

$$Z_B = Z \frac{K}{K-1}$$

6.7 MILLER'S DUAL THEOREM

Miller's theorem also has a dual form: "an impedance (or a branch with impedance) Z where two currents i_1 and i_2 converge, can be equivalently replaced by two impedances (or branches with impedances) Z_A with i_1 flowing and Z_B with i_2 flowing, respectively, the impedances being"

$$Z_A = Z\left(1 + \frac{i_2}{i_1}\right) \text{ and } Z_B = Z\left(1 + \frac{i_1}{i_2}\right)$$

(Figure 6.23).

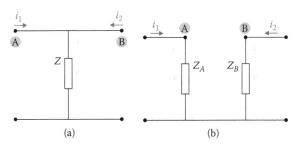

(a)   (b)

FIGURE 6.23   (a) An impedance where two currents converge is equivalent
to (b) two impedances with the two currents flowing separately.

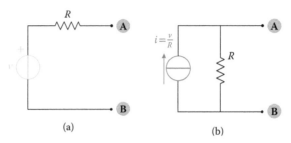

(a)　　　　　　(b)

FIGURE 6.24   (a) A voltage source can be converted into (b) a current source.

To guarantee the equivalence, the voltage v must be unchanged across the elements and the flowing currents i_1 and i_2, so

$$v = \left(i_1 + i_2 \right) Z = Z_A i_1 = Z_B i_2$$

hence the assertion.

6.8 SUBSTITUTION THEOREM

The *substitution theorem* states that *"for branch equivalence, the terminal voltage and current must be the same."* According to the theorem, the components in a network can be interchanged, so long as the terminal current and voltage are maintained.

This theorem can be considered a generalization of the Thévenin or Norton statements.

With the substitution theorem we can, for instance, realize a source conversion (which is sometimes referred to as a separate theorem, the *source conversion theorem*) (Figure 6.24).

So, at the terminals "A"-"B", voltage, current, and resistance are the same in both cases.

A consequence worth noting of the theorem is that when a source is a function of the voltage or current of its branch, then it is equivalent to a passive element, as represented in Figure 6.25.

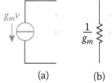

(a)　　　　(b)

FIGURE 6.25   (a) A dependent current source can be equivalent to (b) a resistor.

6.9 IMPEDANCE MATCHING AND BRIDGING

6.9.1 Introduction

Figure 6.26 schematizes a direct connection between a signal voltage source v_s, of internal impedance Z_s, and an impedance load Z_L. The signal source and its internal impedance can be real or the equivalent of a whole circuit, according to Thévenin's theorem.

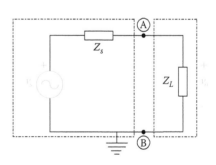

FIGURE 6.26   Direct source-load connection.

We will see that it is somewhat strategic to know the relationship between the values of Z_S and Z_L. So, even if the network is a complex one and is not represented by its Thévenin equivalent, it can be convenient to determine its equivalent impedance anyway.

For this type of connection, it might be necessary to maximize the voltage transferred from source to load, or maximize the transferred current, or maximize the transferred power. These requirements imply different relations between Z_S and Z_L.

6.9.2 Maximum Power Transfer Theorem

In this section, we will consider the conditions we must satisfy to transfer the maximum active power from the source to the load, so to minimize reflections from the load to the source, realizing what is referred as *impedance matching*.

6.9.2.1 Resistive Impedances for Both Source and Load

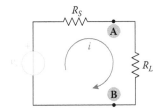

FIGURE 6.27 Source with resistive impedance and load with resistive impedance.

Let us consider a simple circuit as in Figure 6.27, composed of a voltage source v_S with its internal resistance R_S. Across a load resistance R_L. We want to find the value of R_L to realize the maximum transfer of active power from the source to the load.

We study this circuit using the method of phasors (see Section 1.5 in Chapter 1). The output power is expressed as

$$P_A = \frac{1}{2} Re\left[\overline{V}_{AB}\overline{I}^*\right] = \frac{1}{2} R_L Re\left[\overline{I}\overline{I}^*\right] = \frac{1}{2} R_L I^2 = \frac{1}{2R_L} Re\left[\overline{V}_{AB}\overline{V}_{AB}^*\right] = \frac{1}{2R_L} V_{AB}^2$$

For this circuit the following relations are valid:

$$I = \frac{V}{R_S + R_L} ; V_{AB} = \frac{R_L}{R_S + R_L} V_S$$

from which we get

$$P_A = \frac{1}{2} \frac{R_L}{\left(R_S + R_L\right)^2} V_S^2$$

Imposing the condition of maximum transfer of active power, the derivative of power with respect to R_L must be null, and the second derivative must be negative:

$$\frac{dP_A}{dR_L} = \frac{V_S^2\left[\left(R_S + R_L\right)^2 - 2R_L\left(R_S + R_L\right)\right]}{2\left(R_S + R_L\right)^4} = \frac{V_S^2}{2} \frac{\left(R_S^2 - R_L^2\right)}{\left(R_S + R_L\right)^4} \overset{!}{=} 0$$

which implies

$$R_S = R_L$$

and we also have

$$\left.\frac{d^2 P_A}{dR_L^2}\right|_{R_S = R_L} = -\frac{v_S^2}{2}\frac{1}{8 - R_L^3} < 0$$

as we requested.

So, as a conclusion, the maximum power transfer theorem states that, to obtain maximum transferred power from a source with a finite internal resistance to a load, the resistance of the source and the resistance of the load must be equal.

As a consequence, this means that as a maximum, just 50% of the available power from the source can be delivered to the load.

6.9.2.2 Complex Impedance for Source and Resistive Impedance for Load

Let us now consider the case in which the source impedance has a complex value, that is, a resistive and a reactive part such as $Z_S = R_S + jX_S$, and we want to find the condition of maximum power transfer on a purely resistive load R_L (Figure 6.28).

The power applied to the load can be expressed as

$$P_A = \frac{1}{2}Re\left[\bar{V}_{AB}\bar{I}^*\right] = \frac{1}{2}R_L Re\left[\bar{I}\bar{I}^*\right] = \frac{1}{2}R_L I^2$$

but

$$\bar{I} = \frac{\bar{V}_S}{(R_S + R_L) + jX_S} ; I = \frac{V_S}{\sqrt{(R_S + R_L)^2 + X_S^2}}$$

Furthermore, the expression of the power on a resistive load R_L is

$$P_A = \frac{1}{2}R_L I^2 = \frac{1}{2}\frac{R_L V_S^2}{(R_S + R_L)^2 + X_S^2}$$

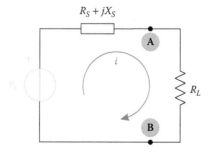

$$R_S + jX_S$$

FIGURE 6.28 Source with complex impedance and resistive load.

But imposing the condition of null derivative

$$\frac{dP_A}{dR_L} = \frac{1}{2}\frac{V_S^2\left(R_S^2 + X_S^2 - R_L^2\right)}{\left[\left(R_S + R_L\right)^2 + X_S^2\right]^2} \overset{!}{=} 0$$

which is satisfied if and only if

$$R_L = \sqrt{R_S^2 + X_S^2}$$

and the second derivative, being negative, confirms the maximum condition.

Furthermore, we have maximum power transfer when the load resistance R_L is equal to the module of internal source impedance $R_S + jX_S$ of the source v_S.

6.9.2.3 Complex Impedances for Both Source and Load

Let us consider the case in which the source impedance is Z_S and the load impedance Z_L. We get (Figure 6.29)

$$\bar{I} = \frac{\bar{V}_S}{\left(R_S + R_L\right) + j\left(X_S + X_L\right)}; \; I = \frac{V_S}{\sqrt{\left(R_S + R_L\right)^2 + \left(X_S + X_L\right)^2}}$$

$$P_A = \frac{1}{2}Re[Z_L]I^2 = \frac{1}{2}R_L I^2 = \frac{1}{2}\frac{R_L V_S^2}{\left(R_S + R_L\right)^2 + \left(X_S + X_L\right)^2}$$

The condition of maximum power transfer is obtained by maximizing P_A with respect to X_L and R_L. Observing the expression of P_A we easily deduce that the condition regarding X_L is given by

$$X_S = -X_L$$

In this case for the active power we have the expression

$$P_A\big|_{X_S=-X_L} = \frac{1}{2}\frac{R_L V_S^2}{\left(R_S + R_L\right)^2}$$

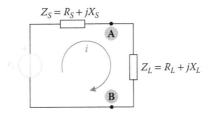

FIGURE 6.29 Source and load both with complete impedance.

The second maximum condition relevant to R_L is obtained considering

$$\frac{d}{dR_L}\left(\frac{V_S^2}{2}\frac{R_L}{(R_S + R_L)^2}\right) = \frac{V_S^2}{2}\frac{R_S - R_L}{(R_S + R_L)^3} \overset{!}{=} 0$$

from which we get

$$R_L = R_S$$

In conclusion, the two conditions of maximum transfer of active power impose

$$Z_L = Z_S^*$$

that is, a complex conjugate matching.

Finally, if we consider the case of load impedance with a constant reactive part, we have

$$P_A = \frac{1}{2}R_L I^2 = \frac{1}{2}\frac{R_L V_S^2}{(R_S + R_L)^2 + (X_S + X_L)^2}$$

$$\frac{dP_A}{dR_L} = \frac{1}{2}\frac{\left[R_L^2 - R_S^2 + (X_S + X_L)^2\right]}{\left[(R_S + R_L)^2 + (X_S + X_L)^2\right]^2} \overset{!}{=} 0$$

This equation is satisfied if and only if

$$R_L = \sqrt{R_S^2 + (X_S + X_L)^2}$$

This is why the maximum transfer of power is obtained when R_L is equal to the absolute value of all the internal impedance of the source (which given Thévenin's theorem can also be considered as an equivalent resistance of the whole circuit).

Curiosities

The violin bridge is necessary to transmit the vibration of the strings to the body of the instrument. Its form and material allow the matching of the acoustic impedance (related to the sound pressure generated by the vibration of molecules of air) of the strings, which has a low value, to the acoustic impedance of the top plate, which has a high value.

So the violin bridge must have with a low impedance near the anchor points of the strings, and a high impedance near the body of the instrument, with a smooth transition between these extreme values. That is the reason for its "strange" shape (Figure 6.30).

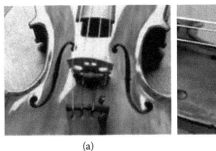

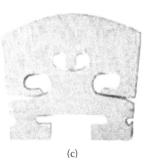

(a) (b) (c)

FIGURE 6.30   (a) Violin and (b) (c) particulars of its bridge.

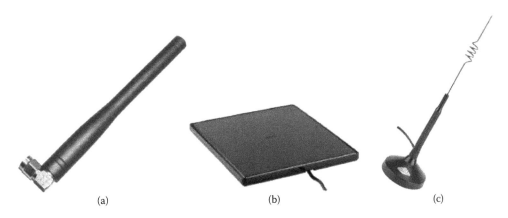

(a) (b) (c)

FIGURE 6.31   Antennas: (a) $900\,MHz$ SMA, (b) flat digital passive indoor, (c) *CB*.

The role of the antenna is to match circuit/air impedances. Without the antenna the signal cannot propagate out of the circuit and is reflected back, and the antenna impedance matches the transmitter output impedance (Figure 6.31).

6.9.3 Maximum Voltage or Current Transfer

Here we will consider the conditions we must satisfy to transfer the maximum voltage value from the source to the load, realizing what is referred to as *impedance bridging*, since it is a bridging connection if the Z_L device does not appreciably load the Z_S component, so that no power is transferred.

An example of impedance bridging comes from the audio amplifiers for which the driver amplifiers has a lower output impedance with respect to the load (the loudspeakers); they usually have 4, 8, or 16 ohm impedance values.

Referring to Figure 6.26, we can write output voltage (from a voltage divider) as

$$v_o = \frac{Z_L}{Z_S + Z_L} v_S = \frac{1}{1 + \dfrac{Z_S}{Z_L}} v_S$$

To maximize the module of voltage at the terminals of the load we need

$$\left|\frac{Z_S}{Z_L}\right| \ll 1 \Leftrightarrow |Z_L| \gg |Z_S|$$

and consequently $v_o \cong v_s$.

We have perfect impedance bridging, that is, a maximum output voltage with respect to input one, when one of the following conditions is true:

$$\begin{cases} Z_s = 0; & Z_L \neq 0 \\ Z_s \neq 0; & Z_L = \infty \end{cases}$$

An important technique in networks is to bridge low source impedance to high load impedance to transfer the most voltage possible between components.

Let's consider an unmatched example: if you connect an electric bass guitar or a synthesizer, devices with high impedance, to a microphone, a device usually with a low impedance, the result is that the low frequencies in the signal will roll off so that the bass will sound thin. An impedance matching network (a sort of "box" connected at the output of the guitar and at the input of the microphone) can solve the problem.

It is not a good idea to combine two or more sources into a single load. In fact, the low output impedance of one source will load down the output of the other source, and vice versa. This can cause level loss and distortion.

To prevent problems, it is convenient to insert a series resistor with each source before combining them, so as to prevent the loading down effect.

To find conditions of perfect current matching, that is, of maximum output current value with respect to that of input, it is convenient to refer to a Norton's equivalent model for the source (a current source with a parallel internal impedance z_S, Figure 6.18b), loaded by an impedance Z_L in parallel.

With similar consideration as the previous illustrations, it follows that we must satisfy the conditions

$$\begin{cases} Z_s \neq 0; & Z_L = 0 \\ Z_s = \infty; & Z_L \neq 0 \end{cases}$$

Note that these conditions are "opposite" to those for perfect voltage matching.

6.9.4 Efficiency

To take into account the net amount of electric power delivered to load, the parameter called *electric power efficiency* or, simply, *efficiency* is commonly adopted.

Efficiency is defined as the ratio between the useful output power P_L delivered to the load, and the total power furnished by the electric source(s) P_S:

$$\frac{P_L}{P_S}$$

Therefore, in the case of maximum power transferred to the load, we obtain 0.5, which is the maximum obtainable.

6.9.5 Matching Networks

When the matching conditions (impedance matching or bridging) are not satisfied, we can utilize the so-called *matching network* (or *impedance-matching adapter*), usually a 2-port inserted between the source and the load, as in Figure 6.32.

In such a manner, the source "sees" the input port of the network and not the load directly, and the load "sees" the exit port of the network and not the source directly.

The network interposed is chosen with appropriate values for its own input and output impedances with the aim of achieving the maximum transfer condition desired (power, voltage, or current).

6.10 WHO DECIDES WHAT

A crucial aspect of electronics never underlined enough is *who decides what*, and *resistance* is one of the key aspects of this. The resistance can be the property associated to a real resistor or the inner one, even a parasite, of any other component.

When we have a voltage divider supplied by a voltage source, as represented in Figure 6.33, the value of the current i flowing through it is mostly "decided" by the resistance of higher value between R_S and R_L, since $i = v_S/(R_S + R_L)$. So, when $R_L \gg R_S$ so that R_S is negligible, $i \cong v_S/R_L$.

Now, let's think about the voltage source v_S and the resistor R_S representing a Thévenin equivalent of a network, and the resistor R_L representing the load. Now

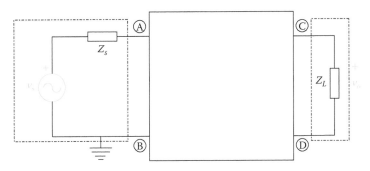

FIGURE 6.32 Matching two-port network.

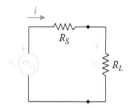

FIGURE 6.33 Voltage divider supplied by a voltage source.

the voltage across R_L is the one "decided" by the voltage source only if $R_L \gg R_S$, since

$$v_L = \frac{R_L}{R_L + R_S} v_s$$

So when R_S is negligible, $v_L \cong v_s$. But, when the contrary occurs, that is, $R_L \ll R_S$, the load "experiences" a lower voltage than the one of the source, "reducing" it, until the extreme that a load of no resistance, $R_L = 0$, "experiences" no voltage at all from the source.

6.11 KEY POINTS, JARGON, AND TERMS

→ An *electric circuit* is a system of conductors and devices interconnected to each other in a way to allow the flow of electrical current, to elaborate and/or transfer electric signals or electric power from one point to another.

→ We consider the conductors that connect the elements as ideal, and therefore equipotential.

→ We consider all components as *lumped elements*, that is, discrete objects that can exchange energy with other objects, with internal physics described by terminal relations, and size smaller than wavelength of the signal.

→ The way the circuit devices are connected (series, parallel, cascade, triangle, star, bridge) is called circuit *topology*.

→ A *closed circuit* refers to an electric circuit without interruptions; otherwise it is an *open circuit*.

→ A *short circuit* occurs when there is a very low-resistance connection, usually accidental, between two points of the circuit at different voltages.

→ If a circuit containing a voltage source is broken, the full voltage of that source will appear across the points of the break.

→ A linear circuit is made only of linear components.

→ A *node* is a point at which two or more components are connected together with shorts; a *branch* is a part of circuit consisting of one or more components connected in series, between two nodes; a *mesh* is any closed circuit.

→ *KVL*: "Sum of voltages around a closed loop is zero" or equally: "Sum of voltages from point A to point B is the same regardless of the path taken."

→ *KCL*: "Sum of currents into a node (or area) is zero."

→ An *equivalent circuit* or *component* is made by a theoretical electric network with exactly the same electrical characteristics of the circuit or component it is intended to replace.

→ *Thévenin's theorem*: "Any linear network, no matter how complex, made of a combination of independent (current and/or voltage) sources and resistors, from the viewpoint of any pair of terminals, is equivalent to a single voltage source and a single series resistor."

→ *Norton's theorem*: "Any linear network, no matter how complex, made of a combination of independent (current and/or voltage) sources and resistors, from the viewpoint of any pair of terminals, is equivalent to a single current source with a single parallel resistor."

→ *Superposition theorem*: "If in a linear network more than one voltage and/or current source operates at the same time, the voltage between any two nodes in the network (or the current in any branch) is the algebraic sum of voltages (or currents) due to each of the independent sources acting one at a time."

→ *Miller's theorem*: "In any given linear two-port network having the output signal proportional to the input one, a shunt impedance directly connected between input and output ports can be replaced by two equivalent impedances placed in the only input and output ports of the network respectively."

→ The *maximum active power transferred* from the source to the load is when the load impedance is equal to the complex conjugate of the source impedance. The process used to make the impedances equal is called *impedance matching*.

→ When the load impedance is much greater than the source impedance, it is called a *bridging impedance*, which results in *maximum voltage transfer* from the source to the load.

→ The matching conditions can be satisfied by means of a *matching network*.

6.12 EXERCISES

EXERCISE 1

A car battery is no longer working since it provides only $V_{B1} = 8V$ across its terminals. To charge it, we connect in parallel a spare battery that provides the necessary $V_{B2} = 12V$. Considering that each battery has an internal resistance of $R_1 = R_2 = 10$, what is the current flowing in the circuit? (Figure 6.34)

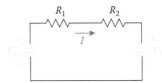

FIGURE 6.34 Two non-ideal batteries connected in parallel.

ANSWER

$$V_{B2} = R_1 I + R_2 I + V_{B1}$$

hence $I = 0.2 [A]$.

Please notice that as the voltage of the car battery increases during the charging process, the current delivered to it decreases.

EXERCISE 2

For the bridge configuration circuit shown in Figure 6.35, use the open/short-circuit approach to derive the Thévenin equivalent at the port "1-2."

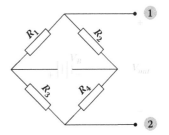

FIGURE 6.35 Bridge topology with one source battery and four resistances.

ANSWER

According to Thévenin's theorem, the previous bridge topology is equivalent to a voltage source V_{TH} in series with a resistor R_{TH}. We have to calculate them (Figure 6.36).

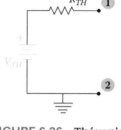

FIGURE 6.36 Thévenin equivalent of the bridge topology.

V_{TH} is the voltage "seen" at the output "1-2" port, so we can determine it as the difference between the voltage at the node "1" V_1, and the one at the node "2" V_2.

Considering the node connecting V_B, R_2, and R_4 as a reference, we have

$$V_1 = \frac{R_2}{R_1 + R_2} V_B$$

and

$$V_2 = \frac{R_4}{R_3 + R_4} V_B$$

therefore

$$V_{TH} = \left(\frac{R_2}{R_1 + R_{42}} - \frac{R_4}{R_3 + R_4} \right) V_B$$

To determine the value of the resistor R_{TH}, we need to nullify the effect of the voltage source (a short-circuit across it) (Figure 6.37). So

$$R_{TH} = R_1 \,||\, R_2 + R_3 \,||\, R_4$$

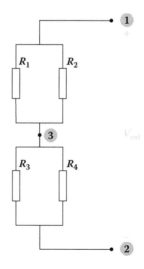

FIGURE 6.37 The bridge topology with the battery being short-circuited.

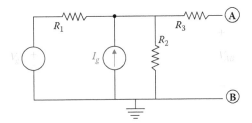

FIGURE 6.38   Network to be solved with the Thévenin and the Norton equivalents.

EXERCISE 3

For the network in Figure 6.38, derive the Thévenin and the Norton equivalents.

   The component values are $V_g = 15V$; $I_g = 0.2\,A$; $R_1 = 200$; $R_2 = 200$; and $R_3 = 500$.

ANSWER

For the Thévenin equivalent circuit, we have to determine the voltage source V_{TH} in series with the resistor R_{TH}.

$$V_{TH}$$

   The Thévenin equivalent voltage is the open-circuit voltage at the "A-B" port.

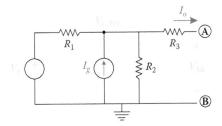

Therefore the current I_o is zero

$$\frac{V_{x,TH} - V_g}{R_1} + \frac{V_{x,TH}}{R_2} = i_g \Rightarrow V_{x,TH} = V_{AB} = 27,5\,V$$

$$R_{TH}$$

The Thévenin equivalent resistance is the one "seen" at the "A-B" port, after that we open-circuited the current source (Figure 6.39).
   Then

$$R_{TH} = R_3 + R_1 \,||\, R_2 = 600\,\Omega$$

FIGURE 6.39   The current source has been open-circuited.

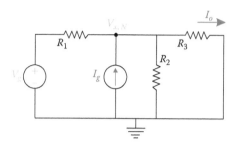

FIGURE 6.40   "A-B" port is short-circuited.

Likewise, for the Norton equivalent circuit, the Norton equivalent current is the short-circuit current at the "A-B" port (Figure 6.40).

$$\frac{V_{x,N} - V_g}{R_1} + \frac{V_{x,N}}{R_2} + \frac{V_{x,N}}{R_3} = i_g \Rightarrow V_{x,N} = \frac{R_2 R_3 V_g + R_1 R_2 R_3 I_g}{R_1 R_2 + R_2 R_3 + R_1 R_3} \Rightarrow I_N = \frac{V_{x,N}}{R_3} \cong 0,046A$$

The same result would be obtained with the equivalent relationship

$$I_N = \frac{V_{TH}}{R_{TH}}$$

$$R_N = R_{TH}$$

EXERCISE 4

For the network in Figure 6.41, determine the load resistance R_L that allows the maximum power transfer from the battery V_B.

Data: $V_B = 20V$, $R_1 = 1k\Omega$, $R_2 = 2.7k\Omega$.

ANSWER

It is useful to replace the given network with its Thévenin equivalent at the **A** and **B** nodes. To this aim we have to remove the load R_L and evaluate the voltage V_{TH}:

$$V_{TH} = \frac{R_2}{R_1 + R_2} V_B$$

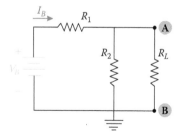

FIGURE 6.41   One battery, three resistors, one is the load.

Again, according to Thévenin's theorem, the equivalent resistance R_{TH} must be evaluated nullifying the independent, so we have to replace the battery V_B with a short-circuit, obtaining

$$R_{TH} = R_1 \mid\mid R_2 = \frac{R_1 R_2}{R_1 + R_2}$$

The result is the scheme seen in Figure 6.42.
Numerically: $V_{TH} \cong 14.6V$; $R_{TH} \cong 730\Omega$.
To obtain the maximum transfer of active power

$$R_L \overset{!}{=} R_{TH} = \frac{R_1 R_2}{R_1 + R_2} \cong 730\Omega$$

and the maximum transferred power to the load P_L is

$$P_L = V_L I_L$$

with

$$V_L = \frac{V_{TH}}{2}$$

$$I_L = \frac{V_{TH}}{2R_{TH}}$$

Therefore

$$P_L = \frac{V_{TH}}{2} \frac{V_{TH}}{2R_{TH}} = \frac{V_{TH}^2}{4R_{TH}}$$

Numerically, $P_L \cong 73mW$.

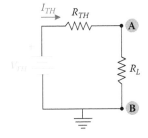

FIGURE 6.42 The Thévenin equivalent at the "A-B" port.

Even with the imposed conditions dictated by the maximum power transfer theorem, the power delivered to the load is not 50% of the source power P_s available from the battery.
In fact, let's calculate P_s:

$$P_s = V_B I_B$$

V_B is already known, while for I_B we can use the equivalent scheme (Figure 6.43) where

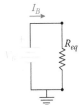

FIGURE 6.43 Equivalent circuit of the given network.

$$R_{eq} = R_1 + R_2 \mid\mid R_L = R_1 + \frac{R_2 R_L}{R_2 + R_L}$$

Therefore, the current flowing from the battery is

$$I_B = \frac{V_B}{R_{eq}} = \frac{V_B}{R_1 + \dfrac{R_2 R_L}{R_2 + R_L}} \cong 13\,mA$$

and the power is

$$P_L = V_B I_B \cong 254\,mW$$

This value is more than double the $P_L \cong 73\,mW$ available on the load resistor, so the maximum of the optimal conditions of transferred power is not reached. That maximum is obtained only in special cases.

Here

$$\frac{P_L}{P_S} \cong 0.29$$

so only 29% of the power from the battery is furnished to the load, and this is for optimal conditions!

EXERCISE 5
Suppose we have a network with voltage and current sources as represented in Figure 6.44.

Determine the current I_{R2} *flowing into* R_2*, adopting the superposition theorem.*

ANSWER.
According to the superposition theorem, we start by turning off the current source I_g, and calculate the current $I_{R2}^{(V_s)}$ flowing into R_2, but due to V_s. The second step is to turn off the voltage source V_s and to calculate the current $I_{R2}^{(I_s)}$

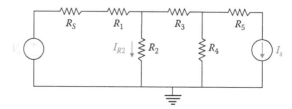

FIGURE 6.44 One voltage and one current source supply a resistive network.

flowing into R_2, but due to I_s. Finally we sum the results.

Figure 6.45 is the first equivalent network with $I_s = 0$ and the current from the voltage source is

$$I_{V_S} = \frac{V_S}{R_g + R_1 + R_2 \,||\, (R_3 + R_4)}$$

Therefore

$$I_{R2}^{(V_S)} = \frac{R_3 + R_4}{R_2 + (R_3 + R_4)} I_{V_S}$$

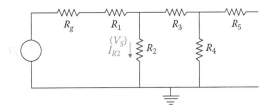

FIGURE 6.45   First equivalent network with $I_s = 0$.

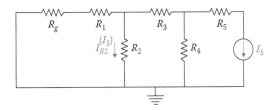

FIGURE 6.46   Second equivalent network with $V_s = 0$.

Figure 6.46 illustrates the second equivalent network with $V_s = 0$ and now the current I_{R3} flowing into the resistor R_3 is

$$I_{R3} = -\frac{R_4}{R_4 + R_3 + R_2 \,||\, (R_g + R_1)} I_s$$

As a consequence

$$I_{R2}^{(I_S)} = \frac{R_g + R_1}{R_2 + (R_g + R_1)} I_{R3}$$

The requested current is then

$$I_{R2} = I_{R2}^{(V_S)} + I_{R2}^{(I_S)} = \frac{R_3 + R_4}{R_2 + (R_3 + R_4)} I_{V_S} + \frac{R_g + R_1}{R_2 + (R_g + R_1)} I_{R3}$$

$$= \frac{R_3 + R_4}{R_2 + (R_3 + R_4)} \frac{V_S}{R_g + R_1 + R_2 \,||\, (R_3 + R_4)}$$

$$- \frac{R_g + R_1}{R_2 + (R_g + R_1)} \frac{R_4}{R_4 + R_3 + R_2 \,||\, (R_g + R_1)} I_s$$

EXERCISE 6

The network in Figure 6.47 is made with a real current dependent source (an ideal dependent one hv_i, depending on the input voltage v_i, with its resistance R_{cs} in parallel), a series input resistor R, and a parallel output resistor R_L. It is known that input and output voltages are directly proportional as $v_2/v_1 = K$. According to Miller's theorem, replace the resistor R between the nodes "A" and "B" with the equivalent resistors R_A and R_B at the input and output port, respectively.

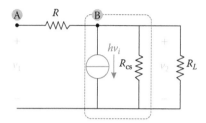

FIGURE 6.47 A real current source with input and output resistors.

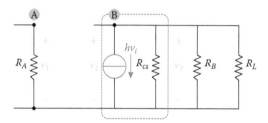

FIGURE 6.48 Application of Miller's theorem on the circuit of Figure 6.47.

ANSWER.

The network can be equivalently redrawn as schematized in Figure 6.48, with

$$R_A = R \frac{1}{1-K}$$

$$R_B = R \frac{K}{K-1} = R \frac{1}{1 - \frac{1}{K}}$$

Observations

It is easy to verify that V_2 depends on R_B. But R_B depends on K which depends on V_2. It seems to be a *cul-de-sac*. The result is that it makes sense to adopt Miller's theorem if the K value, that is, the ratio v_2/v_1, is known in advance.

The resistance values for resistors R_A and R_B will be positive only if $K < 0$. If $K > 0$ one or both resistances will be negative and it would be represented not just by a resistor but by a source to determine the correct values for current and voltage.

7

Frequency Domain

7.1 INTRODUCTION

We saw in Section 4.6 of Chapter 4 that the equations for voltage and current for capacitors are

$$i(t) = C\frac{dv(t)}{dt}; \quad v(t) = \frac{1}{C}\int i(t)\,dt + v_0$$

and in Section 4.6 for inductors

$$v(t) = L\frac{di(t)}{dt}; \quad i(t) = \frac{1}{L}\int v(t)\,dt + i_0$$

Such equations depend on time, which is the *domain* of our functions.

However, thanks to a mathematical tool, an integral transform, it is possible to change the domain of functions and operate not in *time* but in *frequency* or, more generically, in the domain of the complex argument s by the Laplace transform, denoted $L\{f(t)\}$. In such a way, the equations of capacitors and inductors are no longer given by derivatives and integrals, which can be difficult to treat mathematically, but become easy expressions of the type $v = Xi$, in which X is named *reactance*.

In particular, it is called *capacitive reactance* X_C, if it concerns the capacitor, and *inductive reactance* X_L, if it concerns the inductor.

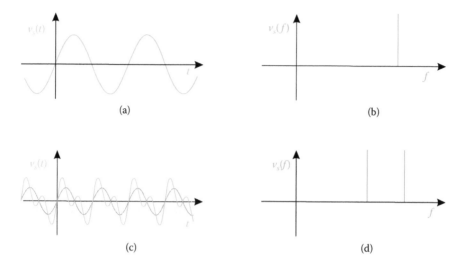

(a)

(b)

(c)

(d)

FIGURE 7.1 (a) A pure sinusoidal wave in the time domain becomes a (b) Dirac delta function in the frequency domain. (c) A periodic function (blue) in the time domain made from two basic sine functions (green and gray), becomes two (d) Dirac delta functions in the frequency domain.

Observation

The behavior of resistances in a circuit is described by Ohm's law. Given that this law does not depend on the time, the results are not varied by the integral transform.

The pure sinusoidal function we are dealing with in the time domain now becomes a Dirac delta function in the frequency domain, as represented in Figure 7.1.

7.2 RESISTANCE, REACTANCE, IMPEDANCE

It is useful to refer to the Laplace transform expressed in the domain of the complex argument $s = \sigma + j\omega$.

In the new domain s, the voltage to current ratio might not be a simple real number, and it is not defined as *resistance* as done previously but, more generically, *impedance*.

For reasons of dimensional coherence, impedance is measured in $ohm\,[\Omega]$ and it is generically a complex number. However it can be reduced to the real number already known as *resistance* (if it lacks the imaginary part), or an imaginary number (if it lacks the real part) called *reactance*.

$$Impedence = resistance + j * reactance$$

In symbols:

$$Z = R + jX$$

The inverse of the impedance is named *admittance* $Y\ [S\ or\ mho\ or\ \Omega^{-1}]$.

In analyzing a circuit, it is often preferred to work in the frequency domain respect to the time one, and we will see that the frequency f is highly correlated to the variable s. This is why we need to have equations with the circuit elements directly in the domain of s.

7.2.1 Capacitive Reactance

In the case of a capacitor the relations expressed in s assume the following forms (the demonstrations are beyond the scope of our discussion):

$$i(t) = C\frac{dv(t)}{dt} \quad \overset{transform}{\Rightarrow} \quad I(s) = sCV(s) - Cv(0)$$

$$v(t) = v(0) + \frac{1}{C}\int_0^t i(t)\,dt \quad \overset{transform}{\Rightarrow} \quad V(s) = \frac{v(0)}{s} + \frac{1}{sC}I(s)$$

or, if the initial conditions can be considered null, more simply they are

$$i(t) = C\frac{dv(t)}{dt} \quad \overset{transform}{\Rightarrow} \quad I(s) = sCV(s)$$

$$v(t) = \frac{1}{C}\int_0^t i(t)\,dt \quad \overset{transform}{\Rightarrow} \quad V(s) = \frac{1}{sC}I(s)$$

The impedance of the capacitor is therefore expressed as

$$Z_C = \frac{1}{sC}$$

In the case of a *steady state* (i.e., when the *transient* described in Section 4.6.4 in Chapter 4 has run out) the real part of the variable s is null ($\sigma = 0$), so $s = j\omega$ and the term $X_C = 1/\omega C$ is called *capacitive reactance*, so we can write

$$V = \frac{1}{j}X_C I = -jX_C I = -j\frac{1}{\omega C}I$$

(with $1/j = -j$) This equation is reminiscent of Ohm's law, whose characteristics it must maintain: dimensionally a voltage $[V]$ is equal to a current $[A]$ multiplied by a resistance $[\Omega]$, consequently X_C has to be measured in Ohm too. But Ohm is tied to the concept of electrical *resistance* and the capacitor, differing from the resistor, does not present power consumption, so the term X_C is called *reactance* (i.e., "*reactive resistance*").

If the voltage applied at the terminals of the capacitor is sinusoidal, $v(t) = V_M\sin(\omega t)$, the current becomes 90° phase shifted,

$$i(t) = C\frac{dv(t)}{dt} = C\omega V_M \cos(\omega t)$$

This means that the voltage/current ratio is between two values that are 90° phase shifted. This ratio is still measured in Ohms, but to keep in mind the fact that the values have a difference in time (because of phase shifting), this ratio is called *reactance* (and not *resistance*). The reactance we talk about concerns a capacitor, hence the name *capacitive reactance*.

7.2.2 Inductive Reactance

Similarly to what happens with a capacitor, for an inductor we have

$$v(t) = L\frac{di(t)}{dt} \overset{\textit{transform}}{\Rightarrow} V(s) = sLI(s) - Li(0)$$

$$i(t) = i(0) + \frac{1}{L}\int_0^t v(t)\,dt \overset{\textit{transform}}{\Rightarrow} V(s) = \frac{i(0)}{s} + \frac{1}{sL}V(s)$$

or, if the initial conditions can be considered null, more simply we have

$$v(t) = L\frac{di(t)}{dt} \overset{\textit{transform}}{\Rightarrow} V(s) = sLI(s)$$

$$i(t) = \frac{1}{L}\int_0^t v(t)\,dt \overset{\textit{transform}}{\Rightarrow} I(s) = \frac{1}{sL}V(s)$$

The impedance of the inductor is then expressed as

$$Z_L = sL$$

In the case of a *steady state* the real part of the variable s is null ($\sigma = 0$), so $s = j\omega$ and the term $X_L = \omega L$ is called *inductive reactance*, so we can write

$$V = jX_L I = j\omega L I$$

7.3 ELECTRONIC FILTERS

Electronic filters (or *frequency-selective networks*, or simply *filters*) are essential circuits to the operation of most electronic networks. They process any given input signal, and can (totally or partially) remove, reduce, or phase shift some (or all) of its frequency components. In such a manner, given an input signal with multiple frequency content, some of these frequencies pass to the output unchanged in amplitude and phase, while others undergo more or less amplitude reductions and phase shifting.

Please be aware that a filter will not add new frequencies, nor will it change the component frequencies of the input signal, but it will change the relative amplitudes and/or phase of the various frequency components.

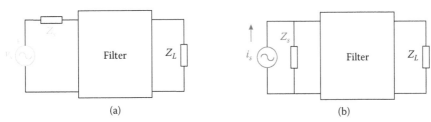

(a)   (b)

FIGURE 7.2   Filter network with (a) voltage source and (b) current source at its input port.

This is the circuital way we have to "select" some frequencies of any arbitrary electric signal.

The filter can be generally described by a two-port network (Figure 7.2), with the real voltage or real current signal source, v_s or i_s, at its input port and at its output port an arbitrary load impedance Z_L.

Here we will deal with basic analog, passive, linear filters, made of linear components, combinations of resistors, capacitors, and inductors.

Curiosity

Filtering is applied not just in electronics, but practically in every technological field.

A water filter removes particulate matter by means of fine physical barriers. An optical filter (Figure 7.3b) selectively blocks or absorbs or transmits one or more colors (depending on their wavelenghts). The filtration process in radiology corresponds to attenuating a beam of x or γ rays. An IVC (Inferior Vena Cava) filter (Figure 7.3a) is an anchor device implanted by vascular surgeons to selectively stop pieces of clots, but not blood of course, to prevent them from

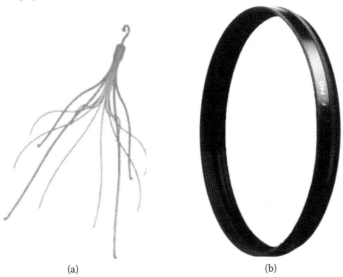

(a)   (b)

FIGURE 7.3   (a) IVC filter implanted in a vein selectively stops pieces of clots; (b) yellow filter that optically reflects the yellow light and lets the other colors pass unchanged.

getting to the heart or lung. In computer programming, a filter can be used to remove or insert headers or control characters in data. In graphic applications, a filter can be applied to a picture to create various effects.

7.3.1 Low-Pass RC Filter

The circuit described here is an example of a *low-pass filter*, so called because it discriminates in frequency in such a way that low-frequency sinusoidal signals pass unchanged, while high-frequency signals undergo greater and greater reductions in amplitude as the frequency value increases.

The simplest realization of such a filter is the RC series circuit, so called because it is composed of a resistor R and a capacitor C, the resistor in series with a load, and the capacitor parallel with the load. The input signal v_s is applied across the series of the two components, while the output signal v_o is measured at the terminals of the capacitor. Given the circuit of Figure 7.4, let us suppose a voltage source v_s in the input having the ideal ability to generate sinusoidal functions of all frequencies, from zero frequency, which corresponds to DC, to an infinite frequency. We will see how this circuit discriminates in frequency.

We have to start by determining the output voltage function v_o. In fact, since filters are defined by their frequency-domain effects on signals, it makes sense to develop the analytical and graphical descriptions of the filter's output function v_o versus amplitude and phase at each working frequency.

Applying the voltage divider formula we get

$$v_o = \frac{-jX_C}{-jX_C + R} v_s$$

$$v_o = \frac{\dfrac{1}{j\omega C}}{\dfrac{1}{j\omega C} + R} v_s = \frac{\dfrac{1}{j\omega C}}{\dfrac{1 + j\omega RC}{j\omega C}} v_s = \frac{1}{1 + j\omega RC} v_s$$

Hence the ratio v_o / v_s, which represents the *transfer function* (also known as the *system function* or *network function*) of the circuit is expressed by the complex value $1/1 + j\omega RC$.

We need to study the transfer function because it defines the filter's response to any input signal.

Its magnitude, called the *amplitude response*, and its phase, called the *phase response* are fundamental.

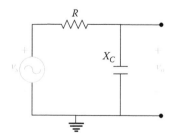

FIGURE 7.4 RC series circuit.

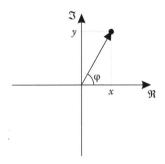

FIGURE 7.5 Representation of the complex value in the complex plane.

The amplitude response corresponds to the effect of the filter on the amplitudes of sinusoidal input signals at various frequencies. The phase response indicates the ammount of phase shift the filter introduces on input signals as a function of frequency.

The amplitude means the absolute value or modulus, i.e., the distance of the complex number of the transfer function from the origin of the axes varying ω, while the phase or argument means the angle the function forms with the x-axis, as depicted in Figure 7.5.

7.3.1.1 Amplitude Response

The absolute value of a complex number expressed in the form $z = x + jy$ is easily determined given that, as is shown in Figure 7.5, it is derived by the Pythagorean theorem applied to the right triangle, which has as its legs (catheti) the real value x and the imaginary value y, and as hypotenuse the very absolute value we are looking for.

In order to derive the absolute value we rationalize the transfer function, multiplying numerator and denominator by the complex conjugate of the denominator (i.e., the same denominator with its imaginary part opposite in sign):

$$\frac{1}{1+j\omega RC} = \frac{1}{1+j\omega RC}\frac{1-j\omega RC}{1-j\omega RC} = \frac{1-j\omega RC}{1+\omega^2 R^2 C^2} = \frac{1}{1+\omega^2 R^2 C^2} - j\frac{\omega RC}{1+\omega^2 R^2 C^2}$$

Therefore, with reference to Figure 7.6, the module is

$$\left|\frac{v_o}{v_s}\right| = \sqrt{\left(\frac{1}{1+\omega^2 R^2 C^2}\right)^2 + \left(\frac{\omega RC}{1+\omega^2 R^2 C^2}\right)^2} \Rightarrow \left|\frac{v_o}{v_s}\right| = \sqrt{\frac{1}{1+\omega^2 R^2 C^2}}$$

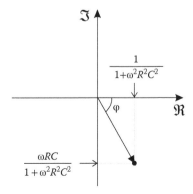

FIGURE 7.6 Representation of the transfer function in the complex plane.

It is useful to evaluate the expression in the following special cases:

$$\omega^2 R^2 C^2 \ll 1 \Leftrightarrow \omega \ll \frac{1}{RC} \Rightarrow \left|\frac{v_o}{v_s}\right| \to 1$$

$$\omega^2 R^2 C^2 = 1 \Leftrightarrow R = \frac{1}{\omega C} = X_C \Leftrightarrow \omega = \frac{1}{RC} \Leftrightarrow \left|\frac{v_o}{v_s}\right| = \sqrt{\frac{1}{2}}$$

$$\omega^2 R^2 C^2 \gg 1 \Leftrightarrow \omega \gg \frac{1}{RC} \Rightarrow \left|\frac{v_o}{v_s}\right| \to 0$$

The particular frequency equal to the value $1/2\pi RC$ ($\Leftrightarrow \omega = 1/RC$), when the output signal is reduced by $\sqrt{1/2} \cong 0.707$ with respect to the input signal, is called *the cutoff frequency* of the low-pass filter, f_{c-lp} (or *corner frequency, breakpoint frequency, –3 dB frequency*, or *critical frequency*). At this special frequency f_{c-lp} the capacitive reactance $X_C = 1/\omega C$ equals the resistance R.

Finally, it is demonstrable that around f_{c-lp} the transfer function has a slope of $-20\,dB/Decade$ or, equivalently, $-6\,dB/Octave$.

Example

If, for the circuit in Figure 7.4 we choose as component values $R = 1\,k\Omega$ and $C = 1\,nF$, with voltage input source of peak $V_M = 1V$, the graph of the absolute value of the transfer function $(= |v_o/v_s|)$ with respect to frequency results:
With the value of the cut-off frequency equal to

$$f_c = \frac{1}{2\pi RC} \cong \frac{1}{6.28 * 10^{-3} * 10^{-9}} \cong 160\,kHz$$

So, for a sinusoidal input function of $V_M = 1V$ peak amplitude and $f = 100\,Hz$ frequency, we would be in the condition of $\omega \ll 1/RC \Rightarrow |v_o/v_s| \cong 1$, therefore the output would be a sinusoidal function of the same input amplitude.

For a sinusoidal input function of $V_M = 1V$ peak amplitude and $f = 150\,MHz$ frequency, we would be in the condition of $\omega \gg 1 \Rightarrow |v_o/v_s| \cong 0$, therefore the output would be practically of null amplitude.

For a sinusoidal input function of $V_M = 1V$ peak amplitude and $f = 160\,kHz$ frequency (i.e., the cutoff frequency in our example), the output would be yet another sinusoid but with a reduced amplitude of $1/\sqrt{2}\,V_M \cong 0.707V$.

All these behaviors are represented in Figure 7.7.

Keeping in mind the definition of *decibel* we can see that just at the cutoff frequency the output is reduced by $3\,dB$ with respect to input. In fact, naming

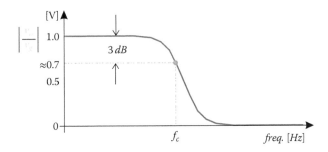

FIGURE 7.7 Absolute value of transfer function vs. frequency for a low-pass RC filter with R and C of given values.

G_{dB} the gain of the network in dB (it is a negative gain, that is, a loss), P_i the input power and P_o the output power, we can write

$$G_{dB} = 10\log\left(\frac{P_o}{P_i}\right) = 10\log\left(\frac{V_{RMS-o}^2 / R_o}{V_{RMS-i}^2 / R_i}\right)$$

and, as for the circuit in question $R_i = R_o = R$:

$$G_{dB} = 10\log\left(\frac{V_o^2}{V_i^2}\right) = 20\log\left(\frac{V_o}{V_i}\right) = 20\log\left(\frac{V_i / \sqrt{2}}{V_i}\right) = -3_{dB}$$

7.3.1.2 Phase Response

The phase φ of the complex number can be derived if we keep in mind that for a right triangle the following relation is true: $y = x\tan(\varphi)$, so $\varphi = arctan(y/x)$. It is useful then to express the relation $1/(1 + j\omega RC)$ in the form $x + jy$.

From Figure 7.6 we deduce that the angle of phase shifting is equal to

$$\varphi = \tan^{-1}\left(\frac{-\dfrac{\omega RC}{1 + \omega^2 R^2 C^2}}{\dfrac{1}{1 + \omega^2 R^2 C^2}}\right) = \tan^{-1}(-\omega RC)$$

Also in this case it is useful to study the function for certain special cases:

$$\omega RC \ll 1 \Rightarrow \varphi \to 0°$$
$$\omega RC = 1 \Rightarrow \varphi = -45°$$
$$\omega RC \gg 1 \Rightarrow \varphi \to -90°$$

which are useful to draw the complete phase response reported in Figure 7.8.

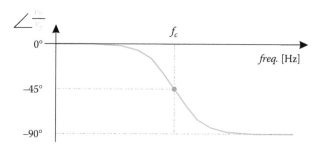

FIGURE 7.8 The phase response curve of the low-pass RC filter.

Example

If, for the circuit in Figure 7.4, which has as its input a voltage signal of $V_M = 1V$ peak amplitude, we choose once more as component values $R = 1\,k\Omega$ and $C = 1\,nF$, the circuit phase response with respect to the frequency is as seen in Figure 7.9

So, for a sinusoidal input function of $V_M = 1V$ peak amplitude and $f = 100\,Hz$ frequency, we would have $\omega RC \ll 1 \Rightarrow \varphi \cong 0$; therefore the output would be a sinusoidal function without outphasing with respect to input.

For a sinusoidal input function of $V_M = 1V$ peak amplitude and $f = 150\,MHz$ frequency, we would have $\omega RC \ll 1 \Rightarrow \varphi \cong -90°$, so the output voltage lags $90°$ the input in phase.

For a sinusoidal input function of $V_M = 1V$ peak amplitude and $f = 160\,kHz$ frequency (i.e., the cutoff frequency in our example), the output signal voltage would be yet another sinusoid that lags the input voltage by a phase angle of $45°$.

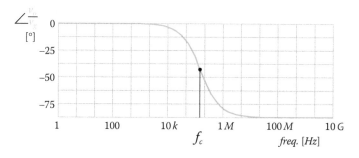

FIGURE 7.9 Phase response curve for example RC filter.

7.3.1.3 Group Delay Response

Another important aspect of electronic filters to deal with is the so called *group delay response* (or simply *group delay*) τ, which can be critical especially in audio and communication systems. It is mathematically defined as

$$\tau(\omega) = -\frac{\delta\varphi(\omega)}{\delta\omega}$$

that is, it is the negative of the rate of change of phase with angular frequency, which has units of time. Practically τ furnishes the way to determine the delay of signals through the filter, being the amount of time a signal takes to pass through the filter. So given the transfer function previously determined

$$\frac{v_o}{v_s} = \frac{1}{1 + j\omega RC}$$

we have

$$\tau(\omega) = \frac{RC}{\omega^2 R^2 C^2 + 1}$$

therefore the group delay varies with the frequency. For the special case of $\omega = 0$ we have $\tau = RC$, while τ reduces with frequency. Ideally, we would have that all frequencies of the input signal should have the same time delay, so to avoid distortions but, as we demonstrated, in simple analog filters this is not the case.

7.3.1.4 Considerations

The important considerations that arise from the analysis of the aforementioned transfer function are

→ For "low" frequencies (in particular for zero frequency, i.e., DC), output and input coincide in absolute value ($|v_o/v_s| = 1 \Leftrightarrow v_o = v_s$) and in phase (no outphasing), so the circuit *does not* influence the signal, which comes out unchanged.

→ For "high" frequencies the output signal is more and more reduced in absolute value with frequency, until it is completely nullified ($v_o|_{f=\infty} = 0$, so the circuit *does not* let input signals pass at all) and outphasing tends to −90°.

→ At the frequency $f_c = 1/2\pi RC$ the output signal is about 70% in absolute value with respect to the input signal and outphasing is −45°.

When we talk about "high" and "low" frequencies we refer to those sufficiently far "upwards" or "downwards" from the frequency considered "central," $f_c = 1/2\pi RC$.

Figure 7.10 shows the interesting effect of "cleaning up" that the application of a low-pass filter can have on two "dirty" signals, i.e., signals containing unwanted components of "high" frequency.

7.3.2 High-Pass RC Filter

The *high-pass filter* discriminates in frequency in such a way that high-frequency sinusoidal signals pass unchanged, while low-frequency signals undergo greater and greater reductions in amplitude as the frequency decreases.

The simplest realization of such a filter is the *CR series* circuit, i.e., composed of the series of a capacitor C and a resistor R. The source signal v_s is applied across the two components, while the output signal v_o is collected at the terminals of the resistor, as in Figure 7.11.

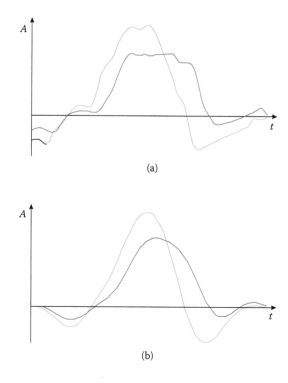

(a)

(b)

FIGURE 7.10 Two signals (a) before and (b) after the application of a low-pass filter.

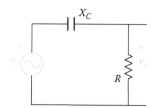

FIGURE 7.11 High-pass RC circuit.

Applying the voltage divider formula we get

$$\frac{v_o}{v_s} = \frac{R}{R + \dfrac{1}{j\omega C}} = \frac{1}{1 + \dfrac{1}{j\omega RC}} = \frac{j\omega RC}{j\omega RC + 1}$$

Rationalizing, we obtain the transfer funtion of the form $z = a + jb$:

$$\frac{v_o}{v_s} = \frac{\omega^2 R^2 C^2}{1 + \omega^2 R^2 C^2} + j\frac{\omega RC}{1 + \omega^2 R^2 C^2}$$

with the real part

$$a = \frac{\omega^2 R^2 C^2}{1 + \omega^2 R^2 C^2}$$

and the coefficient of the imaginary part

$$b = \frac{\omega RC}{1 + \omega^2 R^2 C^2}$$

7.3.2.1 Amplitude Response

The absolute value of the transfer function is

$$\left| \frac{v_o}{v_s} \right| = \omega RC \sqrt{\frac{1}{1 + \omega^2 R^2 C^2}}$$

The cutoff frequency f_{c-hp} of the high-pass filter is the frequency for which the amplitude is reduced by $1/\sqrt{2}$ so, imposing it, we obtain

$$\omega_{c-hp} = \frac{1}{RC} \Rightarrow f_{c-hp} = \frac{1}{2\pi RC}$$

It is useful to evaluate the absolute value also in certain special cases:

$$\omega^2 R^2 C^2 \ll 1 \Leftrightarrow \omega \ll \frac{1}{RC} \Rightarrow \left| \frac{v_o}{v_s} \right| \to 0$$

$$\omega^2 R^2 C^2 \gg 1 \Leftrightarrow \omega \gg \frac{1}{RC} \Rightarrow \left| \frac{v_o}{v_s} \right| \to 1$$

The complete graph of the absolute value of the transfer function is illustrated in Figure 7.12.

This filter passes frequencies that are higher than the cutoff frequency f_{c-hp} and rejects those that are lower than f_{c-hp}. In the transition band the transfer function has a slope of $-20\,dB/Decade$ or, equivalently, $-6\,dB/Octave$.

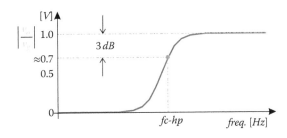

FIGURE 7.12 Absolute value of transfer function vs. frequency for a high-pass RC filter.

7.3.2.2 Phase Response

From the transfer function we have that the angle of outphasing is equal to

$$\varphi = \tan^{-1}\left(\frac{\dfrac{\omega RC}{1+\omega^2 R^2 C^2}}{\dfrac{\omega^2 R^2 C^2}{+\omega^2 R^2 C^2}}\right) = \tan^{-1}\left(\frac{1}{\omega RC}\right)$$

In this case, too, it is useful to study the function in the following occurrences of

$$\omega RC \ll 1 \Rightarrow \varphi \to 90°$$
$$\omega RC = 1 \Rightarrow \varphi = 45°$$
$$\omega RC \gg 1 \Rightarrow \varphi \to 0°$$

The graph of the phase response is shown in Figure 7.13
At the cutoff frequency the phase shift is 45°.

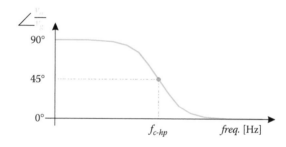

FIGURE 7.13   Typical phase response curve for a high-pass RC filter.

Example

Figures, 7.14, 7.15, and 7.16 show the effective input (blue) and output (red) signals recorded by an oscilloscope for a real high-pass RC circuit, with values $R = 10\,k\Omega$ (tolerance $\pm 5\%$) and $C = 10\,nF$ (tolerance $\pm 10\%$).

Figure 7.14 considers the situation at the cutoff frequency $f_{c-hp} = 1650\,kHz$:·
The output signal is reduced by about 70% in amplitude and leads the input 45° $(= \pi/4)$ in phase.

Figure 7.15 represents the situation for $f < f_{c-hp}$, in particular for $f = 1000\,kHz$. At this frequency, which is lower than the cutoff frequency f_{c-hp}, the output signal is reduced even more in amplitude and leads the input more than 45° the input in phase.

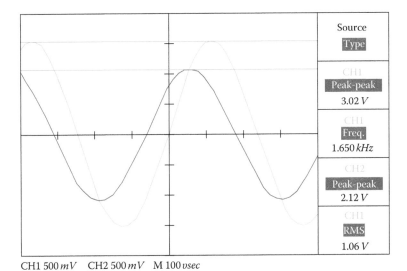

CH1 500 *mV*   CH2 500 *mV*   M 100 *vsec*

FIGURE 7.14   Input (in blue) and output (in red) signals
for the high-pass RC circuit at $f = f_{c\text{-}hp}$.

The third figure regards the situation for $f > f_{c-hp}$, in particular for $f = 3000\,kHz$. As we move away from the cut-off frequency f_{c-hp} in the direction of higher frequencies, the output signal looks more and more like the input signal both in absolute value and in phase.

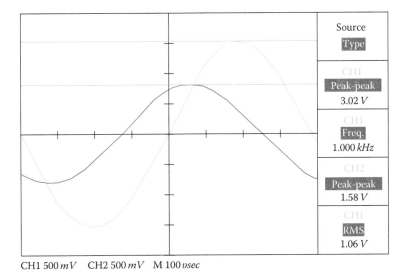

CH1 500 *mV*   CH2 500 *mV*   M 100 *vsec*

FIGURE 7.15   Input and output signals for the high-pass RC circuit at $f < f_{c\text{-}hp}$.

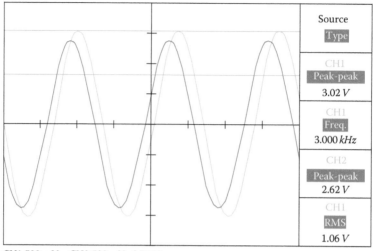

CH1 500 *mV*   CH2 500 *mV*   M 100 *vsec*

FIGURE 7.16   Input and output signals for the high-pass RC circuit at f > f$_{c-hp}$.

7.3.3 LC Filter

The *LC circuit*, made of the series of an inductor and a capacitor, is an ideal network, because it does not consider the unavoidable energy dissipation due to resistance. The resistors are usually present in any network, because their function is to limit or control the amount of flowing current. Even when no resistor is utilized, the resistance of the network is provided in any event by the resistance of wires and by the physical soldering between components (both usually ignored), even possible corrosion around terminal points, or faulty insulations, etc. In any case, just for didactic purposes, we refer here to a pure *LC circuit* (Figure 7.17).

We know that capacitive reactance $X_C = 1/2\pi fC$ decreases with increasing frequency, while inductive reactance $X_L = 2\pi fL$ increases with increasing frequency f. It means that there must exist just one frequency f_r, called the *resonance frequency*, for which these two reactances will be equal (Figure 7.18):

$$\frac{1}{2\pi f_r C} = 2\pi f_r L$$

$$\Rightarrow$$

$$f_r = \frac{1}{2\pi\sqrt{LC}}$$

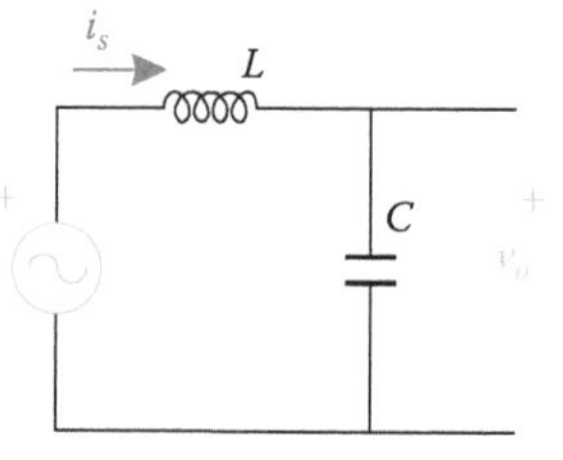

FIGURE 7.17   LC filter.

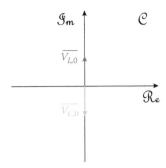

FIGURE 7.18 Capacitive and inductive reactances equal each other at one working frequency.

At this special frequency, capacitor and inductor periodically exchange the electro-magnetic energy, similarly to what occurs with the kinetic and potential energy of a pendulum swinging back and forth.

Let's calculate the transfer function:

$$\frac{v_o}{v_s} = \frac{\dfrac{1}{j\omega C}}{j\omega L + \dfrac{1}{j\omega C}} = \frac{1}{1 - \omega^2 LC}$$

Also, let's consider the special cases:

$$\omega^2 LC \ll 1 \Leftrightarrow \omega \ll \frac{1}{\sqrt{LC}} \Rightarrow \left|\frac{v_o}{v_s}\right| \to 1$$

$$\omega^2 LC = 1 \Leftrightarrow \omega = \frac{1}{\sqrt{LC}} \Rightarrow \left|\frac{v_o}{v_s}\right| \to \infty$$

$$\omega^2 LC \gg 1 \Leftrightarrow \omega \gg \frac{1}{\sqrt{LC}} \Rightarrow \left|\frac{v_o}{v_s}\right| \to 0$$

The complete graph of the absolute value of the transfer function is shown in Figure 7.19.

The total impedance of the circuit $Z = X_C + X_L = (1/j\omega C) + j\omega L$, at the special condition $\omega = 1/\sqrt{LC} \Leftrightarrow X_C = X_L$, increases to infinity, so (theoretically) output voltage v_o is infinite, and zero current comes from the AC power source. This is an unreal situation of course, but we get closer and closer to it, as the real resistance is more and more reduced.

The resonance effect of the LC circuit has many important applications.

As an example, radio stations transmit their signals each around a carrier fre-quency (see Section 1.8 in Chapter 1). To tune one particular radio station, in the radio receiver an LC circuit is set at resonance for the particular carrier frequency

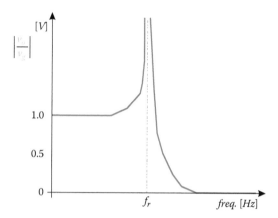

FIGURE 7.19 Absolute value of transfer function vs. frequency for an LC circuit.

of that station, and the remaining carrier frequencies of the other stations can be filtered and rejected.

Another example is the utilization of the LC circuit as an impedance matching network when a maximum voltage transfer is requested (see Section 6.9.3 in Chapter 6).

For all the above considerations, we can understand why the *LC circuit* is also referred to as a *resonant, tuned,* or *tank circuit.*

7.3.4 RLC Filter

The real situation for the LC filter of the previous section is when a resistor is added. Here, in particular we analyze the series topology (Figure 7.20).

Now, the total impedance Z is

$$Z = R + j(X_L - X_C) = R + j\left(\omega L - \frac{1}{\omega C}\right)$$

So, with V_{RMS} the RMS value of the input voltage signal v_s, the flowing current is

$$\bar{I} = \frac{V_{RMS}}{R + j\left(\omega L - \frac{1}{\omega C}\right)}$$

(" $\bar{\ }$ " is phasor notation). The absolute and phase values of the current are

$$|I| = \frac{V_{RMS}}{\sqrt{R^2 + \left(\omega L - \frac{1}{\omega C}\right)^2}}$$

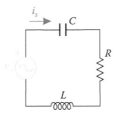

FIGURE 7.20   RLC series circuit

$\angle I = \tan^{-1}(-\omega L - (1/\omega C)/R)$ where $\overline{V_{RMS}}$ is the reference vector.
The condition $X_L = X_C$ is realized at the resonant frequency

$$2\pi f_r L = \frac{1}{2\pi f_r C}$$

(see Figure 7.21) so

$$f_r = \frac{1}{2\pi\sqrt{LC}}$$

At this special frequency the impedance Z has its minimum value

$$Z = R$$

while the current has its maximum value:

$$I_r = \frac{V_{RMS}}{R}$$

Without outphasing,

$$\angle I_r = 0°$$

The complete plot of current versus frequency starts near zero for low frequencies, reaches its maximum I_r value at resonant frequency, and then drops again to nearly zero as f becomes infinite, as illustrated in Figure 7.22.

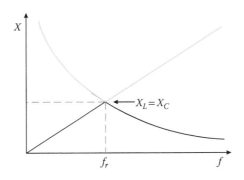

FIGURE 7.21   Reactance X of the series RCL circuit as a function of frequency f.

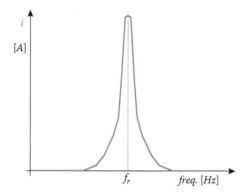

FIGURE 7.22   Current vs. frequency for a series RLC filter.

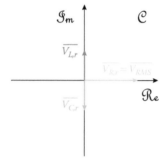

FIGURE 7.23   Inductance, resistance, and capacitance phasors at the resonance frequency.

Regarding the voltages across each component (resistor, inductor, and capacitor), at the resonant frequency they are, respectively,

$$V_{R,r} = V_{RMS} = RI_r$$

$$V_{L,r} = \omega_r L I_r$$

$$V_{C,r} = \frac{1}{\omega_r C} I_r$$

A *quality factor Q* for a series RLC resonance circuit is defined as the ratio that occurs between the voltage across the reactive elements and the resistive one, at resonant conditions:

$$Q \underset{def}{=} \frac{V_{L,r}}{V_{R,r}} = \omega_r \frac{L}{R} \underset{\omega_0 L = \frac{1}{\omega_0 C}}{=} \frac{1}{\omega_r RC} = \frac{V_{C,r}}{V_{R,r}} = \frac{V_{C,r}}{V_{RMS}}$$

In particular consider how the lower R is, the higher Q becomes.

The impedance Z depends on frequency and its amplitude and phase are

$$|Z| = \sqrt{R^2 + \left(\omega L - \frac{1}{\omega C}\right)^2}$$

$$Z = \tan^{-1}\left(-\frac{\omega L - \frac{1}{\omega C}}{R}\right)$$

which correspond respectively to the graphs in Figures 7.25 and 7.26.

Observation

The value of the Q factor is fundamental to discriminate the carrier frequency of just one radio station, with respect to the nearest ones. Figure 7.24 represents the situation of a high vs. low Q factor, and demonstrates how a low value lets frequencies of carriers of adjacent stations pass, so that the final result is to hear more than one radio station at the same time.

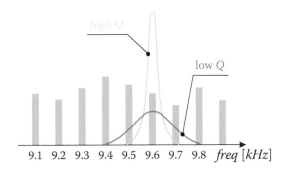

FIGURE 7.24 Effects of high vs. low Q factor.

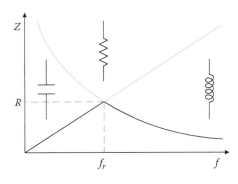

FIGURE 7.25 Z vs. frequency for a RLC series circuit.

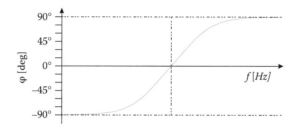

FIGURE 7.26   The phase characteristics of I versus f/f_r.

At zero frequency, the impedance Z is only due to the capacitive reactance X_C, with $Z|_{f=0} = X_C = \infty$, and i leads v_s, so $\varphi|_{f=0} = -90°$. At resonant frequency, the impedance Z is only due to the resistance $Z = R$, with $X_C|_{f_r} = X_L|_{f_r}$, so $\varphi|_{f=f_r} = 0°$. At infinite frequency, the impedance Z is only due to the inductive reactance X_L, with $Z|_{f=\infty} = X_L = \infty$, and i lags v_s, so $\varphi|_{f=\infty} = +90°$.

We define a lower frequency value f_{c-l} and an upper frequency value f_{c-h} at which the current value drops to $1/\sqrt{2}$ ($-3\,dB$) of the maximum value in resonant conditions. These are the *cutoff* frequencies. The difference between these two values defines the so called *bandwidth* $BW = f_{c-h} - f_{c-l}$.

Finally, it is possible to demonstrate that

$$f_r = \sqrt{f_{c-l}f_{c-h}} = Q * BW$$

7.3.5 Ideal Filters

There are five basic filter types: *low-pass, high-pass, bandpass, notch* (or *band-reject* or *band-stop*), and *all-pass*. Their ideal and real behavior does not match perfectly. Let's discuss this in detail.

The *low-pass filter* behaves in a way to distinguish three areas of function in frequency:

→ In frequencies much lower than the cutoff frequency ($f \ll f_{c-lp}$), in which the amplitude of the output signal is practically the same as that of the input signal.

→ In frequencies much higher than the cutoff frequency ($f \gg f_{c-lp}$), in which the amplitude of the output signal is practically null.

→ Around the cutoff frequency f_{c-hp}, called the transition band, where the transfer function has a slope of $-20\,dB/Decade$ or, equivalently, $-6\,dB/Octave$.

This is the behavior of a *real* filter.

A low-pass filter is defined *ideal* insofar if it behaves as described in the first and second points above, but not as in the third point, since the ideal transition band is null, resulting in an abrupt change from one situation to another (Figure 7.27a).

The abrupt transition (null transition band) defines the *ideal* condition of every type of filter.

Similarly to the ideal low-pass filter, but complementary to it, we define an *ideal high-pass filter* as one that stops the frequency lower than the cutoff f_{c-hp} and lets pass the frequencies higher than the same cutoff f_{c-hp} (Figure 7.27b).

If a high-pass filter is in parallel/series with a low-pass filter, we can obtain a *passband/band-reject filter*, when $f_{c-hp} < f_{c-lp}$ or $f_{c-hp} > f_{c-lp}$.

We define the *ideal passband filter* as one that lets pass a band of frequencies between a lower cutoff frequency f_{c-hp} and an upper cutoff frequency f_{c-lp} (Figure 7.27c), but completely rejects signals out of that interval.

An *ideal band-stop filter* (or *band-rejection filter*) is defined as the complement to the bandpass filter. Now, the band-stop admits frequencies below f_{c-lp} and above f_{c-hp} (Figure 7.27d). Usually a band-stop filter with a "narrow" stopband is called a *notch filter*.

Finally, the *all-pass filter* lets pass all the frequencies, so it finds applications just for phase shaping or equalization.

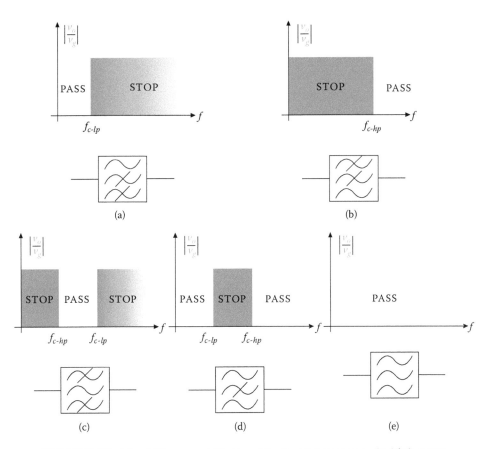

FIGURE 7.27 Ideal filters and their symbols: (a) low-pass, (b) high-pass, (c) bandpass, (d) band-reject (notch), (e) all-pass.

The passband filter separates signals, passing those of interest, and attenuating the unwanted frequencies. An example of application is in a radio receiver or in a TV set, where the filter allows only the signals having the frequencies of your favorite radio station to pass, while heavily attenuating the rest of the signals of all the other stations coming from the antenna.

Curiosity

The so called *equalizers* that we find in some radios and hi-fi systems are typically made by a series of bandpass and band-stop filters and are usually in octave bands (an octave is when the upper cutoff frequency is double than the lower one).

7.4 SOURCES OF PHASE SHIFT

We saw how the reactive elements (capacitor and inductors) cause a phase shift between voltage and current (refer to Sections 4.6.8 and 4.7.9 in Chapter 4).

In particular the capacitor, for which the result is

$$v = \frac{1}{j\omega C} i = -j \frac{1}{\omega C} i$$

ideally produces a phase shift between voltage and current in a way that voltage is −90° out of phase with the current.

The opposite is true for the inductor, for which the result is $v = j\,Li$; it *ideally* causes a phase shift in a way that the current lags the voltage by 90°.

The *ideality* comes from the fact that in any arbitrary circuit, resistive elements are always present, even when no "real" resistors are inserted. The cause is the non-null resistances of conductors and wires, resistances owed to bonds, etc. These resistive elements, together with the reactive ones, prevent the phase shift caused by a single reactance from ever reaching the theoretic level of 90°.

In general we will have a non-null phase shift between voltage and current whose entity depends on the values of the reactive elements *together* with the resistive ones.

To clarify, let us suppose that a circuit is ohmic-inductive and is made of two parallel arms (R and L), as in Figure 7.28

Here the total current $i(t)$, from the source, splits into $i_R(t)$ and $i_L(t)$ respectively for the resistor and the inductor, so that

$$i(t) = i_R(t) + i_L(t)$$

The current $i_R(t)$ has no phase shift with the voltage $v(t)$; the current $i_L(t)$ is late by 90° (theoretical) with respect to the voltage $v(t)$; the total output current $i_o(t)$ that results from these two contributes is phase shifted by an angle φ with respect to voltage $v(t)$, with φ that depends on the numeric values of the inductance L of the inductor and the resistance R of the resistor. Figure 7.29 represents the assertion.

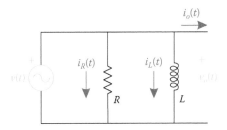

FIGURE 7.28 Parallel RL circuit.

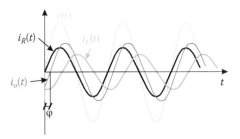

FIGURE 7.29 Response curves of the components of the current.

7.5 RC INTEGRATOR

Due to the nature of the capacitive element, the charge that is established at the plates is equal to the time domain integral of the current. Therefore the voltage across the capacitor can be written as

$$v = \frac{1}{C} \int i dt$$

So, we can use a capacitor to circuitally realize the mathematical operation of integration.

Curiosity

The operation of integration is so important that the acronyms used for the first computers in history were *ENIAC* (1943), which stands for *Electronic Numerical Integrator and Computer*, and *MANIAC* (1952), which stands for *Mathematical Analyzer Numerator Integrator and Computer*.

One method to obtain the charge of a capacitor through a current is by means of the circuit proposed in Figure 7.30

This type of circuit, however, has the disadvantage that it functions only with a constant current, fixed by a current source i_s. Another possibility is to utilize a low-pass RC circuit (Figure 7.31).

In fact, from the analysis of the circuit we have

$$v_o = \frac{\dfrac{1}{j\omega C}}{R + \dfrac{1}{j\omega C}} v_s = \frac{1}{1 + j\omega RC} v_s$$

FIGURE 7.30   First example of an integrator.

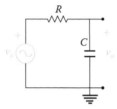

FIGURE 7.31   Second example of an integrator.

so we obtain an output voltage that is proportional to the input integral (remember the Laplace transform from time to frequency domain: $1/j\omega \Leftrightarrow \int dt$) when

$$|jwRC| = \omega RC \gg 1 \Rightarrow w \gg \frac{1}{RC} \Rightarrow f \gg \frac{1}{2\tau RC}$$

Therefore this RC circuit performs the function of integration better and better as the frequency is larger than $1/2\pi RC$ (the gray area in Figure 7.32), which is exactly the cutoff frequency of the circuit.

Analyzing the graph, we gather that the circuit behaves as an integrator only when the output voltage is a fraction of the input one. It would be useful then to combine an amplification circuit (i.e., a circuit that increases the signal amplitude) to the passive integrator one.

Such considerations allow us to establish an optimal band of frequencies for the *differentiator* circuit (complementary with respect to the integrator), which occurs for $f \ll f_c = 1/2\pi RC$.

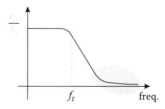

FIGURE 7.32   Transfer function of the RC circuit. The highlighted area represents the frequency band where the circuit performs better as an integrator.

7.6 KEY POINTS, JARGON, AND TERMS

→ The traditional way of observing a signal is to view it in the time domain, i.e. its behavior vs. time. But a periodic waveform can be generated by adding up sine waves, each with its amplitude, frequency, and phase. So, the same periodic signal can be considered both in time and in the *frequency domain*.

→ The opposition to current flow produced by a capacitor/inductor is called capacitive/inductive *reactance*.

→ The *impedance* is a complex number that furnishes a measure of opposition to AC current and generalizes the concept of resistance. The impedance is composed of resistance, inductive reactance, and capacitive reactance.

→ Capacitors and inductors can control signals depending on the frequency.

→ *Filters* provide frequency selectivity and/or phase shaping.

→ An ideal filter will have an amplitude response that is unity for the frequencies of interest and zero everywhere else. The transition frequency is referred to as the cutoff frequency.

→ There are five basic filter types: *low-pass, high-pass, bandpass, notch* (or *band-reject*), and *all-pass*.

7.7 EXERCISES

EXERCISE 1

A series resonance RLC circuit consisting of a resistor $R = 1\,k\Omega$, an inductor $L = 50\,mH$, and a capacitor $C = 100\,nF$ is connected across a sinusoidal supply voltage that has a constant output $V_s = 12\,V$ at all frequencies.

Calculate the voltages respectively across the resistor $V_{R,r}$, the inductor $V_{L,r}$, and the capacitor $V_{L,r}$ at resonance.

ANSWER

At resonance, the angular frequency ω_r for a series RLC circuit is equal to

$$\omega_r = \sqrt{\frac{1}{LC}}$$

so, numerically

$$\omega_r = 14142.14 \left[\frac{rad}{sec}\right]$$

The quality factor Q is defined as

$$Q_{def} = \frac{V_{L,r}}{V_{RMS}} = \frac{V_{L,r}}{V_{R,r}} = \omega_r \frac{L}{R}$$

therefore $Q \cong 0.707$.

Finally we can calculate the requested values of voltage:

$$V_{L,r} = V_{C,r} = QV_s = \omega_r \frac{L}{R} V_s$$

$$V_{R,r} = V_r$$

Numerically, $V_{L,r} = V_{Cr0} \cong 8.48\,V$; $V_{R,r} = 12\,V$.

EXERCISE 2

A hi-fi stereo system has two channel outputs with one tweeter and one woofer speaker each. To not waste low-frequency signal power into the tweeters (dedicated only to high-frequency signals), which is the easiest way to modify the schematic depicted in Figure 7.33?

ANSWER

It is sufficient to insert one capacitor (with the right value of capacitance) in series with each of the tweeters, as in Figure 7.34.

To prevent high-frequency signals from being wasted into the woofers, inductors can be placed in series with the woofers.

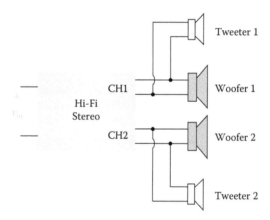

FIGURE 7.33 A hi-fi stereo with two output channels and tweeter + woofer for each channel.

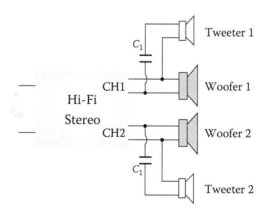

FIGURE 7.34 The hi-fi stereo with low-frequency filtered tweeters.

EXERCISE 3
A series low-pass RL filter consists of an inductance L = 10 mH and a resistor R = 4.7 kΩ, and the output voltage is taken across the resistor.
  Determine the cutoff frequency f_c of the circuit.

ANSWER
The cutoff condition is verified at the frequency f_c for which the absolute value of the transfer function $\left| v_o/v_s \right|$ is equal to

$$\left| \frac{v_o}{v_s} \right| = \frac{1}{\sqrt{2}}$$

with v_s the voltage source across the RL series.
  So, let's determine the transfer function:

$$\left| \frac{v_o}{v_s} \right| = \left| \frac{R}{R + j\omega L} \right| = \left| \frac{R}{R + j\omega L} \frac{R - j\omega L}{R - j\omega L} \right| = \left| \frac{R^2}{R^2 + \omega^2 L^2} - j \frac{\omega R L}{R^2 + \omega^2 L^2} \right| = R\sqrt{\frac{1}{R^2 + \omega^2 L^2}}$$

which is equal to $1/\sqrt{2}$ when $\omega_r = R/L$.

  Therefore,

$$f_r = \frac{1}{2\pi} \frac{R}{L} \cong 74.8\,MHz$$

EXERCISE 4
A RC low-pass filter is terminated with a load resistor R_L, as represented in Figure 7.35, with R = 1.2 kΩ, C = 2.2 μF, R_L = 8.2 kΩ.
  Determine its cutoff frequency f_c.

ANSWER
The transfer function is

$$\frac{v_o}{v_s} = \frac{R_L \, || \, \dfrac{1}{j\omega C}}{R_L \, || \, \dfrac{1}{j\omega C} + R} = \frac{R_L}{R_L + R + j\omega R_L RC}$$

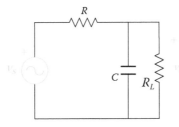

FIGURE 7.35 Terminated RC low-pass filter.

But, if we define

$$R_{||} = R \,||\, R_L$$

we can rewrite the transfer function as

$$\frac{v_o}{v_s} = \frac{R_{||}/R}{1 + j\omega R_{||}C}$$

The absolute value of the transfer function is

$$\left|\frac{v_o}{v_s}\right| = \left|\frac{\left(R_{||}/R\right)}{1+\omega^2 R_{||}^2 C^2} - j\frac{\omega\left(R_{||}^2/R\right)C}{1+\omega^2 R_{||}^2 C^2}\right| = \sqrt{\frac{\left(R_{||}/R\right)^2}{1+\omega^2 R_{||}^2 C^2}} = \frac{R_{||}}{R}\sqrt{\frac{1}{1+\omega^2 R_{||}^2 C^2}}$$

this expression is similar to the one obtained for the unterminated, also known as *unloaded*, low-pass series RC filter, with the difference of a multiplier factor $R_{||}/R$ and the unique resistor R now being replaced by $R_{||}$.
   Imposing the condition $\left|v_o/v_s\right| = 1/\sqrt{2}$, we obtain

$$f_c = \frac{1}{2\pi}\sqrt{\frac{2R_{||}^2 - R^2}{R_{||}^2 R^2 C^2}}$$

Substituing the values we get $f_c \cong 49.9\,Hz$.

Observation

Loading the filter resulted in a shift of the cutoff frequency. In fact, solving the same exercise with no load, the result is $f_c \cong 60.3\,Hz$.

8

Semiconductor Components

Semiconductor components, or devices, are basic elements for all modern electronic products. Therefore, detailed study of them is fundamental, starting from the basic material that they are made of, that is, the semiconductors (Section 3.8 in Chapter 3), and analyzing the artificial treatment they undergo, that is, doping, in order to define their practical use.

8.1 DOPING OF SEMICONDUCTORS

A *pure* matter consists of only one component with definite physical and chemical properties.

Curiosity

Pure water, H_2O, consists of two parts hydrogen, 11.1888% by weight, and one part oxygen, 88.812% by weight. It is free from contaminants like particulate, minerals, salts, bacteria, and so on. Be aware that pure water is likely to be found only in laboratories and only under ideal conditions. Distillation and deionization are the common methods of purifying water.

Going against common sense, pure water is not a good conductor of electricity at all, but the conductivity rapidly increases as water dissolves ionic species.

Typical *conductivity* σ_{H_2O} values are for ultra pure water $5.5*10^{-6}[S/m]$, for drinking water $0.01[S/m]$, and for sea water $5[S/m]$. A way to measure the "purity" of water is by means of its conductivity.

When an arbitrary matter is added with foreign elements that should not be part of the material itself, we talk about *impurity*. When the foreign elements have been artificially and intentionally added to the material, then we use the term *doping*.

Basically, doping is necessary to change the electrical properties of the pure material. More exactly, to dope means to replace some lattice atoms by suitable doping atoms without changing anything else.

For our purposes, the material used to dope must be a semiconductor, the choice of which, among all possibilities, is conditioned by its mechanical and electrical properties.

The mechanical properties must allow processing at as low a cost as possible, while the electrical properties are strictly bound to the number of electrons in the outer atomic shell, which take part in the formation of covalent bonds. These are the so-called *valence electrons* with which the energies of the *valence band* compete.

Obviously, the choice of foreign elements to use in doping cannot be random, since there is a relationship to the type of semiconductor to dope and the electrical properties we want to obtain by doping, in particular, the increase or decrease of the number of charge carriers.

The typical semiconductors used in electronics are silicon (*Si*), germanium (*Ge*), gallium arsenide (*GaAs*), indium phosphide (*InP*), etc. (some of them have already been detailed in Section 3.8.2 in Chapter 3), but silicon particularly meets all the electrical and mechanical criteria required from a generic semiconductor.

So, let's consider now the doping of silicon.

Each silicon atom has four valence electrons (it is named *tetravalent*) useful to form bonds with four more nearby atoms (Figure 8.1).

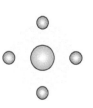

An atom that is bound with four others completes its octet according to the *octet rule* (Section 3.1.3 in Chapter 3), and each of the four bonds is formed by a pair of electrons as schematized in Figure 8.2 (to simplify the graph we have a 2D representation but, obviously, the bonds are in 3D).

FIGURE 8.1 Schematization of the silicon atom with its four valence electrons.

Here we have what we call a *covalent bond*, that is, the type of bond that takes place through pairs of common electrons, in which each atom participates by donating one electron.

By doping, a certain number of silicon atoms are substituted with an equivalent number of atoms of a material that has *pentavalent* atoms (i.e., with five valence electrons) or *trivalent* atoms (i.e., with three valence atoms).

Obviously, whether doping is done with pentavalent or with trivalent elements, silicon remains neutral, due to the equality of the electrical charges, positive

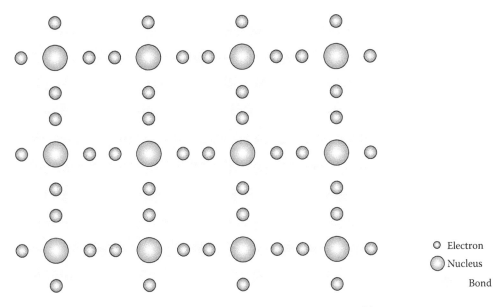

FIGURE 8.2   Representation of the bonds among silicon atoms.

and negative (the silicon atom is complete, just like the atoms of the doping element).

Observation

We could increase the number of carriers simply by adding only electrons to the material, instead of complete atoms. However, this is not done because it would mean deneutralizing the charge of the doped material.

Another way to increase the number of carriers is increasing temperature but, clearly, this method is not very practical.

8.1.1 N-Type Silicon

When the doping process is conducted with pentavalent atoms, four of the five valence electrons will be necessary to saturate the bonds with the adjacent silicon atoms, while the fifth electron will not saturate any covalent bond (as in Figure 8.3).

From the point of view of the energy band diagram, the introduction of pentavalent doping atoms results in a new discrete energy level E_d, termed *donor*, within the energy gap (empty in pure material), just below the minimum of the conduction band (Figure 8.4).

The (usually small) distance between the new energy level E_d and the lower level of the conduction band E_c depends on the choice of pentavalent doping element, which belongs to the "V" group of the periodic table of elements. Given the proximity of the new energetic level E_d to the conduction band, the environmental

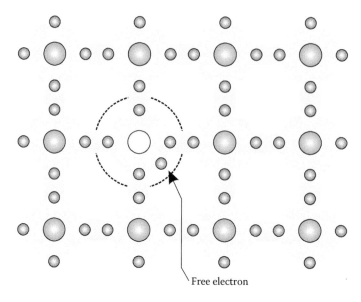

Free electron

FIGURE 8.3   Silicon doping with pentavalent material.

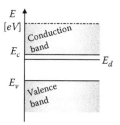

FIGURE 8.4   Energy level E_d introduced with pentavalent doping.

heat energy is enough to cause the fifth electron, not constituting a bond, to make the leap to the conduction band (Figure 8.5).

The elements of the "V" group are called *donors*, seeing as they "donate" an electron that does not create a bond. This electron is therefore useful to conduction. The silicon doped with donor atoms is called *n-type*, where "n" stands for the *"negative"* added charges.

Theoretically, all the atoms belonging to the "V" group of the periodic table (*N, P, As, Sb, Bi*) are functionally adequate to be a pentavalent dopant, as they are on the right column with respect to silicon (Table 8.1).

In any case, arsenic (*As*) or phosphorus (*P*) is almost exclusively preferred. The main reason is to avoid excessive "lattice stress" caused to silicon if doped with atoms that have a typical lattice constant too different from it (Section 3.2 in Chapter 3).

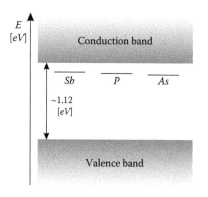

FIGURE 8.5   Donor ionization energy levels for three different dopants. The picture is not to scale. $E_d(Sb) \cong 0.039\,eV$, $E_d(P) \cong 0.044\,eV$, $E_d(As) \cong 0.049\,eV$.

TABLE 8.1
Periodic Table of the Elements in Which We Can Highlight the Possible Candidates for Silicon Doping (to the Left and Right of the Column Where Silicon Belongs)

Group	1		2	3	4	5	6	7	8	9	10	11	12	13	14	15	16	17	18
Period	IA		11A	IIIA	IVB	VB	VIB	VIIB		VI11B		IB	IIB	IIIA	IVA	VA	VIA	VIIA	VIIIA
1	1																Non-metals		2
	H													5	6	7	8	9	He
2	3		4											B	C	N	O	F	10
	Li		Be											13	14	15	16	17	Ne
3	11		12				Transition metals							Al	Si	P	S	Cl	18
	Na		Mg																Ar
4	19		20	21	22	23	24	25	26	27	28	29	30	31	32	33	34	35	36
	K		Ca	Sc	Ti	V	Cr	Mn	Fe	Co	Ni	Cu	Zn	Ga	Ge	As	Se	Br	Kr
5	37		38	39	40	41	42	43	44	45	46	47	48	49	50	51	52	53	54
	Rb		Sr	Y	Zr	Nb	Mo	Tc	Ru	Rh	Pd	Ag	Cd	In	Sn	Sb	Te	I	Xe
6	55		56	71	72	73	74	75	76	77	78	79	80	81	82	83	84	85	86
	Cs		Ba ↓	Lu	Hf	Ta	W	Re	Os	Ir	Pt	Au	Hg	Tl	Pb	Bi	Po	At	Rn
7	87		88	103	104	105	106	107	108	109									
	Fr		Ra ↓	Lr	Rf	Db	sg	Bh	Hs	Mt									

Inner transaction elements

		57	58	59	60	61	62	63	64	65	66	67	68	69	70
Laninanoias	↑	La	Ce	Pr	Nd	Pm	Sm	Eu	Gd	Tb	Dy	Ho	Er	Tm	Yb
		89	90	91	92	93	94	95	96	97	98	99	100	101	102
Aciinoias	↑	Ac	Th	Pa	U	Np	Pu	Am	Cm	Bk	Cf	Es	Fm	Md	No

Noble gases	Alkali metals	Other metals

Observation

The table of the elements by Mendeleev (Dimitrij Mendeleev, Russian chemist, 1834–1907, Figure 8.6) is called *periodic* because the chemical properties of the elements are periodic functions of their atomic numbers. Each horizontal line of the table is called a *period*.

FIGURE 8.6   D. Mendeleev.

Arsenic, for example, is a pentavalent element, that is, with five valence electrons (more specifically at the level M, $n = 3$, it has two electrons in the subshell s and three in the subshell p). Therefore it has 1 electron more than necessary to satisfy the octet rule of the subshell p of the level $n = 3$ of silicon. This is why arsenic and silicon do not form a natural chemical bond, but it is artificially made by *implantation*. Added arsenic atoms furnish four valence electrons each to complete the valence level of silicon atoms, plus one extra electron not necessary for valence bonds, so useful to become a current carrier. Doping, in this case, lowers the electric resistance of silicon.

8.1.2 P-Type Silicon

In the case of the silicon doping process using trivalent atoms, the covalent bond is not completed, missing one electron. This absence of an electron in the valence band is termed a *hole*. The hole is generally considered as a positive charge insofar as it attracts and can be "filled in" by an electron (Figure 8.7).

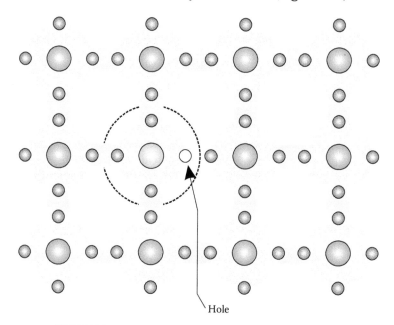

Hole

FIGURE 8.7   Silicon doping with trivalent material.

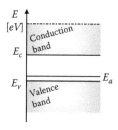

FIGURE 8.8   Energy level E_a introduced with trivalent doping.

The introduction of trivalent doping atoms results in a new discrete energy level E_a, termed the *acceptor* level, within the energy gap (empty in the pure material), just above the maximum value of the conduction band (Figure 8.8).

Given that the new energy level E_a is next to the valence band, valence electrons can easily acquire sufficient heat energy to be promoted to the acceptor level (Figure 8.9).

The elements of group "III" used to dope silicon trivalently are called *acceptors*, because they can "accept" one electron for each atom artificially introduced to silicon. The silicon doped with acceptor atoms is named *p-type* (where "*p*" stands for a "*positive*" charge, given the absence of an electron).

Although all trivalent atoms belonging to group "III" of the periodic table (*B*, *Al*, *Ga*, *In*, *Tl*) are functionally adequate for p-type doping, we practically never use tallium and prefer boron for technological reasons similar to the ones regarding pentavalent doping.

By doping silicon with trivalent elements we obtain an electric current owed not to the transport of electrons in the conduction band (negative charge), but to the transport of the so-called *holes*, assumed to be of positive charge, in the valence band.

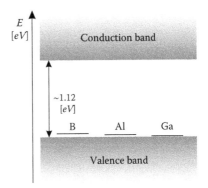

FIGURE 8.9 Acceptor ionization energy levels for three different dopants. The picture is not to scale. $E_a(B) \cong 0.045\,eV$, $E_a(Al) \cong 0.057\,eV$, $E_a(Ga) \cong 0.065\,eV$.

8.2 CHARGE CARRIERS

8.2.1 Electrons and Holes

The main difference between the current flowing in a conductor and the one flowing in a semiconductor is the existence in the latter of a motion of charges within the lattice.

In conductors, just as in semiconductors, the movement of the free electrons in the conduction band contributes, as can be guessed, to the current. Less intuitive, but equally fundamental, is the fact that in semiconductors, the holes also contribute to the overall electric current. When there is an incomplete bond, that is a hole, an electron might be "ripped off" from a nearby atom (due to thermal agitation or other reasons) and go and fill the hole. This phenomenon can be considered a movement of the electron or equally, a movement of the hole. Although it is not a physical particle in the same sense as an electron, a hole can be considered to pass from atom to atom in the opposite direction an electron moves, as if it was a positive charge. To an electron moving in a certain direction inside the lattice, we can associate a hole that "seems" to be moving in the opposite direction, as schematized in Figure 8.10.

Effectively, in semiconductors the total current is the sum of a current due to the movement of free electrons, so being within the conduction band, plus the current due to the movements of electrons within the lattice, so being within the valence band.

Let us underline here an aspect that in most textbooks is not highlighted enough: the current of holes affects the energy levels of the valence band seeing as the electrons move from one atom to the other *not* becoming free electrons, therefore not occupying the conduction band. On the contrary, the current of electrons affects the conduction band and concerns only the flow of free electrons.

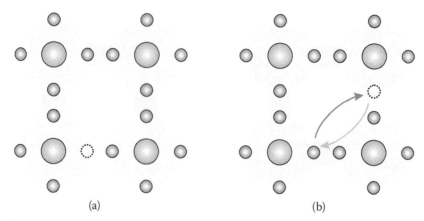

FIGURE 8.10 An electron moving in a direction (green arrow) corresponds to the movement of a hole in the opposite direction (red arrow).

If an electron from the conduction band goes down to the valence band (because of energy loss), this motion cannot be considered current, and we have one electron-hole pair less. In this case we talk about the effect of *electron-hole pair recombination*.

The electron-hole pairs, apart from nullifying each other by recombination, can also be generated. In fact, room heat energy can be absorbed by the crystal, achieving a vibration of the crystalline lattice structure. The heat energy, however, is not distributed uniformly, so that the lattice might break at some points. As a consequence some electrons will become free charges and the places they left will correspond to holes. In this case we talk about *electron-hole pair generation*.

8.2.2 Majority and Minority Carriers

Both electrons and holes are present in any semiconductor. Conventionally, electrons flow from lower to higher voltage (minus to plus), and holes flow in the opposite direction (from plus to minus). Both electrons and holes contribute to the overall current and this is why they are generically called *carriers*.

When a material is pure (not doped) it is generically called *intrinsic* and the density of its electrons (negative charge carriers) n_i equals that of the hole p_i (positive charge carriers):

$$n_i = p_i$$

If we are in absolute zero conditions, and therefore have no heat energy, the valence band is completely full and the conduction band is completely empty, as there is no energy to promote charges overcoming the energy gap. By increasing temperature we increase the possibility of breaking covalent bonds and consequently find electrons in the conduction band. The mean number of carriers is strictly connected to the temperature and to the energy gap (hence to the type of material), according to the law (already mentioned in Section 3.8.2.2 in Chapter 3)

$$n_i = AT^{1.5}e^{-\frac{E_g}{2kT}}$$

in which

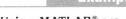

→ A: constant that depends on the material ($A_{Si} = 8.70225 * 10^{24}$)
→ T: absolute temperature [$kelvin$]
→ E_g: value of the energy gap [eV] $E_{g(Si)} = 1.1 eV$
→ k: Boltzmann's constant [$1.38*10^{-23} J/K$ or $8.62*10^{-5} eV/K$]

Example

Using MATLAB® we can graph the amount of carriers in the intrinsic silicon with respect to temperature:

```
% electronics concentration vs T
A = 8.70225e24;
k = 8.62e-5;
for i = 1:10
t(i) = 273 + 10*(i-1);
eg(i) = 1.17 - 4.37e-4*(t(i)*t(i))/(t(i) + 636);
t32(i) = t(i).^1.5;
ni(i) = A*t32(i)*exp(-eg(i)/(2*k*t(i)));
end
semilogy(t,ni)
title('Electronic concentration vs. T','FontSize',12)
xlabel('Temperature, K','FontSize',11)
ylabel('Electronic concentration, cm-3','FontSize',11)
```

The graph is shown in Figure 8.11.

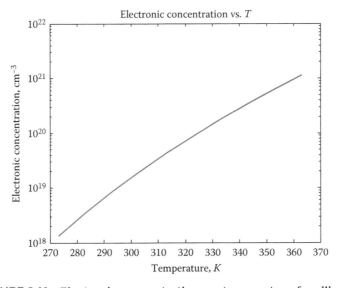

FIGURE 8.11   Electronic concentration vs. temperature for silicon.

As we can note, at room temperature the concentration of carriers for intrinsic silicon is about $n_i|_{T=300k,Si} \cong 10^{10} \, [electrons/cm^3]$, with respect to an atomic density greater than $10^{22} \, atoms/cm^3$.

For a semiconductor at room temperature the density of its free charges is less than its atomic density, unlike for conductors for which each atom contributes one charge. On the other hand, when a material is intentionally doped, it is called *extrinsic*, and the balance of electrons/holes density is altered on purpose.

By definition an *n-type* doped material is extrinsic and has electrons "in excess," meaning that there are more electrons than holes (remaining neutral, seeing as the number of protons of each donor atom is equal to the number of electrons added by doping). Because of what we said before we can basically consider all the electrons in excess as being in conduction band.

For the crystal of n-type doped silicon there are both electrons supplied by the dopant (in excess), and pairs of electrons-holes generated by the breaking of bonds (because of heat energy). Conventionally the density of negative electric charge in excess is indicated by $N_D \, [electrons/cm^3]$ (i.e., "number of donors"), while the density of the electrons generated thermally (equal to the one we have in the intrinsic material) is indicated with $n_i [electrons/cm^3]$ and that of the holes with $p_i[holes/cm^3]$. It becomes obvious that $N_D \gg n_i$ and $n_i = p_i$; this is why N_D represents the concentration of the so-called *majority carriers* (in this case electrons), while n_i and p_i represent the concentration of the so-called *minority carriers* (both electrons and holes).

As with n-type material, p-type doped material is also extrinsic but what it has "in excess" are the holes. In this case, too, when we say "in excess" we mean that there are more holes than electrons (but also in this case the doped material remains neutral).

For the crystal of the p-type doped silicon there are both the holes supplied by the dopant (in excess) and pairs of electrons-holes generated by the breaking of bonds (because of heat energy). Conventionally the density of positive electric charge in excess is indicated with $N_A \, [holes/cm^3]$ (i.e., "number of acceptors"), while the density of the electrons generated thermally is indicated with

$$n_i \, [electrons/cm^3]$$

and that of the holes with $p_i \, [holes / cm^3]$. It becomes obvious that $N_A \gg p_i$ and $n_i = p_i$; this is why N_A represents the concentration of the so-called *majority carriers* (in this case holes), while n_i and p_i represent the concentration of the so-called *minority carriers* (both electrons and holes).

Summing up, n-type silicon has negative free carriers (electrons) of practically equal density to the density N_D of the donors introduced by doping. It can be said that N_D is usually much bigger than the n_i density of free carriers naturally present in silicon at room temperature. In addition, it contains an equal number of fixed positive ions. Therefore, $n = N_D + n_i$, but $N_D \gg n_i$ so $n \cong N_D$.

On the other hand, *p*-type silicon has a density of free holes (therefore "ficti-tious" free positive charges) equal to the density N_A of acceptors introduced by doping. It can be said that N_A is normally much bigger than the n_i density of free carriers naturally present in silicon at room temperature. In addition, it con-tains an equal number of fixed negative ions. Therefore, $p = N_A + p_i$, but, $N_A \gg p_i$, so $p \cong N_A$.

8.2.3 Law of Mass Action

When we dope, increasing the concentration of one type of carrier (*n* or *p*), the consequence is that the other type of carrier will reduce in concentration (*p* or *n*). In fact, the probability of recombination of the hole-electron pairs is related to the product *np* of the two concentrations, while the probability of generation of new pairs because of thermal agitation is related to the concentration of non-ion-ized bonds, and therefore remains constant even after doping. This equilibrium is summed up in the *law of mass action*:

$$np = n_i^2$$

with n_i being the intrinsic carrier density.

So, the product of the electron *n* and hole density *p* is equal to the square of the intrinsic carrier density n_i and not only for intrinsic semiconductors.

Observation

The law of mass action states equilibrium conditions, so it is valid in many fields other than electronics. For instance, in chemistry it is related to the com-position of a reaction mixture at equilibrium.

According to what was mentioned in the previous section, for the n-type doped semiconductor we have $p = n_i^2/n \cong n_i^2/N_D$, and for p-type doped semiconduc-tor $n = n_i^2/P \cong n_i^2/N_A$ (Figure 8.12).

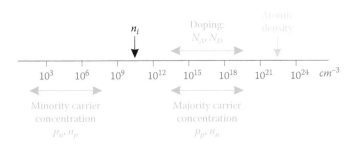

FIGURE 8.12 Carrier concentration.

Example

Suppose we have n-type doped silicon, with doping in the $10^{15} - 10^{19}$ $atoms/cm^3$ range. Using MATLAB® we graph the value of hole concentration (in this case minority carriers) at room temperature:

```
% concentration of holes in n-type doped silicon
nd = logspace(15,19);
n = nd;
ni = 1e10;
ni_sq = ni*ni;
p = ni_sq./nd;
semilogx(nd,p,'b')
title('Hole concentration ','FontSize',12)
xlabel('Dopant concentration, cm-3','FontSize',11)
ylabel('Hole concentration, cm-3','FontSize',11)
```

We get the following graph:

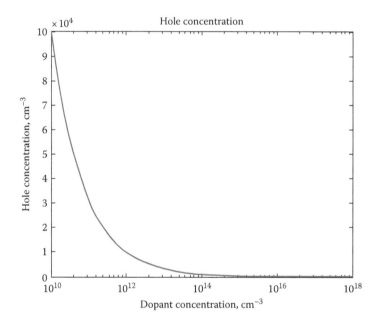

As we can note the hole concentration is reduced as the n-type dopant increases.

8.3 CURRENT IN SEMICONDUCTORS

8.3.1 Drift Current

In Section 2.6 of Chapter 2, the density of the electrical current through any given conductor was defined as

$$J = nq\vec{v}_d$$

in which n is the total number of electrical charges (typically only electrons, negative charges, hence "n"), q is the value of unit charge, and $\vec{v}_d$ is the drift velocity.

This equation now needs to be adapted for a semiconductor for which the current is due to both electrons and holes. Since the holes move in the opposite direction of electrons, but have opposite value of charge, their contribution to the total current is *summed* to that of the electrons. Associating the subscript "n" to the electrons and the subscript "p" to the holes we can write

$$J = J_n + J_p = nq_n\vec{v}_n + pq_p\vec{v}_p = nq\vec{v}_n + pq\vec{v}_p$$

for $q_n = q_p = q$, where n is the number of electrons and p that of holes.

The movement of electrical charges that determines this current is due to an electrical field $\vec{E}$ applied across the material, but it also depends on the *mobility*, $\mu[m^2/V * sec]$ (please refer to Section 2.9 in Chapter 2):

$$\mu_n = \frac{\vec{v}_n}{\vec{E}} ; \; \mu_p = \frac{\vec{v}_p}{\vec{E}}$$

So the total *drift current* can be written as

$$J = J_n + J_p = (nq\mu_n + pq\mu_n)\vec{E}$$

The mobility of electrons and holes in the case of silicon gives the result

$$\mu_{n(Si)} = 0.14\frac{m^2}{Vsec} ; \; \mu_{p(Si)} = 0.05\frac{m^2}{Vsec}$$

Generally, electrons have greater mobility than holes (see Table 8.2 for the most important semiconductors), so if there is an equal number of electrons and holes ($n = p$) the biggest contribution to current is given by the electrons in the conduction band (as expected).

TABLE 8.2
Electron and Hole Mobility Values for Some Semiconductors

	$\mu_n\left[\dfrac{cm^2}{Vsec}\right]$	$\mu_p\left[\dfrac{cm^2}{Vsec}\right]$
Si	1417	471
Ge	3900	1900
GaAs	8800	400

8.3.2 Diffusion Current

We saw how in semiconductors the transport of charge takes place as a drift motion of the carriers because of an electric field (drift current), in a similar way to metals. Apart from this mechanism of charge transport we can have a second one, irrelevant in the case of metals (because of Gauss's law), which takes place in the absence of an applied electrical field, since it is due to a *diffusion* mechanism.

It appears strange that a charged particle can move even if no force is applied to it. But consider that there is always a source of energy within the semiconductor lattice, that is, heat. In normal conditions, the heat action results in a completely random movement for the charges, without a specific direction, so that the overall average current is zero. Normal conditions mean particles uniformly distributed within all of the semiconductor. But for certain reasons this cannot be true, and the particles can be more concentrated in one region of the semiconductor as compared to another. If so, the heat will cause the charges to randomly move across the entire semiconductor, finally resulting in a uniform distribution. This is a way to maximize the so called *entropy* of the system.

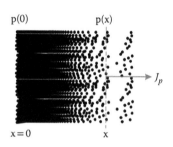

p(0)   p(x)

J_p

x = 0   x

FIGURE 8.13 Non-uniform distribution of charges in one direction.

In particular the diffusive effect produces a flow of particles in an orthogonal direction to the surface of equal concentration to the particles themselves. This flow proceeds from high to low concentrations with an intensity bound to gradient level (Figure 8.13).

So, the drift current is caused by an electric field acting on electric particles having charge, while the diffusion current is caused by thermal energy acting on electric particles having mass. Drift and diffusion currents can be in the same or opposite direction.

Observation

The *diffusion* is generally a physical transport effect that naturally occurs under different circumstances, always moving particles from higher to lower concentration. As an example, when two or more gases are put together, their atoms and molecules move—rearranging to form a homogeneous mixture without any chemical change. Diffusion can happen in liquids too. A little bit of paint into a glass of water will spread slowly through the water until uniformity is obtained.

The rate at which diffusion occurs is termed *diffusivity* and measured by *diffusion coefficients*, D_n for electrons, and D_p for holes, in cm^2/s. The diffusivity depends on the speed of carriers and on the distance between *scattering events*, which are interactions with host atoms in the lattice that cause charges to suddenly move into a different state and direction (Figure 2.19, Section 2.7 in Chapter 2).

Assuming, for simplicity, the diffusion occurs unidimensionally in direction x, there are a number of charge carriers crossing the unit area per unit time, which is termed *flux* ϕ_n and ϕ_p respectively for electrons and holes. The flux is proportional to the change of the concentration gradient:

$$\phi_n = -D_n \frac{dn(x)}{dx}; \; \phi_p = -D_p \frac{dp(x)}{dx}$$

(in the case of 3D flow we should consider gradients instead of simple derivatives) where

$$D_n = \frac{L_n^2}{\tau_n} \left[\frac{cm^2}{sec} \right]; \; D_p = \frac{L_p^2}{\tau_p} \left[\frac{cm^2}{sec} \right]$$

with L_n and L_p being the *mean diffusion lengths*, and τ_n and τ_p the *mean free times* for electrons and holes, between two scattering events.

Between diffusion coefficients and mobility the following relations by Einstein hold:

$$D_n = \mu_n \frac{k}{q} T; \; D_p = \mu_p \frac{k}{q} T$$

with k the Boltzmann's constant ($k = 1.35 * 10^{-23} [J/K]$).

Typical values at room temperature ($T \cong 300 K$) of the diffusion coefficient for electrons and holes in silicon and germanium are

$$D_n \big|_{Si} = 35 \left[\frac{cm^2}{sec} \right]; \; D_p \big|_{Si} = 13 \left[\frac{cm^2}{sec} \right]$$

$$D_n \big|_{Ge} = 100 \left[\frac{cm^2}{sec} \right]; \; D_n \big|_{Ge} = 50 \left[\frac{cm^2}{sec} \right]$$

Multiplying the flows ϕ_n and ϕ_p by the charge of electrons and holes we get the density of the diffusion current J_n and J_p:

$$J_n = -q \left(-D_n \frac{dn(x)}{dx} \right) = q D_n \frac{dn(x)}{dx} \left[\frac{A}{m^2} \right]$$

$$J_p = -q D_p \frac{dp(x)}{dx} \left[\frac{A}{m^2} \right]$$

(equations that, for a non-null gradient along the axes y and z, would have to be written as $J_n = q D_n * grad(n)$ and $J_p = -q D_p * grad(p)$).

8.3.3 Total Current

The total current density J in a semiconductor, which results from the motion of charge carriers, can be the sum of four components depending on the type of carriers, electrons, and/or holes, and the causes, drift or diffusion. Summarizing,

$$J = J_n + J_p$$

with

$$J_n = J_n\big|_{drift} + J_n\big|_{diffusion} = nq\mu_n E + qD_n \frac{dn(x)}{dx}$$

$$J_p = J_p\big|_{drift} + J_p\big|_{diffusion} = pq\mu_p E - qD_p \frac{dp(x)}{dx}$$

8.3.4 Leakage Current

It is somewhat important to also define the so called *leakage current.* In electronics leakage current represents a *parasitic* and *uncontrolled* current, which flows across regions of a material, not specifically a semiconductor, where no current should be flowing or flowing in the opposite direction. An example is a usually negligible current flowing through the body or over the surface of an oxide in which, by definition, no current should be present.

8.4 P–N JUNCTION

When in the same semiconductor, by means of complex technological processes, an abrupt transition is created between n-type and p-type doped regions, we have an *p–n junction.* This is the building block of most semiconductor devices, and so of the electronic age, which is why it is fundamental to understand its working principles.

Initially both n-type and p-type regions are neutral, seeing as doping introduces a balanced number of charges, and every (positive or negative) *free charge* is together with a *fixed charge* opposite in sign (we talk about *free* and *fixed* charge with obvious meaning). Conventionally, we represent a fixed charge closed in a circle, while the free one is out of the circle, as seen in Figure 8.14.

The junction can therefore be schematized as in Figure 8.15.

(a)   (b)

FIGURE 8.14   (a) Fixed positive charge and free negative charge; (b) fixed negative charge and free positive one.

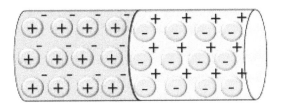

FIGURE 8.15   p–n junction.

We are dealing here with the so-called *homojunction,* that is, made by two regions of the same doped material (silicon in our case), so with the same energy gap. In contrast, a p–n junction can be *heterojunction,* that is, made with two different materials, therefore with unequal band gaps.

8.4.1 Built-In Electric Field

Conventionally, but with no scientific value, free charges close to the junction are called *nearest charges;* free charges a little further away are called *next-nearest charges,* and so on for *third-nearest, fourth-nearest,* etc.

The distribution of charges is due to the doping so that there are large numbers of mobile electrons on the n-type side, acting as majority carriers, but very few mobile electrons on the p-type side, acting as minority carriers. It is the opposite for the holes.

Due both to electrostatic attraction and different concentration, a current will start flowing, so that free positive charges move across the junction, from the p-type to the n-type region, and vice-versa for negative charges.

This initial current, though, is soon stopped for reasons that we are about to analyze.

The *nearest free charges* cross the junction and finding opposite free charges in the other region, they recombine. Due to recombination, the nearest zone is now without free charges and exclusively occupied by fixed charges which set up a uniform electric field, called *built-in* $\vec{E}_{b-i}$ (Figure 8.16), right at the junction between the n-type and p-type material (see Figure 8.39 in Section 8.6.1 as a reference).

The built-in electric field causes some of the electrons and holes to flow in the opposite direction to the flow caused by electrostatic attraction and diffusion. These

(a)

(b)

FIGURE 8.16   Depletion region with (a) E_{b-i} between fixed charges and (b) its schematization.

opposing flows eventually reach an equilibrium deleting each other and the motion of second, third, etc. nearest free charges is stopped, no current will flow anymore, and a situation of stability is obtained. If the carriers had no charge, a current could result for diffusion anyway, but electrons and holes have electric charges so the built-in electric field is capable of stopping them and the current is zeroed.

Figure 8.17 schematizes this situation, and the built-in field $\vec{E}_{b-i}$ is represented by one and only vector outside the junction (even though that field is a homogenous one, made of so many force lines between the fixed charges "positive → negative"). The volume across the junction where there are no free charges is named the *depletion region*, as it is "depleted" of mobile charges, so as to be highly resistive.

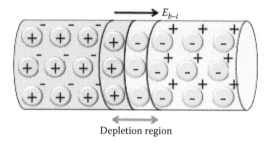

Depletion region

FIGURE 8.17 E_{b-i} in the depletion region prevents the recombination of second, third, etc., nearest free charges.

The width of the depletion region has an inverse proportion with respect to the amount of doping: the greater the doping the narrower the depletion region and vice-versa. In the examples of the previous figure, we assumed the same amount of doping for both regions, so the total depletion region around the junction is symmetric. If, however, the dopings of the two regions were different, the symmetry is lost and the region with the higher doping level will become narrower.

Although there is no net flow of current anymore, across the junction there has been established an electric field, which is the basis of the operation of many fundamental electronic devices.

But, be aware that the built-in electric field prevents *majority* charges from moving, but also moves *minority* charges (electrons in the p-type region, holes in the n-type region), creating a so-called *reverse current* I_0. This current, proportional to the poor number of minority charges and present when the junction is within a device in a circuit, is negligible for the most part of applications, and can be considered as a sort of leakage current (see Section 8.3.4).

Observation

In the case doping heavy enough to have a *degenerate material* (where we can no longer talk about a doped semiconductor, but basically about a different material similar to a conductor), the depletion region is so restricted as to become 2D. This is exactly what we get in conductors for which free charges are only arranged along the surface and not inside the material.

8.4.2 Built-In Potential

We have seen how in a semiconductor the current is possible both through drift and diffusion, and both for electron and hole carriers:

$$J_n = J_n\big|_{drift} + J_n\big|_{diffusion} = nq\mu_n E + qD_n \frac{dn(x)}{dx}$$

$$J_p = J_p\big|_{drift} + J_p\big|_{diffusion} = pq\mu_p E - qD_p \frac{dp(x)}{dx}$$

When the junction is in balance and there is no electrical field applied externally, the total charge motion must be null, so with a zero current density we get

$$E = -\frac{1}{n}\frac{D_n}{\mu_n}\frac{dn(x)}{dx}$$

$$E = \frac{1}{p}\frac{D_p}{\mu_p}\frac{dp(x)}{dx}$$

The electric field is different from zero only in the depletion region, so the voltage in that region V_{dr} is calculated for electrons (see Figure 8.18):

$$V_{b-i} = -\int_{x_p}^{x_n} (E_x)\,dx = -\int_{x_p}^{x_n}\left(-\frac{1}{n}\frac{D_n}{\mu_n}\frac{dn(x)}{dx}\right)dx = \frac{D_n}{\mu_n}\int_{n_p}^{n_n}\frac{1}{n}\,dn(x) = V_T \ln\left(\frac{n_n}{n_p}\right)$$

and equally for holes

$$V_{b-i} = V_T \ln\left(\frac{p_p}{p_n}\right)$$

These equations can be rewritten as

$$\frac{n_n}{n_p} = e^{\frac{qV_{b-i}}{kT}}$$

$$\frac{p_p}{p_n} = e^{\frac{qV_{b-i}}{kT}}$$

which represent the *Maxwell–Boltzmann distribution*, that does not apply to metals (James Clerk Maxwell, Scottish mathematical physicist, 1831–1879).

Considering the mass action law and keeping in mind that in the depletion region $n_n \cong N_D$ and $p_p \cong N_A$, we get

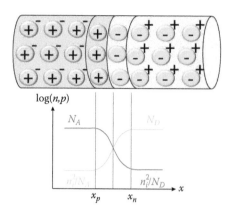

$$\frac{n_n}{n_p} = \frac{N_D}{n_i^2 / N_A} = \frac{N_D N_A}{n_i^2} = \frac{N_A}{n_i^2 / N_D} = \frac{p_p}{p_n}$$

so the value obtained from the voltage V_g considering electrons or the one obtained considering holes is the same, as was obvious. However, as we know, minor charges represent only a very low percentage of the mobile charges available (Section 8.2.2), so the drift current favored by $\vec{E}_{b-i}$ will be of a very low, often negligible, value.

FIGURE 8.18 Concentration of carriers within the depletion region.

Example

Let us determine the value of the built-in potential across the depletion region of a junction formed by two silicon regions doped with concentrations $N_A = 5*10^{15} acceptors/cm^3$ and $N_D = 1*10^{16} donors/cm^3$ respectively. Let us assume a temperature of 20°C.

Keeping in mind that for silicon $n_i = p_i \cong 10^{10} carriers/cm^3$, we get

$$V_{b-i} = \frac{kT}{q} \ln\left(\frac{N_A N_D}{n_i^2}\right) = \frac{1.38*10^{-23} \times 293.16}{1.6*10^{-19}} \ln\left(\frac{5*10^{15} \times 10^{16}}{\left(10^{10}\right)^2}\right) \cong 0.68\,V$$

8.5 DIODE

A *diode* is essentially a p–n junction conveniently packaged and with terminals made of negligible resistance ohmic junctions. It is characterized by a nonlinear voltage-current function. Side *p* of the diode is called an *anode*, while side *n* is called a *cathode*. Figure 8.19 shows the scheme and symbol adopted for this device, and Figure 8.20 shows photos of some commercial diodes.

The main characteristic of the ideal diode is that it minimizes current flow in one direction, while easily carrying current in the other direction. This characteristic is called the *rectifying property* of the diode.

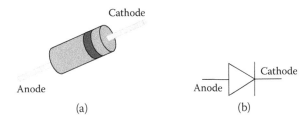

FIGURE 8.19 (a) Scheme and (b) diode symbol. In the scheme, the dark band marks the cathode, in the diode symbol the arrow shows the direction in which the current can flow.

FIGURE 8.20 Different commercial diodes.

Curiosities

The name *diode* comes from di*electrode*s, that is, two electrodes.

Braun (Karl Ferdinand Braun, German inventor, 1850–1918, Figure 8.21) in 1899 patented a device made of a single metal wire, wittily called a *cat's whisker*, soft contacting a semiconductor crystal. The device realized the fundamental rectifying effect.

In 1909, Braun shared the Nobel Prize in Physics with Guglielmo Marconi.

FIGURE 8.21 K. Braun.

A diode can be considered the electrical version of a mechanical valve and early diodes were actually called *valves*. But this was not very accurate, since mechanical valves are devices by which the flow of a gas or a liquid can be started, stopped, diverted, and/or regulated. The diode works only as a specific one-way valve (Figure 8.22).

FIGURE 8.22 Traffic sign: one way directional arrow.

A better example would be *cardiac valves*, which control one-way flow of blood. These are the *tricuspid valve* (located between the right atrium and the right ventricle), the *pulmonary semilunar valve* (lies between the right ventricle and the pulmonary artery), the *mitral valve* (between left atrium and left ventricle), and the *aortic semilunar valve* (between the left ventricular outflow tract and the ascending aorta) (Figure 8.23).

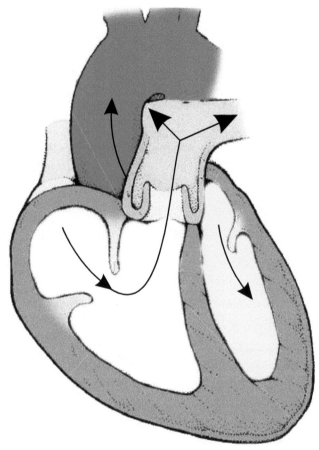

FIGURE 8.23   Blood circulation in the heart. The directions
are controlled by the cardiac valves.

There are several types of diodes, differing in geometric scaling, doping level, and electrodes. But emphasizing different physical aspects, their v-i function can be changed accordingly. As a consequence, there are diodes with electrical behavior that differs from the standard design. We refer to *avalanche diodes, Esaki* or *tunnel diodes, thermal diodes, photodiodes, point-contact diodes, PIN diodes, Schottky diodes, super barrier diodes, step-recovery diodes, varicaps* or *varactor diodes, Peltier diodes, gun diodes, IMPATT diodes,* and so on. In the following, we will treat two of the most important ones: the *Zener diode* (Section 8.5.4) and the *LED* (Section 8.5.6).

Observation

The diode has numerous applications in electronics. We'll detail some of the most important in the following sections.

But now, let's consider that the diode protects any electronic equipment (a radio, a recorder, etc.). In the event that the power source (battery or mains electricity) is connected the wrong way round, the diode prevents the current from flowing, so avoiding damaging or even destroying electronic devices.

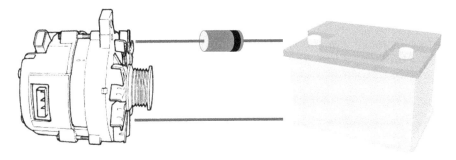

FIGURE 8.24   Alternator–car battery circuit, with a forced one-way current due to the diode.

Another example is in our cars: the *alternator* charges the battery when the engine is running, but when the engine stops, a diode prevents the battery from discharging through the alternator (Figure 8.24).

8.5.1 The Bias

The resistance of the depletion region can be modified by adding a fixed electric field to the built-in one. This operation is termed *biasing*, that is, the application of an external DC voltage, typically by means of a battery, across the junction. The added electric field can be in the same or opposite direction as the built-in electric field, determining an increase or a decrease of the resistance of the depletion region. So the depletion region can be considered to operate as a voltage-controlled resistor.

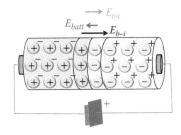

FIGURE 8.25   Forward-biased diode: the battery field is smaller and opposite to the built-in one.

8.5.1.1 Forward Bias

In *forward bias* the battery supplies an electric field $\vec{E}_{batt}$ opposite to the built-in one $\vec{E}_{b-i}$ (positive battery voltage applied to the p-type junction side), so that the two fields are subtracted. This subtraction gives rise to a field $\vec{E}_{tot}$ whose direction is the same as the built-in one when $|\vec{E}_{batt}| < |\vec{E}_{b-i}|$, so the depletion region is thinner and less resistive, but not nullified. This means that majority carriers will still be prevented from crossing the junction and there will be no current yet (Figure 8.25).

But, if the applied voltage is large enough, so that $|\vec{E}_{batt}| > |\vec{E}_{b-i}|$, the resulting field $\vec{E}_{tot}$ will be in the same direction as the battery field and the depletion region's resistance becomes negligible. This means that majority carriers can cross the junction now (Figure 8.26).

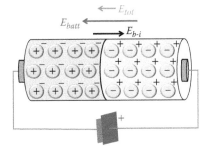

FIGURE 8.26   Forward-biased diode: the battery field is bigger and opposite to the built-in one.

In the particular condition when the built-in field and the battery field are equal in the module, $|\vec{E}_{batt}| = |\vec{E}_{b-i}|$, but opposite in sign, the battery has reached its *threshold voltage* V_γ. For the typical doping values of silicon diodes, this occurs at about $V_\gamma = 0.6 - 0.7\,V$ forward bias.

Observation

Given the typical threshold voltage value of silicon diodes, $V_\gamma \cong 0.7\,V$, it is not accidental that on the market we find $1.5\,V$ batteries to bias electronic devices containing a *p–n* junction. About half of its value, the threshold voltage, is necessary to set the depletion region to zero, just to allow a poor diffusion current, while the other half will determine drift current.

8.5.1.2 Reverse Bias

When the diode is *reverse biased*, the majority carriers cannot cross the junction because the battery and built-in electric fields add, and the resulting field E_{tot} will make the depletion region even larger and resistive than without biasing (Figure 8.27). In reality some current can flow even with these conditions, both because the resistance can be very high but never infinite, so that a really small percentage of the majority carriers can pass, and because there are minority carriers that move just because of E_{tot}. The amount of this current, called *reverse saturation current*, I_s (which is a *leakage current*, Section 8.3.4), is considered negligible for most applications and is related to temperature variation, since it was observed that it doubles as the temperature increases by $10°C$.

The order of magnitude of I_s is nA for silicon and μA for germanium relative to diodes used for "small signal" applications (so not for so-called "power" diodes). The difference is related to the smaller energy gap E_{gap} of germanium, since

$$I_s \approx e^{-\frac{E_{gap}}{kT}}$$

where k is Boltzmann's constant and T the kelvin temperature.

In any case, for a huge range of applications, a reverse-biased diode is considered an open circuit.

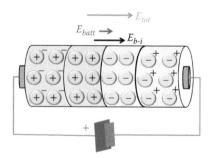

FIGURE 8.27   Reverse-biased diode: the battery field is of the same direction as the built-in one, and the depletion region is enlarged.

8.5.2 I–V Characteristic and Threshold Voltage

In conditions of forward biasing, electrons flow from the n-type region, where they are majority carriers, to the p-type region, where they are minority carriers. The results are similar for holes that flow in the opposite direction.

Once the electrons cross the junction and become minority carriers, they continue to move further and further away from the junction, but their number, and therefore their concentration, exponentially reduces because of the recombination process, until a balanced value is reached.

Near the junction, the system is in a non-balanced condition, and given the injection of minority carriers, we have $n_n p_n \gg n_i^2$ in the n-type region and $n_p p_p \gg n_i^2$ in the p-type region. This situation, maintained by the external battery field, is represented in Figure 8.28.

The parameters in the figure are

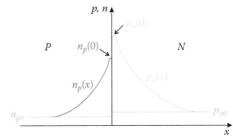

FIGURE 8.28 Concentration of minority carriers, forward biasing. We suppose a different amount of doping in the two regions, otherwise a symmetry must be assumed.

→ n_{p0} : balanced concentration of electrons in the p-region
→ p_{n0} : balance concentration of holes in the n-region
→ $n_p(0)$: electron concentration in the p-region close to the junction
→ $p_n(0)$: hole concentration in the n-region close to the junction
→ $n_p(x)$: electron concentration in the p-region at point x
→ $p_n(x)$: hole concentration in the n-region at point x

The exponential decay can be expressed as

$$n_p(x) = n_{p0} + \left(n_p(0) - n_{p0}\right) e^{\frac{x}{L_n}}$$

$$p_n(x) = p_{n0} + \left(p_n(0) - p_{n0}\right) e^{-\frac{x}{L_n}}$$

To determine the $i_D - v_D$ relationship of the diode, let us hypothesize that

→ The applied voltage v_D is almost exclusively across of the depletion region.
→ The Maxwell–Boltzmann distribution

$$\text{(Section 8.4.2:} \quad \frac{n_n}{n_p} = e^{\frac{qV_{b-i}}{kT}} , \quad \frac{p_p}{p_n} = e^{\frac{qV_{b-i}}{kT}})$$

valid only in the case of thermodynamic equilibrium, is considered valid, with a good approximation, also in the case of a stationary state near the equilibrium.

These conditions allow us to rewrite the Maxwell–Boltzmann distribution as

$$\frac{n_n}{n_p(0)} = e^{\frac{q(V_{b-i}-v_D)}{kT}}$$

$$\Rightarrow \quad n_p(0) = n_{po}e^{\frac{v_D}{V_T}}$$

where $V_T = kT/q$ $[V]$ is defined as the *thermal voltage*, that is, an electrostatic voltage across the junction.

Observation

Let us determine the value of V_T at room temperature (20°C):

$$V_T = \frac{kT}{q} = \frac{\left(1.38{*}10^{-23}\right)(273.16+20)}{1.6{*}10^{-19}} \cong 25.3mV$$

We must underline that it is *not* always a negligible value.

Keeping in mind the expression of the density of diffusion current and combining the last equations:

$$J_n\Big|_{x=0} = qD_n\frac{dn_p}{dx}\Big|_{x=0} = \frac{qD_nn_{po}}{L_n}\left(e^{\frac{v_D}{V_T}}-1\right)$$

$$J_p\Big|_{x=0} = -qD_n\frac{dp_n}{dx}\Big|_{x=0} = \frac{qD_pp_{no}}{L_p}\left(e^{\frac{v_D}{V_T}}-1\right)$$

The diffusion current of minority carriers close to the junction is transformed gradually, due to recombination, in the drift current of the majority carriers. The total value of the current is constant and equals the sum of the contribution of diffusion and drift:

$$i_D = A\left(J_n + J_p\right)$$

with A as junction area. Then

$$i_D = I_S\left(e^{\frac{v_D}{V_T}}-1\right)$$

an expression known as the Shockley diode equation (William Bradford Shockley, English-American physicist and inventor, 1910–1989), in which

$$\rightarrow I_S = qA\left(\frac{p_{n0}}{L_p}D_p + \frac{n_{p0}}{L_n}D_n\right) = qA\left(\frac{D_p}{N_D L_p} + \frac{D_n}{N_A L_n}\right): \text{reverse saturation current}$$

$\rightarrow k$: the Boltzmann constant $(1.38*10^{-23}J/K)$

$\rightarrow q$: fundamental charge $(1.6*10^{-19}C)$

$\rightarrow \eta$: empirical constant (1 for germanium, 2 for silicon), introduced to keep track of the approximations made during the argumentation

A typical plot of i_D versus v_D is given in Figure 8.29.

For $v_D < 0.6\,V$ the current is quite small. For $v_D > 0.6\,V$ the current increases rapidly with v_D, and the $V_\gamma = 0.6\,V$ fundamental value is termed the *cut-in* or *threshold voltage*, that is the voltage at which the diode "appears to begin" conducting.

Please, be aware that in reality the $0.6\,V$ value can range between 0.6–$0.7\,V$, depending on the particular considered diode.

With respect to the values obtainable by the Shockley equation we need to consider some corrections as follows:

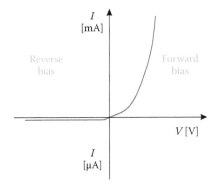

FIGURE 8.29 Typical I–V characteristics of a p–n junction diode.

→ In forward bias conditions, the voltage drop is slightly higher because of the ohmic resistance of the metal contacts relevant to the diode terminals.
→ The reverse saturation current is higher because of a so-called leakage current along the surface of the semiconductor material.
→ The reverse saturation current increases to significant values for v_D beyond a certain negative value because of a *breakdown* phenomenon that we will define later.

Therefore, the Shockley equation is quite accurate in forward bias conditions but not accurate enough for I_s and does not consider the phenomenon of *breakdown*.

8.5.3 The Resistance

In treating resistors, capacitors, and inductors we saw that the voltage to current ratios for them is constant $(v = ri, v = -jX_C i, v = jX_L i)$; they are always equations of lines passing through the origin of the axes. Therefore, for an increase (reduction) of the voltage applied, there is always a corresponding increase (reduction) in linear proportion of the current running through it.

Differently, the diode offers a value of resistance that does not always have the same value but depends on the value of the voltage applied. Therefore the diode is a *nonlinear* device. The resistance value depends on the voltage to current ratio at the point of the graph i, v where the diode is supposed to work. Unlike linear devices, however, we can define two "resistances": a *static resistance* and a *dynamic resistance* (or *incremental resistance*).

8.5.3.1 Static Resistance

Studying the characteristic of the diode, we can define a so-called *static resistance* and a so-called *dynamic resistance*. Dynamic in the sense that it refers to incremental quantities, and static in the sense that it refers to constant quantities.

Given a working voltage-current pair of the diode, represented by one, and only one, point in the characteristic curve, the static resistance of the diode R_D is defined as the voltage drop to flowing current ratio. The resistance is thus represented by the inverse slope of the segment that links that point with the origin of the axes:

$$R_D = \frac{V_D}{I_D}$$

By definition, the static resistance can assume values comprised in an interval that is too large to be a practical parameter. In fact it is from the order of the $G\Omega$ assumed in the reverse bias region, through to the $M\Omega$ until the $m\Omega$ assumed in the forward bias region.

8.5.3.2 Dynamic Resistance

Given a working point of the diode, which is a voltage-current pair on its characteristic curve, the dynamic resistance r_D is defined as the ratio of the variations of voltage to current around that point. It is therefore represented by the inverse of the angular coefficient of the tangent line at that point:

$$r_D = \frac{dV_D}{dI_D}$$

Recalling the Shockley equation we have

$$\frac{1}{r_D} = \frac{dI_D}{dV_D} = \frac{d\left(I_S\left(e^{\frac{V_D}{V_T}} - 1\right)\right)}{dV_D} = \frac{I_S}{V_T} e^{\frac{V_D}{V_T}} = \frac{I_D + I_S}{V_T}$$

When $V_D > V_\gamma$, we can assume that $I_D \gg I_S$, so

$$\frac{1}{r_D} \cong \frac{I_D}{V_T}$$

In direct biasing conditions the diode has low values of dynamic resistance. Therefore, the diode is not a device capable of limiting the current and we can reach power values that could damage it. It is preferable to prevent this by arranging in series to the diode a resistance of an appropriate value.

8.5.4 The Capacitance

A junction diode has two types of capacitances associated with it, a *junction* or *depletion capacitance* and a *diffusion* or *transit time capacitance*. The first is a weak function of the applied reverse voltage, the second is proportional to the diode current.

8.5.4.1 Junction or Depletion Capacitance

We consider a capacitor to be made of two conductors with a dielectric in between. Now, a diode in reverse bias mode is made of two doped semiconductors with a depletion region in between, so there is a "similar-capacitance" that occurs. In fact, outside of the depletion region the materials have no net charge, since fixed charges are with their mobile counterparts, but inside the depletion region the lack of electrons, in the *n*-side, and holes, in the *p*-side, yields charged volumes. Therefore, the junction has a *depletion* (region) or *junction capacitance* C_j, which is due to charges of dopants, which is equal to

$$C_j = \varepsilon \frac{S}{w}$$

where ε is the dielectric permittivity of semiconductor material, S is the junction area, and w is the thickness of the depletion region, which is a function of the applied voltage $w = w(v)$.

It can be demostrated that for a step-abrupt junction, the depletion capacitance can be written as

$$C_j = \sqrt{\frac{\varepsilon q}{2} \frac{N}{(V_D - v)}}$$

with $1/N = 1/N_A + 1/N_D$. It has a nonlinear behavior since it depends on the applied reverse voltage, which is the reason why it is fundamental to determine its dynamic characteristic rather than the static one.

8.5.4.2 Diffusion or Transit Time Capacitance

In forward bias conditions, there is a net current of electrons from the *n*-side to the *p*-side in the conduction band, and vice-versa for holes in the valence band. This current promotes an accumulation of charges in the two regions, therefore the diffusion capacitance C_d is due to free carriers.

It can be demonstrated that

$$C_d = \frac{dQ}{dv} = \frac{d}{dv}\left[\tau I_s e^{\frac{v_D}{V_T}}\right] = \frac{\tau I_s}{V_T}$$

having assumed $\tau = \tau_n = \tau_p$ is the lifetime for minority carriers (both electrons and holes), and knowing that $Q = \tau I$ as the total charge at the junction (Figure 8.30).

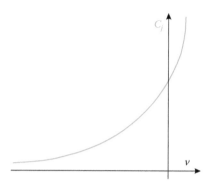

FIGURE 8.30 Capacitance–voltage characteristics for a diode in forward-biased condition.

8.5.5 Zener

As already seen, a reverse-biased diode lets a negligible amount of current run through so we can say that we have approximately the equivalent of an open circuit. However, the reverse voltage cannot be of any value; in fact, beyond a certain limit a standard diode is subject to a phenomenon called *breakdown* due to *avalanche* that causes such a significant increase of current as to damage the device irreversibly. To avoid the damage, the current must be limited by the circuit resistances in which the diode is used.

The *avalanche breakdown* phenomenon causes a significant increase in the number of carriers that takes place when an electric field of relatively "high" energy accelerates charges to such a point to dislodge other charge carriers, creating hole-electron pairs (the phenomenon of *impact ionization*). In turn these new carriers are accelerated and can similarly produce other new pairs iteratively, which is why it is considered the avalanche phenomenon.

The current increase that follows causes the damage to the device, as it is not able to tolerate electrical powers beyond a limit.

Curiosity

Avalanche phenomenae are also possible in other fields of physics. For example, when gamma rays (electromagnetic waves with a higher frequency than that of the light) collide with atoms in the atmosphere, they release energy and we have the creation of electron-positron pairs (the latter are particles, similar to electrons but with an opposite sign). These pairs of particles collide with other atoms, generating new electron-positron pairs, etc., until we get the so-called *electron swarm*. The result is a form of light called *Cherenkov radiation* (Pavel Alekseyevich Cherenkov, Soviet physicist, Nobel laureate, 1904–1990) that is manifested in the form of flashes of light sometimes visible in the sky at night.

FIGURE 8.31 Circuital symbol of the zener diode.

The so-called *zener diode* (Clarence Melvin Zener, American physicist, 1905–1993) is obtained, by means of appropriate doping values, in a way to obtain control over breakdown (Figure 8.31).

When the reverse bias reaches a certain value, called zener voltage V_Z, again we have a significant increase of the current, but at the same time we have a great reduction of the dynamic resistance value, which goes down to a just a few Ohms. The voltage value for which this happens is practically independent from the current value.

This property makes the zener diode a useful device as a *voltage reference*. In fact, its voltage-current characteristic in reverse bias has a practically constant voltage value in correspondence with a big interval of current variation, as shown in Figure 8.32.

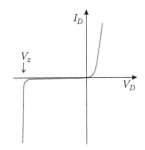

FIGURE 8.32 Typical I–V curve of a zener diode.

The phenomenon of breakdown in the zener diode is due to an effect called *quantum tunneling* of the carriers through the junction. This phenomenon takes place when both *p*-type and *n*-type regions have a very high doping value. *Quantum tunneling* (*tunneling*, or *tunnel effect*) is the crossing of the potential barrier, at the *p-n* junction, by carriers that, according to classical physics would not otherwise have enough energy to do so. Conversely, it can occur according to quantum physics, when the barrier width is reduced to just a few nanometers (remember that the width of the depletion region is inversely proportional to the amount of doping) and the biasing value can supply a high energy value to the carriers.

8.5.6 LED

LED stands for *Light-Emitting Diode* (Figures 8.33 and 8.34). As the name suggests its characteristic is that it emits light (visible or not) when directly biased. The first practical visible-spectrum LED was invented in 1962 by Holonyak (Nick Holonyak Jr., Electrical Engineer, 1928–). Let's analyze its working principle.

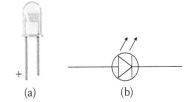

(a) (b)

FIGURE 8.33 LED: (a) scheme, (b) circuit symbol.

When an electron recombines with a hole, it experiences a loss of energy; in fact, the electron moves from the conduction band, with a greater energy, to the valence band, with less energy. The energy lost by the electron has a wavelength that is radiated in the visible or infrared spectrum. The whole phenomenon is known as *electroluminescence*.

The *color* of light that a LED can emit depends on the energy gap value crossed by the electron. As a consequence, seeing as the value of the energy gap depends on the type of semiconductor used, the color emitted by the LED varies depending on semiconductor material (Table 8.3 and Figure 8.35).

According to the equation $E = hf$ (Section 3.3 in Chapter 3), and remembering that $f = c/\lambda$ (Section 1.2 in Chapter 1), we can calculate the energy gap E_g of the material according to the color we want to obtain (Table 8.4):

$$E_g = h\frac{c}{\lambda}$$

TABLE 8.3

Some Materials with Which We Can Make LEDs and Their Typical Radiation

Material	Symbol	λ[nm]	Radiation
Gallium arsenide	GaAs	870	Infrared
Aluminium arsenide	AlAs	590	Orange
Gallium phosphide	GaP	550	Yellow, Green
Indium phosphide	InP	930	Infrared

TABLE 8.4

According to the Color We Want, We Choose the Energy Gap, and Therefore the Material

Color	λ[nm]	E[eV]
Red	750	1.65
Orange	600	2.07
Yellow	580	2.14
Green	490	2.53
Blue	450	2.75
Purple	390	3.17

(with $h \cong 6.6*10^{-34}[J*sec]$ or $h \cong 4.14*10^{-15}[eV*sec]$ given that $1[eV] \cong 1.6*10^{-19}[J]$), as reported in Table 8.4.

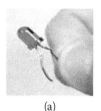 

(a)   (b)

FIGURE 8.34   LED: (a) comparison of dimensions, (b) working on printed circuit.

FIGURE 8.35   Some types of LEDs.

Silicon does not belong to the group of semiconductors used in the construction of an LED. This is because its transition between the conduction and valence bands is "indirect," in the sense that the electrons recombined undergo a variation of impulse and not just energy. This fact does not favor the emission of *photons*, so the energy is released in the form of heat rather than light.

For the construction of LEDs we normally use gallium arsenide, indium phosphide, aluminium arsenide, etc.

LED emits *cold* light unlike common filament bulbs, which work in high temperatures and have a significant thermal inertia. Thanks to this characteristic LEDs can be used to light in high frequencies, greater than *MHz*.

Furthermore, thanks to the fact that the light emitted is proportional to the current, these devices can be applied to the transmission of signals through luminous intensity modulation, as for example when used to impose a signal in optical fiber.

LEDs were first applied as indicators of state in electronic circuits, as transmitters in infrared remote controls, and as seven-segment displays that, according to the segments "lit" show a number between 0 and 9 (Figure 8.36).

Later the fields of application were multiplied and include traffic lights, stoplights, *LCD* (*Liquid Crystal Display*), backlight displays, panel displays, placards, etc. (Figure 8.37).

FIGURE 8.36 Seven-segment display.

FIGURE 8.37 LED panel display.

Curiosities

An *OLED* is a type of LED; the acronym stands for Organic LED. So it is a light-emitting diode but uses electroluminescent *organic* materials. The organic materials have the advantage of a low-cost process, such as inkjet printing on plastic substrates.

An OLED is made by either two or, to improve the efficiency, more layers of electroluminescent organic materials sandwiched between (transparent) anode and (transparent or not) cathode terminals, all deposited on a substrate—usually glass. *ITO* (indium tin oxide) is generally used for the anode, while *barium* or *calcium* is used for the cathode. The organic materials can be small organic or large polymer molecules, the latter more suitable for large-screen displays.

When a voltage is applied across the terminals, the injected electrons and holes recombine in the organic materials, thus emitting light of a frequency that depends on the value of the HOMO-LUMO energy gap crossed.

There are different types of OLEDs: PMOLEDs (passive-matrix OLEDs), AMOLEDs (active-matrix OLED), transparent OLEDs, top-emitting OLEDs, foldable OLEDs, and white OLEDs.

The glass substrate and encapsulation layers of the standard OLED design can be replaced with flexible polymer sheets. This is the way to realize a plastic-based OLED that can be shatterproof, lighter, thinner, more durable, and even flexible compared to a glass-based OLED.

A useful parameter to define the quality of a TV set is the *contrast ratio*, that is the ratio of the luminance of the white to that of the black colors. The higher the ratio, the better the system. For a television with a screen made of OLEDs, this parameter loses significance. In fact for a traditional TV (the old *cathode-ray tube* or *CRT, plasma, LCD*) the black color is really never "black" because its *pixel (picture element)* is always more or less "lit up," but this does not occur for an OLED-based TV set, so that the *constrast ratio* is practically infinite.

Recently, flexible OLEDs for *flexible displays* by Samsung were trademarked as "YOUM."

I expect that the new large TVs will be "flat" when OFF but "curved" when ON, so that the viewer will perceive all parts of an image at the same distance from his/her eyes, for a better viewing experience.

8.6 BJT

8.6.1 Introduction

Along with the diode, the *transistor* is also a semiconductor device. Transistors were born relatively few years ago, and have replaced dated *valves* (Figure 8.38) in most applications.

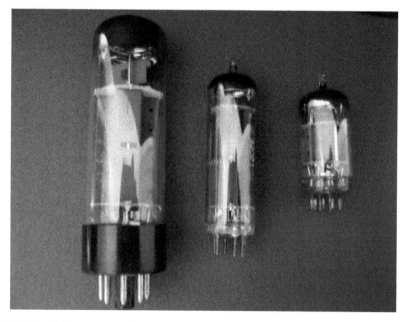

FIGURE 8.38   Valves. Once they performed the tasks assigned today to transistors.

FIGURE 8.39   Some types of BJTs.

There are different types of transistors, such as the *HBT* (Heterojunction Bipolar Transistor), *Schottky transistor, avalanche transistor, IGBT* (Insulated-Gate Bipolar Transistor), *photo transistor,* and so on.

We will deal here with the *BJT (Bipolar Junction Transistor)* and the *MOSFET (Metal Oxide Semiconductor Field Effect Transistor,* Section 8.7), as they account for the vast majority of silicon devices produced today.

BJT is one of the electronic devices that influenced technology the most and, as a consequence, the history of the twentieth century. Since its invention, in 1947 at Bell Laboratories, there have been many improved versions that brought the transistor to a higher efficiency compared to the prototype. In Figure 8.39a–d, we illustrate the main types in use today.

Curiosities

We owe the invention of the transistor to Shockley (William Bradford Shockley, English physicist and inventor who worked in the United States, 1910–1989), the creative visionary, Bardeen (John Bardeen, American physicist and electric engineer, 1908–1991), the excellent theoretician, and Brattain (Walter Houser Brattain, American physicist, 1902–1987), the outstanding experimentalist, who demonstrated it could work on December 23, 1947. The three researchers were granted the Nobel Prize for it in 1956. The transistor was commercialized on May 10, 1954 by Texas Instruments.

William Shockley founded, in 1955, "Shockley Semiconductor" in Palo Alto, California. The company is heralded as the first Silicon Valley business, but it was never profitable and within only a few years it was sold. Moore (Gordon Earle Moore, Ph.D. in chemistry and physics, the author of the famous Moore's Law, 1929–) and Noyce (Robert Norton Noyce, physicist, inventor, and entrepreneur, nicknamed "the Mayor of Silicon Valley," 1927–1990) left Shockley's company to co-found Intel Corporation.

Shockley has taken out more than 50 U.S. patents for his inventions.

John Bardeen won the Nobel Prize in physics twice, for the invention of the transistor and for the "BCS" theory (about conventional superconductivity) applied in Magnetic Resonance Imaging (MRI).

Another of Walter Brattain's chief contributions to solid state physics was the discovery of the photo-effect at the free surface of a semiconductor.

Nick Holonyak, inventor of the LED, was John Bardeen's first student, and later his friend.

It is fairly unknown that another form of transistor, called the *transistron*, was developed in the same years at Westinghouse. It was invented by Mataré (Herbert Franz Mataré, a German physicist, 1912–2011) and Welker (Heinrich Johann Welker, a German theoretical and applied physicist, 1912–1981) with the patent US2673948, "Crystal device for controlling electric currents by means of a solid semiconductor," dated August 13, 1948. The known transistor by Shockley, Bardeen, and Brattain was registered with the patent US2502488 "Semiconductor Amplifier," dated September 24, 1948.

8.6.2 Working Principles

A BJT has three terminals for connection to an external circuit. The letter "B" in BJT stands for *bipolar*, that is, the device is based on two types of carrier polarity (electrons and holes); the letter "J" stands for *junction* meaning that it is formed by three alternate doping regions, which is why we have *npn*– or *pnp*–type BJTs (Figures 8.40 and 8.41).

The three alternate doping regions are called the *emitter* (E), *base* (B), and *collector* region (C). In the same way the terminals used to connect the BJT with other devices are called *emitter terminal, base terminal,* and *collector terminal,* respectively. For reasons that will become clearer later, the construction of a transistor is achieved normally with a small doping value in the base region compared to the doping values of the emitter and the collector.

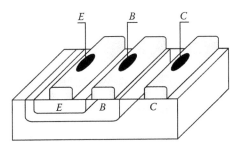

FIGURE 8.40   Scheme of a BJT.

The *collector* owes its name to the fact that it "collects" the carriers supplied by the *emitter*, which actually "emits."

The terms *emitter* and *collector* are due to the first vacuum tubes (the historical *thermionic valve, tube,* or *valve*), while the term *base* derives from the first transistor with a *point contact* for which the central region worked as a real mechanical base.

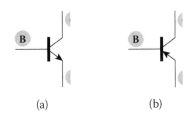

(a)   (b)

FIGURE 8.41   Symbols of (a) *npn*- and (b) *pnp*-type BJT.

8.6.3 Forward-Active Mode

We shall analyze here the flow of the *npn*-type BJT currents but what we will say is obviously valid also for the *pnp*-type, just inverting the biasing signs and the types of carriers.

For reasons of simplicity, from now on we will schematize the BJT with the three regions next to each other along a horizontal axis as in Figure 8.42, although actually they are arranged as in Figure 8.40.

In order for the electrons, majority carriers in the emitter region, to be able to pass into the base region, we need to forward bias the emitter-base junction. In fact, as it was for the diode, when the forward bias V_{EE} is greater than the threshold voltage V_γ, we experience a great reduction of the depletion region, and a so-called *injection* of electrons from the emitter to the base and an injection of holes from the base to the emitter. Once the electrons, majority carriers in the emitter region, cross the junction, they become minority carriers in the base region (where holes are the majority carriers). To conclude the path and bring the electrons in the collector region, we need a reverse bias with a V_{CC} voltage across the base-collector junction. In fact, the reverse bias is necessary to let the minority carriers (electrons in the base) flow.

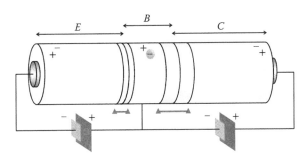

FIGURE 8.42   *npn*-type BJT with *E-B* forward biased and *B-C* reverse biased.

Given an *npn*-type BJT, when the emitter-base junction is forward biased, and the base-collector junction is reverse biased, the BJT is said to work in *forward-active mode*.

For the reasons already discussed in Section 6.1.4 of Chapter 6, it is useful to set a terminal to ground, as a reference. In the case we are now examining, we chose the base terminal for this purpose (but we could also choose the emitter or collector terminals as well).

Observation

Referring to Section 2.11 of Chapter 2, the voltage across the emitter-base junction should have been indicated with the notation V_{EB} and that across the base-collector with V_{BC}. These notations simply mean a constant voltage across two points. On the other hand, notations with the same letter doubled in subscript (V_{EE} and V_{CC}) are adopted to indicate that the specific voltage value is due to a constant voltage source such as a battery. Furthermore, seeing as a battery terminal is always grounded, the second letter of the subscript is not necessary and the first is doubled.

As said before, the base region is lightly doped, less than the emitter and the collector. This is to ensure that the number of electrons that are coming from the emitter and must reach the collector is not drastically reduced due to recombination with the holes, majority carriers in the base. For the same reason the base region can also be made "thin," in such a way that the electrons crossing it have a lower probability of recombination with the holes, as they have a shorter path to cover.

The percentage of electrons that are recombined in the base, deliberately very low, must necessarily be replaced. In fact, assimilating the BJT to a *node* (Section 6.1.4 in Chapter 6), according to KCL (Section 6.2.1), the results are $I_E = I_B + I_C$. But the current is proportional to the number n of charges, and the number of electrons that flow in the emitter region is different from the number in the collector region. So, necessarily, additional charges must flow in the base region, and these charges are furnished by the batteries.

The reverse bias V_{CC} across the base-collector junction produces a current of minority carriers (thermally generated), which corresponds to a reverse saturation current I_s, as seen in the case of the diode. However, the value of V_{CC}, at least within certain limits, does not determine a meaningful variation of I_s.

We can "visualize" such a situation if we imagine a waterfall in a river: the increase of the waterfall height (that corresponds to the increase of V_{CC}) does not determine an increase in the quantity of water (that corresponds to the value of I_s).

Considering electrons as carriers (this is also symmetrically valid

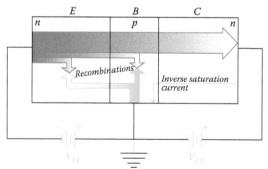

FIGURE 8.43 Currents in an *npn*-type BJT. In red: electron currents; in blue: hole currents.

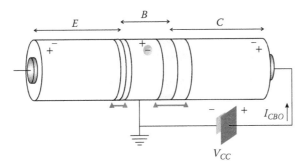

FIGURE 8.44 *B-C* reverse biased and *E-B* left open.

for holes), the reverse saturation current flows from the base to the collector, exactly the same path as the electrons that come from the emitter. The total current I_C that reaches the collector (still considering only electrons for simplicity reasons) is therefore the sum of two components: the part of the electrons injected in the base from the emitter, and the part of electrons thermally generated just within the base (Figure 8.43).

The amount of electrons coming from the emitter that reach the collector is αI_E, in which α is slightly less than the unit, to take into account that a (very low) percentage of electrons recombined in the base. Typical values are $0.95 < \alpha < 0.995$. But how can we evaluate only the current due to minority carriers? Simply by avoiding bias across the emitter-base junction (Figure 8.44).

In this case, we do not have carriers coming from the emitter and the only current that reaches the collector is that of the minority carrier electrons thermally generated. This current is called I_{CBO}, which stands for current between collector "C" and base "B" with the emitter terminal open-circuited ("O").

Now we can write current I_C as the sum of the two terms:

$$I_C = \alpha I_E + I_{CBO}$$

In the case of the diode, the reverse saturation current doubles in value for every increase by 10°C (Section 8.5.1.2). The same happens with I_{CBO} so that great temperature variations can cause problems in the functioning of a BJT. But for normal conditions I_{CBO} is negligible, so we assume

$$I_C \cong \alpha I_E$$

where parameter α is termed the *common-base current gain*.

In the previous equation we can substitute I_E as the sum of currents ($I_E = I_B + I_C$) obtaining

$$I_C = \alpha\left(I_B + I_C\right) + I_{CBO}$$

from which

$$I_C = \frac{\alpha}{1-\alpha} I_B + \frac{1}{1-\alpha} I_{CBO}$$

and defining

$$\beta = \frac{\alpha}{1-\alpha}$$

we get

$$I_C = \beta I_B + (\beta+1)I_{CBO}$$

It can be useful, with an obvious physical meaning, to define the parameter

$$I_{CEO} = (\beta+1)I_{CBO}$$

so the current I_C can be written as

$$I_C = \beta I_B + I_{CEO}$$

Considering, as normally happens, I_{CEO} is negligible,

$$I_C \cong \beta I_B$$

so β represents approximately the ratio of the DC collector current to the DC base current in forward-active mode, and it is termed the *common-emitter current gain*, often simply referred to as *current gain* or *transistor gain* even if this is not exactly true.

With α being near the unit value, β is consequently about one hundred (typical values $20 < \beta < 200$). But β links the collector current I_C to the base current I_B, which means that if we consider the collector terminal as the output and that of the base as the input of an electric signal, we will have a small current I_B, which controls another current I_C, much greater in amplitude. We have, therefore, what we call *signal amplification*.

Thus, we can say that BJT, together with biasing batteries, can supply a greater amplitude output current controlled by a lower amplitude input current.

The consequence is that a BJT in forward-active mode behaves as a controlled source.

Observations

"Amplifying" means creating an output signal with a greater amplitude than the input signal. We can therefore have an output signal greater in energy content than the input signal. It is crucial to observe that this is possible only because of the presence of batteries, which supply the necessary energy and not the BJT, which is a passive element.

Despite technological progress, to this day we do not have sufficient technological control to define a replicable β value. This is the reason why the data sheets of BJTs supply not a single value of β but a minimum value β_{min}, and a maximum one β_{max}.

8.6.4 Cutoff Mode

The emitter-base junction is not forward biased if $V_{BE} \leq V_\gamma \cong 0.7V$. In particular both junctions are reverse-biased (or no biased at all), both I_E and I_C currents are in the order of the reverse saturation current and, for the most part of applications, can be neglected.

In such conditions, the BJT behaves approximately as an open-circuit (Section 6.1.1 in Chapter 6).

Given an *npn*–type BJT, when the emitter-base junction is not forward biased, namely reverse biased or no biased at all, and the base-collector junction is reverse biased, the BJT is said to work in *cutoff mode*.

8.6.5 Saturation Mode

With both junctions forward biased, the result is a carrier injection from both emitter and collector into the base. The I_C current can be of an apprecible value, even if a small voltage value exists across the base-collector junction. In this mode the result is $V_{CE} = V_{CE(SAT)} \leq 0.2\,V$.

In such conditions, the BJT behaves approximately as a short-circuit (Section 6.1.1 in Chapter 6).

Given an *npn*–type BJT, when the emitter-base junction is forward biased, and the base-collector junction is forward biased too, the BJT is said to work in *saturation mode*.

8.6.6 Modes and Models

The two junctions of BJT, base-emitter and base-collector, can be either forward or reverse biased. The BJT operates in different modes depending on the junction biases (Table 8.5).

Essentially, a BJT is utilized in active mode for analog applications, especially for amplifier networks. On the other hand, cutoff and saturation modes are utilized for digital and switching applications.

The *inverse* or *reverse-active mode* is quite unsuitable, and very rarely used, because of its smaller current gain as compared with the *forward-active mode*.

In general, a BJT can be defined as a three-terminal, nonlinear element that can furnish amplification if used in the correct mode in a network. It operates with an input signal, which is applied to two of its terminals, the *input port*, and supplies an output signal at two other terminals, the *output port*. This necessarily implies that one of the three terminals of the transistor will have to be in common between the input and output port. The very choice of the common terminal will give the name to the circuit configuration adopted (Figure 8.45).

TABLE 8.5

BJT's Modes of Operation

Mode	Base-Emitter	Base-Collector
Cutoff	Reverse	Reverse
Active	Forward	Reverse
Inverse	Reverse	Forward
Saturation	Forward	Forward

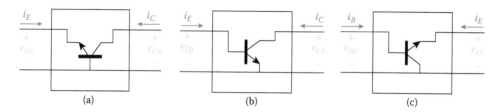

FIGURE 8.45 (a) Common-base, (b) common-emitter, and (c) common-collector configuration.

The *common-base* configuration (C.B.) is the one where the base terminal is in common between the input and output meshes, considering the AC signal analysis. For the C.B. configuration, the input current is the emitter and the output is the collector, so the main parameter is $\alpha \cong I_C/I_E$, which has a value near the unit.

The *common-emitter* configuration (C.E.) is the one for which the emitter terminal is in common between the input and output meshes, considering the AC signal analysis. For the C.E. configuration, the main parameter is parameter is $\beta \cong I_C/I_B$, typically between about ten and some hundreds.

The *common-collector* configuration (C.C.) is the one for which the collector terminal is in common between the input and output meshes, considering the AC signal analysis. For the C.C. configuration, the main parameter is the ratio $\beta/\alpha \cong I_E / I_B$, and given that $\alpha \cong 1$ we have $\beta/\alpha \cong \beta$.

Observation

The name of the configuration should derive from the common terminal established by a dynamic AC analysis (in the presence of electric signals) and not from a static DC analysis (i.e., relative to constant electric quantities, such as those supplied from batteries).

Actually, even if it is not correct, the two conditions of static and dynamic are used without distinction practically all the time to establish the configuration type of the transistor (C.E., C.B., C.C.).

For a two-terminal device, such as the diode, we have only one voltage and only one current, so a 2D Cartesian graph, with the voltage on one axis and the current on the other, is enough to describe its characteristics. For a three-terminal device such as the BJT this is not possible, because there are three voltages $\left(v_{BE}, v_{BC}, v_{CB}\right)$ and three currents (i_B, i_E, i_c) to take into account. To describe graphically the sum of characteristics of a three-terminal device we would need a 6D graph, for which each axis would represent an electric quantity. It is obvious that this is impossible, so what we do is draw graphs including only the quantities of greater interest.

Input characteristics are the curves that establish the link between voltage and current of the input circuit, parameterized by an electric output quantity. Whereas *output characteristics* are curves that establish the link between voltage and current of the output circuit, parameterized by an electric input quantity.

The characteristic curves of the BJT are slightly different for each configuration, C.B., C.E., or C.C., in which it is used.

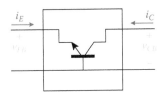

FIGURE 8.46   BJT in C.B. configuration.

8.6.7 I–V Graph for C.B. Configuration

8.6.7.1 Input Characteristics (C.B.)

Let us consider the *base* terminal as a reference for the emitter and collector terminals, that is, the input and output voltages have the base as their reference: V_{EB}, V_{CB} (Figure 8.46).

We then have a C.B. configuration type.

Observation

For DC analysis, the voltage V_{EB} would be V_{EE} (see Figure 8.43) and V_{CB} would be V_{CC}. Actually, as we will see more in detail further on, it makes more sense to define a configuration considering the AC analysis.

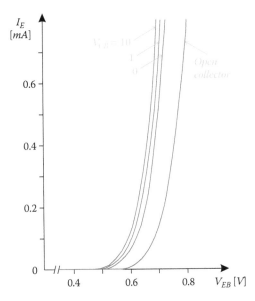

FIGURE 8.47   Typical input curves of CB configuration.

Seeing as the input voltage V_{EB} is across the terminals of a forward-biased junction (emitter-base), it is reasonable to expect a graph of the input current I_E, with respect to input voltage V_{EB}, analogous to the one seen in the case of the diode in forward-biased conditions.

But, unlike the diode, here the current I_E is not just a function only of V_{EB}, but also of the output parameter V_{CB}. This is because the increase of V_{CB} implies an enlargement of the base-collector depletion region and, as a consequence, a reduction of the base region useful for the recombination of charges. To make the concept clearer consider that the physical width of the base region is obviously fixed by the construction parameters of the BJT, but the part of the base region occupied by free charges (potentially useful for recombination with free charges of the opposite sign) depends on how much of it remains occupied exclusively by fixed charges (without mobile charges, missing due to reverse bias).

Summing up, the input current I_E is a function of the input voltage V_{EB}, but parameterized to the value of the output voltage V_{CB}.

This is shown graphically in Figure 8.47.

The reverse-biasing V_{CB} affects the direct current I_E, but not the reverse saturation current I_S, as discussed in Section 8.6.2.

The voltage V_{CB} does not affect the input curves very much. In fact, even for great variations of V_{CB} the current does not vary significantly. This is the reason why the effect of output voltage is often ignored in the characteristic input curves.

Note once again that this discussion, and as a consequence the characteristic curves shown, concerns an *npn*-type BJT. For a *pnp*-type BJT the discussion is similar, as long as we invert all the signs of the electric values.

8.6.7.2 Output Characteristics (C.B.)

As already seen (Section 8.6.3),

$$I_C = \alpha I_E + I_{CBO}$$

so when $I_E = 0$ we get for collector current

$$I_C\big|_{I_E=0} = I_{CBO}$$

As I_E increases, I_C also increases according to the parameter α. However, a greater V_{CB} value corresponds to a greater α value and therefore I_C value (Figure 8.48).

Given the typical values that parameter α can take on and the typical order of magnitude for I_{CBO}, basically we always have $I_C < I_E$, so the C.B. configuration does not have a current amplification effect.

Note how the slope (or gradient) of each characteristic output curve represents the output impedance of the BJT for a specific value of the input current I_E.

For negative V_{CB} values, such that $V_{CB} < -V_\gamma$ (V_γ is the threshold voltage), the base-collector junction is forward biased and therefore cannot be crossed by minority carriers (the electrons in the base coming from the emitter). As a consequence, no current I_C is obtained.

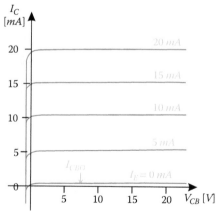

FIGURE 8.48 Typical output characteristics of BJT in C.B. configuration.

The characteristic output curves start from a basically null I_C value, and grow rapidly for small increases of V_{CB}. For this aim V_{CB} must not forward-bias the base-collector junction, so $V_{CB} < -V_\gamma$. Given that every curve is parametrized by a fixed value of I_E, this means that with I_C we also have the increase of the ratio I_C/I_E, so the α value is not constant but starts from zero and increases with V_{CB}. The reason for this behavior is due to charges that from emitter do not reach the collector until a reverse bias does occur across the base-collector junction. If we continue increasing V_{CB} we reach a point (with $V_{CB} \cong 0$) beyond which the current I_C does not grow any more but tends to stabilize, approximating the value of I_E (obviously it cannot be any different given that, at most, when $I_B = 0$ it can be equal to I_E).

In the graph of the output characteristic curves of the BJT we highlight three regions:

→ *Saturation region* (yellow in the graph): Both the emitter-base junction and the base-collector junction are forward biased, current I_C rapidly increases for small increases of the voltage V_{CB}, and parameter α grows from zero to its maximum value.

→ *Active region*: The emitter-base junction is forward biased and the base-collector junction is reverse biased, current I_C tends to be stable for V_{CB} voltage increases, and parameter α is practically stable at its maximum value.

→ *Cutoff region* (green in the graph): Both the emitter-base and the base-collector junctions are reverse biased, and current I_C is practically null to any V_{CB} value, so the transistor is "turned off."

8.6.8 I–V Graph for C.E. Configuration

8.6.8.1 Input Characteristics (C.E.)

Let us consider the *emitter* terminal as a reference for the base and collector terminals, that is, the input and output voltages have the emitter as their reference: V_{BE}, V_{CE} (Figure 8.49).

We then have a C.E. configuration type.

The input characteristic curves for BJT in C.E. configuration are similar to those for the C.B. configuration (Figure 8.47). Basically we have curves regarding a forward-biased junction (in this case the base-emitter junction) parameterized to an electric output quantity. More specifically we will have an I_B versus V_{BE} graph with curves parameterized by V_{CE}.

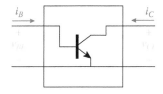

FIGURE 8.49   BJT in C.E. configuration.

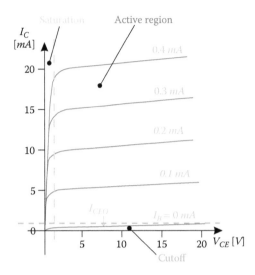

FIGURE 8.50   Typical output characteristics of the BJT in C.E. configuration.

8.6.8.2 Output Characteristics (C.E.)

The output characteristics of a BJT in C.E. configuration show the (output) I_C current as a function of the (output) voltage V_{CE} parameterized to (input) I_B current, as in Figure 8.50.

In the first part the curves are quite linear, with a low resistive value (when V_{CE} is low, typically $\leq 0.2\,V$), after that a short transition curve, and then a rather high resistive linear behavior. We have a transition from an initial situation of low resistance to a final one of high resistance.

Curiosity

The term "transistor" comes from the very fact the device conducts a sort of transfer of resistance values, from a low to a high value, and so is basically a transfer resistor.

As already seen in the C.B. configuration, we distinguish *saturation, cutoff,* and *active* regions.

→ *Saturation region*: On increasing the base current I_B we do not get a proportional increase of the collector current I_C. The value of the collector-emitter voltage V_{CE} is close to short-circuit (normally $V_{CE} \leq 0.2\,V$). To be in the saturation region, both the emitter-base and base-collector junctions are forward biased. When the emitter-base junction is forward biased we consider $V_{BE} \cong 0.7\,V$. Then, the collector-base junction is reverse biased for any $V_{CE} > 0.7\,V$ (so $V_{CE} - V_{BE} = V_{CB} > 0$) and consequently the saturation region lies on the left of $V_{CE} = 0.7\,V$. Commonly, the saturation value considered for the voltage across the collector-emitter terminals is $V_{CE(Sat)} = 0.2\,V$.

→ *Cutoff region*: The base current I_B and collector current I_C are close to zero and in any case negligible. The value of the collector-emitter voltage V_{CB} is high enough to be assimilated to an open circuit. To be in the cutoff region, both the emitter-base and base-collector junctions are reverse biased.

→ *Active region*: The transistor behavior is similar to that of a variable resistor controlled by input current. In order for it to be in the active region, the emitter-base junction is forward biased, while the base-collector junction is reverse biased.

As discussed in Section 8.6.6, the fundamental parameter of this configuration is $\beta \cong I_C/I_B$. This parameter is found by establishing a specific point Q in the graph of

the output characteristics, reading the relative values of I_C and I_B, and calculating their ratio.

As we can easily verify, β is not constant but varies according to where the point Q is chosen. Only in the active region, where BJT will be used for our purposes, the parameter β is practically constant.

We saw in the output characteristics of the C.B. configuration that in the active region the curves of I_C are substantially parallel in the x-axis, so the collector current is practically constant to V_{CE} variations. BJT behaviour in the C.E. configuration is slightly different and, as we can see in the previous figure, the curves in the active region have a certain (though slight) slope. This is due to the so-called *Early effect* (discussed in Section 8.6.8.3).

Observation

In the *active* region the BJT characteristics, in mathematical jargon, *saturate* (i.e., do not grow significantly any more). Do not be fooled by this because, from the point of view of electronics, *saturation* is the condition where we have a great passage of current for a very low applied voltage. Therefore, in electronics, the saturation region is not that of almost parallel curves to the x-axis (mathematics) but that for which I_C changes rapidly even when V_{CE} is low.

8.6.8.3 Early Effect

We have seen how in *p-n* junctions the width of the depletion region depends inversely on doping and, specifically, the higher the doping the smaller the depletion region (Section 8.4.1). However, the width of the depletion region could increase by reverse-biasing the junction (Section 8.5.1.2). This phenomenon has a particular effect when the base-collector junction is reverse biased. More specifically, we want to evaluate what happens only in the base region. Its physical width is obviously fixed and we can think of it as the sum of a depletion region (i.e., without any mobile carriers) and a non-depletion region (with mobile carriers). When a BJT works in normal operative conditions, in the non-depleted region of the base we can have recombinations of carriers coming from the emitter, so here it is established how many carriers can pass, without recombinations, and

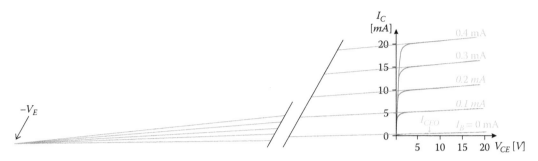

FIGURE 8.51   The ideal extension of the characteristic curves leads to a unique intersection with a negative voltage value called "Early voltage," -V_E.

get to the collector (basically parameter α depends on this). Evidently a narrow non-depleted region implies a reduced number of recombined carriers (therefore a higher I_C current for a constant I_E). Conversely, a larger non-depleted region implies a greater number of recombined carriers (therefore a smaller I_C current for a constant I_E).

This phenomenon is known as the *Early effect* (James M. Early, American engineer, 1922–2004), and it is the reason why the characteristic curves of the BJT in the C.E. configuration are in the active zone with a certain slope and not parallel to the *x*-axis (Figure 8.51).

The ideal extensions of the rectilinear parts of the curves in the active region, in the direction of negative voltages, all meet in one point called the *Early voltage* V_E, whose typical values are in the range $-50 - -100\,V$.

8.6.9 I–V Graph for C.C. Configuration

8.6.9.1 Input Characteristics (C.C.)

Let us consider the *collector* terminal as a reference for the base and emitter terminals, that is, the input and output voltages have the collector as their reference: V_{BC} (Figure 8.52).

We then have a C.C. configuration type.

The input characteristics for the BJT in C.E. configuration are similar to what we have already seen both for the C.B. configuration (Figure 8.47) and for the C.E. configuration. Basically, they are curves regarding a forward-biased junction (in this case the base-collector one) parameterized by an electric output quantity. More specifically we will have I_B versus V_{BC} curves, parameterized by V_{EC}.

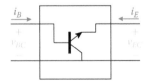

FIGURE 8.52   BJT in C.C. configuration

8.6.9.2 Output Characteristics (C.C.)

The output curves for the BJT in C.C. configuration are similar to those of the C.E. configuration (Figure 8.50). More specifically we have I_E versus V_{EC} curves, parameterized by I_B.

8.7 MOSFET

8.7.1 Field Effect Transistors

The BJT analyzed so far are only one possible transistor type. In fact, the characteristic of *transfer resistance* exists also because of other principles differing from those on which the BJT is based.

Considering the most interesting transistors with the biggest number of applications we can list the following:

→ The JFET (Junction Field-Effect Transistor) uses an reversely biased p-n junction to separate a gate terminal from the rest of the device.
→ The HEMT (High Electron Mobility Transistor) is based on energy-gap modulation.
→ the MODFET (Modulation-Doped Field Effect Transistor) uses a so-called quantum hole for electron confinement.
→ The MOSFET (Metal-Oxide-Semiconductor Field-Effect Transistor) uses a pseudo-capacitive structure to modulate the width of a conductive path to allow flowing charges.

There are others in addition to those on this list. A certain number among the various types can be classified under the acronym FET (Field Effect Transistor), that is, devices that owe their transistor characteristic to an *electric field effect*.

The field-effect transistor basically relies on an electric field to control the shape and hence the conductivity of a channel of one type of charge carrier in a semiconductor material. FETs are unipolar transistors with respect to the dual-carrier-type operation of BJTs.

One of the devices belonging in the FET family is the MOSFET, which has very interesting characteristics and a huge number of applications.

The MOSFET is characterized by a relatively low production cost, really compact dimensions that allow it to be integrated easily, and interesting versatility, applicable to both analog and digital electronics (Figure 8.53).

FIGURE 8.53   Typical commercial MOSFETs.

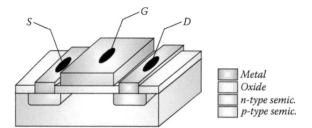

FIGURE 8.54 MOSFET scheme. The three terminals are Source, Gate, and Drain.

8.7.2 Working Principles

The MOSFET can be made on a main structure of a *n*–type doped semiconductor or a *p*–type doped semiconductor. Let us analyze the structure on a *p*–type doped body as it is the one with the greatest number of applications. What we will say can be also applied, with the same principles, for the structure on an *n*–type doped body.

The MOSFET structure is made up of a lightly doped *p*–type substrate (often referred to as the *bulk*) where two heavily doped *n*–type regions, named *source* S and *drain D*, are implanted. On top of these regions conductive layers provide terminal contacts.

A thin layer of insulating material is grown over the region between source and drain, and this layer is covered by a conductive material, forming the third terminal contact named *gate G*. In this way a central structure of metal (M) on oxide (O) on semiconductor (S) is created, which is so fundamental that it became part of the name of the device (MOS: Metal Oxide Semiconductor) (Figure 8.54).

The MOS structure is what we call a "similar-capacitor." We know that a real capacitor is constructed of two metallic layers separated by an insulator. However, in this case, a doped semiconductor is substituted for one of the two metallic layers. In the case of a real capacitor, applying a voltage across its terminals we have an accumulation of charges that are arranged on the *surfaces* of the two metallic plates. In the case of our similar-capacitor, on applying a voltage across its terminals, the charges on the side of the doped semiconductor are not arranged along a surface but on a *volume*, so a 3D volume rather than a 2D plane. The width of this volume (its "third dimension") increases as the doping decreases; it is inversely proportional to it.

Observation

To better realize how an inverse proportionality is possible, we can think about the possibility of a semiconductor with so much doping that the number of atoms of the dopant can be compared with the number of atoms of the doped material. In this case, we have a *degenerate semiconductor*, that is, a material that cannot be called a *doped semiconductor* but has become a compound. It has similar properties to those of a real metal, so if it is part of a capacitor, its charges will be disposed on such a thin volume that it will be assumed as a plane.

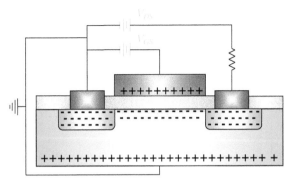

FIGURE 8.55   Formation of the "inversion layer," or "channel," below the oxide.

The MOS type transistor is based on the modulation of the charge concentrations in the semiconductor region below the layer of oxide of the gate, thanks to a voltage applied to the gate terminal above the oxide.

In the case of a p-type semiconductor (with hole density equal to N_A) a positive voltage V_G applied to the gate reduces the concentration of holes in the semiconductor under the oxide, and increases the concentration of electrons. For a sufficiently high value of V_G, the concentration of negative charge carriers below the gate exceeds that of positive charges, and the so-called *inversion layer* is formed. This is a *channel* in the p-type doped semiconductor, but made mainly of electrons (Figure 8.55).

The voltage V_G must be applied across the terminals of the similar-capacitor, so between the metal of the gate and the substrate, that is, the p-doped body of the whole structure. For convenience, the body is usually grounded, and to avoid voltage spikes, we also ground the source terminal. So, we call the voltage across the gate-substrate V_{GS} in reference to the ground.

For the BJT we have seen how the current flows from the emitter to the collector terminals, being modulated (controlled) via the base terminal. In a similar way in the MOSFET, the current must be able to flow from source to drain terminals, being modulated via the middle gate terminal by the gate voltage. Let us analyze how this is possible. To have a current, it is obviously necessary to apply a voltage V_{DS} across source and drain. However this is not enough. The source region charges that can constitute carriers for a current are electrons. In order for them to reach the drain region they must cross underneath the central region through the gate channel. But the gate region is constituted by free holes in excess, therefore the electrons coming from the source do not reach the drain because they are totally recombined in the gate channel. The only way for electrons to reach the drain is by inverting the polarity of carriers in the gate, in a way to avoid the recombination of the electrons. This is exactly the reason for the gate voltage V_{GS}. In fact, it is possible to create the aforementioned layer of inversion that constitutes an actual 3D channel of transit for electrons through the similar-capacitor MOS structure. It is clear that the channel dimension will increase as the gate voltage increases, so for a constant voltage V_{DS}, the bigger the gate voltage, the bigger the flowing current.

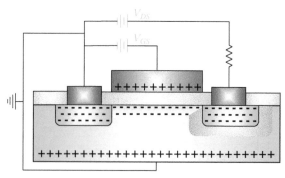

FIGURE 8.56   The channel does not have a constant depth but gets narrow near the drain region.

However, the current is linearly proportional with the voltage up to a limit, above which the current saturates, remaining more or less constant. To understand why, we can consider typical values for the voltages applied. Let us consider for example $V_{GS} = 3V$ and $V_{DS} = 5V$, therefore, $V_{DG} = V_{DS} - V_{GS} = 5 - 3 = 2V$. So V_{GS} forward-biases the gate-source junction, and V_{DG} reverse-biases the drain-gate junction. As a consequence, we have a negligible depletion region at the gate-source junction, and a significant depletion region at the drain-gate junction. The latter opposes the formation of the channel, narrowing it near the drain region. Therefore, the channel will no longer have constant depth below the gate oxide, as schematized in Figure 8.56.

The depletion region of the drain-gate junction would, in fact, tend to close the channel, but the closure condition does not take place. In fact, V_{DG} falls across the junction itself rather than the remaining an ohmic part between the gate and drain terminals (see for example Figure 8.54). But, if the channel closed (a condition called *pinch-off*) the current would cease to flow and, as a consequence, the voltage across the junction (responsible for the charge movement) would be reduced to zero, and the battery voltage distributed elsewhere. But if it was reduced to zero the reverse bias of the junction would also drop to zero, so there would be no reason for the channel to close and therefore it would open. On opening again the current would flow again and we would be in the same conditions as before, which would tend to close the channel again. This situation tends to a stationary state in which the channel is neither completely open nor completely closed.

8.7.3 Ohmic Condition

What we previously discussed at a qualitative level can be now analytically transferred at a quantitative level. Applying a voltage V_{GS} at the gate, higher than the *threshold voltage* V_{th} (it is called threshold in the case of MOSFET but it is equivalent to the V_γ already discussed for the diode), we have an accumulation of volumetric charge Q_n along the channel with a uniform voltage. The MOS region has the geometry of a plane capacitor, so indicating the oxide thickness as d and the dielectric constant as ε, we have

$$Q_n = \frac{\varepsilon}{d}\left(v_{GS} - V_{th}\right)$$

Now applying a voltage V_{DS} across drain and source terminals, the voltage along the channel formed previously will not be uniform any more, but will depend on the position x between source ($x = x_s = 0$) and drain ($x = x_D$).

The equation that expresses the charge Q_n, now a function of x, becomes

$$Q_n(x) = \frac{\varepsilon}{d}\left[v_{GS} - V_{th} - v(x)\right] = C_S\left[v_{GS} - V_{th} - v(x)\right]$$

where C_S is the capacitance per surface unit.

Associated to voltage $V(x)$ we have an electric field E_x, through which it is useful to express the value of the superficial density of current J_x:

$$J_x = \mu_n Q_n(x) E_x$$

As the channel is homogeneous along the direction y, and orthogonal to x and to the plane of the previous figure, by indicating with W the channel width, the total current can be written as

$$i_{DS} = -\mu_n Q_n(x) W E_x$$

in which, conventionally, i_{DS} and E_x have opposite directions, hence the reason for the negative sign.

We must note how the current i_{DS} does not depend on the coordinate x, given that the only direction in which the charges can flow is between source and drain (towards the gate the charges cannot flow given the presence of oxide and towards the substrate we make sure that only one, practically negligible, current can run through).

Making the necessary substitutions, the previous equation can be written as

$$i_{DS} = -\mu_n C_S\left[v_{GS} - V_{th} - v(x)\right] W E_x$$

and integrating along the channel

$$\int_0^{x_D} i_{DS} dx = \mu_n C_S W \int_0^{x_D}\left[v_{GS} - V_{th} - v(x)\right]\frac{dv}{dx}dx$$

But i_{DS} is constant and $v(x_D) = v_{DS}$, so

$$i_{DS}L = \mu_n C_S W\left(v_{GS} - V_{th} - \frac{v_{DS}}{2}\right)v_{DS}$$

$$i_{DS} = \mu_n C_S \frac{W}{L}\left(v_{GS} - V_{th} - \frac{v_{DS}}{2}\right)v_{DS} = K_n\left[2(v_{GS} - V_{th}) - v_{DS}\right]$$
$$v_{DS} = K_n\left[2(v_{GS} - V_{th})v_{DS} - v_{DS}^2\right]$$

with

$$K_n = \frac{\mu_n C_S}{2}\frac{W}{L}\left[\frac{A}{V^2}\right]$$

termed the *transconductance parameter*, a constant that depends on the process technology used to fabricate the MOSFET.

The current i_{DS} is therefore a function both of the geometric dimensions of the channel and the applied voltages. The relationship of i_{DS} with the voltages is not linear. In particular it is quadratic with v_{DS}.

This situation describes the so called *ohmic*, or *triode*, condition of the MOSFET.

Observation

Be aware that the transconductance parameter K_n, here defined as

$$K_n = \frac{\mu_n C_S}{2} \frac{W}{L} \left[\frac{A}{V^2} \right]$$

is sometimes considered to be $\mu_n C_S$ (W/L), which is double what is considered here.

8.7.4 Pinch-Off Condition

The density of the inversion charge vanishes at the drain when the voltage v_{DG} is higher than the threshold voltage of the drain-gate junction, that is, for $v_{DG} \geq V_{th}$. This borderline condition, called *pinch-off*, corresponds to having the channel in contact with the drain at only one point (as illustrated in Figure 8.57) precisely because $v_{DG} = V_{th}$. This condition imposes a null charge value $Q_n(x = x_D)$ at the contact point:

$$Q_n(x_D) = C_S[v_{GS} - V_{th} - v_{DS}] = 0$$

so

$$v_{DS} = v_{GS} - V_{th}$$

and the aforementioned equation for i_{DS} becomes

$$i_{DS} = K_n(v_{GS} - V_{th})^2$$

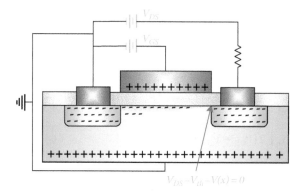

FIGURE 8.57 Pinch-off condition.

8.7.5 Saturation Condition

Beyond the pinch-off limit, the voltage v_{DS} further increases as the device enters the so-called *saturation region*. The current i_{DS} remains almost constant, increasing very slowly as voltage increases:

$$i_{DS} = K_n \left(v_{GS} - V_{th} \right)^2 \left(1 + \lambda v_{DS} \right)$$

where λ is the *channel length modulation coefficient*, due to a *channel length modulation effect*, similar to that of the *Early effect* already considered for the This determines the small amount of slope of the current versus voltage graph, as schematized in Figure 8.58. With the position of $\lambda \cong 0$ we neglect the modulation effect of the channel.

Observation

Advantages of MOSFET with respect to the BJT are the relative simplicity of technological structure, the fact that they are "driven" by practically null power (given that there is no gate current flow), and the high density of integration in *LSI (Large Scale of Integration)* and *VLSI (Very Large Scale of Integration)* scales.

8.7.6 I–V Graph

What we have analyzed so far can be summed up graphically through a MOS transistor output characteristics plot of i_{DS} versus v_{DS} for different values of i_{GS} (Figure 8.58).

We distinguish three working regions:

$$\rightarrow \text{Cutoff region} \begin{cases} condition: & v_{GS} < V_{th} \\ current: & i_{DS} = 0 \end{cases}$$

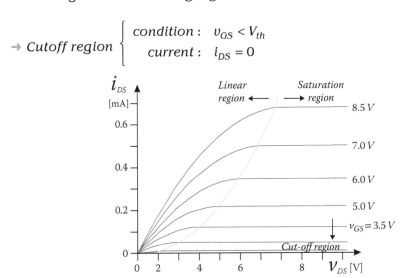

FIGURE 8.58   Example of MOSFET output characteristics i_{DS} vs. v_{DS} parametrized by v_{GS}.

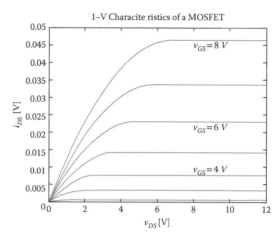

FIGURE 8.59   I–V characteristics of a MOSFET as simulated by a MATLAB® routine.

$$\rightarrow \text{Ohmic or triode, or linear region} \begin{cases} condition: \quad v_{GS} - V_{th} > v_{DS} > 0 \\ current: \quad i_{DS} = K_n\left[2(v_{GS} - V_{th})v_{DS} - v_{DS}^2\right] \end{cases}$$

$$\rightarrow \text{Saturation region} \begin{cases} condition: \quad 0 < v_{GS} - V_{th} < v_{DS} \\ current: \quad i_{DS} = K_n(v_{GS} - V_{th})^2(1 + \lambda v_{DS}) \end{cases}$$

The blue line in Figure 8.58 for which $v_{GS} - V_{th} = v_{DS}$ separates ohmic and saturation regions.

Observation

The λ parameter is very often ignored due to its low value, which depends on the channel length and is typically $10^{-3} \le \lambda \le 10^{-1}$ [1 / v].

The following is a MATLAB® routine to graph an example of the I–V characteristic of a MOSFET (Figure 8.59), neglecting the overall contribution of λ, and assuming $K_n = 10^{-3}[A/V^2]$, $V_{th} = 1.2[V]$:

```
% I-V characteristics of mosfet
Kn = 1e-3; Vth = 1.2;
vds = 0:0.5:12; vgs = 2:1:8;
lvds = length(vds); lvgs = length(vgs);
for i = 1:lvgs
  for j = 1:lvds
        if vgs(i) < Vth current(i,j) = 0;
        elseif (vgs(i)-Vth)>vds(j) current(i,j) = Kn*(2*(vgs(i)-
        Vth)*vds(j)-vds(j)^2);
        elseif (vgs(i)-Vth)< = vds(j) current(i,j) = Kn*(vgs(i)-Vth)^2;
     end
  end
end
```

```
plot (vds,current (1,:), 'r',  vds,current (2,:), 'r',  vds,current (3,:), 'r',
vds,current (4,:), 'r',  vds,current (5,:), 'r',  vds,current (6,:), 'r',
vds,current (7,:), 'r')
xlabel ('v ds [V]'); ylabel ('i ds [A]')
title ('I-V Characteristics of a MOSFET')
text (8, 0.009, 'Vgs = 4 V'); text (8, 0.022, 'Vgs = 6 V'); text (8, 0.045,
'Vgs = 8 V')
```

8.7.7 Subthreshold Current

As mentioned, the condition $v_{GS} < V_{th}$ corresponds to no current flowing in the channel, that is, $i_{DS} = 0$. However, it can happen that an undesirable leakage current (Section 8.3.4), named the *subthreshold current* I_{off}, can flow between the drain and the source. It can have a quite low value, of the order of tens of nA, so it can be neglected for many applications of the MOSFET. But, there are circuits for which even such a low value can be a problem. Let's consider, for instance, a cell-phone chip containing one hundred million MOSFETs. In the so-called *idle mode*, the standby current it consumes corresponds, as a first approximation, to the sum of the subtreshold currents of all the MOSFETs, something around $1\,A$. As a consequence, the battery will discharge very fast! The problem is even worse when it is necessary to scale down the MOSFET's dimensions to increase the level of miniaturization, because the leakage current can increase dramatically in sub-$100\,nm$ processes. This is one of the reasons why static power consumption is now one of the crucial parameters in designing electronic circuits.

8.7.8 Types of MOSFETs

The transistor described to this point is called an *n-channel enhancement* MOSFET or, equivalently, *NMOS enhancement FET*. This is because its working principle depends on the creation (and therefore *enhancement*) of an *n-type channel* thanks to a positive voltage applied across the gate-ground terminals.

Another possibility is to make a MOSFET with inverted doping, so on an *n-type* substrate that works thanks to the enhancement of a *p-type channel*. In this case we talk about *p-channel enhancement* MOSFETs or *PMOS enhancement FETs*.

There are also MOSFETs for which the channel is already formed by construction and for it the control voltage across gate and ground does not serve to *enhance* the channel but, on the contrary, to *deplete* it. Unlike the type we described before this type of transistor allows a drain current applying a voltage v_{DS} even if $v_{GS} = 0$. Now, applying a control voltage between gate and ground is useful to deplete (or even eliminate) the channel, so this transistor is called a *depletion* MOSFET. As far as the type of channel charges is concerned, there are *n-channel depletion* MOSFETs, or *NMOS depletion FETs*, and *p-channel depletion* MOSFETs, or *PMOS depletion FETs*.

The schematic representations of the four types of MOSFETs are shown in Figure 8.60. The channel is represented by a continuous line for depletion MOSFETs, and a dotted line for enhancement MOSFETs. The gap at the gate is for the oxide layer. The arrow points from p to n and is with the source terminal.

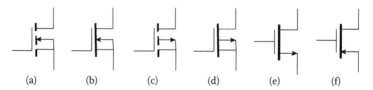

FIGURE 8.60 MOSFET: (a) *n*-channel enhancement mode, (b) *n*-channel depletion mode, (c) *p*-channel enhancement mode, (d) *p*-channel depletion mode, (e) *n*-channel enhancement mode no-bulk, (f) *p*-channel enhancement mode no-bulk.

The bulk (or body) can be either connected or not, and when it is connected to the source it is usually not represented, as in the last two symbols in Figure 8.60.

Observations

As previously mentioned, the mobility of electron and hole carriers in a silicon semiconductor are respectively $\mu_{n(Si)} = 0.14 \ m^2/Vsec$ and $\mu_{p(Si)} = 0.05 \ m^2/Vsec$ (Section 8.3.1). Proportionally a *p*–type MOSFET will conduct only around 35% of the current of a *n*–type MOSFET, which is why the NMOS is often preferred.

The drain current has been named i_{DS} until now, since the subscripts exactly specify that this current, flowing through the channel, goes from the drain to the source of an arbitrary MOSFET. But it can be referred simply as i_D too.

Curiosity

MOSFETs can be subject to damage from electrostatic discharge. So to prevent damage these transistors have to be shipped and stored in conductive foam. It is also advisable to not apply signals while the DC supply is off.

8.8 BJT VERSUS MOSFET

MOS technology offers higher layout densities and lower power requests as compared to BJT technology. But BJTs are preferred in applications for which higher speed and gains are necessary. In any case, the choice between a BJT or a MOSFET strictly depends on applications we want to realize. Here are some fundamental differences:

→ A BJT has an emitter, collector, and base, while a MOSFET has a gate, source, and drain.
→ A BJT is based on bipolar carriers (electrons *and* holes), while a MOSFET is based on unipolar charges (electrons *or* holes).
→ A BJT is a current-controlled device, while a MOSFET is a voltage-controlled device, so (theoretically) no power is requested to drive it.
→ The bulk mobility of a BJT is always better than the surface mobility of a MOSFET.
→ A BJT is generally less expensive than a MOSFET.
→ A BJT has a low input resistance, which can often be a problem, while a MOSFET has a very high input resistance, typically more than $10^7 \Omega$.

→ A BJT has an input current of the order of mA, while a MOSFET has a theoretically zero input current (practically a few pA) so it is not a source of DC offset errors, but the gate leakage current increases rapidly with temperature.

→ A MOSFET has a smaller size, lower power dissipation, and is easier to fabricate as compared to a BJT.

8.9 NUMBERING AND CODING SCHEMES

The identification of semiconductor devices is done through codes stamped on them. These codes respect normative standards that can vary according to the international association of reference. Here is a list of some of the most important associations:

→ *JEDEC* (*Joint Electron Device Engineering Council Solid State Technology Association*, www.jedec.org), created by the EIA (Electronic Industries Alliance)

→ *Pro Electron* (united to the "European Electronic Component Manufacturers Association" since 1983, www.eeca.org)

→ *JIS* (*Japanese Industrial Standards*, www.jisc.go.jp/eng)

There can also be other types of codes of identification depending on the manufacturer.

8.9.1 JEDEC

According to the JEDEC standards devices are identified by a prefix and a suffix. Prefixes:

→ 1N → diode
→ 2N → transistor
→ 4N-6N → optoelectronic device

The suffix identifies the single component and its technical features. We can simply have a stamped number (from two to four digits), or in the case of diodes, a code in colored lines, where a number corresponds to each color, as shown in Table 8.6.

TABLE 8.6
Standard JEDEC Color Code for
Semiconductor Devices

0	black	1	brown
2	red	3	orange
4	yellow	5	green
6	blue	7	purple
8	grey	9	white

With the color code, the cathode can be identified from the first colored line, which is larger compared to the others, or from the fact that the colored lines are grouped together on the side of the cathode.

As an example, the code "1N4001" identifies a rectifying diode for generic application with a maximum current of $1\,A$.

8.9.2 Pro Electron

Pro Electron identifies semiconductor devices usually through two letters followed by a numeric or alphanumeric code of three letters/numbers and, in the case of transistors, an optional suffix.

The *first* of the two letters identifies the material that constitutes the active part of the device:

→ A: germanium or other material with a band gap ranging from $0.6–1.0\,eV$
→ B: silicon or other material with a band gap within $1.0–1.3\,eV$
→ C: gallium arsenide or other material with a band gap greater than $1.3\,eV$
→ R: compounds

The *second* letter indicates the function for which the device has been designed, as shown in Table 8.7.

The *code* that follows can be composed of three numbers (100–999) or a letter and two numbers (10–99). The first type (three numbers) is used for consumer equipment devices and the second type (a letter X, Y, Z, etc., and two numbers) is used for devices designed for industrial and/or professional equipment.

The *suffix* (optional) indicates the gain (h_{fe}) of the device:

→ A: low gain
→ B: medium gain
→ C: high gain
→ No suffix: not catalogued

Examples

BC107: BJT, *Si*, *n*-type, $50\,V$, $0.2\,A$, $0.3\,W$, $250\,MHz$

BC117: BJT, *Si*, *n*-type, $120\,V$, $50\,mA$, $0.3\,W$, $>60\,MHz$

8.9.3 JIS

The *JIS* identifies semiconductor devices through a code of the following type: number, two letters, serial number, suffix (optional).

The first number and the first letter identify the type of component. For example 2S is for transistors. This code (2S) is sometimes omitted so, for example, if the device should have "2SC733" as an identification code we can find it simply as "C733."

The two letters indicate the field of application of the device, as follows:

TABLE 8.7
Standard Pro Electron Codes

A: Signal diodes, low power	B: Variable capacitance diodes	C: Low-power transistors, audio frequency
D: High-power transistors, audio frequencies	E: Tunnel diodes	F: Low-power, high-frequency transistors
G: Generic devices (example: oscillators)	H: Magnetically sensitive diodes	L: High-frequency, high-power transistors
N: Photocouplers	P: Radiation detectors (ex. photo-transistors)	Q: Radiation sources (ex. LED, LASER)
R: Controllers/switchers (ex. thyristors), low power	S: Low-power transistors, switcing	T: High-power controllers/ switchers (ex. thyristors)
U: High-power transistors, switching	W: SAW (surface acoustic waves)	X: Multipler diodes (ex. varactor)
Y: Rectifying diodes, booster	Z: Voltage regulator or reference diodes	

→ *SA*: *pnp*-type BJT, HF (High Frequency)
→ *SB*: *pnp*-thyristors BJT, AF (Audio Frequency)
→ *SC*: *npn*-thyristors BJT, HF (High Frequency)
→ *SD*: *npn*-thyristors BJT, AF (Audio Frequency)
→ *SE*: diode
→ *SF*: thyristor
→ *SG*: *Gunn* diode
→ *SH*: UJT (UniJunction transistor)
→ *SJ*: p-channel FETs
→ *SK*: n-channel FETs
→ *SM*: triac
→ *SQ*: LED
→ *SR*: rectifier
→ *SS*: signal diode
→ *ST*: avalanche diode
→ *SV*: varicap
→ *SZ*: *Zener* diode

The serial number is within 10 and 9999.

The suffix (optional) indicates that the device has been approved by one or more Japanese boards.

8.9.4 Others

Apart from the standards examined (*JEDEC, Pro Electron, JIS*), there are others, depending on the manufacturer of the device. More specifically the prefixes can be

→ *MJ*: Motorola (power devices, metallic package)
→ *MJE*: Motorola (power devices, plastic package)
→ *MPS*: Motorola (low-power devices, plastic package)
→ *MRF*: Motorola (transistors, HF, VHF, microwave)
→ *RCA*: RCA

→ *RCS*: RCS
→ *TIP*: Texas Instruments (power devices, plastic package)
→ *TIPL*: Texas Instruments (planar power transistors)
→ *TIS*: Texas Instruments (small signal transistors, plastic package)
→ *ZT*: Ferranti
→ *ZTX*: Ferranti

8.10 KEY POINTS, JARGON, AND TERMS

→ A *pure* semiconductor has a relatively high electrical resistivity that can be lowered by many orders of magnitude by *doping*.

→ *To dope* means to replace some lattice atoms by suitable doping atoms without changing anything else. *N*–doping means substituting some atoms with neighboring ones that contribute excess electrons. *P*–doping means substituting some atoms with neighboring ones that contribute excess holes.

→ The heavier the doping, the greater the conductivity or the lower the resistance.

→ In silicon there are covalent bonds between atoms. These are loosely bound, so more and more electrons will be free, passing from the valence to the conduction bands, as the temperature increases. In such a manner hole and electron pairs will be created.

→ For semiconductors there are two types of electric carriers: *electrons* and *holes*. For silicon the *donor* dopants will increase the electron concentration, and the *acceptor* dopants will increase the hole concentration.

→ The *Law of Mass Action* states that, at thermal equilibrium, the product of the electron and hole density is equal to the square of the intrinsic carrier density. This is not only for intrinsic semiconductors.

→ *Electron-hole pairs* can be either generated or recombined by means of the thermal effect.

→ Current in semiconductors can be due to two main reasons: *drift* and *diffusion*. Drift occurs as charged particle motion in response to an electric field. Diffusion occurs when particles tend to spread out or redistribute from areas of higher to lower concentration.

→ When *p*-type and *n*-type materials are located next to each other a *p*–*n* junction is formed. This is the way to established an electric field inside a doped semiconductor.

→ A *diode* is an electronic nonlinear two-terminal device capable of restricting current flow chiefly to one direction. Its resistance depends on the biasing conditions.

→ A *transistor* is an electronic nonlinear three-terminal device in which current flowing between two electrodes is controlled by a current/voltage flowing in/across the third electrode/the ground.

→ The BJT is so called because it is *bipolar*, that is, both holes and electrons participate in the conduction of current.

→ The charges in the base of the BJT flow for diffusion, while they flow in the channel of the MOSFET for drift. This is why the MOSFET can operate at higher frequencies.

→ To flow a current, for the BJT it is sufficient to have just one bias voltage (V_{BE}). For the enhancement MOSFET, two bias voltages are required (V_{GS} and V_{DS}).

→ There are two types of BJTs: *npn* and *pnp*. There are four types of MOSFETs: *n*–channel or *p*–channel, enhancement or depletion.

→ The "working regions" for analog applications are named *active region* for the BJT and *saturation region* for the MOSFET, even if they both correspond to similar behavior.

→ Npn-type BJT (for pnp-type, use the opposite signs): the emitter region is "heavily" doped so to furnish lots of electrons to conduct current; the base region is "lightly" doped and "narrow" so to extremely limit the number of electrons that get recombined with the electrons coming from the emitter region.

→ Because the base region is "lightly" doped, the I_B current is "small" compared to I_E and I_C currents.

→ To work in the active region, the emitter-base junction must be forward biased and the collector-base junction must be reverse biased.

→ The equation $I_E = I_B + I_C$ applies always, while $I_C \cong \alpha I_E$ applies only in the active region.

→ With an approximation, we consider the parameters α and β quite "constant" in the active region (where the BJT works to amplify) but this is so only as long as the BJT is in the active region.

→ When non-negligible currents flow in the C.E. configuration, for $V_{CE} < 0.7V$, the BJT is in its saturation region; for $V_{CE} > 0.7V$, it is in its active region.

→ *N*-channel enhancement MOSFET: the metal gate is insulated from the channel by an oxide layer; a positive $V_{GS} \geq 0.7V$ voltage attracts electrons from the substrate to form a "*n*-channel" through which electrons can flow from the source to the drain region, without recombination. The gate current is zero, because any flow is prevented by the oxide insulating layer. Generally the substrate is connected to the gate region.

8.11 EXERCISES

EXERCISE 1

A Si cylindrical sample is 1mm in diameter and 1cm long. It was doped with donor impurities with concentration $5*10^{16}$ [cm^{-3}]. How much is its resistance now?

ANSWER

The resistance is

$$R = \rho \frac{l}{S} = \frac{1}{\sigma} \frac{l}{S}$$

where the conductivity

$$\sigma = nq\mu_{n(Si)}$$

We know that $\mu_{n(Si)} = 1500 \, [cm^2/V*s]$ (Section 3.8.2.2 in Chapter 3) and $q \cong 1.6*10^{-19} \, [C]$ (Section 2.3 in Chapter 2) so the result is $\sigma = 12 \, [1/\Omega cm]$ and $R = 10.61 \, [\Omega]$.

EXERCISE 2

A p-n silicon junction has uniform doping levels of $N_A = 2*10^{16} \, [cm^{-3}]$ and $N_D = 3*10^{18} \, [cm^{-3}]$. Determine the electron and hole concentrations on the two sides.

ANSWER

Defining

n_n : Concentration of electrons on the n-side
p_n : Concentration of holes on the n-side
p_p : Concentration of holes on the p-side
n_p : Concentration of electrons on the p-side

on the n-side we have

$$n_n \cong N_D = 3*10^{18} \, [cm^{-3}]$$

$$p_n \cong \frac{n_i^2}{N_D} = \frac{(1.0*10^{10})^2}{3*10^{18}} = 33.\overline{3} \, [cm^{-3}]$$

while on the p-side

$$p_p \cong N_A = 2*10^{16} \, [cm^{-3}]$$

$$n_p \cong \frac{n_i^2}{N_A} = \frac{(1.0*10^{10})^2}{2*10^{16}} = 5000 \, [cm^{-3}]$$

EXERCISE 3

Determine the i_{DS} current flowing in a MOSFET, when $v_{GS} = 3V$, $v_{DS} = 8V$, if $V_{th} = 1V$ and $K_n = 5 \, mA/V^2$.

ANSWER

Since we have $0 < v_{GS} - V_{th} < v_{DS}$ the MOSFET operates in the saturation region, so the current is

$$i_{DS} = K_n (v_{GS} - V_{th})^2$$

when λ is neglected. Numerically: $i_{DS} = 20 \, mA$.

EXERCISE 4

Design the circuit represented in Figure 8.61 including a npn silicon BJT, so that $V_{CB} = 5V$ and $I_C = 3mA$. The fixed data are $V_{CC} = 12V = -V_{EE}$, $\beta = 120$, $V_{BE} = 0.7V$.

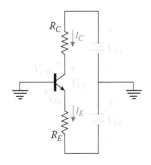

FIGURE 8.61 Double-biased BJT.

ANSWER

To design means to calculate the value of the two resistances.

Since $V_{CB} > 0$ and $V_{BE} = 0.7 = V_\gamma$ the BJT is in its active region, so that

$$R_C = \frac{V_{CC} - V_{CB}}{I_C} = \frac{12 - 5}{3{*}10^{-3}} \cong 2.3\,k\Omega$$

$$I_E = I_B + I_C = \frac{I_C}{\beta} + I_C = \frac{3{*}10^{-3}}{120} + 3{*}10^{-3} \cong 3\,mA \cong I_C$$

$$R_E = \frac{-V_{BE} - \left(-V_{EE}\right)}{I_E} \cong 3.8\,k\Omega$$

EXERCISE 5

A NMOS enhancement FET with parameters $K_n = 0.1\,[mA/V^2]$, $V_{th} = 1.5[V]$, is used in the network shown in Figure 8.62, where $V_{DD} = 12[V]$ and $R_D = 2.2\,k\Omega$.

Determine the output voltage v_o for $v_{GS} = 0.5[V]$ and $v_{GS} = 5[V]$.

ANSWER

The KVL equation of the circuit is

$$V_{DD} = R_D i_D + v_{DS} = R_D i_D + v_o$$

When $v_{GS} = 0.5V$, it results in $v_{GS} < V_{th}$, so the NMOS is in its cutoff region. thus $i_D = 0$ and

$$V_{DD} = R_D i_D + v_o = 12[V]$$

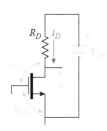

When $v_{GS} = 5V$, it results in $v_{GS} > V_{th}$, so the NMOS can be both in its ohmic or saturation region. Let's assume the hypothesis of the ohmic region. In this case

$$i_D = K_n \left(v_{GS} - V_{th}\right)^2 \cong 1.2\,mA$$

hence

FIGURE 8.62 Simple NMOS circuit.

$$v_o = V_{DD} - R_D i_D \cong 9.3[V]$$

Since $v_{DS} = 9.3 > v_{GS} - V_{th}$, the NMOS is indeed in its active region.

9

Diode Circuits

In this chapter, we will analyze the first *electronic circuits*, in particular (but not exclusively) based on diodes as semiconductor devices. Their function is to process a signal, that is, a time-varying electronic quantity (voltage, current, etc.) to which information is associated.

Diode applications include rectification (Sections 9.1 and 9.2), voltage multiplication (Section 9.3), and wave shaping (Sections 9.4 and 9.5) of the signal.

9.1 RECTIFIERS

Rectifiers are circuits that convert an alternating (including both signs) input signal to an unidirectional output signal (i.e., of a unique sign). So having an AC sinusoidal iso-frequency input signal, the rectifiers supply a DC pulsating multi-frequency output signal, whose harmonic content has more frequencies, including the zero frequency. *Rectification* is a *nonlinear* process implemented by the rectifiers.

Most electronic devices require a plate DC voltage supply to work correctly, but in most cases we have single-phase AC mains. So, rectifiers are the basic circuit to convert AC to DC.

9.1.1 Half-Wave Rectifier

The *half-wave rectifier* is a circuit that, given a pure sinusoidal input voltage, supplies an output positive (or negative) half-sine wave voltage. So, the signal is cut into its negative (or positive) part. Let's consider the sinusoidal voltage signal

$$v_s = V_{sM} \sin(\omega t)$$

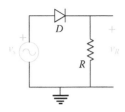

FIGURE 9.1 Half-wave rectifier.

(V_{sM} is its amplitude) as input to the following circuit utilizing a single diode in series with a resistor (Figure 9.1).

The output voltage is measured across the resistor.

In such a circuit, we can only have current when the diode is forward biased beyond its threshold voltage V_γ, so only when $v_s > V_\gamma$. This is also the condition for the current to flow in the resistor and for the output voltage v_R, across the resistor, to be non-zero, as in Figure 9.2.

The different maximum value reached by the two voltages, V_{sM} and V_{RM}, depends on the non-null value of the voltage V_γ that "triggers" the diode.

If (and only if) the amplitude V_{sM} of the input signal is much greater than V_γ, then the delay in the time with which the positive part of v_R starts and how early it ends with respect to v_s can be neglected.

9.1.2 Filtered Half-Wave Rectifier

To the half-wave rectifier one can add a capacitor filter, so as to obtain a circuit that converts an input AC signal to an output DC one. The resulting network is named a *peak rectifier* or *filtered half-wave rectifier*.

The output waveform v_R of the half-wave rectifier was made of several components in frequency, including the zero frequency. A low-pass filter can then be added to "isolate" this zero-frequency component, resulting in a DC output.

Passive filtering can be realized with a shunt capacitor filter, or a series inductor filter, or a choke input LC filter, or a Π CLC (see Figure 4.34b in Chapter 4, but with two capacitances and one inductance replacing the resistors) filter. Here, we will adopt the most common and easiest one, because of its low cost and small size and weight, which is the shunt capacitor filter, that is, a (generally large value) parallel capacitor.

The overall circuit is schematized in Figure 9.3.

During the positive half-wave of the input signal when $v_s > V_\gamma$, the diode is forward biased, so the output signal replicates the input one, and the capacitor is charged up to the voltage $V_{sM} - V_\gamma$. During the negative half-wave of the input signal, the voltage remains below V_γ and the diode is reverse biased, so no current flows from the source, and the charged capacitor supplies current to the resistor. The output voltage is now established by the capacitor discharge.

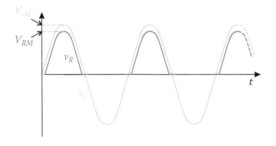

FIGURE 9.2 Input (in blue) and output (in red) waveforms for the half-wave rectifier.

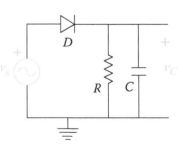

FIGURE 9.3 Half-wave rectifier with filter capacitor.

We establish then two distinct intervals of time to define who "pilots" the output voltage: during the first it is the signal source, during the second it is the capacitor.

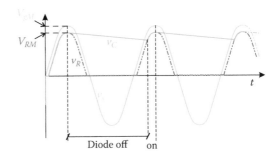

We should not forget that the function that describes the capacitor discharge is exponential (Figure 4.54 in Chapter 4), but if the capacitor discharging time is "low" enough, we can reduce the first part of the exponential curve simply by a fitting line, as shown in Figure 9.4.

FIGURE 9.4 Input and output waveforms for the half-wave rectifier with filter capacitor.

The RC low-pass filter is not ideal (no "abrupt" transition as in Figure 7.27 in Chapter 7), so the zero-frequency component of the signal is not the only one "selected"; there are also a few components in frequencies near to it. The output is not just a constant voltage; there is still an AC "ripple" voltage component, as the voltage is not completely "smoothed." The so-called *ripple factor* γ (or, simply, *ripple*) is defined as the ratio of the RMS value of the AC component to the absolute value of the DC component in the output:

$$\gamma = \frac{\text{RMS value of AC component of output}}{\text{DC value of the component of output}}$$

The ripple can be considered both for the voltage and for the current and, obviously, the smaller the ripple value the better the circuit behavior.

Observation

There are two possible approaches to the transmission of power: 1) using direct current or 2) using alternating current (DC or AC). The system that uses alternating current is more efficient at sending electricity for long distances, so it is used worldwide. However, many common electrical devices require direct current to operate. Thus the need for a device like a filtered rectifier.

9.1.2.1 Choice of Smoothing Capacitor

With reference to Figure 9.5, we define α_1, α_2, α_3 respectively as the angles covered in times t_1, t_2, t_3 (clearly in a whole period T the sinusoidal input voltage $v_s(t)$ covers an angle $\alpha = 360°$).

We have

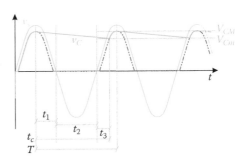

$$V_{Cm} = V_{CM} \sin(\alpha_3) \Rightarrow \alpha_3 = \sin^{-1}\left(\frac{V_{Cm}}{V_{CM}}\right)$$

FIGURE 9.5 Ripple of the filtered half-wave rectifier circuit.

$$t_C = t_1 + t_2 + t_3 = \frac{\alpha_1}{360}T + \frac{\alpha_2}{360}T + \frac{\alpha_3}{360}T$$

The capacitor discharge function, $v_c = V_{CM}e^{-\frac{t}{RC}}$, at time $t = t_C$, when we know that $v_c = V_{cM}$, becomes

$$V_{Cm} = V_{CM}e^{-\frac{t_C}{RC}} \Rightarrow \ln\left(\frac{V_{Cm}}{V_{CM}}\right) = -\frac{t_C}{RC}$$

and finally

$$C = \frac{t_C}{\ln\left(\frac{V_{CM}}{V_{Cm}}\right)R}$$

 Observations

The filtered half-wave rectifier can never have a null ripple (so $V_{CM} - V_{Cm} = 0$) because such a state is only reached for an impossible infinite capacity value.

As previously stated, the current in an inductor cannot change instantaneously, so an inductor tends to resist any change in current flow (Section 4.7.4 of Chapter 4). This means that an inductor can be used as a filter element as well, since it tends to "smooth out" the ripples in the rectified waveforms, but in a "dual form" (see the "duality principle," in Section 4.8 of Chapter 4) with respect to the capacitor.

9.1.2.2 Choice of Diode

Unfortunately, the use of a large capacitor does not offer only the advantage of the reduction of ripple. In fact the presence of the capacitor implies important current and voltage "stresses" for the diode.

When the circuit is initially turned on, a transient time high-value current flows in the diode, much higher than in steady-state conditions. With a input voltage $v_s = V_{sM}\sin(\omega t)$ during the first quarter of cycle, the current that flows through the diode is approximately expressed as

$$i_D(t) = i_C(t) \cong C\left[\frac{d}{dt}V_{SM}\sin(\omega t)\right] = \omega C V_{sM}\cos(\omega t)$$

We have maximum current at the very initial time $t = 0^+$:

$$i_D(t = 0^+) = I_{DM} = \omega C V_{sM}$$

Therefore it depends proportionally on the value of the capacity chosen. Substituting the typical numeric values in the previous expression, we may even have a considerable current value I_{DM}, even of the order of hundreds of Amperes. In most circuits, however, the transient current can be quite lower than the one calculated, because of series resistances in the circuit, which here are ignored. In fact the diode has an internal series resistance and there will also be resistances associated with the primary and secondary coils of an input transformer, when present. The resistance of the secondary coil, even if very low in value, can adequately reduce the maximum current through the diode.

We consider now the steady-state condition. The diode is in conduction only for a very short interval of time t_D during each period, and the current in it assumes the typical approximately pulse shaped behavior shown in Figure 9.6. The area under the curve of I_D (current through the diode) corresponds to the charge that the diode furnishes to the circuit to restore the one lost during the capacitor's discharge.

According to the *principle of charge conservation,* electric charge can neither be created nor destroyed, the charge supplied by the diode must be equal to the one released by the capacitor. Accordingly, we deduce the value of maximum current I_{DM} that flows through the diode.

To determine the total charge Q supplied by the diode, we approximate the area under each pulse with that of the rectangle with base t_D and width I_{DM}, obtaining $Q = I_{DM}t_d$.

To determine the total charge Q lost during the process of capacitor discharge, we consider T as discharge time (assuming t_D negligible with respect to T) and I_{RM} as maximum current (supplied by the capacitor) through the resistance, obtaining $Q = I_{RM}T$.

Matching the two discharge values, there results for the maximum current flowing in the diode

$$I_{DM} = I_{RM} \frac{T}{t_d} = I_{RM} \frac{2\pi}{\omega t_d}$$

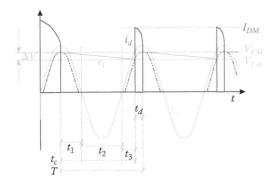

FIGURE 9.6 Voltage across the capacitor
and current that flows in the diode.

However, considering that the already seen equation $V_{Cm} = V_{CM} \sin(\alpha_3)$ can also be written as $V_{Cm} = V_{CM} \sin(90° - \omega t_D) = V_{CM} \cos(\omega t_D)$ (recall the expression of $\sin(\alpha - \beta)$, in Section 1.2.5 of Chapter 1) we have

$$\Delta V = V_{CM} - V_{Cm} = V_{CM} - V_{CM} \cos(\omega t_d)$$

Now substituting the cosinusoidal function with its Taylor series $(\cos(x) = 1 - (x2 / 2!) + (x4 / 4!) - (x6 / 6!) + ...$, Section 1.6 in Chapter 1) truncated, for simplicity, at second order, we have

$$\Delta V \cong V_{CM} \left[1 - \left(1 - \frac{1}{2}(\omega t_d)^2 \right) \right] = \frac{1}{2} V_{CM}(\omega t_d)^2$$

from which we have $\omega t_d = \sqrt{2(\Delta V / V_{SM})}$ which, replaced in the expression of the maximum current through the diode, gives

$$I_{DM} = 2\pi I_{RM} \sqrt{\frac{V_{CM}}{2\,\Delta V}}$$

In conclusion, the maximum current that flows through the diode is higher than the one through the load resistor, and it is inverse with the voltage drop ΔV, which we would like to be as small as possible. We cannot use a capacitor as large as desired, because it increases the current that the diode must support facing a reduction of ripple. But diodes normally do not support high-value currents.

The choice of the diode and of the capacitor must be a trade-off between opposing needs.

9.1.2.3 Observation on Harmonic Content

The output voltage of the half-wave rectifier (Figure 9.1) has a periodic behavior (Figure 9.2) that can be analyzed in terms of a Fourier series expansion. As an example, the easiest base frequency is $1 / 2\pi$ cycles per second, so that in 2π seconds we have a complete cycle, and the complete frequency distribution of this half-wave signal would be as follows:

$$v = V_M \left(\frac{1}{\pi} + \frac{1}{2}\sin(t) - \frac{2}{3\pi}\cos(2t) - \frac{2}{15\pi}\cos(4t) - \frac{2}{35\pi}\cos(6t) - ... \right)$$

This contains

→ A fundamental harmonic with the same frequency as the input signal
→ A series of harmonics set to fractions of the value of the first harmonic
→ A DC component

With the capacitor we aim to low-pass filter to obtain only the DC component of the signal, that is, its zero frequency. But it is not convenient to study the circuit using filter theory because of the nonlinearily of the diode, which changes its resistance (and therefore that of the circuit) according to its biasing. This is why previously we adopted the discharge equation of a capacitor.

9.1.3 Bridge Full-Wave Rectifier

With the half-wave rectifier, the negative half of the input waveform is not utilized. This problem is solved with the *bridge full-wave rectifier*.

The bridge configuration that uses four diodes on its branches is called a *Graetz bridge* (Leo Graetz, German physicist, 1856–1941), and can be used to realize a *full-wave rectifier*, or *bridge rectifier*, a circuit that provides full-wave rectification (Figure 9.7). So both negative and positive half-cycles of an input voltage are made unidirectional during the entire cycle (basically the negative half-wave is positive "flipped" around the *x*-axis).

The four diodes labeled D_1 to D_4 are arranged in "series pairs" with only two diodes conducting current during each half cycle, and the other two remaining in "idle" mode. During the positive half-cycle of the input voltage $v_s = V_{SM} \sin(\omega t)$, when $v_s > 2V_\gamma$, the pair D_1 and D_3 are forward biased and conduct (D_2 and D_4 in idle mode), as in Figure 9.8a. During the negative half-cycle, when

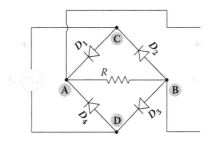

FIGURE 9.7   Full-wave bridge rectifier.

$|v_S| > 2V_\gamma$, the D_2 and D_4 diode pair are in forward bias mode (D_1 and D_3 in idle mode) and conduct, as in Figure 9.8b. So, while one pair of diodes is "on" because they are forward biased, the other set is "off" because they are reverse biased.

But both conduction paths cause current to flow always in the same direction through the load resistor, from node "A" to node "B" as in Figure 9.8, so that the

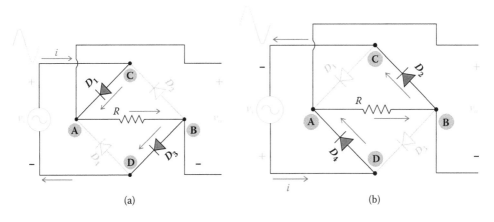

(a)   (b)

FIGURE 9.8   Current flow through the bridge in (a) the D1 and D3 diode pair during the positive half of the input cycle and in (b) the D2 and D4 diode pair during the negative half of the input cycle.

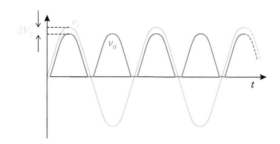

FIGURE 9.9 Output (in red) versus input (in blue) voltage in the full-wave bridge rectifier.

voltage across the resistor always has the same sign, $v_o = v_A - v_B > 0$, for both positive and negative swings of the input source, as represented in Figure 9.9.

The circuit therefore furnishes as output the absolute value of the input sine function.

The output signal has its main component frequency doubled with respect to the input signal.

Observations

To correctly work, the bridge requires four absolutely twin diodes. These diodes can be purchased as a single four-terminal device (Figure 9.10).

Half- and full-wave rectifiers are also frequency multipliers. In fact, focusing on the full-wave rectifier, it rectifies an input $60\,Hz$ signal into an output signal at $120\,Hz$ but also with harmonic components at $240\,Hz$, $360\,Hz$, $480\,Hz$, etc.

FIGURE 9.10 Four diodes connected in bridge topology.

9.1.4 Filtered Full-Wave Rectifier

What we already discussed for the half-wave rectifier can also be replicated here. We can add a filter capacitor to the full-wave rectifier to realize a low-pass filter that can select the DC component of the signal (Figure 9.11).

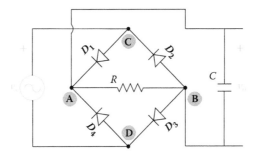

FIGURE 9.11   Filtered full-wave rectifier.

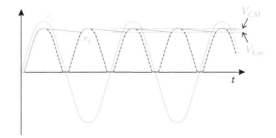

FIGURE 9.12   Input (blue), output (red),
and ripple (green) voltages of the
filtered full-wave rectifier.

However, now the discharge time of the capacitor to "pilot" the load resistance is substantially reduced. Therefore the ripple will be intuitively reduced and closer to the ideal of no ripple (Figure 9.12).

Curiosities

In 1925, the UV-213, started by GE (General Electric, a multinational company), was the first of the commercial full-wave rectifiers. It evolved into the UX-213, marketed through RCA (Radio Corporation of America, an American electronics company in existence from 1919 to 1986), which in turn was redesigned into the UX-280, which remained in continuous production for a longer period than any other tube.

From the RC-32862 data sheet: "Rectron model UX-213 is a full-wave rectifier tube for supplying DC power from an AC source."

A filtered full-wave rectifier is also the "heart" of some commercial defibrillators.

A *defibrillator* is a medical device that "resets" the heart of a person with ventricular fibrillation, by means of a large amount of electrical energy sent through the heart (its public sign is seen in Figure 9.13). This energy is previously stored in an large capacitor, called an energy capacitor, with a typical $C \cong 200\,\mu F$, charged at high voltage, typically $V \cong 2\,kV$. The defibrillation is measured by means of the energy stored in the energy capacitor, $W = (1/2)CV^2$ (Section 4.6.7 of Chapter 4), which can reach hundreds of *Joules* (often $200 \leq W \leq 360\ J$).

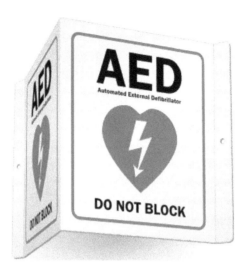

FIGURE 9.13   An AED (Automated External
Defibrillator) sign in a public space.

The capacitor is discharged through the patient creating a characteristic *RC*–circuit discharging behavior, *R* being the resistance offered by the patient to the flowing of the current. The discharge can be done by means of a *H-bridge* (Section 4.3.3 in Chapter 4) topology to obtain a bi-phasic defibrillation waveform.

The capacitor's charger can be a high-voltage, patient-isolated circuit full-wave rectifier that charges an energy capacitor to the correct voltage (Figure 9.13).

Observation

 Neither the half-wave nor the full-wave rectifier is suitable for applications that require an output DC voltage less than the threshold voltage drop V_γ of the diode. To overcome this problem, we can adopt the so-called *precision rectifier*, also termed *super diode*, not treated here.

9.2 DC POWER SUPPLY

The requirement of a *power supply* is to provide a (relatively) ripple-free source of DC voltage from an AC "wall" source, usable for low-voltage electronics, replacing batteries when their usage is unsustainble for power consumption.

The power supply can be described by functional blocks as represented in Figure 9.14.

The transformer, the first block, converts the voltage from the wall outlet (mains) to a (usually) lower voltage necessary as input to the second block and, at the same time, creates an electrical insulation between the mains and the rest of the circuit via inductive coupling.

The rectifier, the second block, converts the sine input signal (with zero DC value), into an unipolar (unidirectional) signal whose harmonic content includes the zero frequency (pure DC).

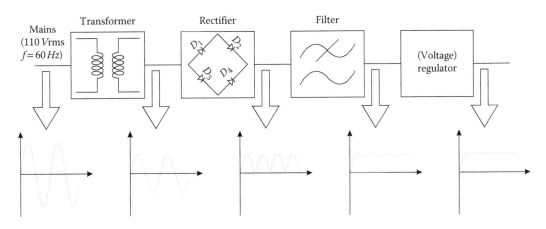

FIGURE 9.14   Block diagram of a DC power supply.

The filter, the third block, smooths the output signal of the rectifier preserving only its very low frequency parts. The smoothing cannot be perfect because the filter cannot furnish only the requested DC component, and the *ripple* factor "measures" this non-ideal scenario.

The group of blocks described so far (transformer, rectifier, and filter) constitutes a power supply circuit that is *not stabilized*. For a power supply that requires extremely good stabilization, an electronic *regulation* can be added, furnished by a *regulator*.

To the *regulator*, the fourth block, is assigned the task to drastically reduce (or cancel) the ripple, to produce a nearly pure and steady DC voltage, even irrespective of "anomalies" of the main or load variations or simply thermal effects.

The overall scheme, with all four blocks, is known as a *stabilized* (or *regulated*) *power supply* (Figure 9.15).

On the market there are many types of stabilized power supplies that differ in characteristics, performance, and costs. The rated output voltage that can be fixed, or could be determined by the user, can be maintained practically constant to the load variations (within certain limits) (Figure 9.16a–c and Figure 9.17).

FIGURE 9.15   Simple stabilized power supply made by students of the course "Laboratory of Electronics" held at University of Rome Tor Vergata.

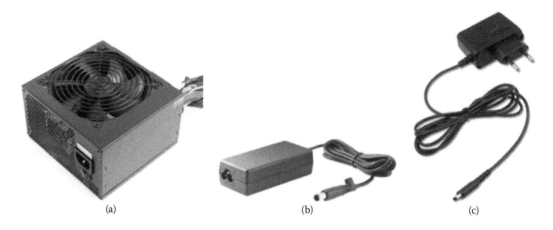

(a) (b) (c)

FIGURE 9.16 Some commercial power supplies (a) for desktop computers, (b) for notebooks, and (c) for cellular phones.

FIGURE 9.17 A common power supply inside of its box: we can clearly see the transformer (with the red cable coils), the four diodes (that form the Graetz bridge), and the capacitor.

9.3 VOLTAGE DOUBLER CIRCUITS

In a power supply circuit, sometimes it may be necessary to maintain low transformer peak voltage; therefore a *voltage multiplier circuit* is used. Different types of multipliers exist, depending on the output voltage: voltage doublers (Figure 9.18), voltage triplers, voltage quadruplers, etc.

Here, we'll deal with the simplest form, that is the *half-wave voltage doubler*, as in Figure 9.18, which produces a DC voltage almost twice the RMS value of the input AC voltage. Assuming an AC input voltage of $v_s = V_{sM}\sin(\omega t)$, we have that during the positive input half-wave, the diode D_1 is forward biased, and so "on," and the diode D_2 is reverse biased, and so "off." Therefore

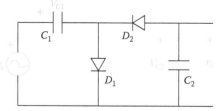

FIGURE 9.18 Half-wave voltage doubler.

the capacitor C_1 charges to V_{sM}. During the negative input half-wave, the diode D_1 is reverse biased, and so "off," and the diode D_2 is forward biased, and so "on." Therefore the capacitor C_2 charges to $+2V_{sM}$; in fact $-V_{C_2} + V_{C_1} + V_{sM} = 0$, so $V_{C_2} = v_o = V_{C_1} + V_{sM} = 2V_{sM}$, from which results

$$v_o = 2V_{sM}$$

Anyway, this result is possible when the capacitors keep their peak voltage value, not discharging through the load.

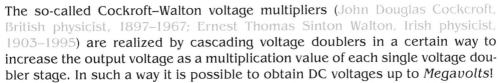

Curiosities

The so-called Cockroft–Walton voltage multipliers (John Douglas Cockcroft, British physicist, 1897–1967; Ernest Thomas Sinton Walton, Irish physicist, 1903–1995) are realized by cascading voltage doublers in a certain way to increase the output voltage as a multiplication value of each single voltage doubler stage. In such a way it is possible to obtain DC voltages up to *Megavolts*!

As an example of application, the Cockroft–Walton voltage multipliers are adopted to realize ion-generator circuits, for which it is requested to generate DC voltages greater than few *kVs*, with a 120 *V* AC input. As a result, a load wire electrode will produce a high electric field that can ionize the surrounding air.

Cockroft and Walton shared the Nobel Prize in Physics for their work in "atom-smashing"; they were the first in history to artificially split the atom.

9.4 CLAMPER CIRCUITS

The *clamping circuits* or *clampers* (also known as *DC restorers*) are used to shift a waveform to a different DC level (termed *offset*) without changing its appearance in steady-state conditions. The clampers, which add a positive DC value, are called *positive clampers; the* addition of a negative DC value is for *negative clampers* (Figure 9.19).

A clamping circuit is made of a capacitor C, a diode D, and a resistor R (a real one or a load output resistor). To keep a constant voltage on the capacitor over the period of the input, the RC time constant must be greater than the half-period of the input signal (a design rule of thumb suggests an RC at least five times greater).

We can realize a *positive* or *negative* clamper circuit (Figure 9.20a and b), here we will deal with the positive one.

For the sake of simplicity, let's consider an ideal diode that when it conducts acts like a short circuit ($V_\gamma = 0$, no inner resistance when forward biased). In the first positive half-wave of the input source voltage $v_s = V_{sM} \sin(\omega t)$, the diode D is reverse biased (acting as an open circuit), the current flows through the resistor R_L, and the capacitor C is charged to the peak value V_{sM}. Hence, we get a positively clamped output voltage $v_o = +V_{sM} + V_{sM} \sin(\omega t) = V_{sM}[\sin(\omega t) + 1]$ (in a real application we need to subtract $V_\gamma = 0.7V$ from the result).

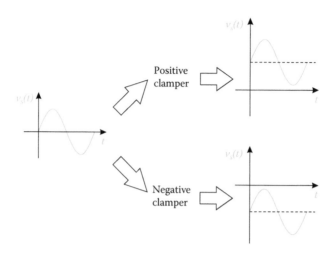

FIGURE 9.19 Positive and negative clamping
effects. The output signal is an exact replica of
the input signal but shifted by a DC value.

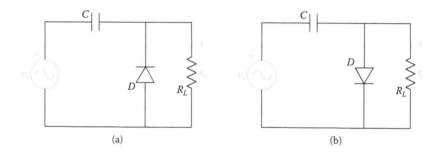

(a)   (b)

FIGURE 9.20 (a) Positive and (b) negative clamper.

Observation

During TV signal transmission, a signal can "lose" its DC level. A clamper circuit
is then adopted in the TV receiver to restore the original signal.

9.5 CLIPPER CIRCUITS

Clipping circuits, also referred to as *clippers*, clip off a fraction of the input signal,
above or below a specified level, keeping the remaining part unchanged. They are
also known as *voltage limiting circuits*, because they limit the input voltage to
certain minimum and maximum values, so as to provide protection to an arbitrary
electronic circuit, preventing component damage if the voltage tends to a value
higher than the designed one.

A clipper can be made with resistors and diodes, as in the following schemes
where *negative* (i.e., a negative portion of the input signal is clipped) and *positive* (i.e., a positive portion of the input signal is clipped) clippers are shown in
Figure 9.21.

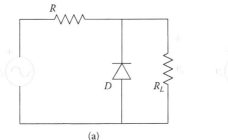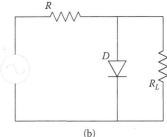

FIGURE 9.21   (a) Negative and (b) positive clipper circuit diagrams.

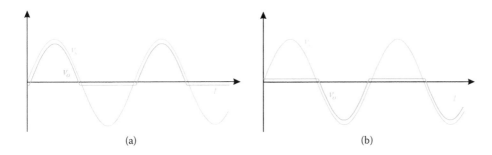

(a)   (b)

FIGURE 9.22   Results of a (a) negative and (b) positive clipping
action. Input signals are in blue and output ones in green.

We start by considering a *negative clipper* (Figure 9.21a) with ideal conditions for the diode, which acts like a short or a open circuit (neglecting its threshold voltage V_γ).

During the positive half-cycle of the input signal v_s the diode is reverse biased, so it is to be considered an open circuit and does not conduct. So, the output voltage is

$$v_o = \frac{R_L}{R_L + R} v_s$$

a replica of the input signal but reduced in amplitude.

During the negative half-cycle of the input signal v_s the diode is forward biased, so it is to be considered a short circuit, that is conducting without limiting the current. So the output voltage result is $v_o = 0$, and the negative half-cycle of the input wave is shorted out.

Taking into account a a non-negligible threshold voltage of the diode, the output voltage will be shifted accordingly (Figure 9.22a).

In a *positive* clipper (Figure 9.22b), the diode is connected in a direction opposite to that of a positive clipper. The resulting output waveform is reported in Figure 9.22b.

Observation

 The half-wave rectifier (Section 9.1.1) can also be considered a clipper circuit since voltages below a certain level (imposed by V_γ) are eliminated.

9.6 KEY POINTS, JARGON, AND TERMS

→ A *rectifier* converts alternating input signals into unidirectional output ones.

→ A *filtered rectifier* converts AC signals into "similar-DC" signals, containing *ripple*.

→ A *regulated filtered rectifier* makes a *power supply*, which is a device that converts a AC input voltage into a constant output voltage under changing load conditions.

→ Characteristics of *half-wave rectifier*: The diode is off for almost the entire period T; the peak-to-peak ripple voltage, $V_{CM} - V_{Cm}$, typically is about 5% less than V_{CM}.

→ Characteristics of *full-wave rectifiers* (bridge version): There is reduced ripple value γ with respect to the half-wave circuit version. The current flows in the opposite direction in two half-cycles, so its net DC component is zero. The four diodes must be identical twins, otherwise the two half-waves are not symmetrical. Every half-cycle two diodes are forward biased, so in the output a $2V_\gamma$ voltage drop must be taken into consideration. In a full-wave rectifier, C discharges for approximately half as long as in the half-wave rectifier. Thus only half of the capacitance is required for a given ripple value.

→ A *voltage doubler* rectifies the input AC voltage into a DC output voltage with double the peak input value.

→ A *clamper* shifts a waveform to a different DC level without changing its appearance.

→ A *clipper* furnishes an output voltage limited in amplitude.

9.7 EXERCISES

EXERCISE 1

A half-wave rectifier as in Figure 9.1 has an input voltage signal $v_s = V_M \sin(\omega t)$.

Assuming an ideal diode ($V_\gamma = 0$, zero resistance in forward biasing condition), calculate the following:

- The average output voltage $V_{0,DC}$ across the resistor
- The average current $I_{0,DC}$ flowing through the resistor
- The RMS value of the output voltage $V_{0,RMS}$ across the resistor
- The RMS value of the current $I_{0,RMS}$ flowing through the resistor
- The form factor $F_{F,1s}$
- The ripple factor γ

ANSWER

$$V_{o,DC} = \frac{V_M}{\pi}; I_{o,DC} = \frac{V_M}{\pi R}; V_{o,RMS} = \frac{V_M}{2}; I_{o,RMS} = \frac{V_M}{2R}; F_{F,1s} \cong 1,57; \gamma = 1.21$$

EXERCISE 2

A half-wave rectifier circuit, as in Figure 9.1, has an input sinusoidal voltage source $v_s = V_M \sin(\omega t)$, with $V_M = 10\,V$, $R = 500$. Assuming an ideal diode ($V_\gamma = 0$, zero resistance in forward biasing condition) calculate the following:

- The maximum current value I_M
- DC component of current I_{DC}
- RMS value of current I_{RMS}
- DC component of voltage in output $V_{o,DC}$
- DC component of power delivered to the load $P_{o,DC}$
- Power value supplied by the source P_s
- The ripple value γ

ANSWER

$$I_M = 20\,mA\,; I_{DC} = 6.3\,mA\,; I_{RMS} = 10\,mA\,; V_{DC} = 6.3\,V\,; P_{o,DC} \cong 0.04\,W\,; P_s = 100\,mW\,;$$

$$P_{DC} = 39.7\,mW\,; \gamma \cong 1.23$$

EXERCISE 3

A half-wave rectifier has an input stage with a transformer, turns ratio n = 20:1, and a load resistance R = 50Ω. The input voltage source has an RMS value of $V_{s,\,RMS} = 220\,V$ (Figure 9.23).

Let's determine the following:

- DC component of voltage in output $V_{o,DC}$
- DC component of current in output $I_{o,DC}$
- RMS value of voltage in output $V_{o,RMS}$
- RMS value of current in output $V_{o,RMS}$
- The ripple value γ

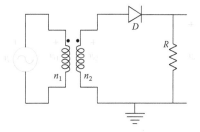

FIGURE 9.23 Half-wave rectifier with a transformer as the first stage.

ANSWER

$$V_{o,DC} \cong 9.9\,V\,; I_{o,DC} \cong 198\,mA\,; V_{o,RMS} \cong 15.56\,V\,; I_{o,RMS} \cong 0.31\,A\,; \gamma \cong 1.21.$$

Note that this ripple value is quite high, so the half-wave rectifier is a poor AC to DC converter.

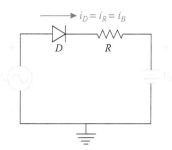

EXERCISE 4

A half-wave rectifier network is used as a battery charger (Figure 9.24)

When the source voltage is higher than that of the battery, the current flows from the source to the battery, which charges. Reverse conditions are not possible because of the presence of the diode. Assuming a voltage source of $v_s = 150 \sin(\omega t)$, an ideal diode ($V_\gamma = 0$ and a

FIGURE 9.24 A simple battery charger by means of a half-wave rectifier.

short circuit in forward bias condition, an open circuit in reverse bias conditions), and a 75 V battery with a charge current of 750 mA, determine the
resistance value of the resistor R.

ANSWER

$$R \cong 22\,\Omega$$

Observations

A *battery's capacity* $C[Ah]$ refers to the stored electric charge that can be delivered in an amount of time at room temperature (77°F or 25°C). A $500[Ah]$ rated
battery can supply $1\,A$ for $500\,h$, or $5\,A$ for $100\,h$, or $10\,A$ for $50\,h$, or $100\,A$ for $5\,h$.
But the capacity is not the perfect parameter to give a real measure for a battery
because it depends on the discharge conditions: the current's value (not necessary constant), the value of the voltage, the temperature, and the discharging rate.

The *C-rate* (or *charge-rate* or *hourly-rate*) of a battery specifies the discharge
rate, as a multiple of the *capacity*. So, for example, a battery with a capacity
$C = 1.5[Ah]$ and a $C/10$ rate, delivers $1.5 / 10 = 0.15[A]$; a $1\,C$ rate means that
the battery discharges entirely in $1[h]$.
    This is similar for the *E-rate* but refers to the power, not the current.

EXERCISE 5
*A battery with C = 1200 mAh, V_B =
5 V, and C−rate = 10 must be charged
by a half-wave rectifier with a
transformer at its input port. The
AC voltage source $v_s = V_M \sin(\omega t)$
has a RMS voltage of $V_{s,\ RMS} = 220\,V$
(Figure 9.25).*
    *Assuming a diode with $V_\gamma = 0.74\,V$,
calculate the following:*

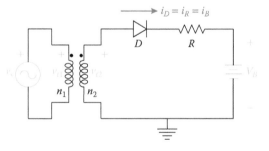

FIGURE 9.25   Half-wave rectifier with
a transformer at its input port.

- The average charging current, I_B
- The value of the resistance R
  necessary to limit the current
- The RMS current flowing through the battery, I_{RMS}
- The power P_R dissipated by the resistor
- The power P_B delivered to the battery
- The charging time, t_B
- The efficiency of the half-wave rectifier

$$= \frac{Power\ delivered\ to\ the\ battery}{Total\ power} = \frac{P_B}{P_{tot}}$$

ANSWER

$$I_B = 0.12\,A; R \cong 34\,\Omega; I_{RMS} \cong 0.21\,A; P_R \cong 1.42\,W; P_B 0.6\,W; t_B = 10\,h; = \eta \cong 0.30$$

Observation

A *constant voltage (C-V) charging* method refers to the method by which a charger sources current into the battery to force its voltage to a preset value, named the *set(-point) voltage*. When this value is reached, the charger sources only the current to maintain the battery at the same constant voltage level.

It is the opposite for the *constant current (C-I) charging* method.

Some rechargeable batteries need to be *C-V* while others are *C-I* charged. Some chargers are *C-V* while others are the *C-I* type; that's why generally a charger can be used to charge some batteries and not others. Universal chargers are those that have both *C-V* and *C-I* characteristics.

EXERCISE 6

The bridge full-wave rectifier has better performance than the half-wave rectifier in terms of form factor and ripple. Consider a bridge full-wave rectifier with an input transformer, as in Figure 9.26.

Calculate the following:

- DC component of voltage in output $V_{0,DC}$
- DC component of current in output $I_{0,DC}$
- RMS value of voltage in output $V_{0,RMS}$
- RMS value of current in output $I_{0,RMS}$
- The form factor $F_{F,2s}$
- The ripple value γ

ANSWER

$$V_{0,DC} = \frac{2V_M}{\pi}; I_{0,DC} = \frac{2V_M}{\pi R}; V_{0,RMS} = \frac{V_M}{\sqrt{2}}; I_{0,RMS} = \frac{V_M}{R\sqrt{2}}; F_{F,2s} \cong 1,1; \gamma = 0.482.$$

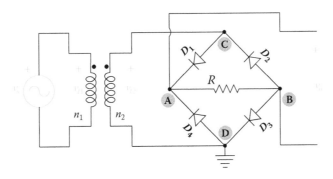

FIGURE 9.26   Bridge full-wave rectifier
with input transformer.

EXERCISE 7

Given a bridge full-wave rectifier as in Figure 9.26, with diodes that have a threshold voltage $V_\gamma = 0$ and a resistance $R_{on} = 10\,\Omega$ when "on," sourced by a sinusoidal voltage signal of amplitude $V_{sM} = 30\,V$, and loaded by a resistance $R_L = 1\,k\Omega$, calculate the following:

- *Maximum, average, and RMS values of the output current: $I_M, I_{o,DC}, I_{o,RMS}$*
- *Average and RMS values of the output voltage: $V_{o,DC}, V_{o,RMS}$*
- *Average value of the output power, $P_{o,DC}$*
- *Average value of the input power, P_i*

ANSWER

$$I_M = 29.4\,mA\,; I_{o,DC} \cong 18.7\,mA\,; I_{o,RMS} \cong 20.8\,mA\,; V_{o,DC} = 18.7\,V\,; V_{o,RMS} = 20.8\,V\,;$$
$$P_{o,DC} \cong 0.3\,W\,; P_i \cong 0.41\,W$$

Observation

A lower ripple requires a higher capacitance, but lower current spikes require lower capacitance. The trade-off is often with a capacitance of the order of a few thousand µF.

EXERCISE 8

A resistance load requires a DC voltage $V_{DC} = 12\,V$ and current $I_{DC} = 0.5\,A$. To this aim a bridge full-wave rectifier is utilized, sourced by an AC line with a frequency $f = 50\,Hz$. Calculate the value of the smoothing capacitance for a 10% ripple (Figure 9.27).

ANSWER

$$C \cong 1.2\,mF$$

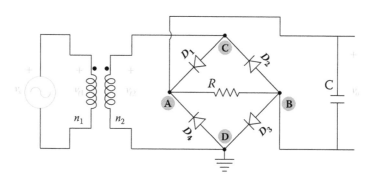

FIGURE 9.27 Filtered bridge full-wave rectifier.

10

Amplifiers

10.1 DEFINITIONS AND CLASSIFICATIONS

In electronics, *amplification* concerns the production, by means of an appropriate circuit, of an output signal that has the same form as the input signal but with an increased level of voltage and/or current and/or power (Figure 10.1a and b).

The input signal is provided by the *signal source*, and we need to amplify it when the source available is not powerful enough to pilot a load. If we consider an audio system, the AC signal source can be a CD or DVD player, the microphone in a telephone handset, the antenna of a radio receiver, etc. The related voltage signal can range between a few μV and a few mV, and the current can range between μA and mA values, so a very low level of power, not enough to pilot many loads (for example, a system of speakers needs a power that can reach even hundreds or thousands of watts).

Amplification concerns a huge number of signal sources very different from each other, including many we really do not imagine as, for example, a fuel-level sensor in an automobile gas tank or a smoke detector in a fire alarm system.

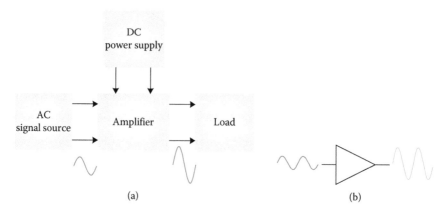

(a)

(b)

FIGURE 10.1 (a) Logical scheme and (b) circuit symbol of an amplifier.

Curiosities

In electronics, the possibility to amplify certain electric signals is fundamental, but the concept of amplification is obviously not exclusive to it. Let us think about our sense of hearing: a physical event that perturbs the air (the classic stone in the pond) produces around it a phenomenon of air rarefaction/compression (sound) that, propagating, reaches our ear. The acoustic wave causes our *eardrum* (the tympanic membrane) to vibrate and the vibration is sent to the three little ossicles (the *auditory ossicles*: *malleus*, *incus*, and *stapes*, *Figure 10.2*) contained within the middle ear space, which behave practically as amplifiers. In fact, the stapes forwards the acoustic waves to the internal ear where the waves arrive amplified about 180 times. Therefore, we perceive sounds with greater amplitude than they actually have!

In the internal ear, sensory cells are predisposed to transform sounds into electric signals that arrive in the brain, which interprets such signals as "sound sensations."

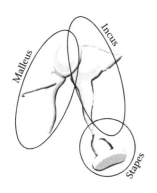

FIGURE 10.2 Malleus, incus, and stapes in inner ear.

Amplification is fundamental in radio equipment. Motorola was one of the first companies to make radios using transistors. Originally, the transistors were made with germanium as the semiconductor, but it could happen that if a car was parked in the really hot sun, the radio no longer worked. This was because the *Ge* had become *intrinsic* (no longer *extrinsic*) with temperature. So Motorola replaced *Ge* with *Si* semiconductors.

Through amplifiers we can control a great variation of an output signal by means of a small variation of an input signal. There are four main types of amplifiers, which as distinguished by the input and output signals, as in the scheme shown in Table 10.1.

TABLE 10.1
Different Types of Amplifiers

Input signal	Output signal	Type of Amplifier
Voltage	Voltage	Voltage Amplifier
Voltage	Current	Transconductance Amplifier
Current	Voltage	Transresistance Amplifier
Current	Current	Current Amplifier

An amplifier can be considered a functional block, that is, a two-port network to which we apply an input signal of a source and from which we get an output signal furnished to a load.

In Figure 10.3, the input source is thought of as a real source of voltage V_s, with its own internal series resistance R_s, and the load as a simple resistor R_L. Obviously, this is just for illustration, because we can have any type of source as the input signal (which we can bring back, according to Thévenin's theorem, to an equivalent real source).

Equally, there can be any type of load in the output, whether of purely real or complex value (so with resistive and capacitive/inductive parts), or another functional block (but with its own equivalent input resistance/impedance).

Let us consider, as an example, a voltage amplifier. Seeing as we said that an amplifier's task must simply be to repeat an output signal v_0 equal to the input one v_s but amplified, ideally we have the output-input voltage relationship as

$$v_o = A_v v_s$$

where A_v is the value of voltage amplification. Regarding the value of A_v the possibilities are as follows:

→ A_v negative: The amplifier is called *inverting* because the output signal has an opposite sign to the input one, that is, it is 180° out of phase, but if $|A_v| > 1$ the output amplitude is larger than the input one.
→ A_v positive: The amplifier is called *non-inverting*, because the output signal maintains the same sign as the input one.
→ $|A_v| < 1$: The output signal has a smaller amplitude than the input one, so the functional block actually reduces rather than increases the amplitude of the input signal (therefore we *do not* have amplification but *attenuation*).

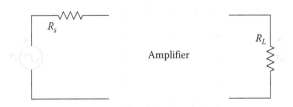

FIGURE 10.3 Amplifier as a functional block.

But what does a 180° phase shift mean? For a video signal we will have the negative of an image (black instead of white and vice versa); for a non-stereo audio signal there is no problem at all.

10.2 KEY PARAMETERS

There are several key parameters to take into account in designing an amplifier. Here, we'll define and describe the most relevant for our purposes.

10.2.1 Gain

The amplifier *gain*, or *amplification factor*, is a measure of the degree to which a signal has been strengthened. The gain can be expressed in terms of voltage, current, and power.

The *voltage gain* A_v is defined as the ratio of the amplitudes of the output voltage v_o furnished to the load by the amplifier, and the input voltage v_s provided by the source:

$$A_v \overset{\text{def}}{=} \frac{v_o}{v_s}$$

The *current gain* A_i is defined as the ratio of the amplitudes of the output current i_o through the load and the input current from the source i_s:

$$A_i \overset{\text{def}}{=} \frac{i_o}{i_s}$$

The *power gain* A_p is defined as the ratio of output power p_o to input power p_i:

$$A_p \overset{\text{def}}{=} \frac{p_o}{p_i}$$

By definition, a gain does not have an SI unit, as it is a ratio of homogenous units.

Ideally all types of gains of an arbitrary amplifier have to be constant at all frequencies. For real amplifiers this cannot be true. The *gain flatness* indicates the variations of the amplifier's gain characteristic, over the frequency range of amplification. It is expressed as half of the difference between the maximum and minimum gain within the frequency range of amplification.

10.2.2 Impedances

The electrical impedance Z, already defined in Section 7.2 of Chapter 7, describes the combined effects of the overall resistance R, capacitive reactance X_c, and inductive reactance X_L in an AC circuit. According to its definition, the impedance cannot be a constant value, since it depends on X_c and X_L, which vary with frequency. Neverthless, typically in audio systems, you find it printed near the input and output sockets, expressed in *Ohms*, typically with values of 2Ω, 4Ω, 6Ω, or 8Ω. The reason is that, as a standard, it is considered the value of the impedance around the middle log-scale frequency of the network and, for audio systems, it is often referred to at the frequency of $1\,kHz$.

The main impedances to be considered for an arbitrary amplifier are those at its input and output ports, named *input* and *output impedances*, Z_i and Z_o, respectively.

These impedances cannot be "real" components but, since currents flow when voltages are applied across these ports, impedances must occur, which can be represented as a two-terminal component Z_i at the input port and a two-terminal component Z_o at the output port.

Depending on the type of source with its internal impedance at the input port, on the type of amplifier with its input and output impedances, and on the type of impedance of the load, sometimes it can be necessary to design matching networks (Section 6.9 in Chapter 6), to be inserted between the source and amplifier and/or amplifier and load, to guarantee maximum voltage, current, or power transfer (Section 6.9.5 in Chapter 6).

In the next chapter, we'll describe the methodology to design an amplifier, but the input and output impedances, Z_i and Z_o, will be reduced to resistances, R_i and R_o. This is because we'll design the amplifiers intending for them to work in their middle band of frequencies, not considering the out-of-band questions, which is outside the scope of a book about the fundamental aspects of electronics. In such a way, the capacitive and inductive reactances will have just zero or infinite values (treated as short or open circuits).

Observations

When the output of an amplifier is connected to the input of a headphone, we have to consider the respective values of the impedances. Generally speaking, the lower the headphone's electrical impedance, the easier it is to get higher volume.

The impedance varies with frequency, so it is an AC property of the network. In that sense, it cannot be simply measured by an ohmmeter, which measures just the DC resistance.

10.2.3 Efficiency

The *efficiency* of an amplifier represents the amount of RMS AC power $P_o|_{AC}$ delivered to load, with respect to the power supplied by the DC source $P_i|_{DC}$:

$$\eta = \frac{P_o|_{AC}}{P_i|_{DC}}$$

$$\eta(\%) = \frac{P_o|_{AC}}{P_i|_{DC}} * 100\%$$

According to the definition, an amplifier with 100% efficiency furnishes an output power equal to the input power, without any loss (no Joule effect). It is evident that this is not possible. So, any arbitrary amplifier has an efficiency lower than 100% for sure. Lower efficiency means higher levels of heat dissipation.

10.2.4 Class

The amplifier efficiency depends on the so-called "class" of the amplifier. We distinguish four classes in analog designs: A, B, AB, and C. Differences are related to the amount of variation of the output signal for a full cycle of input signal, as in Figure 10.4.

The classes are defined as follows:

→ *Class A*: The output signal varies for a full 360° of the cycle of the input waveform.
→ *Class B*: The output signal varies over one-half, that is, 180°, of the input signal cycle.
→ *Class AB*: The output signal swings between 180° and 360° of the input waveform.
→ *Class C*: The output signal varies less than 180° of the input signal cycle.

The four different behaviors are related to both the amplitude of the input signal and the *quiescent working point* (or *quiescent point* or *bias-point*) Q of the transistor, a fundamental part of the amplifier. The Q-point corresponds to the DC behavior of the circuit, so all voltage and current levels are fixed, and named *quiescent voltages* and *quiescent currents*. The Q-point is determined by the biasing conditions, due to the imposed DC sources and their divisions among circuit components. Once the quiescent point Q is fixed, AC perturbations will be determined by the input signal swinging around Q.

→ When there is a "small" input signal and a Q-point in the "center" of the output characteristics of the transistor, the output signal will swing for a full 360° of the cycle of the input waveform, as in Figure 10.5a, where the input/output characteristics of a BJT are adopted as an example.
→ When the point Q is chosen closer to the cutoff region, that is the quiescent current is reduced, the output signal will swing between 180° and 360° of the input waveform, as in Figure 10.5b.
→ When the point Q is in the cutoff region no bias is applied at all, and the output signal will swing for 180° of the input signal waveform, as in Figure 10.5c.
→ When the active device is biased beyond the cutoff, that is, negative biasing conditions, the amplifier operates for less than half of the input waveform, as in Figure 10.5d.

Please note that all classes amplify only the portion of the input signal that is within the operating region, that is, the active one, of the transistor.

The four different classes of operations are for different purposes, each with its own distinct advantages and disadvantages.

Class A keeps the signal waveform always within the linear region of the transistor's characteristics, avoiding running out in nonlinear zones. Therefore the main advantage is that the output signal is always an exact amplified reproduction of the input one. A drawback is poor efficiency, because the DC quiescent current flows during the whole waveform cycle, differing from zero, even if there is no input signal to amplify. Therefore class A is used when high fidelity is requested, without distortion, in low to medium power output stages (typically 2 W and below),

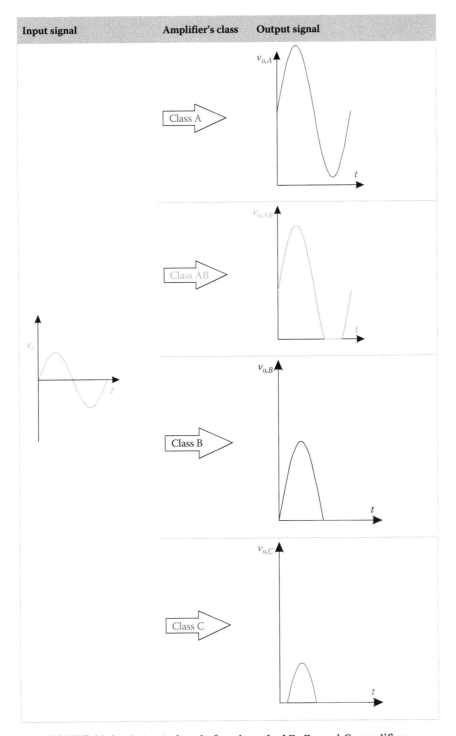

Input signal	Amplifier's class	Output signal

FIGURE 10.4   Output signals for class A, AB, B, and C amplifiers.

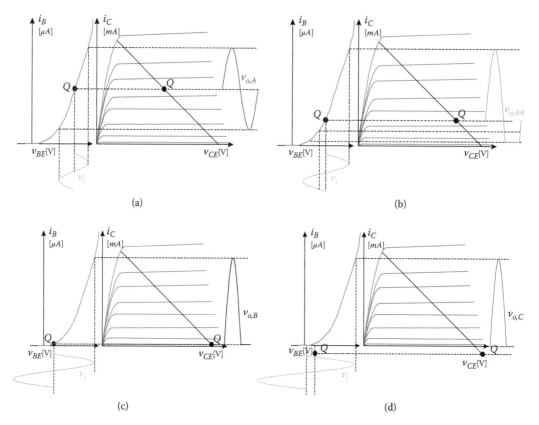

FIGURE 10.5 The transistor supplies current to the load during (a) the entire swing, (b) the output signal swings between 180° and 360° of the input waveform, (c) the output signal varies during 180° of the input signal cycle, and (d) the output signal varies less than 180° of the input signal cycle.

but it is not often used for higher power amplification stages. So applications include (but are not limited to) radio or TV receivers and headphone amplifiers.

Class B is utilized because the DC component of the output current is zero when the input signal to the amplifier is zero; therefore, no power is dissipated in the quiescent condition, and the efficiency is increased with respect to class A. But, since the output reproduces the amplification for only a half-cycle of the input waveform, a two-transistor configuration is necessary. In fact, one transistor is required to amplify one half of the input waveform cycle, and the other transistor amplifies the other remaining half, with the two resulting amplified halves joined together at the output terminal.

Class AB is utilized as a compromise between the advantages and drawbacks of the class A and class B amplifiers.

The *class C* amplifier is not useful for audio applications, but is commonly used in radio-frequency circuits where a resonant output circuit keeps the sine wave going during the non-conducting portion of the input cycle.

The power efficiency improves with class, getting higher as we move from class A to class C.

Even if a transistor is correctly biased, and the Q-point is located at the center of its output characteristics, the amplifier cannot work as a class A if the amplitude of the input signal is too large. That is the reason we talk about a "small" input signal, or *small signal conditions* (Figure 10.6).

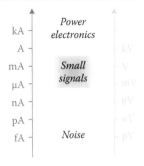

FIGURE 10.6 Order of magnitude for small signals.

10.2.5 Stability

An amplifier is configured by one or more transistors, with one or more DC sources, together with other components useful to to define a predictable and stable operating Q-point (see Section 10.2.4). This network is then sourced by an input AC signal to be amplified linearly around the Q-point, and the amplified signal is provided to an external load.

The linearity, and the class of the amplifier, is determined by the perfect knowledge of the exact position of the Q-point. That's the reason why it is fundamental that the Q-point remains always fixed, even changing the operative conditions of the overall system. So, the *stability* of an amplifier states that the DC output current remains as constant as possible with variations in operation conditions and parameters of components (not only the transistor).

Specific configurations will be further examined (Section 11.9 in Chapter 11) to assure a good stability in the design steps.

10.2.6 Bandwidth

An amplifier is said to be *linear* when it is capable of delivering more power into a load, while mantaining a constant output-to-input signal amplitude ratio, no matter the source or load. So, *linearity* entails the ability to produce only output signal frequencies that are already present in the applied input signal, without added harmonics. An input signal usually contains several frequencies; all of them must be amplified by the same amount of amplification to avoid changes in the output waveform with respect to the input one.

An *ideal* amplifier mantains linearity even with respect to the frequency. Actually this is not true for a *real* amplifier, for which the gain is always referred to as the *mid-frequency gain*, that is, the uniform gain it provides within a band of frequencies. In particular, with the same aforementioned meaning given for electronic filters (Section 7.3 in Chapter 7), it is possible to distinguish *low-pass, bandpass,* and *high-pass* amplifiers, and variants like *wide-band* or *narrow-band* amplifiers.

The differences are related to the so-called *frequency response* of the amplifier, namely its behavior with frequency (which applies only to linear networks).

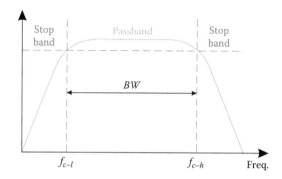

FIGURE 10.7 Amplifier bandwidth.

We are interested here in particular with the *low-pass* amplifiers, commonly called *audio* or *video* amplifiers, depending upon the final applications.

Stricly speaking, a real low-pass amplifier is rarely designed, because it needs to amplify the DC component of the signal, too. The amplification of the zero frequency component is neither convenient, nor practical, so it is commonly avoided. This is why the low-pass amplifiers usually also amplify the very low frequencies but avoid the zero component.

For similar and evident reasons, the design of a real high-pass amplifier is impracticable due to an infinite frequency component, and an upper limit frequency will be determined.

Therefore, a band of frequencies exists where the amplifier gain can be considered practically constant, but also a lower value of frequency f_{c-l} and an upper value of frequency f_{c-h} at which the voltage or current gain drops to $1 / \sqrt{2}$ ($-3\,dB$) of the mid-band value, or the power gain drops to one-half the mid-band value. These are the *cutoff* (or *band-edge*) frequencies.

This is defined as *bandwidth BW*, the amplified frequencies within the higher and lower cut-off frequencies (Figure 10.7):

$$BW = f_{c-h} - f_{c-l}$$

In general the *real* capacitors impact the behavior of the lower frequencies of the amplifier, so to determine its lower cutoff frequency f_{c-l}, while the *parasitic* capacitors (intrinsic in doped semiconductor devices, due to junction/depletion or diffusion/ transit time capacitances, Section 8.5.4 in Chapter 8) impact the behavior of the higher frequencies of the amplifier, so to determine its higher cutoff frequency f_{c-h}.

Curiosity

Our life has a bandwith too! The longer in time, the shorter in frequency and vice versa.

An ideal amplifier must present a *flat* frequency response within its passband, that is, it has to reproduce output signals with no emphasis or attenuation of a particular frequency. In real amplifiers this may not be true. In demanding applications, for instance professional video, it is desiderable to maintain a relatively flat bandwidth, so that we have to refer to the *bandwidth flatness* expressed in *dB*. A 0.5 *dB* flatness means that there is no more than 0.5 *dB* ripple within the amplifier's range of amplified frequencies.

When buying earphones or headphones, an unskilled person takes into consideration the design, the price, the comfort, the weight, even the colors of the plastic envelope. That's makes no sense if other, more important parameters, are neglected. Those who know about electronics pay attention in particular to some specifications, among which is the bandwidth. The audible bandwidth for most healthy human ears is around 20 *Hz*–20 *kHz*, so a good earphone must respect that range. A smaller bandwidth prevents you from hearing bass and/or high tones, while a larger bandwidth is simply a waste unless you are a sound professional. Ultra-low frequencies less than 20 *Hz* and ultra-high frequencies higher than 20 *kHz* are more felt than heard.

Roughly speaking, an audio-frequency amplifier has a 20–20 *kHz* bandwidth, a radio-frequency amplifier has a a bandwidth from 20 *kHz* to many *MHz*, and a video-frequency amplifier has a bandwith within the range 50–100 *MHz*.

The bandwidth of an arbitrary amplifier can be determined by means of its *square wave response*, which is the experimental application of an input square wave signal with the verification of the amplifier's output response.

The square signal has the shape shown in Figure 10.8 (Section 1.2.2 in Chapter 1).

This comes from adding more and more odd-integer harmonics to a sine wave, in a way that the more harmonics are added, the closer to a square shape it becomes, until there is a perfect square shape for an infinite number of harmonics. This is evident by recalling Figure 1.18 in Chapter 1, limited to the seventh harmonic function (Figure 10.9).

Since the square wave is made of infinite sinusoidal functions, starting from zero to infinite frequency, the amplifier can amplify without distortion only those withinits bandwidth. So, only an amplifier with infinite bandwidth can replicate an

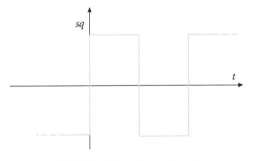

FIGURE 10.8   Step signal.

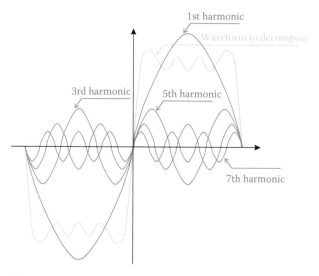

FIGURE 10.9   An similar-square wave periodic function.

amplified undistorted replica of an input square wave. But any real amplifier cannot properly amplify frequencies lower that its lower cutoff frequency f_{c-l} and higher than its higher cutoff frequency f_{c-h}, that is, out of its BW.

The results are schematized in Figure 10.10a and b. Figure 10.10a shows the amplifier's response when low frequencies below f_{c-l} are not amplified. Figure 10.10b shows the result when frequencies higher than f_{c-h} are not amplified.

 Observation

Please pay close attention to the forms of the Figure 10.10 compared to those shown in Figure 4.55 om Chapter 4, relative to capacitor's charge/discharge.

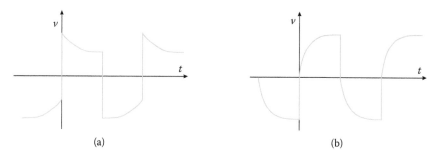

FIGURE 10.10   The amplifier's response when (a) low frequencies and (b) high frequencies are not amplified.

10.2.7 Slew rate

The *slew rate SR* of an amplifier is the maximum rate of change of its output voltage. Accordingly, it is expressed in V/s or, more commonly, $V/\mu s$.

With an input sine wave signal

$$v(t) = V_M \sin(\omega t) = V_M \sin(2\pi ft)$$

with V_M its peak value, the maximum rate of change at the zero crossing is

$$\frac{dv(t)}{dt} = \omega V_M = 2\pi f V_M$$

The amplifier is asked to perform with an *SR* at the same rate of that input signal, as a limit, or faster, which is preferable.

The *full-power bandwidth (FPBW)* is defined as the maximum frequency at which slew rate does not represent a limit for rated voltage output:

$$FPBW = \frac{SR}{2\pi V_M}$$

10.2.8 Distortion

Differences from the linear behavior of an amplifier lead to *distortion*, which indicates the differences between the output and input waveforms except for a scaling factor (the amplification). According to what we defined in Section 10.2.4, the higher the class of an amplifier, the higher the distortion (class A: practically no distortion, class C: the highest).

Distortion is the event that occurs whenever the output signal of an amplifier differs from the input source waveform.

We distinguish three main types of distortion: *frequency distortion, phase (or delay) distortion,* and *amplitude distortion* (others, including *transient distortion, scale or volume distortion,* and *frequency modulation distortion* are not considered as much).

Frequency and *phase* distortions occur because the amplifier's gain cannot be constant at all frequencies within its bandwidth. *Amplitude* distortion, subdivided into *harmonic* and *intermodulation* distortion, occurs because the amplifier itself introduces new frequencies other than those of the input signal.

Observation

We are dealing with *distortion* provided by an amplifier. But any arbitrary circuit can distort a signal. An *RC* high-pass filter, such as the one reported in Section 7.3.2 in Chapter 7, causes distortion to a square waveform, as in Figure 10.11.

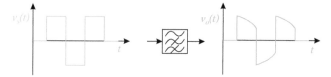

FIGURE 10.11   RC high-pass filter causes distortion to a square signal.

The output signal has a reduced high-frequency content with respect to the input one.

Frequency distortion is unequal amplification of all frequencies of a multi-frequency input signal. For example, if the fundamental frequency of the input signal falls within, while the third harmonic falls outside, the amplifier's bandwidth, the results is a different waveform in the output, and thus distortion. At mid-frequencies there is no frequency distortion, and no new frequencies are produced with this type of distortion.

Phase distortion is when the reactive parts of the amplifier (both the effective ones and the parasistic ones, even of the transistor itself) shift the phase of the components of a multi-frequency signal with respect to each other. The result is that some parts of the output signal will be delayed with respect to some others.

Amplitude distortion (also known as *nonlinear distortion)* occurs when the amplification doesn't take place over the whole signal cycle. This can be due to incorrect biasing (the Q-point not in the middle of the output characteristics of the transistor) or an input signal "larger" than that allowed to assure linearity (the input is overdriven and the transistor operates on nonlinear parts of its characteristics), so that the signal is clipped-off as in Figure 10.5b through Figure 10.5d.

This type of distortion generates new harmonic components of the signals and greatly reduces the efficiency of an amplifier. Odd harmonics (third, fifth, seventh, etc.) result from symmetrical nonlinearities, while even harmonics (second, fourth, sixth, etc.) result from asymmetrical nonlinearities, and an amplifier can have both symmetrical and asymmetrical characteristics.

The amplitude distortion can generate harmonic distortion and intermodulation distortion.

Harmonic distortion adds harmonics to a one-tone pure sinusoidal waveform fed to the amplifier, caused by a clipping effect mainly due to the transistor, as represented in Figure 10.12.

Given an iso-frequency input signal as

$$v_s(t) = V_{sM} \cos(\omega t)$$

with V_{sM} being its peak value, the output signal v_o will have the fundamental and harmonic components

$$v_o(t) = V_{o1} \cos(\omega t) + V_{o2} \cos(2\omega t) + V_{o3} \cos(3\omega t) + \dots$$

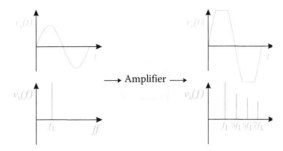

FIGURE 10.12 A single-frequency sinosoidal input is clipped by the amplifier and new (harmonic) frequencies are produced in the output signal. Here a symmetrical clipping is proposed, so that only odd harmonics are generated. If the output spectrum also presents even harmonics, the distortion is not symmetrical.

The harmonic distortion of the *n*th harmonic frequency component is defined as the ratio of the RMS value of the harmonic of interest (second, third, fourth, ..., *n*th) to the RMS signal level:

$$D_n = \frac{|V_{0,n-RMS}|}{|V_{0,1-RMS}|}$$

which can be also expressed as a percentage or in *dB*.

The overall distortion, called *total harmonic distortion (THD)*, which takes into account all the harmonics added geometrically, is defined as the ratio of the root-sum-square value of all the harmonics (second, third, fourth, etc.) to the RMS signal level:

$$THD = \frac{\sqrt{V_{o,2-RMS}^2 + V_{o,3-RMS}^2 + \dots}}{V_{o,1-RMS}}$$

In general, only the first five or six harmonics are significant.

Curiosities

Indeed, some guitarists prefer the sound to be highly distorted (*overdriven*) by intentionally clipping the output waveform. But some people do not tolerate distortion at all; some others consider an even-order distortion to be *sweet* or *harmonically rich*, the rest define an odd-order distortion as *bright* or *detailed* It's up to you!

High-order harmonics can be more "audible" even when with very low amplitude. This is because they are "far" from the fundamental frequency that cannot "mask" them.

The lower detectable value of harmonic distortion for a human ear is difficult to evaluate, but it is considered to be around 0.3%. So, for practical aims an amplifier producing a THD higher than that value does not "guarantee" hearing a "cleaner" sound.

Some recent studies demonstrated that many people consider "more musical" a signal with 0.5% second harmonic distortion added.

Intermodulation distortion (IMD) occurs when two or more input frequencies are applied in a nonlinear amplifier. Intermodulation is intended as the production of new components, termed *intermodulation products*, having frequencies corresponding to undesired sums and differences of the fundamental and harmonic frequencies of the applied waves. Specifically, when the sum of two sine waves is applied to the input of a non-ideal amplifier, it produces new harmonic components that mix to the input sine waves, resulting in an output signal containing sine wave components at frequencies other than those of the input signals.

Curiosities

IMD can occur in many systems, even unimagined, other than amplifiers. For example, if a connector on a transmission feed becomes corroded, it can behave as a nonlinear diode junction, adding IMD to the propagating signal.

In audio amplifiers, it is recommended to avoid IMD, since it produces disagreeable, harsh sounds. IMD can mask musical detail such that, with orchestral music, the instruments seem covered by a "veil."

Filtering cannot be a solution to reduce or eliminate IMD, since IMD is close in frequency to the desired signal.

Class A amplifiers, using the relatively linear part of the transistor's characteristics, are no more subject to distortion than other bias classes.

10.2.9 Noise

Noise is defined as any electrical disturbance, that is, any current or voltage not the signal. So, it does not carry any information, but determines the acceptability of an output signal from an amplifier. If the signal is, for instance, an iso-frequency sine-wave, any frequency components other than its frequency is noise. Evaluating noise is important for different practical reasons, but mainly because it limits the perception of small changes of the signal, and hence the signal resolution we can obtain. An important request is that, in analog signal processing, the amplifier should not limit the overall resolution.

Interference can be classified as a particular noise. It can be considered an *artificial noise* since it is generated by other signals within the same or in other circuits. The signal itself can generate interference. Examples come from an imperfect matching on a transmission line, or from two signals coming from two different local radio stations but which broadcast on the same frequency band.

Both noise (in a general sense) and interference (in a particular sense) cause problems in receiving the desired signal properly, by causing errors in TV (analog or digital) video signal reception or objectionable sounds in the background of a radio program (Figure 10.13a and b).

10.2.9.1 Types

Noise can be mainly grouped into *thermal noise, shot noise, flicker noise, avalanche noise,* and *burst noise.*

Thermal noise, or *Johnson–Nyquist noise,* or *Johnson noise,* or *Nyquist noise* (John Bertrand Johnson, Swedish-born American electrical engineer and physicist, 1887–1970; Harry Theodor Nyquist, Swedish-born American physicist, 1889–1976), is caused by the random motion of charge carriers, in other words the differences in charge density within a matter, due to thermal energy. It is both material-independent and frequency-independent as it is the same through the frequency spectrum (at least up to the frequencies adopted here). It is measured by means of the RMS value of the voltage it generates

$$v_{th-n} = \sqrt{4kTR\Delta f}$$

and, accordingly, the noise current

$$i_{th-n} = \sqrt{\frac{4kT\Delta f}{R}}$$

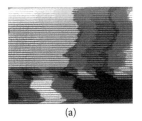
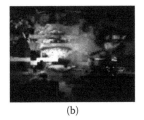

(a)　　　　　　　　(b)

FIGURE 10.13　(a) Analog and (b) digital TV images with added noise.

with k being Boltzmann's constant, T the absolute temperature of the resistance R in which the noise is generated (it can be the resistance of a conductor as well as of a semiconductor from which the transistor is made), and Δf the considered bandwidth. So to take into account the thermal noise in a real resistor, we can assume an idealized resistor together with a fluctuating voltage generator $v_{th\text{-}n}$ at its terminals.

Shot noise is due to the corpuscular nature of matter, and it is always associated with a DC flow. Since the current is due to motion of charged particles that are discrete and independent, at nano scale and in instant time, the current's variation is unpredictable. The number of carriers passing through a surface varies instant by instant, even if it can be considered constant in average over a certain net time. These instantaneous fluctuations are noise. This noise in a *p–n* junction is generated by current flowing and is a function of the bias current I_{DC} and the electron/hole charges, and we can write for the related current

$$i_{s-n} = \sqrt{2qI_{DC}\Delta f}$$

with q being the fundamental charge value and Δf the considered bandwidth. Conductors *do not* experience shot noise, because there are no potential barriers in them, as in a *p-n* junction.

Flicker noise, or *1 / f noise*, or *low-frequency noise*, or *excess noise*, was discovered in 1925 by Johnson. It occurs in so many fundamentally different systems that its explanation is subject to controversal debates and no satisfactory explanation has been developed, since it is highly improbable that an identical mechanism causes noise in all of them. It is termed *flicker* because of the anomalous strange "flicker" that was observed in the plate current of vaccum tubes early in 1925, and $1 / f$ because its spectrum varies as $1 / f^{\alpha}$. The expression of its spectral density can be furnished by the current as

$$i_{fl-n} = \sqrt{K_1 \frac{I_{DC}^{\beta}}{f^{\alpha}} \Delta f}$$

with K_1 and β being dependent on the device and the application.

Flicker is especially evident in semiconductors, such as MOSFETs, which conduct currents by transporting carriers near to the device surface, which is generally rife with imperfections.

Curiosities

It is really interesting to note that $1 / f^{\alpha}$ signals, generally with $0.5 < \alpha < 1.5$, can be found practically in every field: physics, astrophysics, geophysics, biology, psychology, economics/financials, music, and so on.

It has been recorded within an EEG (electroencephalography) brain signal, in a human heart signal, at musical concerts, in the rate of traffic flow on highways, in loudness and pitch of speech, etc., and of course in conductors and semiconductors.

The noise can be referred to by colors. A *white noise* is not related to the frequency, so it is the same overall the frequency spectrum, a *blue noise* is proportional to the frequency f, a *pink noise* goes as $1 / f$, a *violet noise* as f^2, and a *brown noise* as $1 / f^2$.

Avalanche noise is produced by a junction diode operating in reverse conditions at the onset of avalanche breakdown (Section 8.5.5 in Chapter 8). When a "strong" electric field creates additional electron-hole pairs causing electrons to dislodge other charge carriers creating hole electron pairs, the resulting effect, being cumulative, iterates, and a random noise spike can be observed. Avalanche diodes can generate large quantities of RF (radio frequency) noise.

Burst noise, sometimes referred as *popcorn noise* or *random-telegraph-signal noise*, occurs in semiconductor devices, and in amplifiers, and manifests as a noise crackle. it's part of $1 / f$ noise. For a MOSFET device, it can be described as a discrete modulation of the channel current caused by trapping/release of channel carriers. The spectral density can be expressed by means of its current value:

$$i_{b-n} = \sqrt{K_2 \frac{I_{DC}^c}{1+\left(\frac{f}{f_c}\right)^2} \Delta f}$$

with K_2 being a constant depending on a particular component, c a constant, and f_c the cutoff frequency depending on the considered noise process.

10.2.9.2 Measures

Every device has some amount of noise at its output and the importance of this noise is relative to the application. A strong output signal level masks a low level noise, but if the output signal level is small, even a very low noise level can be unacceptable. So the signal has to be related to the noise, and the *signal-to-noise ratio* (*SNR*) is its measure, as a ratio of respective powers:

$$SNR = \frac{P_{signal}}{P_{noise}}$$

or, in *dB*:

$$SNR_{dB} = 10\log_{10}\left(\frac{P_{signal}}{P_{noise}}\right) = P_{signal,dB} - P_{noise,dB}$$

The *noise factor F* allows us to know the additional noise a component or a network will contribute to the noise from the source:

$$F = \frac{SNR_{in}}{SNR_{out}}$$

so ideally $F = 1$.

Finally, the *noise figure NF* expresses F in *dB*:

$$NF = 10\log(F)$$

Observations

The residual AC ripple on the power supply means "hum" in the output of the amplifier, since the "hum" can be treated like a low frequency signal.

The noise can cause distortion when an amplifier is designed with more than one stage hooked up one after another. This is because the later circuit receives noise at its input along with the signals.

For the operation of a MOS transistor, thermal noise and shot noise are physically fundamental and are always present.

A *low noise amplifier* (*LNA*) is a special type of electronic amplifier mainly used in communication systems to amplify very weak signals. This is because it adds a very low noise value, so to affect the signal as little as possible.

Usually an LNA is located close to the receiving antenna and adopted to boost the received signal. It amplifies the signal before additional noise is injected from other electronic circuits.

10.3 TYPES OF AMPLIFIER

Amplifiers can be classified according to a very large number of characteristics: channels (mono, two-, four-, etc.), class rating (A, B, AB, C), frequency range (audio, video, RF, HF, microwave, etc.), adopted technology (monolitic, hybrid, integrated, discrete, etc.). Here we will illustrate the most relevant feature to classify an amplifier, based on its very fundamental characteristic, that is, the ability to amplify.

An ideal amplifier is only a concept, but we will define it to understand how to approach it when we will design one.

10.3.1 Voltage Amplifier

An *ideal* voltage amplifier can be modeled by a voltage-controlled voltage source (VCVS) (see Figure 5.6a in Chapter 5). For a *real* voltage amplifier we have to add to

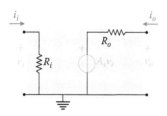

FIGURE 10.14 Model of a real voltage amplifier.

that model an input resistance R_i (or more generally an input impedance Z_i), to take into account the input current the amplifier draws from the signal source, and a series output resistance R_o (or more generally an output impedance Z_o), to form a Thévenin equivalent model to take into account the change in output voltage as the amplifier is called upon to supply output voltage to a load, hence the scheme shown in Figure 10.14.

For this scheme, the gain, termed open-circuit voltage gain A_v (*unitless*) is

$$A_v = \frac{v_o}{v_i}\bigg|_{i_o=0}$$

This is a *voltage amplifier*, so to understand how it behaves we add a voltage source (modeled by a Thévenin equivalent) v_s with a series internal resistance R_s, and a load R_L to which the output voltage is supplied (Figure 10.15).

The resistance R_s of the voltage source and the input resistance R_i of the amplifier form a voltage divider, as well as the output resistance of the amplifier R_o and the resistance of the load R_L. Accordingly we can write

$$v_o = \frac{R_L}{R_L + R_o} A_v v_i$$

and

$$v_i = \frac{R_i}{R_i + R_s} v_s$$

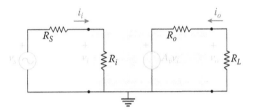

FIGURE 10.15   Model of a real voltage amplifier (blue box) with source and load.

Therefore, the overall voltage gain $\frac{v_o}{v_s}$ can be found as

$$\frac{v_o}{v_s} = \frac{R_L}{R_L + R_o} A_v \frac{R_i}{R_i + R_s}$$

The conclusion is that if we satisfy the conditions

$$\left[\begin{array}{ccc} R_L \gg R_o & \Leftrightarrow & \dfrac{R_L}{R_L + R_o} \to 1 \\[3mm] R_i \gg R_s & \Leftrightarrow & \dfrac{R_i}{R_i + R_s} \to 1 \end{array}\right.$$

the overall voltage gain $\frac{v_o}{v_s}$ roughly equals the open-circuit voltage gain A_v, so that

$$\frac{v_o}{v_s} \cong A_v$$

which approximates the ideal request.

In designing an amplifier we can define its input and output resistances, R_i and R_o, to operate independently from possibly unknown values of source and load resistances, R_s and R_L. Therefore, it would be important to assure the following values:

$$R_i \to \infty; \ R_o \to 0$$

Example

The *ECG* (*electrocardiogram*) is used to detect the heart's electrical activity (Figure 10.16a) to identify eventual heart disorders. It commonly utilizes a pair of electrodes to reveal the ionic potential difference between their respective points of application on the body surface. The electrodes behave as capacitors and/or resistors or somewhere between them, depending on their type. The voltage across the electrodes can be on the order of 1–5 mV. Apart from the electrodes, the second stage of the instrumentation is an amplifier (for example the Analog Devices AD624) because amplification is required to increase the signal amplitude to further process and/or to drive a display. The amplifier is asked to act as a voltage amplifier, with a high input impedance. Generally, the signal of interest resides within the 0.67–40 Hz bandwidth, so an additional stage is made by a filter to reject signals out of that frequency range.

As a curiosity, we can mention the fact that the early ECG recording technique adopted a very curious method to obtain the signals. The subject was asked to put his/her hands into two different pots with saline solutions to enhance electrical conduction (Figure 10.16b). Two signals were then taken from the hands and a third signal from one leg.

Observation

An infinite input resistance $R_i = \infty$ implies that the input current flowing into the amplifier would be ideally zero $i_i = 0$, so that the power flow into the input is zero, which results in an amplifier with infinite power gain. This is another reason to call this amplifier *ideal*.

10.3.2 Current Amplifier

An *ideal* current amplifier can be modeled by a current-controlled current source (CCCS) (see Figure 5.6d). For a *real* current amplifier we have to add to that model an input resistance R_i (or more generally an input impedance Z_i), and an output

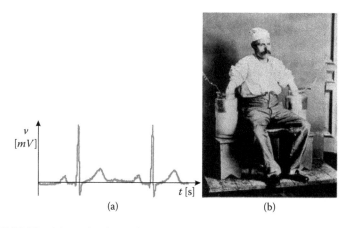

(a)   (b)

FIGURE 10.16   (a) Typical cardiac cycle; (b) early ECG recording technique.

resistance R_o (more generally an output impedance Z_o), to form a Norton equivalent model, hence the scheme shown in Figure 10.17.

For this scheme, the gain, termed the short-circuit current gain A_i (*unitless*), is

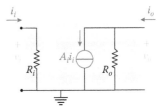

$$A_i = \left.\frac{i_o}{i_i}\right|_{v_o=0}$$

FIGURE 10.17 Model of a real current amplifier.

This is a *current amplifier,* so to understand how it behaves we add a current source (modeled by a Norton equivalent) i_s with its parallel internal resistance R_s, and a load R_L (Figure 10.18).

The resistance R_s of the current source and the input resistance R_i of the amplifier form a current divider, as well as the output resistance of the amplifier R_o and the resistance of the load R_L. Accordingly we can write

$$i_o = \frac{R_o}{R_o + R_L} A_i i_i$$

and

$$i_i = \frac{R_s}{R_s + R_i} i_s$$

Therefore, the overall current gain $\dfrac{i_o}{i_s}$ can be found as

$$\frac{i_o}{i_s} = \frac{R_o}{R_o + R_L} A_i \frac{R_s}{R_s + R_i}$$

The conclusion is that if we satisfy the conditions

$$
\begin{cases}
R_o \gg R_L & \Leftrightarrow & \dfrac{R_o}{R_o + R_L} \to 1 \\[2ex]
R_s \gg R_i & \Leftrightarrow & \dfrac{R_s}{R_s + R_i} \to 1
\end{cases}
$$

the overall current gain $\dfrac{i_o}{i_s}$ roughly equals the short-circuit current gain A_i, so that

$$\frac{i_o}{i_s} \cong A_i$$

which approximates the ideal request.

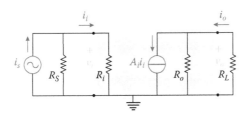

FIGURE 10.18 Model of a real current amplifier (blue box) with source and load.

In designing an amplifier we can define its input and output resistances, R_i and R_o, to operate independently from possibly unknown values of source and load resistances, R_s and R_L. Therefore, it would be important to assure the following values:

$$R_i \to 0; \quad R_o \to \infty$$

Observation

An null input resistance of $R_i = 0$ implies that the voltage across the input port would be ideally zero, $V_i = 0$, so that the power flow into the input is zero, which results in an amplifier with infinite power gain. This is another reason to call this amplifier *ideal*.

Example

A photovoltaic cell acts like a current source when it measures light intensity. A current amplifier can amplify such a current to drive a meter, which requires a current drive.

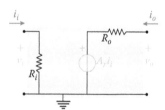

FIGURE 10.19   Model of a real transresistance amplifier.

10.3.3 Transresistance Amplifier

The *transresistance* (or, more generally, *transimpedance*) amplifier converts current to a proportional voltage. So, it can also be called a current to voltage, or *I-to-V*, converter.

An *ideal* transresistance amplifier can be modeled by a current-controlled voltage source (CCVS) (see Figure 5.6b in Chapter 5). For a *real* transresistance amplifier, we have to add to that model an input resistance R_i (or more generally an input impedance Z_i), and an output resistance R_o (more generally an output impedance Z_o), to form a Thévenin equivalent model, hence the scheme shown in Figure 10.19.

For this scheme, the gain, termed the open-circuit transresistence gain A_r [Ω], is

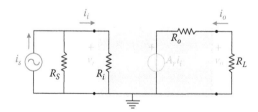

FIGURE 10.20   Model of a real transresistance amplifier (blue box) with source and load.

$$A_r = \left. \frac{v_o}{i_i} \right|_{i_o=0}$$

This is a *transresistance amplifier*. So to understand how it behaves we add a current source (modeled by a Norton equivalent) i_s with its parallel internal resistance R_s, and a load R_L (Figure 10.20):

The resistance R_s of the current source and the input resistance R_i of the amplifier form a current divider, while the output resistance of the amplifier R_o and the resistance of the load R_L form a voltage divider. Accordingly we can write

$$v_o = \frac{R_L}{R_L + R_o} A_r i_i$$

and

$$i_i = \frac{R_s}{R_s + R_i} i_s$$

Therefore, the overall transresistance gain $\dfrac{v_o}{i_s}$ can be found as

$$\frac{v_o}{i_s} = \frac{R_L}{R_L + R_o} A_r \frac{R_s}{R_s + R_i}$$

The conclusion is that if we satisfy the conditions

$$\left\{ \begin{array}{ll} R_L \gg R_o & \Leftrightarrow \quad \dfrac{R_L}{R_L + R_o} \to 1 \\[3mm] R_s \gg R_i & \Leftrightarrow \quad \dfrac{R_s}{R_s + R_i} \to 1 \end{array} \right.$$

the overall transresistance gain $\dfrac{v_o}{i_s}$ roughly equals the open-circuit transresistence gain A_r, so that

$$\frac{v_o}{i_s} \cong A_r$$

which approximates the ideal request.

In designing an amplifier, we can define its input and output resistances, R_i and R_o, to operate independently from possible unknown values of source and load resistances, R_s and R_L. Therefore, it would be important to assure the following values:

$$R_i \to 0; \; R_o \to 0$$

Example

The transresistance amplifier has applications when it is requested to convert an input current to a proportional output voltage. It can be necessary, for instance, utilizing a photodiode, which transduces light energy into electrical energy in a form of a current at high impedance. A photomultiplier or a capacitive sensor or a solar cell are also current-producing devices, but their current can have a very low value. It is possible to convert the input current to an analog output voltage simply by using a load resistor, but the voltage available is often only a few tens of millivolts. A transresistance amplifier can solve the problem.

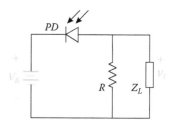

FIGURE 10.21 A photodiode to reveal light radiation for an impedance load.

Let's go deeper. We want to reveal a weak light radiation. A photodiode PD is useful to convert light into a current converted into a voltage by means of a resistor R. This voltage feeds a load impedance Z_L, as represented in Figure 10.21.

But this network does not work properly. In fact we need the output voltage merely related to the light, but this is true only with a load impedance as high as possible, so to satisfy the maximum voltage transfer condition (Section 6.9.3 in Chapter 6). In turn, the higher the load impedance, the lower the current flowing through it. But, a poor current value cannot guarantee that the load is correctly driven. The problem is solved by means of a transresistance amplifier between the resistance R and the impedance Z_L.

Observation

An input resistance of $R_i = 0$ implies that the voltage across the input port would be ideally zero, $v_i = 0$, so that the power flow into the input is zero, which results in an amplifier with infinite power gain. This is another reasons to call this amplifier *ideal*.

10.3.4 Transconductance Amplifier

The *transconductance* (or, more generally, *transadmittance*) amplifier converts voltage to a proportional current. So, it can be also termed as a voltage to current, or *V-to-I*, converter.

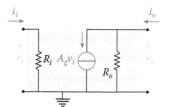

FIGURE 10.22 Model of a real transconductance amplifier.

An *ideal* transconductance amplifier can be modeled by a voltage-controlled current source (VCCS) (see Figure 5.6c in Chapter 5). For a *real* transconductance amplifier we have to add to that model an input resistance R_i (or more generally an input impedance Z_i), and an output resistance R_o (more generally an output impedance Z_o), to form a Norton equivalent model, hence the scheme shown in Figure 10.22.

For this scheme, the gain, termed the short-circuit transconductance gain A_g $\left[\Omega^{-1}\right]$, is

$$A_g = \left. \frac{i_o}{v_i} \right|_{v_o=0}$$

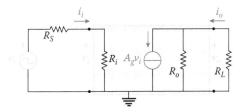

FIGURE 10.23 Model of a real tranconductance amplifier (blue box) with source and load.

This is a *tranconductance amplifier,* so to understand how it behaves we add a voltage source (modeled by a Thévenin equivalent) v_s with its series internal resistance R_s, and a load R_L (Figure 10.23).

The resistance R_s of the voltage source and the input resistance R_i of the amplifier form a voltage divider, while the output resistance of the amplifier R_o and the resistance of the load R_L form a current divider. Accordingly we can write

$$i_o = \frac{R_o}{R_o + R_L} A_s i_i$$

and

$$i_i = \frac{R_i}{R_i + R_s} v_s$$

Therefore, the overall transconductance gain $\dfrac{i_o}{v_s}$ can be found as

$$\frac{i_o}{v_s} = \frac{R_o}{R_o + R_L} A_s \frac{R_i}{R_i + R_s}$$

The conclusion is that if we satisfy the conditions

$$\begin{cases} R_o \gg R_L & \Leftrightarrow & \dfrac{R_o}{R_o + R_L} \to 1 \\[2mm] R_i \gg R_s & \Leftrightarrow & \dfrac{R_i}{R_i + R_s} \to 1 \end{cases}$$

the overall transresistance gain $\dfrac{i_o}{v_s}$ roughly equals the short-circuit transconductance gain A_g, so that

$$\frac{i_o}{v_s} \cong A_g$$

which approximates the ideal request.

In designing an amplifier we can define its input and output resistances, R_i and R_o, to operate independently from possible unknown values of source and load resistances, R_s and R_L. Therefore, it would be important to assure the following values:

$$R_i \to \infty;\ R_o \to \infty$$

Example

The transconductance amplifier has applications when it is requested to drive an output current proportional in magnitude to the input voltage.

The input voltage can be due to voltage sources of a different nature: a thermocouple produces a voltage proportional to a temperature, a light-to-voltage (LTV) sensor provides a linear analog voltage output that is proportional to light, most microphones produce an electrical voltage signal from mechanical vibration, biosignals are recorded as very low level voltages ($1\,\mu V$–$100\,mV$) generated by nerves and muscles, etc. Even the brain waves use electroencephalography (EEG) to furnish microvolt-range potential differences measured across locations on the user's scalp (Figure 10.24).

Some loads need to be driven by voltage, and others by current, as the aforementioned meter. In hi-fi systems, loudspeakers can be voltage or current driven. With voltage driving, the loudspeaker must have low impedance to realize matching conditions. But this means we need high-section copper coils (and therefore we have high weight and poor efficiency in mechanical movements). With current driving conditions, the loudspeaker can be realized with higher impedance to satisfy the same matching conditions. The consequence is lower section copper coils (less weight and higher efficiency). Neverthless current-

FIGURE 10.24   Cap for EEG recording.

drive or transconductance amplifiers for consumer audio applications are very rare, because a typical speaker impedance ranges "only" from 2 to 16 ohms, and loudspeaker systems are mostly optimized for voltage drive.

Observations

A simple resistor can be considered a voltage-to-current converter, since when a voltage is applied across it, a related current will be flowing. So, it appears that the simplest way to realize a transconductance amplifier is to connect a series resistor at the output of a voltage amplifier. The higher the series resistance, the purer the amplifier operates in the current-drive mode. But, as an unavoidable drawback, the available output power would be greatly reduced, so this method is impractical. In addition, a resistor cannot deliver an output resistance that differs from its input resistance.

An infinite input resistance $R_i = \infty$ implies that the input current flowing into the amplifier would be ideally zero $i_i = 0$, so that the power flow into the input is zero, which results in an amplifier with infinite power gain. This is another reason to call this amplifier as *ideal*.

10.3.5 Power Amplifier

A voltage amplifier can amplify a $100\,mV$ signal to $10\,V$ and can drive a $10\,k\Omega$ load, but if the load is reduced to $10\,\Omega$, there is the possibility that the voltage amplifier cannot furnish the current necessary to maintain the same $10\,V$ as output, creating mismatching problems (Section 6.9.3 in Chapter 6).

Similarly, it may occur that a current amplifier amplifies a $10\,\mu A$ signal to $1\,mA$ to drive a load at $1\,V$ but cannot supply the same current at $10\,V$.

In these circumstances, a power amplifier is needed.

A power amplifier has to drive loads sort of that require both voltage and current to be driven, as, for instance, in loudspeakers. Loudspeakers are devices that convert electric energy into mechanical energy at audio frequencies (the sound), and have a maximum power handling capability usually between 50 and 250 watts RMS (home systems). To perform the energy conversion, they move the quite "heavy" *cone* made of inductors with their own inductances and unavoidable resistive losses.

Let's consider an amplifier rating, for example, $P = 100\,W\ @\ R = 8\,\Omega$. Considering that $P = VI$ and $V = RI$, combining these two equations we get that the amplifier delivers a voltage $V \cong 28\,V$ across the loudspeaker, resulting in a $I = 3.5\,A$ current. If we now want to double the cone's movements, we have to double the current to $I_2 \cong 7\,A$, so now the voltage is $V_2 \cong 56\,V$. This means that to have the effect *doubled*, we need an amplifier with *four times* the power $P_2 \cong 396\,W$. The necessary power escalates quite rapidly!

Similar remarks can be made concerning a mobile phone that typically delivers $1\,W$ to a $50\,\Omega$ antenna.

10.4 LOADING EFFECT AND DRIVING PARAMETERS

To properly design an amplifier, there are some key questions to consider. Among the fundamental ones are the following:

→ What type of source is available (current or voltage) and what is the value of its resistance (low, medium, high)?
→ What type of load has to be driven (in current or voltage), and what is the value of its resistance (low, medium, high)?
→ What matching conditions must be satisfied (maximun transfer of current, voltage, or power)?

The answers to these points suggest that the designer choose among the voltage, current, transresistance, transconductance amplifiers, and the proper input/output matching networks..

Let's consider the most common amplifier, which is the voltage amplifier.

An ideal voltage amplifier has an infinite input resistance $R_i = \infty$, and thus a null input current, $i_i = 0$. A real voltage amplifier has a non-zero input current; therefore it has a *loading effect*, that is, the input voltage of the amplifier is smaller than the source voltage.

An ideal voltage amplifier has a null output resistance $R_o = 0$, but a real one has a non-zero value, so that the output of the amplifier will vary if the loading at the output changes. As a result the output can only faithfully amplify if the load resistance value falls within a certain range.

An ideal voltage amplifier can furnish any voltage across the load resistor. A real voltage amplifier cannot exceed the value of the voltage supplied by the DC source, since it cannot "create" energy, but only utilize the energy furnished by the battery or DC power source.

The *loading effect* generally consists of a reduction of the amplifier's gain due to (voltage or current) divider effects.

Therefore the value of the load has an effect on the voltage gain of the amplifier. The lower the load resistance, the lower the voltage gain and, vice versa, the greater the load resistance, the greater the voltage gain, until the highest voltage gain occurs when the load opens.

Curiosity

Did you ever notice that when you turn on your water heater switch, the bathroom bulb's intensity gets reduced? This is just due to the loading effect.

If you design an amplifier that has to furnish an output voltage to a certain load, if you short the output, of course that output voltage cannot be maintained. This is another example of the loading effect.

When you measure the voltage of a battery without any load connected, the measure differs from what you get when the battery is closed on a low resistance load. Again, this is a loading effect problem.

Let's recall the voltage divider scheme of Figure 6.33 in Chapter 6, and the considerations made in the relative section. Now we can affirm that if a resistor load R_L is driven from a high resistance source, this resistor operates under conditions of current drive. Otherwise, if the load is driven from a low resistance source, this load operates under conditions of voltage drive. The terms "high" and "low" are clearly with respect to the load. The same is true from the input side, since a high resistance source can operate under conditions of current drive, while a low resistance source can operate under conditions of voltage drive.

10.5 KEY POINTS, JARGON, AND TERMS

→ A signal source is anything that provides the signal. An amplifier is a system that provides (voltage or current or power) gain. The load is anything we deliver the amplified signal to.

→ To *amplify* refers to the ability of a circuit to produce an output (voltage and/or current) signal larger than the input one. An *amplifier* increases the level of voltage and/or current and/or power of an arbitrary signal, and the operation is called *amplification*. The opposite is *attenuation* when the signal gets reduced, as occurs in a passive filter.

→ A DC amplifier operates when there is no signal but only static voltage or current is at its input. An AC amplifier operates when a signal is at its input port and its output voltage changes with changing voltage drop across a load impedance.

→ The key main parameters for amplifiers are gain, impedances, efficiency, class, stability, bandwidth, slew rate, distortion, and noise.

→ The classes A, B, AB, and C refer to the way the amplifiers are biased in comparison to the amplitude of the input signal. Class A preserves the original wave shape as the transistor is biased using the most linear part of its characteristics. The drawback is poor efficiency. In class B amplifiers there is practically no standing bias current, so there is negligible power consumption without a signal. So, it can be used for more powerful output and it is more efficient than class A. Class AB is slightly less efficient than class B because it uses a small quiescent current to bias the transistors just above cutoff. Class C much improves efficiency to the amplifier, but has a very heavy distortion of the output signal.

→ DC stability for an amplifier means that the Q-point remains always fixed.

→ *Bandwidth* is the frequency band that passes the amplified signal, namely the differences between the cutoff frequencies.

→ The *slew rate* of an amplifier is the maximum rate of change of its output voltage.

→ Differences from the linear behavior of an amplifier lead to *distortion*.

→ *Noise* is defined as any electrical disturbance, that is, any current or voltage not the signal.

→ The main types of amplifier are voltage amplifier, current amplifier, transresistance (or transimpedance) amplifier, and transconductance (or transadmittance) amplifier. A voltage amplifier produces an output voltage larger than the input voltage. A current amplifier produces an output current larger than the input current. A *transresistance amplifier* uses input current to produce a proportional output voltage. A *transconductance amplifier* uses input voltage to produce a proportional output current.

→ A power amplifier drives loads that require both voltage and current to be driven.

→ The *loading effect* generally consists of a reduction of the amplifier's gain due to (voltage or current) divider effects.

10.6 EXERCISES

EXERCISE 1

There are several different types of microphone: carbon, dynamic, crystal, capacitive (electret or condenser). The latter are preferred in portable devices (in mobile phones, tie clips, consumer video cameras, computer sound cards) because of their small size and low production cost. Consider a capacitive microphone with a very low voltage output, $v_s = 1\,mV$, and very high output impedance (due to the electric element itself), $R_s = 1\,M\Omega$. To amplify the signal a voltage amplifier is used with an input

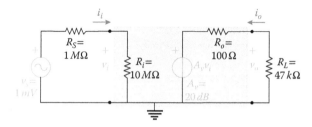

FIGURE 10.25   Voltage amplifier with a mic as source and a load resistor.

resistance $R_i = 10\,M\Omega$, a voltage gain of 20 dB, and an output resistance $R_o = 100\,\Omega$. The resistance load is $R_L = 47\,K\Omega$ (Figure 10.25).
   Calculate the overall voltage amplification v_o/v_s.

ANSWER

$$\frac{V_o}{V_s} \cong 9.1$$

EXERCISE 2

A current amplifier has a gain of 40 dB, an input resistance $R_i = 10\,\Omega$ and an output resistance $R_o = 22\,k\Omega$. Calculate the input signal current $i_o = 1\,A$ required to produce an output signal current $i_o = 1\,A$ in a load resistor $R_L = 50\,\Omega$.

ANSWER

$$i_i = 10\,mA$$

EXERCISE 3

If an amplifier genetares an output voltage $V_o = 10\,V$, having an output internal resistance $R_o = 470\,\Omega$, which voltage could deliver a load resistance $R_L = 50\,\Omega$?

ANSWER

$$v_o = 0.96\,V$$

11

Amplifiers: Basic BJT Configurations

11.1 CONVENTIONAL NOTATIONS

When we indicate a voltage V we need two subscripts that suggest the two nodes across which the voltage is applied (e.g., V_{BE}). We use only one subscript (e.g., V_B) to refer to a voltage applied across the node indicated by that subscript and the (implicit) ground. When only one subscript that is doubled is used (e.g., V_{CC}), we refer to a voltage owing to a DC source, applied across the node indicated by the subscript and the ground.

Examples: V_{CE} indicates the voltage across nodes "C" and "E," V_B indicates the voltage across node "B" and the ground, V_{BB} indicates the voltage owing to a DC source across the node "B" and the ground.

11.2 STEP-BY-STEP FIRST DESIGN

An amplifier consists of one or more transistors together with their biasing circuits and input/output stages. Now, we will deal with single-stage single-transistor class-A amplifiers, starting here to systematically design our first, elementary, scheme.

11.2.1 DC Bias Resistor

The heart of the amplifier is the transistor, and we start by utilizing the BJT type. Power to operate is furnished by one or more batteries, which bias the

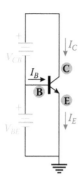

FIGURE 11.1   BJT biased by
means of two batteries.

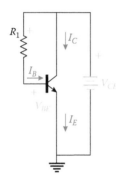

FIGURE 11.2   BJT biased
by means of one battery
and one bias resistor.

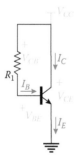

FIGURE 11.3   BJT biased by V_{cc}
and R_1. This network is called
fixed bias or base bias.

BJT's junctions. With biasing, a DC operating point Q is fixed, that is all DC voltages across the BJT's junctions, $V_{BE,Q}$, $V_{CB,Q}$, and $V_{CE,Q}$, and all DC currents flowing into the three BJT regions, $I_{E,Q}$, $I_{B,Q}$, and $I_{C,Q}$. The operating point Q is fundamental to determine the portion of the cycle during which the source current flows through the BJT (Section 10.2.4 in Chapter 10).

To bias the two junctions we start by utilizing two DC voltage sources (by means of two batteries in particular, as here represented, or two DC power supplies in general), an obvious solution. Figure 11.1 schematizes the applied DC voltage sources, V_{CB} and V_{BE}, and the directions of the currents which, conventionally, are those of the holes.

For reasons of cost and size, to bias the two BJT junctions it is more convenient to utilize just one DC source, here the battery V_{CE}, and a so-called bias resistor R_1, as in the scheme shown in Figure 11.2.

The two previous networks are equivalent; let us see why. We can provide V_{CE} equal to $V_{BE} + V_{CB}$ and use a resistance R_1 to obtain a voltage drop V_{BC} at its terminals. Let's assume, for instance, $V_{BE} = 2V$ and $V_{CB} = 3V$ for the two batteries in Figure 11.1, so that the base-emitter junction is forward biased by $2V$ and the base-collector junction is reverse biased by $3V$. Choosing for Figure 11.2 a battery $V_{CE} = 5V$ and a resistance R_1 of such a value to obtain $3V$ across it, then the $5V$ of the battery are divided into $3V$ across the base-collector junction and $2V$ across the base-emitter junction. Therefore, the two situations are equivalent.

In any arbitrary amplifier's scheme, however, the DC source is never represented with its complete loop but, simply, with one of its terminals, with the knowledge that the other is always connected to the ground; the new scheme is shown in Figure 11.3.

According to conventional notations (Section 11.1) the DC source is represented as V_{CC} across the BJT's collector terminal and the ground.

This type of biasing network can be referred to as *fixed bias* or *base bias*.

11.2.2 AC Source

Previously, we considered the DC source and the biasing network. Here, we will deal with the AC source and the signal to be amplified to drive the load. In particular, the AC signal source v_s is connected across the base-emitter junction of the BJT, and the resistor load R_L across the collector-emitter terminals of the BJT, as represented in Figure 11.4.

This is because the source current i_s will flow through the base of the transistor and will condition the collector current i_c, as the BJT a current-controlled current device.

In the BJT itself, there are now both DC and AC currents, due to the DC and AC signal source, respectively:

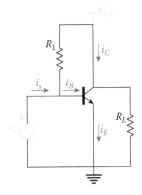

FIGURE 11.4   Biased BJT with signal source and load resistor.

$$\begin{cases} i_B = I_B + i_b \\ i_C = I_C + i_c \\ i_E = I_E + i_e \end{cases}$$

11.2.3 DC Blocking Capacitors

The previous topology, though, still needs to be improved because it offers drawbacks. In fact, the DC values, necessary to bias the BJT's junctions, here affect the signal source (through R_1) and (directly) the load resistance R_L. This is useless and creates problems too. If you consider the AC voltage as provided by a microphone and the load resistor representing the resistance (or the impedance) of an audio speaker, the DC current must be prevented from flowing through the source and through the load, so to avoid possible damage. The load resistor can even represent the input resistance of the next amplifier stage for which the DC loading effect must be avoided.

To solve this, we can use two blocking capacitors, also called coupling capacitors, C_{B1} and C_{B2}, to filter away the DC values (not the AC signal of course). In fact, since $X_C = 1/\omega C = 1/2\pi fC$, the capacitive reactance is practically infinite, representing an open circuit only for DC components. The modified scheme is shown in Figure 11.5.

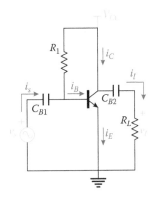

FIGURE 11.5   Amplifier with DC blocking capacitors.

Now the current i_l flowing through the load resistor has AC components only, and no DC current flows toward the signal source.

In our dissertation, we consider the amplifier working within its frequency bandwidth, so that the blocking capacitors are replaced by short circuits. However,

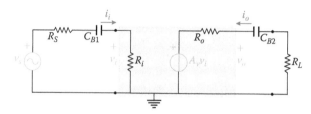

FIGURE 11.6 Model of a voltage amplifier with DC blocking capacitors, signal source, and load.

these capacitors affect the behavior of the lower frequencies of the amplifier, and determine its lower cutoff frequency f_{c-l} (Section 10.2.6). Such a frequency can be easily determined by referring to the general model of the amplifier, which includes C_{B1} and C_{B2}. As an example, we can consider the voltage amplifier model as represented in Figure 11.6.

The capacitor C_{B1} and the resistors R_s and R_i form a high-pass filter at the input port, while the capacitor C_{B2} and the resistors, R_c and R_L form a high-pass filter at the output port. These two filters can be treated as already discussed in Section 7.3.2 in Chapter 7.

The higher cutoff frequency f_{c-h} of the amplifier is determined by parasitic capacitance elements of the BJT (see Section 8.5.4 in Chapter 8) rather than real ones.

11.2.4 AC Collector Resistor

The previous scheme represents an amplifier that cannot actually work. This is because the output signal v_{ce} is across the collector terminal and the ground, but the collector terminal is actually AC grounded by the battery! In fact, if we consider an ideal battery, it has no internal resistance, so that the AC current finds no resistance to flow directly through the ground terminal. Technically, we say that the battery represents a ground for the signal. If the output current i_c "sees" a ground, it will never actually be able to flow through to the load resistance. We can also say that the battery "imposes" its constant voltage value on the collector terminal, not allowing voltage (and current) swings. To solve this fundamental problem we adopt the scheme shown in Figure 11.7, including the resistor R_C.

The aim of this resistor is *to allow an AC voltage drop between the BJT's collector and the positive battery terminal*. The resistor R_C (the subscript "C" stands

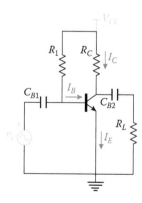

FIGURE 11.7 Amplifier with the AC voltage drop resistor R_c.

for "collector") is also known as an *AC voltage drop resistor*. Thanks to it, now the output signal v_{ce} can swing, no longer being forced to ground.

Given that the battery represents a ground for the signal, the resistors R_C and R_L are dynamically in parallel and have both terminals in common (with the capacitor behaving as a short circuit at the working frequencies).

Observation

We could admit to consider the collector resistor itself as a load. But this has the drawbacks to allow DC values for the load and to admit just few values for the load resistance.

11.3 BJT AS A SIGNAL AMPLIFIER

11.3.1 BJT as a Two-Port Network

Following the description of our first type of amplifier, we need to analyze it to establish the value of its components, the voltage and current gains, and the input and output resistances.

To this aim, we need to solve a set of equations that consider the physical behavior of each component of the network, to find the common voltage and current values. This is relatively easy for all components except the BJT, with nonlinear equations that are not so simple to treat. Therefore, to simplify the problem we adopt its two-port equivalent model.

11.3.2 BJT Small Signal Model

Here, we consider the BJT as *instantaneous*, even though, actually, in a real transistor there are "inertial" mechanisms that are usually modeled using capacitors, here ignored because their effect are on the higher cut-off frequency, and we are now interested only in the mid-band behavior.

From a mathematical point of view, the study of active, nonlinear two-port networks is not an easy task, and simplifications can bring approximate solutions. To solve, we recall the so-called small-signal analysis, treating the BJT just as a linear network, and assuming the signals to be small compared to the rest DC levels, to satisfy the class A conditions (Figure 10.5 in Chapter 10). In such a way, the obtained solutions can be considered pretty much the same as the ones that could be obtained without simplifications.

The BJT's two-port linear network model will be analyzed by means of one of the matrix representations already treated (Section 5.4.1 in Chapter 5) and, in particular, the hybrid one, that is,

$$\begin{cases} v_1 = h_{11}i_1 + h_{12}v_2 \\ i_2 = h_{21}i_1 + h_{22}v_2 \end{cases}$$

This choice is due to many reasons, among which is the ease of measuring the [H] elements and their associated mathematical and physical meaning, treated in the following.

According to the IEEE (Institute of Electrical and Electronics Engineers) standard notation, the double numerical subscripts adopted in the previous set of two equations can be substituted by

$$\begin{cases} 11 = i \ (input) \\ 12 = r \ (reverse \ transfer) \\ 21 = f \ (forward \ transfer) \\ 22 = o \ (output) \end{cases}$$

The reason is clear when we refer to the equivalent network of the hybrid representation as already treated (Figure 5.7c in Chapter 5).

An additional subscript helps to identify the circuit configuration (Section 8.6.6 of Chapter 8):

$$\begin{cases} e \quad (common \ emitter) \\ b \quad (common \ base) \\ c \quad (common \ collector) \end{cases}$$

Therefore, for example, the parameters $h_{ie}, h_{re}, h_{fe}, h_{oe}$ refer to $h_{11}, h_{12}, h_{21}, h_{22}$, respectively, for a common emitter configuration, for which

$$v_1 \rightarrow v_{be}; \ i_1 \rightarrow i_b; \ v_{21} \rightarrow v_{ce}; \ i_2 \rightarrow i_c$$

As a result

$$\begin{cases} v_{be} = h_{ie}i_b + h_{re}v_{ce} \\ i_c = h_{fe}i_b + h_{oe}v_{ce} \end{cases}$$

and the relative circuit scheme is shown in Figure 11.8.

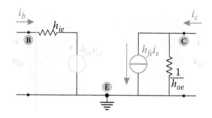

FIGURE 11.8   Equivalent h-parameter model of a BJT in common emitter configuration.

Observations

This equivalent model can be used in place of the real BJT. However, this is only for small signal condition applications and only within the BJT's frequency band, when the transistor does not present reactive phenomena.

The h-parameter model is not the only model used for BJTs in electronics. The so called r-parameter model is also adopted in many circumstances. It is possible to swap between the two models with known relationships among h- and r-parameters.

11.3.2.1 Mathematical Meaning of the Hybrid Parameters

The two-port network model of the BJT derives from the most general case of a two-port network hybrid representation, for which

$$\begin{cases} v_1 = f(i_1, v_2) \\ i_2 = f(i_1, v_2) \end{cases}$$

Keeping in mind that both voltages and current are the sum of a constant (V, I) and variable terms (v, i), the pair of aforementioned equations, when developed in a Taylor series around a point (think of a signal around the quiescent Q-point, far enough from the BJT's cutoff and saturation regions), can be written as

$$v_1 = v_1(I_1, V_2) + \frac{\partial v_1(I_1, V_2)}{\partial i_1}(i_1 - I_1) + \frac{\partial v_1(I_1, V_2)}{\partial v_2}(v_2 - V_2) +$$

$$+ \frac{1}{2}\frac{\partial^2 v_1(I_1, V_2)}{\partial i_1^2}(i_1 - I_1)^2 + \frac{1}{2}\frac{\partial^2 v_1(I_1, V_2)}{\partial v_2^2}(v_2 - V_2)^2$$

$$+ \frac{1}{2}\frac{\partial^2 v_1(I_1, V_2)}{\partial i_1 \partial v_2}(i_1 - I_1)(v_2 - V_2) + \ldots$$

$$i_2 = i_2(I_1, V_2) + \frac{\partial i_2(I_1, V_2)}{\partial i_1}(i_1 - I_1) + \frac{\partial i_2(I_1, V_2)}{\partial v_2}(v_2 - V_2) +$$

$$+ \frac{1}{2}\frac{\partial^2 i_2(I_1, V_2)}{\partial i_1^2}(i_1 - I_1)^2 + \frac{1}{2}\frac{\partial^2 i_2(I_1, V_2)}{\partial v_2^2}(v_2 - V_2)^2 +$$

$$+ \frac{1}{2}\frac{\partial^2 i_2(I_1, V_2)}{\partial i_1 \partial v_2}(i_1 - I_1)(v_2 - V_2) + \ldots$$

Focusing only on signal variations and accepting a first-order approximation, we have

$$\begin{cases} v_1 \cong \dfrac{v_1(I_1, V_2)}{i_1}i_1 + \dfrac{v_1(I_1, V_2)}{v_2}v_2 \\[2ex] i_2 \cong \dfrac{i_2(I_1, V_2)}{i_1}i_1 + \dfrac{i_2(I_1, V_2)}{v_2}v_2 \end{cases}$$

The representation of the two-port network through hybrid parameters will then be valid, defining the following:

$$h_{11} \stackrel{\text{def}}{=} \left.\frac{\partial v_1(I_1, V_2)}{\partial i_1}\right|_{v_2=0} \qquad [\Omega], \text{ named the input impedance with output shorted}$$

$$h_{12} \stackrel{\text{def}}{=} \left.\frac{\partial v_1(I_1, V_2)}{\partial v_2}\right|_{i_1=0} \qquad [\#], \text{ named the voltage feedback ratio with input terminals open}$$

$$h_{21} \overset{\text{def}}{=} \frac{\partial i_2(I_1, V_2)}{\partial i_1}\bigg|_{v_2=0} \qquad \text{[\#], named the current gain with output shorted}$$

$$h_{22} \overset{\text{def}}{=} \frac{\partial i_2(I_1, V_2)}{\partial v_2}\bigg|_{i_1=0} \qquad \text{[S], named the output admittance with input terminals open}$$

11.3.2.2 Physical Meaning of the Hybrid Parameters

For a two-terminal linear resistor the equation $v = Ri$ describes a line passing through the origin of the 2D plane i, v. Similarly, for a four-terminal device the equations

$$\begin{cases} v_1 = h_{11}i_1 + h_{12}v_2 \\ i_2 = h_{21}i_1 + h_{22}v_2 \end{cases}$$

describe a surface in a 4D volume i_1, v_1, i_2, v_2. As it is impossible to visualize this volume, we need to graph one parameterized equation at a time. So, let us consider the equation $v_1 = {}_{h_{11}}i_1 + {}_{h_{12}v_2}$, which can represent a family of parallel lines on the plane i_1, v_1, with a slope h_{11} parameterized by v_2 (Figure 11.9)

When $v_1 = h_{11}i_1 + h_{12}v_2$ represents a nonlinear two-port device, we will not have lines but curves on the plane i_1, v_1 once again parameterized by the value of v_2.

The situation is similar for the equation $i_2 = h_{21}i_1 + h_{22}v_2$, which we can visualize as curves on the plane i_2, v_2 parameterized by the value of i_1.

These considerations help us to understand the physical meaning of the h coefficients; for example, for a BJT in C.E. configuration (but similarly for C.B. and C.C. configurations as well):

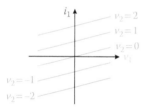

FIGURE 11.9 Lines parameterized by v_2 in the plane i_1, v_1.

$$\begin{cases} h_{11} = h_{ie} \\ h_{12} = h_{re} \\ h_{21} = h_{fe} \\ h_{22} = h_{oe} \end{cases} ; \begin{cases} v_1 = v_{be} \\ i_1 = i_b \\ v_2 = v_{ce} \\ i_2 = i_c \end{cases}$$

$$\Rightarrow \begin{cases} v_1 = h_{11}i_1 + h_{12}v_2 \\ i_2 = h_{21}i_1 + h_{22}v_2 \end{cases} \text{becomes} \begin{cases} v_{be} = h_{ie}i_b + h_{re}v_{ce} \\ i_c = h_{fe}i_b + h_{oe}v_{ce} \end{cases}$$

h_{ie} since $v_{be} = h_{ie}i_b + h_{re}v_{ce}$ we can write

$$h_{ie} = \frac{v_{be}}{i_b}\bigg|_{v_{ce}=0}$$

Therefore h_{ie} is the ratio of v_{be} and i_b variations (Section 2.11 of Chapter 2) when v_{ce} does not vary, that is, it is constant, but not necessarily zero.

Let's consider the BJT input I-V characteristic in Figure 11.10.

The constant output voltage V_{CE} refers to one of the parametric curves of the BJT's input characteristic, here highlighted in blue, and we consider the signal that can swing around the Q–point fixed by a V_{BE}, I_B pair (Figure 11.10).

As we can see the variation of v_{be} is around tenths of Volts, that of i_b is few tens of micro-Ampere, therefore their ratio is typically of the order of 10^3–$10^4\,\Omega$. Of course, for the exact value of h_{ie} we have to refer to the particular BJT in use and to its input I-V graph.

Let's refer to the BJT's model of Figure 11.8.

For a null output voltage v_{ce}, the hybrid BJT model results with output shorted. In this model, h_{ie} expresses a voltage to current ratio, so it represents a series resistance in the input loop, since $v_{be} = h_{ie}i_b + h_{re}v_{ce}$ is a sum of voltages.

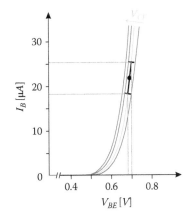

FIGURE 11.10 BJT's input I-V graph; the signal swings around the Q-point, and V_{CE} is fixed.

h_{re} since $v_{be} = h_{ie}i_b + h_{re}v_{ce}$ we can write

$$h_{re} = \left.\frac{v_{be}}{v_{ce}}\right|_{i_b=0}$$

Therefore h_{re} is the ratio of v_{be} and v_{ce} variations (Section 2.11 of Chapter 2) when i_b does not vary, that is, it is constant, but not necessarily zero.

Let's consider the BJT input I-V characteristic in Figure 11.11.

The constant input current I_B refers a fixed value of the y-axis, and we consider a signal that can swing around the Q-point fixed by a V_{BE}, I_B pair (Figure 11.11).

As we can see the variation of v_{be} can be quite small, on the order of $10^{-3}\,V$, and that of v_{ce} on the order of $10^1\,V$; therefore their ratio is typically on the order of 10^{-4}. Of course, for the exact value of h_{re} we have to refer to the particular BJT in use and to its input I-V graph.

It follows that for its typical value, in some applications h_{re} can be ignored.

Let's refer to the BJT's model of Figure 11.8.

For a null input current i_b, the hybrid BJT's model has open input terminals. In this model, h_{re} expresses a voltage-to-voltage ratio. As a consequence $h_{re}v_{ce}$ represents a series VCVS

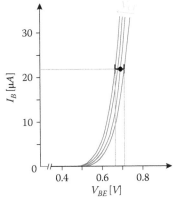

FIGURE 11.11 BJT's input I-V graph; the signal swings around the Q-point, and I_B is fixed.

(Section 5.3 in Chapter 5) in the input loop, since $v_{be} = h_{ie}i_b + h_{re}v_{ce}$ is a sum of voltages.

This VCVS takes into account the part of the output voltage which "influences" the input one, representing a *feedback effect*.

Observation

The feedback effect represents one of the most powerful tools in the design of high-performance networks even with components with low stability values. It has been successfully adopted to design many type of circuits including oscillators and timers, signal comparators, regulators, filters, etc.

However, the feedback effect finds applications in many fields other than electronics. For instance in pedagogy, when teachers give students exercises to challenge their understanding, the results furnish a feedback to teachers to eventually modify their teaching patterns. In computer operating systems, since the software is becoming larger and complex, the so-called β-version is furnished to some selected testers, and they furnish feedback to the developers to change the software accordingly. The Facebook "Like" button is a feature that allows users to provide a feedback to someone, representing an appreciation for content.

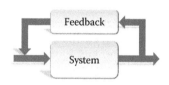

FIGURE 11.12 Representation of the feedback effect.

Practically, feedback represents how the result (i.e., the output) can "go back" and influences the cause (i.e., the input) that produced that result to possibly originate another result (Figure 11.12).

h_{fe} since $i_c = h_{fe}i_b + h_{oe}v_{ce}$ we can write

$$h_{fe} = \frac{i_c}{i_b}\bigg|_{v_{ce}=0}$$

Therefore h_{re} is the ratio of i_c and i_b variations (Section 2.11 in Chapter 2) when v_{ce} does not vary, i.e., it is constant, but not necessarily zero.

Let's consider the BJT output I-V characteristic:

The constant output voltage V_{CE} refers to a fixed value on the x-axis, and we consider a signal that can swing around the Q–point fixed by a V_{CE}, I_C pair (Figure 11.13).

As we can see, the variation of i_C can be on the order of $10^{-2}A$, and that of i_b on the order of $10^{-4}V$; therefore, their ratio is typically on the order of 10^2. Of course, for the exact value of h_{fe} we have to refer to the particular BJT in use and to its output I–V graph.

FIGURE 11.13 BJT's output I-V graph; the signal swings around the Q-point, V_{CE} is fixed.

Let's refer to the BJT's model of Figure 11.8.

For a null output voltage v_{ce}, the hybrid BJT model results with output shorted. In this model, h_{fe} expresses unitless current-to-current ratio. As a consequence $h_{fe}i_b$ represents a parallel CCCS (Section 5.3 in Chapter 5) in output parallel configuration, since $i_c = h_{fe}i_b + h_{oe}v_{ce}$ is a sum of currents.

This CCCS takes into account the part of the input current that "influences" the output one, representing a *feedforward* (reverse of *feedback*) effect.

h_{oe} — since $i_c = h_{fe}i_b + h_{oe}v_{ce}$ we can write

$$h_{oe} = \left. \frac{i_c}{v_{ce}} \right|_{i_b=0}$$

Therefore, h_{oe} is the ratio of i_c and v_{ce} variations (Section 2.11 in Chapter 2) when i_b does not vary, that is, it is constant, but not necessarily zero.

Let's consider the BJT output I–V characteristic:

The constant input current I_B refers to one of the parametric curves of the BJT's output characteristic, here highlighted in blue, and we consider the signal that can swing around the Q-point fixed by a V_{CE}, I_C pair (Figure 11.14).

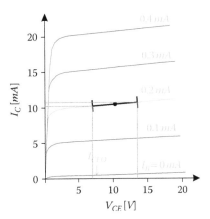

FIGURE 11.14 BJT's output I-V graph; the signal swings around the Q-point, and I_B is fixed.

As we can see the variation of i_c can be of the order of 10^{-3}–$10^{-4}\,A$, and that of v_{ce} on the order of $10^1\,V$; therefore, their ratio is typically of the order of 10^{-4}–$10^{-5}\,\Omega^{-1}$. Of course, for the exact value of h_{oe} we have to refer to the particular BJT in use and to its output I–V graph. It follows that for its typical value, in some applications h_{oe} can be ignored.

Let's refer to the BJT's model of Figure 11.8.

For a null input current i_b, the hybrid BJT's model has open input terminals. In this model h_{oe} expresses a current to voltage ratio, so it represents a parallel conductance in the output parallel configuration, since $i_c = h_{fe}i_b + h_{oe}v_{ce}$ is a sum of currents.

More specifically h_{oe} represents the reciprocal of the slope of the output characteristic curves of the BJT, so it takes into account the Early effect (Section 8.6.8.3 in Chapter 8).

11.3.3 DC and AC Current Gains

In Section 8.6.3 of Chapter 8, we saw the collector-base current relationship:

$$I_C = \beta I_B + (\beta + 1)\, I_{CBO}$$

We have assumed the term inverse saturation current I_{CBO} to be negligible since, for example in the case of silicon, it can be "only" of the order of nA. Regardless of the multiplier factor $(\beta + 1)$, in many practical cases we can set

$$I_C \cong \beta I_B \;\; \Rightarrow \beta \cong \frac{I_C}{I_B}$$

But it is an approximation anyway. Strictly speaking, the collector to base current ratio is exactly defined as

$$h_{FE} = \frac{I_C}{I_B}$$

h_{FE} is known as DC *current gain*.

Parameters β and h_{FE} are definitely similar, so often they are mixed up, but they are *not equal*.

Pay close attention to how the additional parameter h_{fe} has been defined:

$$h_{fe} = \left.\frac{i_c}{i_b}\right|_{v_{ce}=0}$$

So h_{fe} is the *small signal forward current gain* (*small signal current gain* for short) of a bipolar junction transistor. It is relevant to the ratio of incremental measures, so it does not refer to DC components but to AC ones.

11.3.4 Typical H-Parameter Values

After the theoretical approach, let's review some real typical hybrid parameters furnished by manufacturers for some commercial bipolar junction transistors.

The BC107 transistor (Pro Electron standard, Section 8.9.2 in Chapter 8: B for silicon, C for transistor) is an npn BJT, commonly used for low-power audio applications. It has $V_{CE,MAX} = 45V$, supports a max power $P_{MAX} = 300mW$, and has $I_{CBO}|_{T=25°C} = 15nA$. For this transistor and its variants (BC107A, BC107B) refer to Table 11.1.

The BC108 series is an npn-type, commonly used for low-power, general-purpose applications (Table 11.2).

TABLE 11.1

Typical Hybrid Parameters of BJT BC107x Series

BJT	Conditions	h_{ie} [Ω]	h_{ie}	h_{fe}	h_{oe} [Ω^{-1}]
BC107	$I_C = 2mA$, $V_{CE} = 5V$, $f = 1kHz$	4 k	$2.2*10^{-4}$	250	$30*10^{-6}$
BC107A	$I_C = 2mA$, $V_{CE} = 5V$, $f = 1kHz$	3 k	$1.7*10^{-4}$	190	$13*10^{-6}$
BC107B	$I_C = 2mA$, $V_{CE} = 5V$, $f = 1kHz$	4.8 k	$2.7*10^{-4}$	300	$26*10^{-6}$

TABLE 11.2

Typical Hybrid Parameters of the BJT BC108x Series

BJT	Conditions	h_{ie} [Ω]	h_{re}	h_{fe}	h_{oe} [Ω^{-1}]
BC108	$I_C = 2mA$, $V_{CE} = 5V$, $f = 1kHz$	5.5 k	$3.1*10^{-4}$	370	$30*10^{-6}$
BC108A	$I_C = 2mA$, $V_{CE} = 5V$, $f = 1kHz$	3 k	$1.7*10^{-4}$	190	$13*10^{-6}$
BC108B	$I_C = 2mA$, $V_{CE} = 5V$, $f = 1kHz$	4.8 k	$2.7*10^{-4}$	300	$26*10^{-6}$
BC108C	$I_C = 2mA$, $V_{CE} = 5V$, $f = 1kHz$	7 k	$3.8*10^{-4}$	500	$34*10^{-6}$

The BC84x (x: 6,7,8) series is an npn-type BJT, commonly used for low-power audio applications. It has a $V_{CE,MAX} = 30$–$65\,V$, supports a max power $P_{MAX} = 300\,mW$, has a minimum DC current gain $h_{FE,min} = 180 \div 520$ @ $I_C = 2\,mA$, $V_{CE} = 5\,V$, and its voltage across the base-emitter junction, when working in forward-active mode, is $V_{BE,on} = 660\,mV$. The hybrid parameters are shown in Table 11.3.

Series 2N2222 (JEDEC standard, Section 8.9.1 of Chapter 8) is a npn-type BJT, commonly used for high-frequency medium-power applications. It has $V_{CE,MAX} = 30\,V$ (for the 2N2222A version $V_{CE,MAX} = 40\,V$), $I_{C,MAX} = 600\,mA$, $P_{MAX} = 625\,mW$, minimum DC current gain $h_{FE,min} = 50$ @ $I_C = 1\,mA$, and $V_{CE} = 10\,V$. The hybrid parameters are shown in Table 11.4.

The BJT's h-parameter model is typically adopted to design AF (audio frequency) and RF (radio frequency) amplifiers. Therefore, it is evident how important it is to know the four elements of the $[H]$ matrix. Among these elements, the most important is h_{fe}, but it is replaced by h_{FE} in some BJT datasheets. We know that h_{fe} and h_{FE} are not the same, since

$$h_{fe} \overset{\text{def}}{=} \left.\frac{i_c}{i_b}\right|_{v_{ce}=0}$$

which is an AC-to-AC ratio of current with output shorted, and

$$h_{FE} \overset{\text{def}}{=} \frac{I_C}{I_B}$$

which is a DC-to-DC ratio of currents. However, fortunately h_{fe} and h_{FE} have quite similar values for many BJTs when used in "small signal" conditions (Section 10.2.4 in Chapter 10), within a large difference in current values.

TABLE 11.3

Typical Hybrid Parameters of the BJT BC84x Series

BJT	Conditions	$h_{ie}\,[\Omega]$	h_{re}	h_{fe}	$h_{oe}\,[\Omega^{-1}]$
BC846A BC847A BC848A	$I_C = 2\,mA$, $V_{CE} = 5\,V$, $f = 1\,kHz$	2.7 k	$1.5*10^{-4}$	220	$18*10^{-6}$
BC846B BC847B BC848B	$I_C = 2\,mA$, $V_{CE} = 5\,V$, $f = 1\,kHz$	4.5 k	$2*10^{-4}$	330	$30*10^{-6}$
BC846C BC847C BC848C	$I_C = 2\,mA$, $V_{CE} = 5\,V$, $f = 1\,kHz$	8.7 k	$3*10^{-4}$	600	$60*10^{-6}$

TABLE 11.4

Typical Hybrid Parameters of the BJT 2N2222x Series

BJT	Conditions	$h_{ie}\,[\Omega]$	h_{re}	h_{fe}	$h_{oe}\,[\Omega^{-1}]$
2N2222	$I_C = 1\,mA$, $V_{CE} = 10\,V$, $f = 1\,kHz$	5 k		150	$20*10^{-6}$
2N2222A	$V_{CE} = 10\,V$, $f = 1\,kHz$, $I_C = 1\,mA$	5 k	$8*10^{-4}$	175	$20*10^{-6}$
	$V_{CE} = 10\,V$, $f = 1\,kHz$, $I_C = 10\,mA$	0.75 k	$4*10^{-4}$	225	$110*10^{-6}$

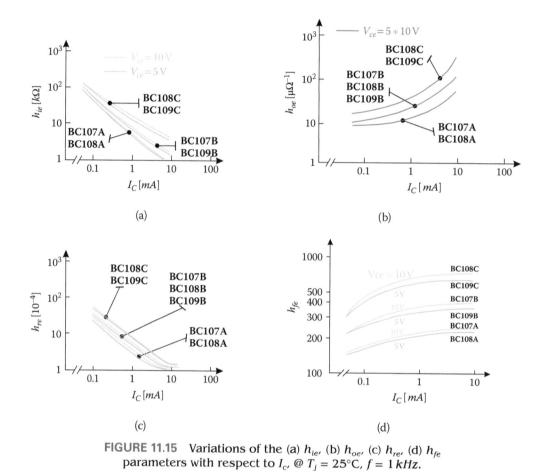

FIGURE 11.15 Variations of the (a) h_{ie}, (b) h_{oe}, (c) h_{re}, (d) h_{fe} parameters with respect to I_c, @ $T_j = 25°C$, $f = 1\,kHz$.

The hybrid parameters are not constant and their values can change depending on the conditions of operation of the particular BJT, that is, its currents and voltages. Figure 11.15 reports changes of the h–parameters with the collector current for some BJTs operating at $f = 1\,kHz$, and $T = 77°F\,(= 25°C)$.

The values of the h-parameters reported in datasheets are commonly considered in the BJT's middle band, that is $f = 1\,kHz$ and, at this particular frequency, they have real values (no imaginary components).

11.3.5 Hybrid Parameter Conversion

In general, the manufacturers furnish h-parameters specified for C.E. configuration. Therefore, if we want to perform analysis for a C.C. or C.B. configuration, it is necessary to convert the C.E. h-parameters into the desired ones. To this aim, you can refer to the conversion Table 11.5.

TABLE 11.5
Hybrid Parameters Conversion Table

C.E.	C.B.	C.C.
h_{ie}	$h_{ib} = \dfrac{h_{ie}}{1+h_{fe}}$	$h_{ic} = h_{ie}$
h_{re}	$h_{rb} = \dfrac{h_{ie}h_{oe}}{1+h_{fe}} - h_{re}$	$h_{rc} = 1 - h_{re}$
h_{fe}	$h_{fb} = -\dfrac{h_{fe}}{1+h_{fe}}$	$h_{fc} = -(1+h_{fe})$
h_{oe}	$h_{ob} = \dfrac{h_{oe}}{1+h_{fe}}$	$h_{oc} = h_{oe}$

Observation

Note the recurring factor $1+h_{fe}$ in Table 11.5.

11.4 C.E. CONFIGURATION

The common emitter, or C.E., amplifier is so called because the BJT shares its emitter terminal for both input and output loops in AC conditions. In DC conditions, this cannot be verified. Be aware that a common terminal does not necessarily have to be grounded, even though this is often preferred.

In our study, we will suppose linear conditions, even though the BJT is not a linear device. This is because we assume that the BJT works in its active region where its characteristics are considered to be "almost linear." Therefore, we will apply the superposition theorem (Section 6.5 in Chapter 6), according to which DC and AC analysis can be done separately, and the relative results will be summed up.

The study of an arbitrary amplifier consists of determining the values of its components, that is, battery, (real) resistors, capacitors, and BJT(s), and the values of its operative conditions, that is, currents, voltages, and input and output equivalent resistances (or, more generally, impedances).

In particular, the DC analysis allows us to determine the *Q-point* and the resistors, while the AC analysis studies the gain(s) and resistances as "seen" at input and output ports (Figure 11.16).

Starting from the real topology of the amplifier, we will derive a network to model and analyze its DC conditions, neglecting the AC conditions, and then we will improve the topology according to detailed considerations. Later, we will analyze the AC performance, neglecting the DC performance. The superposition theorem will help in adding the DC and AC solutions.

$$\text{Amplifier's analysis} \begin{cases} DC\ analysis \begin{cases} calculate\ the\ DC\ Q-point \\ calculate\ the\ values\ of\ the\ resistances \end{cases} \\ AC\ analysis \begin{cases} calculate\ the\ gains\ (voltage\ and\ current) \\ calculate\ the\ resistances\ (input\ and\ output) \end{cases} \end{cases}$$

FIGURE 11.16 Analysis to perform the study of an amplifier.

Our first C.E. configuration is illustrated in Figure 11.17.

We have a DC source V_{CC} that biases the transistor to work in its active region, a sinusoidal voltage source v_s for input, a load resistor R_L for output, and the BJT is a silicon npn-type with output characteristic curves as represented in Figure 11.18.

To operate in class A conditions (Section 10.2.4 in Chapter 10), we require a *Q-point* approximately in the middle of the active region.

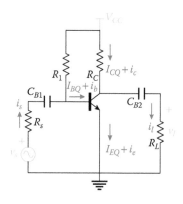

FIGURE 11.17 Simple C.E. configuration.

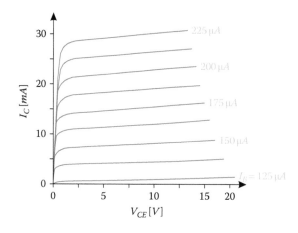

FIGURE 11.18 Output characteristics of the BJT used here.

The DC voltage can be supplied by a battery or a power source, such as the one described in Section 9.2 of Chapter 9.

Although DC and AC conditions are considered separately, note that the AC model is not independent from the DC conditions, since the values of the hybrid parameters depend on the *Q-point*.

11.4.1 DC Analysis

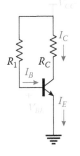

FIGURE 11.19 DC model of the simple C.E. configuration.

To perform DC analysis, the simple amplifier's C.E. configuration in Figure 11.17 can be reduced to only the parts related to DC values, disregarding the AC signals.

Therefore, the capacitors, behaving as filters, have to be replaced by short circuits within the amplifier's frequency range. Then the signal source v_s and the load resistance R_L are "disconnected" from the rest of the amplifier.

11.4.1.1 Fixed Bias

Figure 11.19 illustrated the DC model. This biasing network is known as *fixed bias* or *base bias*.

To operate in class A conditions, the Q-point must be fixed, and according to Figure 11.18, it can be found as pairs $I_{CQ} = 15\,mA$, $V_{CEQ} = 7.5\,V$.

Now, considering currents, applying the known general equations for an arbitrary BJT working in forward-active mode (Section 8.6.3 in Chapter 8) results in

$$I_{BQ} \cong \frac{I_{CQ}}{\beta}$$

$$I_{EQ} \cong (1+\beta)\,I_{BQ}$$

In the DC model, the input loop includes $V_{CC} - R_1 - V_{BE} - ground$, and the output loop includes $V_{CC} - R_C - V_{CE} - ground$.

The KVL applied to the input loop where there is a base-emitter (BE) junction, known as BE-KVL, gives

$$V_{CC} = I_{BQ}R_1 + V_{BEQ}$$

$$\Rightarrow$$

$$R_1 = \frac{V_{CC} - V_{BEQ}}{I_{BQ}}$$

where $V_{BEQ} \cong 0.7V$ (Section 8.6.8.2 in Chapter 8).

For the output loop where there are the collector-emitter (CE) terminals, the Kirchhoff's voltage law (KVL), here called CE-KVL, gives

$$V_{CC} = R_C I_{CQ} + V_{CEQ}$$

$$\Rightarrow$$

$$R_C = \frac{V_{CC} - V_{CEQ}}{I_{CQ}}$$

Therefore, thanks to the DC analysis. The Q-point was determined, along with the quiescent currents I_{CQ}, I_{BQ}, and I_{EQ}, the bias resistor R_1, and the collector resistor R_C.

11.4.1.2 DC Load Line

Only the points belonging to the output characteristic curves represent acceptable pairs of working values $I_C - V_{CE}$ for the BJT, because they satisfy its output equations. However, the output loop of the network reduces these pairs to a subset, because the KVL equation must also be satisfied, obtaining combinations of $I_C - V_{CE}$ for the given collector resistor R_C. The DC load line is a graph representing the subset of possible $I_C - V_{CE}$ pairs for the DC model of the given amplifier, and we can establish its y- and x-intercepts, that is, the saturation current and the cutoff voltage, by drawing the load line simply joining these two points.

From the KVL of the output loop we have

$$V_{CC} = R_C I_C + V_{CE}$$

which represents just the DC load line (Figure 11.20) intercepting the x-axis at $V_{CE} = V_{CC}$ and the y-axis at $I_C = V_{CC}/R_C$.

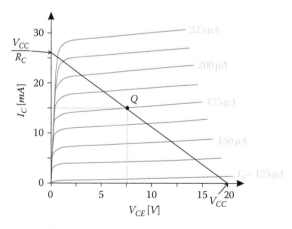

FIGURE 11.20 DC load line superimposed to the output characteristics of the BJT.

The quiescent point Q represented here on the output characteristics has, obviously, its counterpart on the graph of the input characteristics.

For reasons of simplicity, a rule of thumb can help the designer to locate the Q-point roughly at the center of the active region of the BJT. In particular

- When $R_L \gg R_C$, set

$$V_{CEQ} \cong \frac{V_{CC}}{2} + V_{CE(SAT)}$$

(from Section 8.6.5 in Chapter 8, we can always consider $V_{CE(SAT)} \cong 0.2\,V$ as a general rule).

- When $R_L = R_C$, set

$$V_{CEQ} \cong \frac{V_{CC}}{3} + V_{CE(SAT)}$$

- When $R_L \ll R_C$, the network works as a transconductance amplifier (we will see the output resistance of the current amplifier's configuration late in Section 11.4.2.5) so it is fundamental to maximize the output current swing rather than the output voltage swing.

Thermal conditions influence this amplifier's configuration. We know that the BJT's collector current I_C increases by 9% per °F (or °C) for a fixed V_{BE}. If the amplifier was designed to work in class A conditions at a room temperature of $80°F (= 26.7°C)$, it happens that at $91°F (= 32.8°C)$, the current I_C approximately doubles, so that the BJT works now in the saturation region and the class A condition is no longer satisfied.

11.4.1.3 DC Stability

For the current amplifier under study, the Q-point ($I_{CQ} - V_{CEQ}$ pair) directly depends on β. In fact,

$$I_{CQ} = \beta I_{BQ} = \beta \frac{V_{CC} - V_{BEQ}}{R_1}; \quad V_{CEQ} = V_{CC} - \beta \frac{R_C}{R_1}(V_{CC} - V_{BEQ})$$

This dependence represents a problem, given that twin BJTs can have different βs, or the same BJT has β varying with temperature and, therefore, Q can move accordingly along the DC load line. Different βs can have the consequence that the class A specification can be no longer satisfied (Figure 11.21).

The problem, termed *DC stability*, can be solved making Q not β-dependent, or can be drastically reduced if Q is slightly β-dependent. To this aim, one solution can derive from the utilization of an emitter resistor R_E.

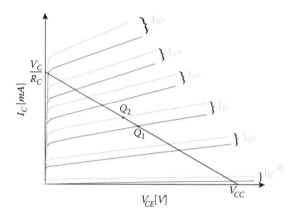

FIGURE 11.21   Q-point can move because of β variations.

11.4.1.4 Emitter Feedback Bias

An emitter resistor R_E provides higher stability of the quiescent point Q, as the DC bias values remain closer to the designed ones even when boundary conditions, such as temperature and the BJT's β, change.

This type of biasing network is a *modified fixed bias*, referred to as *emitter feedback bias* (Figure 11.22).

In particular, R_E improves stability with respect to thermal drifts, and it is referred to as a *swamping resistor*. For example, when there is a rise in the transistor's temperature, the number of carriers increases (Section 3.9 in Chapter 3) and all the currents I_B, I_C, and I_E increase accordingly (with the relative proportions). The higher emitter current I_E causes an increased voltage drop V_E (just for the presence of R_E) and, consequently, a lower voltage drop across the bias resistor R_1, so to respect the KVL at the input mesh $V_{CC} \div R_1 I_B \div V_{BE} \div R_E I_E$. The reduction in voltage across R_1 decreases the base current I_B, and the collector current I_C decreases proportionally (Figure 11.23).

Schematically,

$$T \uparrow \Rightarrow \underbrace{I_C \uparrow \Rightarrow I_E \uparrow \Rightarrow V_{R_1} \downarrow \Rightarrow I_B \downarrow}_{\text{in opposition}} \Rightarrow I_C \downarrow$$

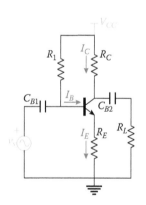

FIGURE 11.22   Amplifier with resistance R_E that allows for a thermal stabilization effect.

However, the advantage of the presence of R_E, in terms of parameter stability to thermal variations, is paid in terms of an increase of the dissipated power (proportional to the sum of all resistances) and of a decrease of the amplifier's gain, as we will see.

To evaluate the obtained advantage, let us perform the DC analysis of the network of the previous figure (the AC source and the load will not be considered).

The KVL at the input loop is

$$V_{CC} = R_1 I_{BQ} + V_{BEQ} + R_E I_{EQ}$$

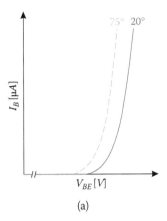

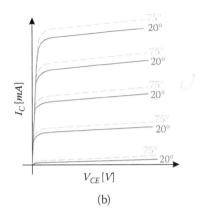

(a)   (b)

FIGURE 11.23   (a) Input and (b) output characteristics with temperature variations.

However, as $I_{EQ} = I_{BQ} + I_{CQ}$ and $I_{CQ} \cong \beta I_{BQ}$, the result is $I_{EQ} \cong (1+\beta) I_{BQ}$; therefore, the KVL equation can be rewritten as

$$V_{CC} \cong R_1 I_{BQ} + V_{BEQ} + R_E (1+\beta) I_{BQ}$$

from which

$$I_{BQ} \cong \frac{V_{CC} - V_{BEQ}}{R_1 + (1+\beta) R_E}$$

and

$$I_{CQ} \cong \beta I_{BQ} \cong \beta \frac{V_{CC} - V_{BEQ}}{R_1 + (1+\beta) R_E}$$

$$I_{EQ} \cong (1+\beta) I_{BQ} \cong \frac{V_{CC} - V_{BEQ}}{R_E + \dfrac{R_1}{(1+\beta)}}$$

If we meet the conditions of $R_1 \ll (1+\beta) R_E$ and $(1+\beta) \cong \beta$ (the latter practically always verified) then

$$I_{CQ} \cong \frac{V_{CC} - V_{BEQ}}{R_E}$$

The result is that I_{CQ} is now practically independent of β. It seems a good solution but it may not be, because $R_1 \ll (1+\beta) R_E$ means that R_1 is low or R_E is large, but R_1 low cannot bias the base-emitter junction correctly, and a large R_E can drastically reduce the amplifier's voltage gain, as we will see.

Slower or faster variations of the output collector current $i_C = (I_{CQ} + i_c)$ are due to thermal effects and to the amplified signal respectively. However, the emitter resistor R_E only has to provide better stability of the Q-point (DC conditions) related to the thermal effects and not to the signals (AC conditions); otherwise, the amplifier's gain will be reduced. So, R_E has to "act" for DC or for very low variable signals, but not within the bandwidth of the amplifier. To this aim, a bypass capacitor C_E can be adopted (Figure 11.24).

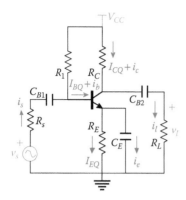

FIGURE 11.24 Amplifier with a bypass capacitor C_E.

The capacitor C_E behaves as an open circuit for DC or very slow signals, and as a short circuit within the amplifier's frequency range. Therefore, the current $i_E = I_E + i_e$ is split between its DC I_E and AC i_e components, through R_E and C_E, respectively.

11.4.1.5 Voltage Divider Bias

The C.E. amplifier network is the most widely used transistor analog amplifier configuration. Here is the complete version, with a voltage divider bias made from the two resistors R_1 and R_2, a collector resistor R_C (Section 11.2.4), an emitter resistor R_E (Section 11.4.1.4) bypassed by a capacitor C_E, two DC blocking capacitors (Section 11.2.3), a voltage signal source with its resistance (Section 4.5.3 in Chapter 4, Section 6.3 in Chapter 6, Section 10.1 in Chapter 10), and a load resistor (Section 10.1) (Figure 11.25).

Curiosity

It is fascinating to visualize the amplifier as an orchestra (Figure 11.26). The signal source could be the conductor, the transistor could represent the soloist, and the electronic components the instrumental ensemble (containing different sections, the string, bass, percussion, etc., as the resistors, capacitors, inductors, etc.); the DC source can represent the sound mixer for setting the reference for volume and tone of sounds. The amplifier's output will be, of course, the orchestral music.

A smaller size orchestra exists as well—the chamber orchestra; the amplifier can also be analogized to this simplified version (Figure 11.17). However, a large full orchestra performs better, as does a complete amplifier (Figure 11.25).

We will analyze now the reason to adopt a voltage divider biasing network, starting from an exact analysis and then an approximate one.

Exact Analysis

In DC analysis, the DC impedance of capacitors is infinite, so they block DC currents and represent open circuits. Therefore, the AC voltage source v_s is

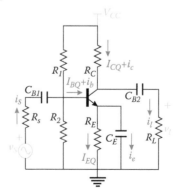

FIGURE 11.25 Complete C.E. amplifier topology.

FIGURE 11.26 The conductor leads the orchestra as the signal source drives the amplifier.

"disconnected" by C_{B1} (for DC conditions this source becomes a short anyway), the resistor load R_L is "disconnected" by C_{B2}, and the emitter resistor R_E is without any parallel impedance since C_E equals an open circuit.

In this occurrence, the DC model is shown in Figure 11.27.

This can be modeled and replaced by its Thévenin equivalent (Figure 11.28), obtaining

$$R_B = R_1 \parallel R_2$$

$$V_{BB} = \frac{R_2}{R_1 + R_2} V_{CC}$$

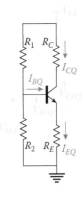

FIGURE 11.27 DC model of the complete C.E. configuration.

From KVL analysis of the input BE (base-emitter) loop, named BE-KVL, we obtain

$$V_{BB} = R_B I_{BQ} + V_{BEQ} + R_E I_{EQ}$$

and recalling for currents

$$I_{EQ} = I_{BQ} + I_{CQ} = I_{BQ} + \beta I_{BQ} = (1 + \beta) I_{BQ}$$

the quiescent I_{BQ} current is

$$I_{BQ} = \frac{V_{BB} - V_{BEQ}}{R_B + (1 + \beta) R_E}$$

and the quiescent I_{EQ} current is

$$I_{EQ} = \frac{V_{BB} - V_{BEQ}}{R_E + \dfrac{R_B}{(1 + \beta)}}$$

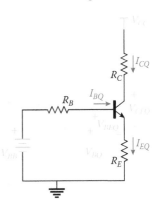

FIGURE 11.28 Amplifier represented by Thévenin equivalent at its input loop.

To guarantee that I_{EQ} is independent enough from β, so that DC stability conditions can be satisfied, the condition is $R_E \gg R_B/(1+\beta)$ or, since we always have $\beta + 1 \cong \beta$, the condition simplifies to $R_E \gg R_B/\beta$, which is considered to be true when

$$\beta R_E \geq 10 R_B$$

Approximate Analysis

If the previous inequality is true, the current I_{BQ} flowing into the base is much lower than the one flowing into the resistor R_2, and this allows us to perform the DC analysis in a simpler way, considering the voltage at the BJT's base terminal as due to a simple voltage divider made by R_1 and R_2, since we can admit $I_{R1} \cong I_{R2}$.

With this assumption, the Q-point is determined as

$$V_{BQ} = \frac{R_2}{R_1 + R_2} V_{CC}$$

$$V_{EQ} = V_{BQ} - V_{BEQ}$$

(and we know $V_{BEQ} \cong 0.7V$ for a silicon BJT)

$$I_{EQ} = \frac{V_{EQ}}{R_E}$$

and since the I_{BQ} value has been considered negligible:

$$I_{CQ} \cong I_{EQ}$$

Also, considering the collector-emitter (CE) loop, the CE-KVL can be written as

$$V_{CC} = R_C I_{CQ} + V_{CEQ} + R_E I_{EQ}$$

but, again, $I_{CQ} \cong I_{EQ}$, so

$$V_{CEQ} \cong V_{CC} - (R_C + R_E) I_{CQ}$$

Please note that none of the above equations that fix the Q-point depend on β, exactly what we were looking for.

Observations

Obviously even with the presence of R_2, we still have the thermal stabilization effect due to the resistor R_E. If temperature increases, the number of carriers increases accordingly, so that the base current I_B increases and proportionally so does the collector current I_C and the emitter current I_E as well. The latter current causes a rise in the emitter voltage V_E which, compared to the steady base voltage V_B, causes a reduction of the base-emitter voltage V_{BE}. Consequently, there is a drop in the collector current I_C:

$$I_C \uparrow \Rightarrow V_E \uparrow \Rightarrow V_{BE} \downarrow \Rightarrow I_B \downarrow \Rightarrow I_C \downarrow$$

The presence of R_2 creates, as we have seen, an undeniable advantage. On the other hand, R_2 creates a path toward ground for the input signal, and as a consequence it loses a part that would have been destined to the transistor input.

The DC analysis allowed us to determine the Q-point, and we learned how it is important for it to remain fixed. We returned to the voltage divider bias network and to the emitter resistor just to stabilize the Q-point with regard to variations of the BJT's parameters and temperature. However, what about changes of V_{CC}? We supposed it is fixed by the battery or by a general DC power source. But any battery discharges with time, and the power source can have a not-negligible ripple value, so the DC load line tends to "shift" and the Q-point moves or "swings" accordingly. This is the reason why even a well-designed radio receiver, working perfectly in class A conditions, tends to generate an unpleasant sound when the battery discharges.

The voltage divider bias network works correctly if the BJT does not load the divider. Therefore, the base voltage must be approximately the same if measured with the BJT inserted or removed from the circuit.

We treat here voltage bias design, but current bias design can be performed too. Since the latter is more complicated (for instance, it requires other BJTs to be designed) it will be ignored here.

11.4.2 AC Analysis

To perform AC analysis, the amplifier's C.E. configuration in Figure 11.25 can be reduced to only those parts related to AC values, disregarding the DC ones. To this aim we consider the following:

→ The signals "see" the DC supply as a zero voltage (being a constant), the same as the ground. So the DC source is killed (in general it is to replace DC voltage sources with short circuits, and DC current sources with open circuits).
→ The DC blocking capacitors operate as a short circuit at the amplifier's working frequencies.

So,

→ Resistance R_1 is between the BJT base terminal and the ground.
→ Resistance R_C is between the BJT collector terminal and the ground.

→ Resistance R_L is between the BJT collector terminal and the ground, and parallel to resistance R_C.
→ The signal source, with its internal resistance, is between the BJT base terminal and the ground, so in parallel with resistance R_1.

Observation

The battery, considered an ideal source of constant voltage, has zero resistance (Figure 4.3c in Chapter 4).

As a result, the equivalent AC model layout looks like the signals shown in Figure 11.29.

This configuration is confirmed to be a C.E. as the emitter is evidently shared between the AC input and output loops.

Before determining the circuit's equations, it is useful to recall the formulas of the h-parameter network, useful to model the AC behavior of the BJT when it operates in C.E. configuration in the center of the forward-active region (Section 11.3.2):

$$v_{be} = h_{ie}i_b + h_{re}v_{ce} \quad (eq.I\,C.E.)$$

$$i_c = h_{fe}i_b + h_{oe}v_{ce} \quad (eq.II\,C.E.)$$

Now we can replace the BJT with its hybrid parameter small signal model; the result is shown in Figure 11.30.

For convenience, resistors R_1 and R_2 can be combined into $R_{12} = R_1 \,||\, R_2 = \dfrac{R_1 R_2}{R_1 + R_2}$.

11.4.2.1 AC Load Line

Similar to the DC load line determined in DC conditions, we can also determined an AC load line, effective in AC conditions.

KVL applied at the collector-emitter loop gives us

$$R_{CL}I_C + V_{CE} = 0$$

with $R_{CL} = R_C \,||\, R_L$.

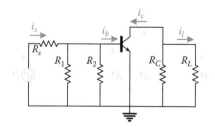

FIGURE 11.29 AC equivalent circuit of the C.E. configuration.

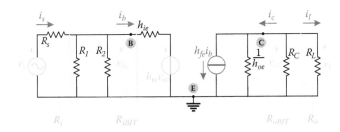

FIGURE 11.30   AC equivalent circuit of the C.E. amplifier
with the BJT's h-parameter model.

Therefore, the AC load line has a slope equal to $-1/R_{CL}$ and intersects the origin.

However, taking into account the superposition theorem (Section 6.5 of Chapter 6) and the DC findings, the AC load line is "shifted," intersecting the Q-point, as here represented in Figure 11.31.

To this line belongs all the possible $i_c - v_{ce}$ AC working pairs of the network.

The AC signal i_b (due to the voltage source and swinging around zero), which enters to the base of the BJT, is added to the DC biasing current I_{BQ}, so that swings around the Q-point along the AC load line.

If the Q-point is located at the center of the active region and the amplitude of i_b is "small enough" (small signal condition), the output signal i_c will just be an amplified copy of i_b (class A conditions, Section 10.2.4 in Chapter 10), as represented in Figure 11.32.

If, on the other hand, the biasing circuit has not been properly chosen, so that the Q-point results near the cutoff region, the output signal will be "cut" in its lower part (clipping effect), and thus distorted, as represented in Figure 11.33.

Similarly, this also happens if the Q-point is near the saturation region (Figure 11.34).

However, we could have a clipping effect even with a correct bias network when the input signal i_b is too large (no small signal conditions) (Figure 11.35).

Because of what we have seen so far, we understand how the choice of Q-point, determined in DC analysis, is fundamental also for AC conditions. In fact, it defines the maximum swings for the input signal

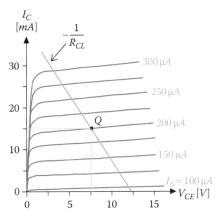

FIGURE 11.31   AC load line superimposed to the output characteristics of the BJT.

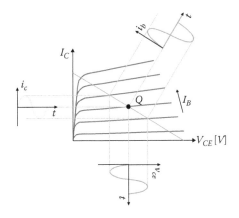

FIGURE 11.32   Class A conditions.

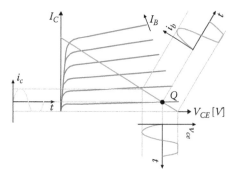

FIGURE 11.33 *Q*-point near the cutoff region; the output signal is cut in its lowest part.

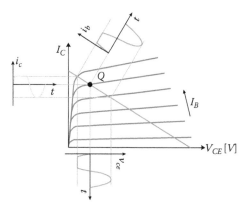

FIGURE 11.34 *Q*-point near the saturation area; output signal is cut in its upper part.

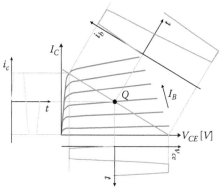

FIGURE 11.35 Large signal conditions.

without distortions in output. This is why the *Q*-point should be "centered" in the active region, so assuring the maximum swings for the input signal.

If, regardless of an optimally centered *Q*-point, the input signal is too large, we can adopt a larger battery value V_{CC}. This will shift upward the DC load line, so to take advantage of a larger area in the active region. This expedient, however, works only if the increased DC power can be tolerated by the BJT without damaging it.

11.4.2.2 C.E.: Input Resistance

We want to determine the value of the input resistance of the amplifier in the mid-band frequency regime, as already considered in Section 10.2.2 of Chapter 10. In particular, we define as R_{iBJT} the input small-signal resistance looking in at the base of the BJT, and as R_i the overall small-signal input resistance as "seen" by the AC source, so that

$$R_{iBJT} \stackrel{\text{def}}{=} \frac{v_{eb}}{i_e}$$

$$R_i \stackrel{\text{def}}{=} \frac{v_{eb}}{i_s}$$

as already represented in Figure 11.30.

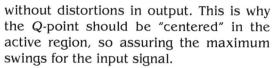

The knowledge of the input resistance is essential to quantify the loading effect (Section 10.4 in Chapter 10) and to find dimensions for the input DC blocking capacitor (Section 11.2.3).

In general, the resistance "seen" at two nodes is determined according to the Thévenin or Norton statements, that is, removing all power independent sources (voltage independent sources shorted, current independent sources open) and calculating the resistance between the open connection nodes. Here, the nodes are the ones of the input port, and at that port, no independent sources are present when looking toward the load.

We can distinguish two input resistances: the first, called R_{iBJT}, is the one at the input port of the BJT, and the second, called R_i, is the one "seen" by the AC source at the input port of the overall amplifier. This distinction can be useful to consider separately the effect of reduction to the input current due to the BJT itself.

For convenience, resistors R_C and R_L can be combined into $R_{CL} = R_C \| R_L$ $= R_C R_L / (R_C + R_L)$.

The results are

$$R_{iBJT} \quad\quad R_{iBJT} \overset{\text{def}}{=} \frac{v_{be}}{i_b}$$

The voltage across the collector-emitter terminals can be written as $v_{ce} = -R_{CL} i_c$, an equation that inserted into the *eq. II C.E.* gives $-v_{ce} / R_{CL} = h_{fe} i_b + h_{oe} v_{ce}$. This is useful to determine a new expression for v_{ce}, which can be substituted into the *eq. I C.E.* so that we obtain

$$v_{be} = h_{ie} i_b - h_{re} \frac{h_{fe}}{\dfrac{1}{R_{CL}} + h_{oe}} i_b, \text{ from which } R_{iBJT} = h_{ie} - \frac{h_{re} h_{fe} R_{CL}}{1 + h_{oe} R_{CL}}$$

or, equivalently

$$R_{iBJT} \overset{\text{def}}{=} \frac{v_{be}}{i_b} = \frac{h_{ie} + \left(h_{ie} h_{oe} - h_{re} h_{fe} \right) R_{CL}}{1 + h_{oe} R_{CL}} = \frac{h_{ie} + \Delta h * R_{CL}}{1 + h_{oe} R_{CL}}$$

where $\Delta h = h_{ie} h_{oe} - h_{fe} h_{re}$.

Observations

When h_{re} can be considered practically zero, we have the following relevant result:

$$R_{iBJT} \cong h_{ie}$$

$$R_i \overset{\text{def}}{=} \frac{v_{be}}{i_s}$$

$$R_i = R_{12} \| R_{iBJT}$$

Note that in general R_i depends on R_L. This is due to the feedback effect (Section 11.3.2.2).

11.4.2.3 C.E.: Current Gain

We want to determine the value of the current gain of the amplifier in the mid-band frequency regime, as already defined in Section 10.2.1 of Chapter 10. It generally refers to the output current to input current ratio. Here the output current flowing through the load is i_l and the input current coming from the source is i_s, therefore

$$A_i \overset{\text{def}}{=} \frac{i_l}{i_s}$$

The equation can be rewritten as

$$A_i = \frac{i_l}{i_c} \frac{i_c}{i_b} \frac{i_b}{i_s} = \alpha_{i,o} A_{iBJT} \alpha_{i,i}$$

So, it has been conveniently separated into three terms: $\alpha_{i,o} = i_l/i_c$, $A_{iBJT} = i_c/i_b$ and $\alpha_{i,i} = i_b/i_s$. The term A_{iBJT} can be considered the current gain due to the BJT itself, because of the ratio of its output and input currents. The terms $\alpha_{i,o}$ and $\alpha_{i,i}$ will reduce the overall current gain A_i because this takes into account current divider effects due to output and input loops, as we will see, so these terms furnish an attenuation effect.

$$\boxed{A_{iBJT}} \quad A_{iBJT} \overset{\text{def}}{=} \frac{i_c}{i_b}$$

The expression of the output voltage $v_{ce} = -R_{CL}i_c$ can be substituted into Equation II C.E. resulting in $i_c = h_{fe}i_b - h_{oe}R_{CL}i_c$, rewritten as $i_c(1 + h_{oe}R_{CL}) = h_{fe}i_b$; therefore

$$A_{iBJT} = \frac{h_{fe}}{1 + h_{oe}R_{CL}}$$

The same result would be obtained by writing the output current divider equation

$$i_c = \frac{\dfrac{1}{h_{oe}}}{\dfrac{1}{h_{oe}} + R_{CL}} h_{fe}i_{b'}$$

from which we obtain the ratio i_c/i_b.

Observation

When h_{oe} can be considered practically zero, the results are

$$A_{iBJT} \cong h_{fe}$$

Wih MATLAB® we can graph A_{iBJT} versus h_{fe} and h_{oe} (Figure 11.36):

```
% AiBJTCE vs. hfe and hoe
clear clf
Rc = 4000; Rl = 10000;
hfe = [50:10:250]; hoe = [0:0.0001:0.001];
[X,Y] = meshgrid(hfe,hoe);
Z = X./(1+Y.*((Rc.*Rl)./(Rc+Rl)));
mesh(X,Y,Z)
xlabel('hfe'); ylabel('hoe'); zlabel('AiBJTCE');
```

Similarly, we can graph A_{iBJT} versus R_C and R_L (Figure 11.37):

```
% AiBJTCE vs. Rc and Rl
hfe = 200; hoe = 0.001;
Rc = [10:100:10000]; Rl = [10:100:10000];
[X,Y] = meshgrid(Rc,Rl);
Z = hfe./(1+hoe.*((X.*Y)./(X+Y)));
mesh(X,Y,Z)
axis([0 10000 0 10000 50 200])
xlabel('Rc'); ylabel('Rl'); zlabel('AiBJTCE');
```

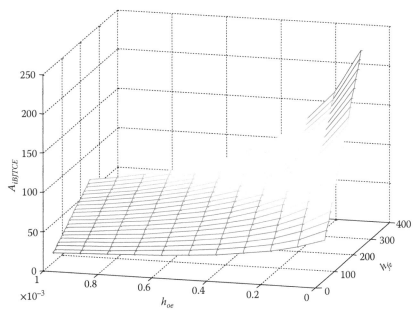

FIGURE 11.36 $A_{iBJT,CE}$ vs. h_{fe} and h_{oe}.

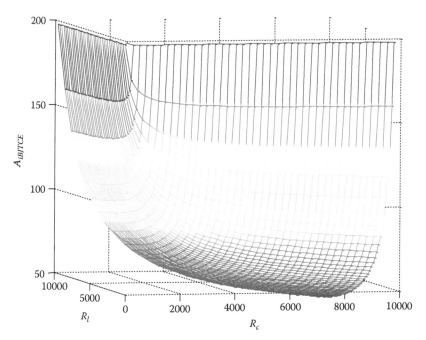

FIGURE 11.37 A_{iBJTCE} vs. R_c and R_l.

| $\alpha_{i,o}$ | $\alpha_{i,o} = \dfrac{i_l}{i_c}$ |

From the output current divider equation $i_l = -\dfrac{R_C}{R_C + R_L} i_c$, therefore,

$$\alpha_{i,o} = -\frac{R_C}{R_C + R_L}$$

Surely $|\alpha_{i,o}| \leq 1$, so the overall current gain will be reduced.

| $\alpha_{i,i}$ | $\alpha_{i,i} = \dfrac{i_b}{i_s}$ |

From the input current divider equation (Figure 11.38):

$$i_b = \frac{G_{iBJT}}{G_{iBJT} + G_{12}} i_s = \frac{\dfrac{1}{R_{iBJT}}}{\dfrac{1}{R_{iBJT}} + \dfrac{1}{R_{12}}} i_s = \frac{R_{12}}{R_{12} + R_{iBJT}} i_s$$

where "G" are the admittances. So,

$$\alpha_{i,i} = \frac{R_{12}}{R_{12} + R_{iBJT}}$$

which is ≤ 1 for sure, and so represents an attenuation.

Therefore, the overall current gain A_i results in

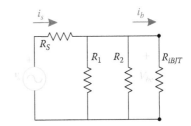

$$\boxed{A_i} \quad A_i = \alpha_{i,o} A_{iBJT} \alpha_{i,i}$$

$$A_i = -\frac{R_C}{R_C + R_L} \frac{h_{fe}}{1 + h_{oe} R_{CL}} \frac{R_{12}}{R_{12} + R_{iBJT}}$$

FIGURE 11.38 Input current divider.

11.4.2.4 C.E.: Voltage Gain

We want to determine the value of the voltage gain of the amplifier in the mid-band frequency regime, as already considered in Section 10.2.1 of Chapter 10. It refers to the output voltage to input voltage ratio. Here the output voltage across the load is $v_l = v_{ce}$ and the input source voltage is v_s, therefore,

$$A_v \overset{\text{def}}{=\!=} \frac{v_l}{v_s} = \frac{v_{ce}}{v_s}$$

This equation can be rewritten as

$$A_v = \frac{v_{ce}}{v_{be}} \frac{v_{be}}{v_s} = A_{vBJT} \alpha_{v,i}$$

So, it has been conveniently separated into two terms $A_{vBJT} = v_{ce}/v_{be}$ and $\alpha_{v,i} = v_{be}/v_s$. The term A_{vBJT} can be considered as the voltage gain due to the BJT itself, because of the ratio of its output and input voltages. The term $\alpha_{v,i}$ will reduce the overall voltage gain A_v because this takes into account the voltage divider effect due to the input loop, as we will see, so this term furnishes an attenuation effect.

$$\boxed{A_{vBJT}} \quad A_{vBJT} \overset{\text{def}}{=\!=} \frac{v_{ce}}{v_{be}}$$

From Equation I C.E. we can extract

$$i_b = \frac{1}{h_{ie}} \left(v_{be} - h_{re} v_{ce} \right)$$

which inserted into Equation II C.E. gives

$$i_c = \frac{h_{fe}}{h_{ie}} \left(v_{be} - h_{re} v_{ce} \right) + h_{oe} v_{ce}$$

But since $i_c = -v_{ce}/R_{CL}$ we can write

$$v_{ce} \left(h_{ie} + h_{ie} h_{oe} R_{CL} - h_{fe} h_{re} R_{CL} \right) = -h_{fe} R_{CL} v_{be}$$

Therefore,

$$A_{vBJT} = -\frac{h_{fe} R_{CL}}{h_{ie} + \left(h_{ie} h_{oe} - h_{fe} h_{re} \right) R_{CL}} = -\frac{h_{fe} R_{CL}}{h_{ie} + \Delta h * R_{CL}}$$

(where $\Delta h = h_{ie} h_{oe} - h_{fe} h_{re}$) or, equivalently,

$$A_{vBJT} = \cfrac{1}{h_{re} - \cfrac{h_{ie} + h_{ie}h_{oe}R_{CL}}{h_{fe}R_{CL}}}$$

Observation

Note the direct proportion between A_{vBJT} and h_{fe}.

We can utilize MATLAB® to draw A_{vBJT} versus the load resistance R_L, keeping the other parameters constant (Figure 11.39):

```
hie = 1500; hre = 0.001; hfe = 200; hoe = 0.001; Rc = 4000;
Deltah = hie*hoe-hfe*hre;
for i = 1000:100000;
Rl(i) = i; Rcl = (Rc*i)/(Rc+i); Avbjt(i) = -(hfe*Rcl)/(hie+Deltah*Rcl);
end;
plot(Rl,Avbjt)
ylabel('AvBJT'); xlabel('Rl'); title('AvBJT vs Rl');
```

$\alpha_{v,i}$	$\alpha_{v,i} = \dfrac{v_{be}}{v_s}$

From the input voltage divider $v_{be} = \dfrac{\left(R_{12} \mid\mid R_{iBJT}\right)}{\left(R_{12} \mid\mid R_{iBJT}\right) + R_s} v_s$, so,

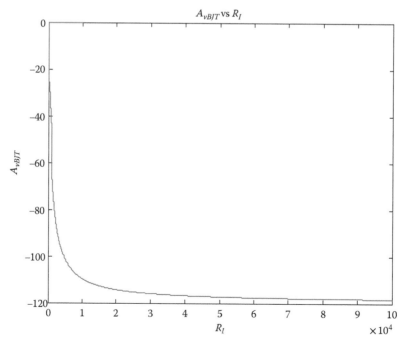

FIGURE 11.39 A_{vBJT} vs. R_1. The absolute value of A_{vBJT} is greater with increasing R_1.

$$\alpha_{v,i} = \frac{(R_{12} \,||\, R_{iBJT})}{(R_{12} \,||\, R_{iBJT}) + R_s}$$

Surely $\alpha_{v,i} \le 1$, so the overall voltage gain will be reduced. The overall voltage gain is then

$$\boxed{A_v} \quad A_v = A_{vBJT}\alpha_v$$

$$A_v = -\frac{h_{fe}R_{CL}}{h_{ie} + \Delta h * R_{CL}} \frac{(R_{12} \,||\, R_{iBJT})}{(R_{12} \,||\, R_{iBJT}) + R_s}$$

11.4.2.5 C.E.: Output Resistance

We want here to determine the value of the output resistance of the amplifier as already defined in Section 10.2.2 of Chapter 10. The output resistance R_o is the one that the amplifier presents to its load. As a consequence, R_o acts as the source resistance for that load.

Similarly to what has already been said for the input resistance, in general a resistance "seen" at two nodes is determined according to the Thévenin or Norton statements, that is, removing all power independent sources (voltage independent sources shorted, current independent sources open) and calculating the resistance between the open connection nodes. Here the nodes are the ones of the output port, and at that port only the AC voltage independent source v_s is present when looking toward the input.

With the load removed and the AC source killed ($v_s \overset{!}{=} 0$), the output resistances R_{oBJT} and R_o, as represented in Figure 11.30, looking back, are as follows:

$$R_{oBJT} \overset{def}{=\!=} \left.\frac{v_{ce}}{i_c}\right|_{R_L \; "open"}$$

$$R_o \overset{def}{=\!=} \left.\frac{v_{ce}}{i_l}\right|_{R_L \; "open"}$$

$$\boxed{R_{oBJT}} \quad R_{oBJT} \overset{def}{=\!=} \left.\frac{v_{ce}}{i_c}\right|_{R_L \; "open"}$$

With the condition $v_s \overset{!}{=} 0$ for the input loop we get the result $h_{re}v_{ce} = -(h_{ie} + R_S \,//\, R_{12})i_b$, from which we can extract i_b, which can be substituted into Equation II C.E. obtaining $i_c = -h_{fe}\dfrac{h_{re}}{h_{ie} + R_S \,||\, R_{12}}v_{ce} + h_{oe}v_{ce}$. Therefore,

$$R_{oBJT} = \frac{1}{h_{oe} - \dfrac{h_{re}h_{fe}}{h_{ie} + R_s \,||\, R_{12}}}$$

R_{oBJT} can be written in the equivalent form

$$R_{oBJT} = \frac{h_{ie} + R_s \mid\mid R_{12}}{h_{oe}\left(R_s \mid\mid R_{12}\right) + \left(h_{ie}h_{oe} - h_{re}h_{fe}\right)} = \frac{h_{ie} + R_s \mid\mid R_{12}}{h_{oe}\left(R_s \mid\mid R_{12}\right) + \Delta h}$$

where $\Delta h = h_{ie}h_{oe} - h_{fe}h_{re}$.

Finally for R_o:

$$\boxed{R_o \quad R_o \overset{\text{def}}{=} \left.\frac{v_{ce}}{i_l}\right|_{R_L\ "open"}}$$

$$R_o = R_{oBJT} \mid\mid R_C$$

Observations

Knowing the output resistance value is fundamental for different reasons, among which are the satisfaction of the matching conditions with the load and the evaluation of the filtering blocking capacitor C_{B2}.

When h_{re} can be considered practically zero, so that its feedback effect is null (Section 11.3.2.2), and when $v_s \overset{!}{=} 0$, it follows that no current will flow through the input loop $i_b = 0$ and, as a consequence $h_{fe}i_b = 0$. Therefore, $R_{oBJT} = \infty$.

For the *C.E.* configuration, the input and output resistances are typically of the same order. This leads to the advantage that more than one BJT can be used in cascade configurations without using impedance-matching networks.

11.5 SWAMPED C.E.

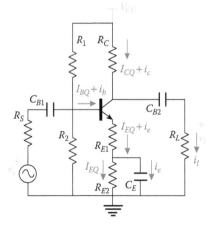

FIGURE 11.40   Swamped C.E. amplifier.

A swamped amplifier refers to a common emitter amplifier having a resistance tied to the emitter of the npn transistor. In particular the emitter resistance R_E is partially bypassed by a capacitor C_E, so that R_E is "split" in two parts, that is R_{E1} and R_{E2}, with only R_{E2} bypassed by C_E, as in Figure 11.40.

A swamped C.E. amplifier has the advantage of stabilizing, that is, reducing, possible variations in voltage gain by increasing the AC resistance of the emitter tail. This higher resistance reduces, in turn, the loading effect of this amplifier to the previous stage. Other positive effects can be the rejection of eventual distortion and noise. As a drawback, the voltage gain will reduce.

The resistor R_{E1} has a feedback effect, because of the voltage across it, proportional to the output current, which will be subtracted from the AC source voltage in the input loop.

11.5.1 DC Analysis

The DC analysis is exactly the same as what we already performed for the C.E. configuration (Section 11.4.1), with the exception of evaluating R_{E1} in place of R_E.

11.5.2 AC Analysis

To draw the equivalent AC model of the swamped C.E. topology we have to consider, as usual, the capacitors as AC short circuits at the amplifier's operating frequencies, and replace the BJT with its hybrid model with equations

$$\begin{cases} v_{be} = h_{ie}i_b + h_{re}v_{ce} & (eq.I\,C.E.) \\ i_c = h_{fe}i_b + h_{oe}v_{ce} & (eq.II\,C.E.) \end{cases}$$

Figure 11.41 shows the AC model representation for which, different from the C.E. configuration, the resistor labeled R_{E1} is not bypassed.

For convenience, resistors R_1 and R_2 can be combined into $R_{12} = R_1 \,\|\, R_2 = R_1 R_2/(R_1 + R_2)$, and R_C and R_L into $R_{CL} = R_C \,\|\, R_L = R_C R_L/(R_C + R_L)$.

11.5.2.1 Swamped C.E.: Current Gain

As usual, the overall current gain A_i of the amplifier can be written as the ratio of output load current i_l and input source current i_s:

$$A_i \overset{\text{def}}{=\!=} \frac{i_l}{i_s}$$

This equation can be rewritten as

$$A_i \overset{\text{def}}{=\!=} \frac{i_l}{i_c}\frac{i_c}{i_b}\frac{i_b}{i_s} = \alpha_{i,o}A_{iBJT}\alpha_{i,i}$$

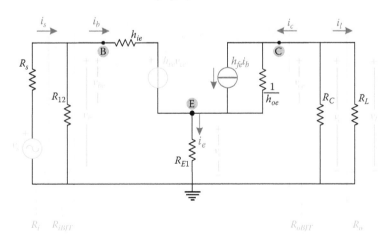

FIGURE 11.41 Swamped C.E. equivalent AC model.

for which $\alpha_{i,o} = i_l / i_c$ and $\alpha_{i,i} = i_b / i_s$ represent current attenuations due respectively to the output and input current divider, and $A_{iBJT} = i_c / i_b$ is the current gain of the BJT itself, being its collector to base current ratio.

A_{iBJT} $\quad$ $A_{iBJT} \overset{\text{def}}{=\!=} \dfrac{i_c}{i_b}$

For the output loop we can write the equation $v_{ce} = v_c - v_e$ $= -R_{CL} i_c - R_{E1} i_e$, in which we can replace v_{ce} with the one obtained from Equation II C.E., and i_e with $i_b + i_c$, obtaining

$$\frac{1}{h_{oe}} \left(i_c - h_{fe} i_b \right) = -R_{CL} i_c - R_{E1} i_b - R_{E1} i_c$$

Rearranging,

$$A_{iBJT} = \frac{h_{fe} - h_{oe} R_{E1}}{1 + h_{oe} \left(R_{CL} + R_{E1} \right)}$$

$\alpha_{i,o}$ $\quad$ $\alpha_{i,o} = \dfrac{i_l}{i_c}$

The output current divider equation is $i_l = -\dfrac{R_C}{R_C + R_L} i_c$, so

$$\alpha_{i,o} = -\frac{R_C}{R_C + R_L}$$

$\alpha_{i,i}$ $\quad$ $\alpha_{i,i} = \dfrac{i_b}{i_s}$

The input current divider equation is $i_b = \dfrac{R_{12}}{R_{12} + R_{iBJT}} i_s$, so

$$\alpha_{i,i} = \frac{R_{12}}{R_{12} + R_{iBJT}}$$

Therefore, the overall current gain of the swamped C.E. amplifier is

A_i $\quad$ $A_i \overset{\text{def}}{=\!=} \dfrac{i_l}{i_c} \dfrac{i_c}{i_b} \dfrac{i_b}{i_s}$

$$A_i = -\frac{R_C}{R_C + R_L} \frac{h_{fe} - h_{oe} R_{E1}}{1 + h_{oe} \left(R_{CL} + R_{E1} \right)} \frac{R_{12}}{R_{12} + R_{iBJT}}$$

11.5.2.2 Swamped C.E.: Input Resistance

We can define two different small-signal input resistances: the first, R_{iBJT}, is the one considered looking in at the base of the BJT with respect to the ground, and the second, R_i, is the overall one as "seen" by the AC source, so that

$$R_{iBJT} \overset{\text{def}}{=\!=} \frac{v_b}{i_b}$$

$$R_i \overset{\text{def}}{=\!=} \frac{v_b}{i_s}$$

as already represented in Figure 11.41.

R_{iBJT} | $R_{iBJT} \overset{\text{def}}{=\!=} \dfrac{v_b}{i_b}$

We can write

$$R_{iBJT} \overset{\text{def}}{=\!=} \frac{v_b}{i_b} = \frac{v_{be} + (v_e)}{i_b} = \frac{v_{be} + (R_{E1}i_b + R_{E1}i_c)}{i_b}$$

Now, from Equation II C.E. we have

$$v_{ce} = \frac{1}{h_{oe}}\left(i_c - h_{fe}i_b\right)$$

which placed into Equation I C.E. gives

$$v_{be} = h_{ie}i_b + \frac{h_{re}}{h_{oe}}\left(i_c - h_{fe}i_b\right)$$

Placing v_{be} into the expression of R_{iBJT} gives

$$R_{iBJT} \overset{\text{def}}{=\!=} \frac{v_b}{i_b} = \frac{h_{ie}i_b + \dfrac{h_{re}}{h_{oe}}(i_c - h_{fe}i_b) + (R_{E1}i_b + R_{E1}i_c)}{i_c}$$

an equation that depends on the input and output currents i_b and i_c. For i_c we can replace what we already found in the previous section with

$$i_c = A_{iBJT}i_b = \frac{h_{fe} - h_{oe}R_{E1}}{1 + h_{oe}\left(R_{CL} + R_{E1}\right)}i_b$$

and after some simple steps, the result is

$$R_{iBJT} = \frac{h_{ie} + \left(h_{ie}h_{oe} - h_{re}h_{fe}\right)R_{CL}}{1 + h_{oe}\left(R_{CL} + R_{E1}\right)}$$

$$+ \frac{\left[h_{oe}R_{CL} + \left(h_{ie}h_{oe} - h_{re}h_{fe}\right) + 1 + h_{fe} - h_{re}\right]R_{E1}}{1 + h_{oe}\left(R_{CL} + R_{E1}\right)}$$

or

$$R_{iBJT} = \frac{h_{ie} + \Delta h * R_{CL}}{1 + h_{oe}\left(R_{CL} + R_{E1}\right)} + \frac{\left[h_{oe}R_{CL} + \Delta h + 1 + h_{fe} - h_{re}\right]R_{E1}}{1 + h_{oe}\left(R_{CL} + R_{E1}\right)}$$

where $\Delta h = h_{ie}h_{oe} - h_{fe}h_{re}$

With $R_{E1} = 0$ we are no longer in a swamped condition and, in fact, the BJT's input resistance R_{iBJT} will be the same as already found for the C.E. configuration.

In many practical cases the BJT's inner feedback effect can be considered as zero, that is $h_{re} \cong 0$, and the Early effect too, that is $h_{oe} \cong 0$. In these occurrencies the BJT's voltage gain reduces to:

$$A_{vBJT} = -h_{fe}R_{CL}/h_{ie} + h_{fe}R_{E1}$$

Now, it can easily result that $h_{ie} \ll h_{fe}R_{E1}$ and, if so, we have the really interesting and relevant consequence that:

$$A_{vBJT} \cong -R_{CL}/R_{E1}$$

Therefore the BJT's voltage amplification does not depend on the BJT itself! This is so relevant because the amplifier's stability (Section 10.2.5 in Chapter 10) exceptionally increases.

Of course, this result can be possible only for small signal conditions, only for those BJTs for which we can neglect two of the four h-parameters and present a quite low input resistance. Neverthless, these requirements are not so stringent after all, so that the relevant result $A_{vBJT} \cong -R_{CL}/R_{E1}$ is quite often utilized. But, we want underline the fact that this result is so *elegant* that it is commonly adopted by novice even when it is wrong because the conditions are not satisfied!

The overall input resistance R_i is easily determined as

$$R_i \quad R_i = R_{12} \,||\, R_{iBJT}$$

$$R_i = R_{12} \,||\, \left\{ \frac{h_{ie} + \Delta h * R_{CL}}{1 + h_{oe}(R_{CL} + R_{E1})} + \frac{\left[h_{oe}R_{CL} + \Delta h + 1 + h_{fe} - h_{re} \right]R_{E1}}{1 + h_{oe}(R_{CL} + R_{E1})} \right\}$$

11.5.2.3 Swamped C.E.: Voltage Gain

The voltage gain A_v of the amplifier in the mid-band frequency regime is the output voltage to input voltage ratio. Here the output voltage across the load is $v_l = v_c$ and the input source voltage is v_s. Therefore,

$$A_v \overset{\text{def}}{=\!=} \frac{v_l}{v_s} = \frac{v_c}{v_s}$$

This equation can be rewritten as

$$A_v = \frac{v_c}{v_b}\frac{v_b}{v_s} = A_{vBJT}\alpha_{v,i}$$

So, it has been conveniently separated into two terms $A_{vBJT} = v_c/v_b$ and $\alpha_{v,i} = v_b/v_s$. The term A_{vBJT} can be considered the voltage gain owing to the BJT itself, because of the ratio of its output and input voltages. The term $\alpha_{v,i}$ will reduce the overall voltage gain A_v because it takes into account the voltage divider effect owing to the input loop, as we will see, so this term furnishes an *attenuation effect*.

$$\boxed{A_{vBJT}} \qquad A_{vBJT} \stackrel{\text{def}}{=\!=} \frac{v_c}{v_b}$$

The base and collector voltages can be written as

$$v_b = v_{be} + v_e = v_{be} + R_{E1}i_e = v_{be} + R_{E1}\left(i_b + i_c\right)$$

$$v_c = v_{ce} + v_e = v_{ce} + R_{E1}i_e = v_{ce} + R_{E1}\left(i_b + i_c\right)$$

and from Equation II C.E.

$$v_{ce} = \frac{1}{h_{oe}}\left(i_c - h_{fe}i_b\right)$$

which, inserted into Equation I C.E., gives

$$v_{be} = h_{ie}i_b + h_{re}v_{ce} = h_{ie}i_b + \frac{h_{re}}{h_{oe}}\left(i_c - h_{fe}i_b\right)$$

Therefore,

$$A_{vBJT} = \frac{v_c}{v_b} = \frac{v_{ce} + R_{E1}\left(i_b + i_c\right)}{v_{be} + R_{E1}\left(i_b + i_c\right)} = \frac{\dfrac{1}{h_{oe}}\left(i_c - h_{fe}i_b\right) + R_{E1}\left(i_b + i_c\right)}{h_{ie}i_b + \dfrac{h_{re}}{h_{oe}}\left(i_c - h_{fe}i_b\right) + R_{E1}\left(i_b + i_c\right)}$$

an equation that depends on both input and output currents i_b and i_c. It is then useful to replace i_c with its previous determined expression

$$i_c = A_{iBJT}i_b = \frac{h_{fe} - h_{oe}R_{E1}}{1 + h_{oe}\left(R_{CL} + R_{E1}\right)}$$

so that in a few steps we have

$$A_{vBJT} = \frac{-h_{fe}R_{CL} + h_{oe}R_{CL}R_{E1}}{h_{ie} + \Delta h * R_{CL} + \left(h_{oe}R_{CL} + \Delta h + h_{fe} - h_{re}\right)R_{E1}}$$

where $\Delta h = \left(h_{ie}h_{oe} - h_{re}h_{fe}\right)$.

With $R_{E1} = 0$ we are no longer in a swamped condition and, in fact, the BJT's voltage gain A_{vBJT} will be the same as already found for the C.E. configuration. In many practical cases the BJT's inner feedback effect can be considered zero, that is, $h_{re} \cong 0$, and the Early effect as well, that is, $h_{oe} \cong 0$. In these occurrences the BJT's voltage gain reduces to

$$A_{vBJT} = -\frac{h_{fe}R_{CL}}{h_{ie} + h_{fe}R_{E1}}$$

Now, it can easily result that $h_{ie} \ll h_{fe}R_{E1}$ and, if so, we have the really interesting and relevant consequence that

$$A_{vBJT} \cong -\frac{R_{CL}}{R_{E1}}$$

Therefore, the BJT's voltage amplification does not depend on the BJT itself! This is relevant because the amplifier's stability (Section 10.2.5 in Chapter 10) exceptionally increases.

Of course, this result can be possible only for small signal conditions—only for those BJTs for which we can neglect two of the four h–parameters and present a quite low input resistance. Nevertheless, these requirements are not so stringent after all, so that the relevant result $A_{vBJT} \cong -R_{CL}/R_{E1}$ is quite often utilized. However, we want to underline the fact that this result is so elegant that novices commonly adopt it even when it is wrong because the conditions are not satisfied!

$\alpha_{v,i}$ $\alpha_{v,i} = \dfrac{v_b}{v_s}$

From the input voltage divider equation $v_b = \dfrac{(R_{12} \| R_{iBJT})}{(R_{12} \| R_{iBJT}) + R_S}v_s$, so

$$\alpha_{v,i} = \frac{(R_{12} \| R_{iBJT})}{(R_{12} \| R_{iBJT}) + R_S}$$

Surely $\alpha_{v,i} \leq 1$, so the overall voltage gain will be reduced.
The overall voltage gain is then

A_v $A_v = A_{vBJT}\alpha_{v,i}$

$$A_v = \frac{-h_{fe}R_{CL} + h_{oe}R_{CL}R_{E1}}{h_{ie} + \Delta h * R_{CL} + (h_{oe}R_{CL} + \Delta h + h_{fe} - h_{re})R_{E1}} \frac{(R_{12} \| R_{iBJT})}{(R_{12} \| R_{iBJT}) + R_S}$$

11.5.2.4 Swamped C.E.: Output Resistance

$$\boxed{R_{oBJT}} \quad R_{oBJT} \stackrel{\text{def}}{=} \frac{v_c}{i_c} = \frac{v_{ce} + v_e}{i_c}$$

With the condition $v_s \stackrel{!}{=} 0$ for the input loop we obtain $h_{re}v_{ce} + v_e = -(h_{ie} + R_s \| R_{12})i_b$, but $v_e = R_{E1}i_e = R_{E1}(i_b + i_c)$, so

$$v_{ce} = \frac{1}{h_{re}}\left[-R_{E1}i_c - R_{E1}i_b - h_{ie}i_b - (R_s \| R_{12})i_b\right]$$

These expressions for v_{ce} and v_e are useful to determine

$$R_{oBJT} = \frac{v_{ce} + v_e}{i_c} = \frac{\frac{1}{h_{re}}\left(-R_{E1}i_c - R_{E1}i_b - h_{ie}i_b - R_s // R_{12}i_b\right) + R_{E1}(i_b + i_c)}{i_c}$$

But, since $i_c = A_{iBJT}i_b = \dfrac{h_{fe} - h_{oe}R_{E1}}{1 + h_{oe}(R_{CL} + R_{E1})}i_b$, we get

$$R_{oBJT} = \frac{\left(1 + h_{fe} + \Delta h + h_{oe}R_s \| R_{12} - h_{re}\right)R_{E1}}{h_{re}\left(h_{oe}R_{E1} - h_{fe}\right)}$$

$$+ \frac{\left[h_{ie} + R_s \| R_{12} + (1 + h_{re})R_{E1}\right]h_{oe}R_{CL} + h_{ie}}{h_{re}\left(h_{oe}R_{E1} - h_{fe}\right)}$$

where $\Delta h = \left(h_{ie}h_{oe} - h_{re}h_{fe}\right)$.

The overall output resistance:

$$\boxed{R_o} \quad R_o = R_{oBJT} \| R_C$$

$$R_o = R_{oBJT} \| \left\{ \frac{\left(1 + h_{fe} + \Delta h + h_{oe}R_s \| R_{12} - h_{re}\right)R_{E1}}{h_{re}\left(h_{oe}R_{E1} - h_{fe}\right)} \right.$$

$$\left. + \frac{\left[h_{ie} + R_s \| R_{12} + (1 + h_{re})R_{E1}\right]h_{oe}R_{CL} + h_{ie}}{h_{re}\left(h_{oe}R_{E1} - h_{fe}\right)} \right\}$$

11.6 C.C. CONFIGURATION (EMITTER FOLLOWER)

The *common collector*, or C.C., amplifier is so called because the BJT shares its collector terminal for both input and output loops in AC conditions. In DC conditions, this cannot be verified.

As we will see, the voltage gain of this configuration is nearly one ($A_v \cong 1$) and the output voltage is in phase with the input voltage, so it is said that the output *follows* the input; this is why an alternative, commonly used name for this configuration is *emitter follower*.

In the following scheme, the resistor R_C is omitted. In the C.E. configuration, this resistor was necessary to obtain an AC voltage drop, which is not necessary in the current case. The output voltage is across emitter and ground, and R_C can be useless for the AC output swing (Figure 11.42).

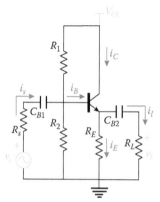

FIGURE 11.42 Emitter follower.

However, in some ways, the roles of resistors R_C and R_E are here "swapped" with respect to the C.E. configuration. The resistor R_C, if used, could have a role in stabilizing the working Q-point against possible variations with temperature, but it has to be bypassed by a capacitor for AC conditions, as in the scheme shown in Figure 11.43.

When the emitter follower configuration is used as a resistance (or impedance) matching network, it can be important to ensure an input resistance as high as possible. This is why an emitter feedback bias (Section 11.4.1.4) is often preferred to the voltage divider bias (Section 11.4.1.5).

11.6.1 DC Analysis

DC operation of the C.C. configuration is determined similarly to that of the C.E. configuration (Section 11.4.1). Capacitors are removed since no DC current flows through them.

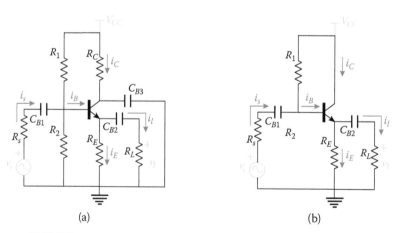

(a)   (b)

FIGURE 11.43 C.C. configuration (a) with collector resistor bypassed by C_{B3} and (b) with an emitter feedback bias.

The known general equations for currents through an arbitrary BJT working in forward-active mode (Section 8.6.3 in Chapter 8) are reported here:

$$\boxed{\text{BJT}}\quad I_E = I_B + I_C$$

$$I_C \cong \beta I_B$$

$$I_E \cong (1 + \beta) I_B$$

$$I_C \cong \frac{\beta}{1+\beta} I_E$$

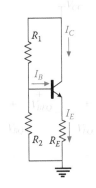

FIGURE 11.44   DC equivalent model for the C.C. configuration.

The DC equivalent model is represented in Figure 11.44. It was drawn considering the capacitors, behaving as filters, replaced by open circuits in DC. The quiescent voltage values are easily obtained:

$$V_{BQ} = \frac{R_2}{R_2 + R_1} V_{CC}$$

$$V_{EQ} = V_{BQ} - V_{BEQ} = V_{BQ} - 0.7$$

$$V_{CEQ} = V_{CC} - V_{EQ}$$

The DC load line can be drawn considering its y- and x-intercepts, that is, the saturation current $I_{C(sat)}$ and the cut-off voltage $V_{CE(off)}$, and connecting them with a straight line:

$$I_{C(sat)} = \frac{V_{CC}}{R_E}$$

$$V_{CE(off)} = V_{CC}$$

11.6.2 AC Analysis

To perform the AC analysis we start with finding the AC equivalent model of our C.C. configuration amplifier. To this aim, in general we consider the following:

→ DC voltage sources are replaced by ground connections and DC current sources by open circuits. Here only one DC voltage source, V_{CC}, is present, so the collector terminal is now grounded.
→ The AC impedance of capacitors, within the amplifier's bandwidth, is zero, so the capacitors represent short circuits. Here the AC voltage source v_s (with its internal resistance R_s) is now directly connected to the amplifier, as well as the resistor load R_L.

This equivalent AC scheme is shown in Figure 11.45.

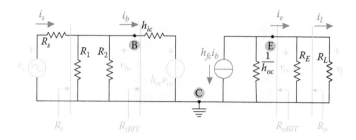

FIGURE 11.45   AC equivalent model for C.C. configuration,
with hybrid parameters in "c" subscript.

In addition, let us define, as usual, $R_{12} = R_1| \,|R_2$ and $R_{EL} = R_E| \,|R_L$

In order to calculate the unknown values, we will use the two hybrid parameter equations of the BJT in "c" subscripts:

$$\begin{cases} v_{bc} = h_{ic}i_b + h_{rc}v_{ec} & (eq.I\,C.C.) \\ i_e = h_{fc}i_b + h_{oc}v_{ec} & (eq.II\,C.C.) \end{cases}$$

The subscript "c" merely indicates that h–parameters are specific for a BJT used in a common-collector amplifier topology.

11.6.2.1 C.C. (C.C. Hybrid Parameters): Input Resistance

$$\boxed{R_{iBJT}} \quad R_{iBJT} \stackrel{\text{def}}{=} \frac{v_{bc}}{i_b}$$

The equation of the output voltage, $v_{ec} = -R_{EL}i_e$, is substituted in Equation II C.C. obtaining

$$-\frac{v_{ec}}{R_{EL}} = h_{fc}i_b + h_{oc}v_{ec}$$

from which we find the v_{ec} expression and substitute it in Equation I C.C.:

$$v_{bc} = h_{ic}i_b - h_{rc}\,\frac{h_{fc}}{\dfrac{1}{R_{EL}} + h_{oc}}i_b$$

thus obtaining

$$R_{iBJT} = h_{ic} - \frac{h_{rc}h_{fc}R_{EL}}{1 + h_{oc}R_{EL}}$$

R_{iBJT} can also be written in the equivalent form

$$R_{iBJT} = \frac{h_{ic} + \left(h_{ic}h_{oc} - h_{rc}h_{fc}\right)R_{EL}}{1 + h_{oc}R_{EL}} = \frac{h_{ic} + \Delta h * R_{EL}}{1 + h_{oc}R_{EL}}$$

where $\Delta h = h_{ic}h_{oc} - h_{fc}h_{rc}$.

R_i

$$R_i = R_{12} \,||\, R_{iBJT}$$

11.6.2.2 C.C. (C.C. Hybrid Parameters): Current Gain

The current gain refers to the output current to input current ratio. Here, the output current flowing through the load is i_l and the input current coming from the source is i_s, therefore,

$$A_i \overset{\text{def}}{=\!=} \frac{i_l}{i_s}$$

This equation can be rewritten as

$$A_i = \frac{i_l}{-i_e}\frac{-i_e}{i_b}\frac{i_b}{i_s} = \alpha_{i,o} A_{iBJT} \alpha_{i,i}$$

So, it has been conveniently separated into the three terms $\alpha_{i,o} = i_l/-i_e$, $A_{iBJT} = -i_e/i_b$, and $\alpha_{i,i} = i_b/i_s$. The term A_{iBJT} can be considered the current gain due to the BJT itself, because of the ratio of its output and input currents. The terms $\alpha_{i,o}$ and $\alpha_{i,i}$ will reduce the overall current gain A_i because this takes into account current divider effects owing to output and input loops, as we will see, so these terms furnish an attenuation effect.

A_{iBJT} $A_{iBJT} \overset{\text{def}}{=\!=} \dfrac{i_c}{i_e}$

We substitute the equation of the output voltage, $v_{ec} = R_{EL}i_e$, in Equation I C.C. obtaining $-i_e = h_{fc}i_b - h_{oc}R_{EL}i_e \Rightarrow -i_e(1 + h_{oc}R_{EL}) = h_{fc}i_b$, so

$$A_{iBJT} = -\frac{h_{fc}}{1 + h_{oc}R_{EL}}$$

Given that typically $h_{fc} < 0$, the value of A_{iBJT} is negative.

$\alpha_{i,o}$ $\alpha_{i,o} = \dfrac{i_l}{-i_e}$

From the output current divider equation $i_l = \dfrac{R_E}{R_E + R_L}i_e$, therefore,

$$\alpha_{i,o} = \frac{R_E}{R_E + R_L}$$

Surely $|\alpha_{i,o}| \leq 1$, so the overall current gain will be reduced.

$$\boxed{\alpha_{i,i}} \quad \alpha_{i,i} \overset{\text{def}}{=} \frac{i_b}{i_s}$$

This term takes into account the current division that occurs in the input circuit, so it is determined by applying the current divider equation

$$i_b = \frac{R_{12}}{R_{12} + R_{iBJT}} i_s$$

and we obtain

$$\alpha_i = \frac{R_{12}}{R_{12} + R_{iBJT}}$$

Given that R_{iBJT} can be quite high, the current gain A_i of a C.C. configuration can be significantly lower than the current gain of the BJT itself A_{iBJT}.

$$\boxed{A_i} \quad A_i = \alpha_{i,o} A_{iBJT} \alpha_{i,i}$$

$$A_i = -\frac{R_E}{R_E + R_L} \frac{h_{fc}}{1 + h_{oc}R_{EL}} \frac{R_{12}}{R_{12} + R_{iBJT}}$$

11.6.2.3 C.C. (C.C. Hybrid Parameters): Voltage Gain

The voltage gain refers to the output voltage to input voltage ratio. Here the output voltage across the load is $v_l = v_{ec}$ and the input source voltage is v_s, therefore,

$$A_v \overset{\text{def}}{=} \frac{v_l}{v_s} = \frac{v_{ec}}{v_s}$$

This equation can be rewritten as

$$A_v = \frac{v_{ec}}{v_{bc}} \frac{v_{bc}}{v_s} = A_{vBJT} \alpha_{v,i}$$

So, it has been conveniently separated into two terms $A_{vBJT} = v_{ec}/v_{bc}$ and $\alpha_{v,i} = v_{bc}/v_s$. The term A_{vBJT} can be considered the voltage gain due to the BJT itself, because of the ratio of its output and input voltages. The term $\alpha_{v,i}$ will reduce the overall voltage gain A_v because this takes into account the voltage divider effect owing to the input loop, as we will see, so the term furnishes an attenuation effect.

$$\boxed{A_{vBJT}} \quad A_{vBJT} \overset{\text{def}}{=} \frac{v_{cb}}{v_{eb}}$$

From Equation I C.C. we obtain

$$i_b = \frac{1}{h_{ic}} (v_{bc} - h_{rc}v_{ec})$$

which inserted in Equation II C.C. furnishes

$$i_e = \frac{h_{fc}}{h_{ic}}\left(v_{bc} - h_{rc}v_{ec}\right) + h_{oc}v_{ec}$$

but

$$i_e = -\frac{v_{ec}}{R_{EL}}$$

so $v_{ec}\left(h_{ic} + h_{ic}h_{oc}R_{EL} - h_{fc}h_{rc}R_{EL}\right) = -h_{fc}R_{EL}v_{bc}.$

Finally

$$A_{vBJT} = -\frac{h_{fc}R_{EL}}{h_{ic} + \left(h_{ic}h_{oc} - h_{fc}h_{rc}\right)R_{EL}} = -\frac{h_{fc}R_{EL}}{h_{ic} + \Delta h \ast R_{EL}}$$

where $\Delta h = h_{ic}h_{oc} - h_{fc}h_{rc}.$

$\boxed{\alpha_{v,i}}$ $\quad \alpha_{v,i} = \dfrac{v_{bc}}{v_s}$

From the input voltage divider $v_{bc} = \dfrac{R_{12} \,||\, R_{iBJT}}{R_s + \left(R_{12} \,||\, R_{iBJT}\right)}v_s$, therefore,

$$\alpha_{v,i} = \frac{R_{12} \,||\, R_{iBJT}}{R_s + \left(R_{12} \,||\, R_{iBJT}\right)}$$

$\boxed{A_v}$ $\quad A_v = -\dfrac{h_{fc}R_{EL}}{h_{ic} + \Delta h \ast R_{EL}}\dfrac{R_{12} \,||\, R_{iBJT}}{R_s + \left(R_{12} \,||\, R_{iBJT}\right)}$

11.6.2.4 C.C. (C.C. Hybrid Parameters): Output Resistance

The C.C. configuration is particularly used in resistance (or impedance) matching applications, so the output resistance is quite important. It is necessary to guarantee the necessary power transfer from the source to the load.

The output resistance can be determined by killing the independent sources

$$(\text{here } v_s \overset{!}{=} 0)$$

and determining the Thévenin resistance at the output port.

$\boxed{R_{oBJT}}$ $\quad R_{oBJT} \overset{\text{def}}{=} \dfrac{v_{ec}}{i_e}$

The independent voltage source is killed, so $v_s \overset{!}{=} 0$, and for the input loop we have $h_{rc}v_{rc} = -\left(h_{ic} + R_s \,||\, R_{12}\right)i_b$, from which we extract i, to be substituted into Equation II C.C., so that $i_e = -h_{fc} \dfrac{h_{rc}}{h_{ic} + R_s \,||\, R_{12}} v_{ec} + h_{oc} v_{ec}$;

therefore,

$$R_{oBJT} = \cfrac{1}{h_{oc} - \cfrac{h_{fc} h_{rc}}{h_{ic} + R_s \,||\, R_{12}}}$$

or, equivalently

$$R_{oBJT} = \frac{h_{ic} + R_s \,||\, R_{12}}{h_{oc}\left(R_s \,||\, R_{12}\right) + \left(h_{ic} h_{oc} - h_{rc} h_{fc}\right)} = \frac{h_{ic} + R_s \,||\, R_{12}}{h_{oc}\left(R_s \,||\, R_{12}\right) + \Delta h}$$

where $\Delta h = h_{ic} h_{oc} - h_{fc} h_{rc}$.

The output resistance of the C.C. configuration is complicated by the fact that the load can be considered including R_E, so we have to take into account $R_{EL} = R_E \,||\, R_L$, or excluding R_E, and considering only R_L, or, even, assuming just R_E to be the real load (avoiding R_L). The choice depends of the final application of the amplifier. It is convenient to consider R_{EL} as a load when the C.C. configuration is adopted as a buffer stage and, to avoid the loading effect, we must guarantee $R_L \gg R_E$, so that the current mostly flows through R_E. Otherwise, it is good to have only R_E as a load in a C.C. configuration to be used as a current amplifier, so that the power level is raised just to correctly drive the load.

To generalize, here we always treat R_{EL}, but be aware of the aforementioned considerations.

$$\boxed{R_O} \quad R_o = R_{EL} \,||\, R_{oBJT}$$

11.6.3 AC Analysis (with C.E. Hybrid Parameters.)

The datasheets can present the h–parameters of the C.E. configuration but not of the C.C. configuration. In this occurrence, we can convert the parameters according to Table 11.5, or directly utilize the furnished parameters, as reported here. The AC equivalent model of the C.C. amplifier results are seen in Figure 11.46.

The h-parameter formulas are rewritten here:

$$\boxed{h-}$$

$$v_{be} = h_{ie} i_b + h_{re} v_{ce} \qquad (eq.I\ C.E.)$$
$$i_c = h_{fe} i_b + h_{oe} v_{ce} \qquad (eq.II\ C.E.)$$

11.6.3.1 C.C. (C.E. Hybrid Parameters): Current Gain

The current gain refers to the output current to input current ratio. Here the output current flowing through the load is i_l and the input current coming from the source is i_s, therefore,

$$A_i \overset{\text{def}}{=\!=} \frac{i_l}{i_s}$$

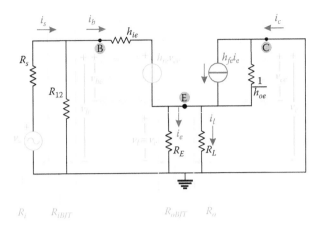

This equation can be rewritten as

$$A_i = \frac{i_l}{i_e}\frac{i_e}{i_b}\frac{i_b}{i_s} = \alpha_{i,o}A_{iBJT}\alpha_{i,i}$$

A_{iBJT} $A_{iBJT} \stackrel{\text{def}}{=\!=} \dfrac{i_e}{i_b}$

For the BJT's current, $i_b + i_c = i_e$, and we replace the i_c value with that from Equation II C.E. $i_b + h_{fe}i_b + h_{oe}v_{ce} = i_e$. But $v_{ce} = -v_e = -R_{EL}i_e$, so $i_b + h_{fe}i_b - h_{oe}R_{EL}i_e = i_e$ from which $(1 + h_{fe})i_b = (1 + h_{oe}R_{EL})i_e$.

So we finally write

$$A_{iBJT} = \frac{i_e}{i_b} = \frac{1 + h_{fe}}{1 + h_{oe}R_{EL}}$$

We can compare this result with the one obtained in Section 11.6.2.2, which is in agreement with the expressions reported in Table 11.5, that is, $h_{fc} = -(1 + h_{fe})$ and $h_{oc} = h_{oe}$.

$\alpha_{i,o}$ $\alpha_{i,o} = \dfrac{i_l}{i_e}$

From the output current divider equation $i_l = \dfrac{R_E}{R_E + R_L}i_e$, therefore,

$$\alpha_{i,o} = \frac{R_E}{R_E + R_L}$$

Surely $|\alpha_{i,o}| \le 1$, so the overall current gain will be reduced.

$\boxed{\alpha_{i,i}}$ $\quad \alpha_{i,i} = \dfrac{i_b}{i_s}$

According to the input current divider expression $i_b = \dfrac{R_{12}}{R_{12} + R_{iBJT}} i_s$, therefore,

$$\alpha_{i,i} = \frac{i_b}{i_s} = \frac{R_{12}}{R_{12} + R_{iBJT}}$$

$\boxed{A_i}$ $\quad A_i = \alpha_{i,o} A_{iBJT} \alpha_{i,i}$

$$A_i = \frac{R_{12}}{R_{12} + R_{iBJT}} \frac{1 + h_{fe}}{1 + h_{oe} R_{EL}} \frac{R_E}{R_E + R_L}$$

11.6.3.2 C.C. (C.E. Hybrid Parameters): Input Resistance

$\boxed{R_{iBJT}}$ $\quad R_{iBJT} \overset{\text{def}}{=} \dfrac{v_b}{i_b}$

Using Equation I C.E., knowing that $v_{ce} = -v_e$, we have

$$R_{iBJT} \overset{\text{def}}{=} \frac{v_{bc}}{i_b} = \frac{v_{be} + v_e}{i_b} = \frac{h_{ie} i_b - h_{re} v_e + v_e}{i_b}$$

but $v_e = R_{EL} i_e = -R_{EL} i_o$; therefore

$$R_{iBJT} = \frac{h_{ie} i_b + h_{re} R_{EL} i_o - R_{EL} i_o}{i_b}$$

However, the output to input current was

$$A_{iBJT} \overset{\text{def}}{=} \frac{i_o}{i_b} = -\frac{1 + h_{fe}}{1 + h_{oe} R_{EL}} \;\Rightarrow\; i_o = -\frac{1 + h_{fe}}{1 + h_{oe} R_{EL}} i_b$$

As a result,

$$R_{iBJT} = h_{ie} - h_{re} R_{EL} \frac{1 + h_{fe}}{1 + h_{oe} R_{EL}} + R_{EL} \frac{1 + h_{fe}}{1 + h_{oe} R_{EL}} \Rightarrow$$

$$R_{iBJT} = h_{ie} - h_{re} R_{EL} \frac{1 + h_{fe}}{1 + h_{oe} R_{EL}} + R_{EL} \frac{1 + h_{fe}}{1 + h_{oe} R_{EL}} \Rightarrow$$

$$R_{iBJT} = \frac{h_{ie} \left(1 + h_{oe} R_{EL}\right) + \left(1 - h_{re}\right)\left(1 + h_{fe}\right) R_{EL}}{1 + h_{oe} R_{EL}}$$

$\boxed{R_i}$ $\quad R_i = R_{12} \,||\, R_{iBJT}$

$$R_i = R_{12} \,||\, \frac{h_{ie} \left(1 + h_{oe} R_{EL}\right) + \left(1 - h_{re}\right)\left(1 + h_{fe}\right) R_{EL}}{1 + h_{oe} R_{EL}}$$

11.6.3.3 C.C. (C.E. Hybrid Parameters): Voltage Gain

$$A_v = \frac{v_l}{v_s} = \frac{v_e}{v_s} = \frac{v_e}{v_b}\frac{v_b}{v_s} = A_{vBJT}\alpha_{v,i}$$

A_{VBJT} | $A_{vBJT} \stackrel{\text{def}}{=} \dfrac{v_e}{v_b}$

From BJT's KCL $i_b + i_c = i_e$, in which we can substitute i_c from Equation II C.E., obtaining $i_b + h_{fe}i_b + h_{oe}v_{ce} = i_e$. But $i_e = v_e/R_{EL}$, so

$$\frac{v_e}{R_{EL}} = i_b h_{fe}i_b - h_{oe}v_e \Rightarrow (1 + h_{fe})R_{EL}i_b = (1 + h_{oe}R_{EL})v_e$$

from which we can extract i_b to be substituted into Equation I C.E.:

$$v_{be} = h_{ie}\frac{1 + h_{oe}R_{EL}}{(1 + h_{fe}R_{EL})}v_e + h_{re}v_{ce} = \left[h_{ie}\frac{1 + h_{oe}R_{EL}}{(1 + h_{fe}R_{EL})} - h_{re}\right]v_e$$

$$A_{vBJT} \stackrel{\text{def}}{=} \frac{v_e}{v_b} = \frac{v_e}{v_{be} + v_e} = \frac{v_e}{\left[h_{ie}\dfrac{1 + h_{oe}R_{EL}}{(1 + h_{fe}R_{EL})} - h_{re}\right]v_e + v_e}$$

As a result,

$$A_{vBJT} = \frac{(1 + h_{fe})R_{EL}}{h_{ie}(1 + h_{oe}R_{EL}) + (1 - h_{re})(1 + h_{fe})R_{EL}}$$

Observation

As we would expect, the A_{vBJT} value is certainly ≤ 1. In particular, the result is that practically A_{vBJT} slightly less than the unity. This is the reason why the C.C. configuration is said to be an emitter follower.

$\alpha_{v,i}$ | $\alpha_{v,i} = \dfrac{v_b}{v_s}$

From the input voltage divider $v_b = \dfrac{R_{12}}{R_{12} + R_s}v_s$, so

$$\alpha_{v,i} = \frac{R_{12}}{R_{12} + R_s}$$

A_v | $A_v = A_{vBJT}\alpha_v$

$$A_v = \frac{(1 + h_{fe})R_{EL}}{h_{ie}(1 + h_{oe}R_{EL}) + (1 - h_{re})(1 + h_{fe})R_{EL}}\frac{R_{12}}{R_{12} + R_s}$$

11.6.3.4 C.C. (C.E. Hybrid Parameters): Output Resistance

This considers resistance "seen" at the output terminals of the BJT:

$$\boxed{R_{oBJT}} \quad R_{oBJT} \overset{\text{def}}{=} \frac{v_e}{i_e} = -\frac{v_{ce}}{i_e} = \frac{v_{ce}}{i_b + i_c}$$

With the condition $v_s \overset{!}{=} 0$ for the input loop we obtain the result $h_{re}v_{ce} + v_e = -(h_{ie} + R_S || R_{12})i_b$, but $v_{ce} = -v_e$ hence

$$v_{ce} = \frac{h_{ie} + (R_S || R_{12})}{1 - h_{re}} i_b$$

Utilizing the latter expression and recalling Equation II C.E.,

$$R_{oBJT} = \frac{v_{ce}}{i_b + (i_c)} = \frac{v_{ce}}{i_b + (h_{fe}i_b + h_{oe}v_{ce})} = \frac{i_b}{i_b + h_{fe}i_b + h_{oe}\dfrac{h_{ie} + (R_S || R_{12})}{1 - h_{re}}i_b}$$

$$= \frac{h_{ie} + (R_S || R_{12})}{1 - h_{re}} \quad \frac{1 - h_{re}}{(1 + h_{fe})(1 - h_{re}) + \left[h_{ie} + (R_S || R_{12})\right]h_{oe}}$$

Therefore,

$$R_{oBJT} = \frac{h_{ie} + R_s || R_{12}}{h_{oe}(h_{ie} + R_s || R_{12}) + (1 + h_{fe})(1 - h_{re})}$$

The overall output resistance is

$$\boxed{R_O} \quad R_o = R_{oBJT} || R_E$$

$$R_o = \frac{h_{ie} + R_s || R_{12}}{h_{oe}(h_{ie} + R_s || R_{12}) + (1 + h_{fe})(h_{re} - 1)} || R_E$$

11.7 C.B. CONFIGURATION

The common base, or C.B., amplifier is configured with the base terminal common to both the input and output loops in AC conditions (Figure 11.47).

The input voltage is applied across the emitter-base terminals, and the output voltage is taken across the collector-base terminals; this is because the base terminal is AC grounded by the capacitor C_B. Here, the input current is the emitter i_E and the output current is the collector i_C. For this configuration, the emitter resistor R_E is strictly necessary to avoid short-circuiting of the AC voltage source. Please observe that the two bias resistors, R_1 and R_2, are totally irrelevant in AC conditions as they are across two grounded AC nodes.

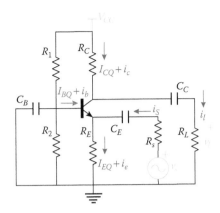

FIGURE 11.47 Complete C.B. amplifier topology.

11.7.1 DC Analysis

The DC analysis for the C.B. configuration is identical to that of the C.E. amplifier. Therefore the DC formulas are identical to those for C.E. already determined (Section 11.4.1).

11.7.2 AC Analysis

As usual, the AC equivalent model of the amplifier is derived by shorting the coupling and bypass capacitors and by setting the DC voltage source to zero (ground). Capacitor C_B practically removes the voltage divider resistors R_1 and R_2 by placing an AC ground at the base of the transistor. The result is represented in Figure 11.48.

Before proceeding further with AC analysis there are some considerations:

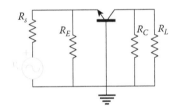

FIGURE 11.48 AC equivalent model of the C.B. configuration; the BJT is not modeled.

→ For the BJT we can adopt its equivalent h–parameter model but with subscript "b," as in Figure 11.49, or with subscript "e," as in Figure 11.50. The AC analysis will be simpler in first case but manufacturers supply exclusive hybrid parameters for BJT in C.E. configurations, as in the equivalent model represented in the second case.

→ Given an arbitrary BJT we know that its emitter current i_e (the input current in the current configuration) is slightly less than its collector current i_c (the output current in the current configuration), so we expect $h_{fb} = \dfrac{i_c}{i_e}\bigg|_{v_{cb}=0}$ to be slightly less than unity.

→ Given that

$$h_{ib} = \frac{v_{eb}}{i_e}\bigg|_{v_{cb}=0}$$

when $i_b \ll i_e$ (which typically always occurs), we get

$$h_{ib} \ll h_{ie} = \frac{v_{be}}{i_b}\bigg|_{v_{ce}=0}$$

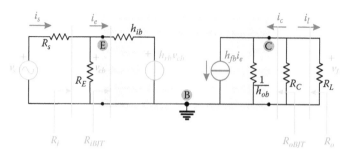

FIGURE 11.49   Equivalent AC model of C.B. configuration,
h-parameters with subscript "b."

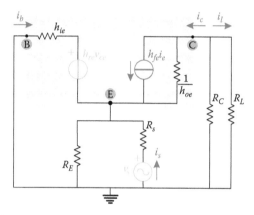

FIGURE 11.50   Equivalent AC model of C.B. configuration,
h-parameters with subscript "e."

⇒ The parameter

$$h_{ob} = \frac{i_c}{v_{cb}}\bigg|_{i_e=0}$$

can often be negligible since it is related to the poor slope of the BJT's output characteristics within the active region.

The AC analysis will here be performed according to Figure 11.49 (the analysis for Figure 11.50 is similar to that already performed in Section 11.5.2).

Before determining the circuit's equations, it is useful to recall the formulas of the h–parameter network, which are useful to model the AC behavior of the BJT when it operates in C.B. configuration in the center of the forward-active region (Section 11.3.2):

$$h - \begin{cases} v_{eb} = h_{ib}i_e + h_{rb}v_{cb} & (eq.I\ C.B.) \\ i_c = h_{fb}i_e + h_{ob}v_{cb} & (eq.II\ C.B.) \end{cases}$$

For convenience, resistors R_C and R_L can be combined into $R_{CL} = R_C\,||\,R_L$.

11.7.2.1 C.B.: Input Resistance

We want to determine the value of the input resistance of the amplifier in the mid-band frequency regime, as already considered in Section 10.2.2 of Chapter 10. In particular, we define R_{iBJT} as the input small-signal resistance looking in at the base of the BJT across nodes E-B, and R_i as the overall small-signal input resistance as "seen" by the AC source, so that

$$R_{iBJT} \overset{def}{=} \frac{v_{eb}}{i_e}$$

$$R_i \overset{def}{=} \frac{v_{eb}}{i_s}$$

as already represented in Figure 11.49.

$$\boxed{R_{iBJT}} \quad R_{iBJT} \overset{def}{=} \frac{v_{eb}}{i_e}$$

The expression of the output voltage $v_{cb} = -R_{CL}i_c$ can be substituted into Equation II C.B. obtaining $-v_{cb}/R_{CL} = h_{fb}i_e + h_{ob}v_{cb}$, so

$$v_{cb} = -\frac{h_{fb}}{\dfrac{1}{R_{CL}} + h_{ob}}i_e$$

This v_{cb} expression can be substituted into Equation I C.B. resulting in

$$v_{eb} = h_{ib}i_e - h_{rb}\frac{h_{fb}}{\dfrac{1}{R_{CL}} + h_{ob}}i_e$$

from which

$$R_{iBJT} = h_{ib} - \frac{h_{rb}h_{fb}R_{CL}}{1 + h_{ob}R_{CL}}$$

or, equivalently

$$R_{iBJT} = \frac{h_{ib} + \left(h_{ib}h_{ob} - h_{rb}h_{fb}\right)R_{CL}}{1 + h_{ob}R_{CL}} = \frac{h_{ib} + \Delta h * R_{CL}}{1 + h_{ob}R_{CL}}$$

where $\Delta h = h_{ib}h_{ob} - h_{fb}h_{rb}$.

Observation

When we can consider $h_{rb} \cong 0$ and $h_{ob} \cong 0$, the input resistance at the BJT's input port is $R_{iBJT} \cong h_{ib}$.

The overall input resistance is then

$$\boxed{R_i} \quad R_i \overset{def}{=} \frac{v_{eb}}{i_s}$$

$$R_i = R_E \,||\, \left(h_{ib} - \frac{h_{rb}h_{fb}}{h_{ob} + \dfrac{1}{R_{CL}}} \right)$$

The low input resistance of the C.B. configuration makes it suitable for utilization when the AC source is provided by microphones with very low output currents.

11.7.2.2 C.B.: Current Gain

We want to determine the value of the current gain of the amplifier in the mid-band frequency regime, as already defined in Section 10.2.1 of Chapter 10. It generally refers to the output current to input current ratio. Here the output current flowing through the load is i_l and the input current coming from the source is i_s, therefore,

$$A_i \overset{\text{def}}{=} \frac{i_l}{i_s}$$

This equation can be rewritten as

$$A_i = \frac{i_l}{i_c}\frac{i_c}{i_e}\frac{i_e}{i_s} = \alpha_{i,o} A_{iBJT} \alpha_{i,i}$$

So, it has been conveniently separated into the three terms $\alpha_{i,o} = i_l/i_c$, $A_{iBJT} = i_c/i_e$, and $\alpha_{i,i} = i_e/i_s$. The term A_{iBJT} can be considered the current gain due to the BJT itself, because of the ratio of its output and input currents. The terms $\alpha_{i,o}$ and $\alpha_{i,i}$ will reduce the overall current gain A_i because this takes into account current divider effects owing to output and input loops, as we will see, so these terms furnish an attenuation effect.

$\boxed{A_{iBJT}}$ $\quad A_{iBJT} \overset{\text{def}}{=} \dfrac{i_c}{i_e}$

The expression of the output voltage $v_{cb} = -R_{CL}i_c$ can be substituted into Equation II C.E. resulting in $i_c = h_{fb}i_e - h_{ob}R_{CL}i_c$, rewritten as $i_c(1 + h_{ob}R_{CL}) = h_{fb}i_e$; therefore,

$$A_{iBJT} = \frac{h_{fb}}{1 + h_{ob}R_{CL}}$$

The same result would be obtained by writing the output current divider equation

$$i_e = \frac{\dfrac{1}{h_{ob}}}{\dfrac{1}{h_{ob}} + R_{CL}} h_{fe}i_b$$

from which we obtain the ratio i_c / i_e.

Since typically $h_{fb} < 0$, we have a negative value for A_{iBJT}.

When we can consider $h_{ob} \cong 0$: $A_{iBJT} \cong h_{fb}$.

$\alpha_{i,o}$ | $\alpha_{i,o} = \dfrac{i_l}{i_c}$

From the output current divider equation $i_l = -\dfrac{R_C}{R_C + R_L} i_c$; therefore,

$$\alpha_{i,o} = -\frac{R_C}{R_C + R_L}$$

Surely $|\alpha_{i,o}| \leq 1$, so the overall current gain will be reduced.

$\alpha_{i,i}$ | $\alpha_{i,i}$ $\qquad \alpha_{i,i} = \dfrac{i_e}{i_s}$

From the input current divider equation (Figure 11.51),

$$i_e = \frac{G_{iBJT}}{G_{iBJT} + G_E} i_s = \frac{\dfrac{1}{R_{iBJT}}}{\dfrac{1}{R_{iBJT}} + \dfrac{1}{R_E}} i_s = \frac{R_E}{R_E + R_{iBJT}} i_s$$

where "G" are the admittances. So

$$\alpha_{i,i} = \frac{R_E}{R_E + R_{iBJT}}$$

which is certainly ≤ 1, and so represents an attenuation.

The overall current gain is then

A_i | $A_i = \dfrac{i_l}{i_c} \dfrac{i_c}{i_e} \dfrac{i_e}{i_s} = \alpha_{i,o} A_{iBJT} \alpha_{i,i}$

$$A_i = -\frac{R_C}{R_C + R_L} \frac{h_{fb}}{1 + h_{ob} R_{CL}} \frac{R_E}{R_E + R_{iBJT}}$$

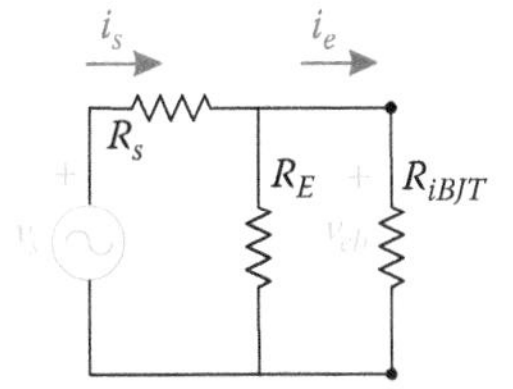

FIGURE 11.51 Detail of the equivalent input network.

11.7.2.3 C.B.: Voltage Gain

We want to determine the value of the voltage gain of the amplifier in the mid-band frequency regime, as already considered in Section 10.2.1 of Chapter 10. It refers to the output voltage to input voltage ratio. Here the output voltage across the load is $v_l = v_{cb}$ and the input source voltage is v_s, therefore,

$$A_v \overset{def}{=\!=} \frac{v_l}{v_s} = \frac{v_{cb}}{v_s}$$

This equation can be rewritten as

$$A_v = \frac{v_{cb}}{v_{eb}} \frac{v_{eb}}{v_s} = A_{vBJT} \alpha_{v,i}$$

Therefore, it has been conveniently separated into two terms $A_{vBJT} = v_{cb}/v_{eb}$ and $\alpha_{v,i} = v_{eb}/v_s$. The term A_{vBJT} can be considered the voltage gain due to the BJT itself, because of the ratio of its output and input voltages. The term $\alpha_{v,i}$ will reduce the overall voltage gain A_v because this takes into account the voltage divider effect owing to the input loop, as we will see, so this term furnishes an attenuation effect.

$$\boxed{A_{vBJT}} \quad A_{vBJT} \overset{def}{=\!=} \frac{v_{cb}}{v_{eb}}$$

From Equation I C.B. we can extract

$$i_e = \frac{1}{h_{ib}} \left(v_{eb} - h_{rb} v_{cb} \right)$$

which inserted into Equation II C.B. gives

$$i_c = \frac{h_{fb}}{h_{ib}} \left(v_{eb} - h_{rb} v_{cb} \right) + h_{ob} v_{cb}$$

However, since

$$i_c = -\frac{v_{cb}}{R_{CL}}$$

we can write

$$v_{cb} \left(h_{ib} + h_{ib} h_{ob} R_{CL} - h_{fb} h_{rb} R_{CL} \right) = -h_{fb} R_{CL} v_{eb}$$

Therefore,

$$A_{vBJT} = -\frac{h_{fb} R_{CL}}{h_{ib} + \left(h_{ib} h_{ob} - h_{fb} h_{rb} \right) R_{CL}} = -\frac{h_{fb} R_{CL}}{h_{ib} + \Delta h * R_{CL}}$$

(where $\Delta h = h_{ib} h_{ob} - h_{fb} h_{rb}$) or, equivalently,

$$A_{vBJT} = \frac{1}{h_{rb} - \dfrac{h_{ib} + h_{ib} h_{ob} R_{CL}}{h_{fb} R_{CL}}}$$

When we can consider $h_{ob} \cong 0$ and $h_{rb} \cong 0$, then $A_{vBJT} \cong \dfrac{h_{fb}}{h_{ib}} R_{CL}$.

$\boxed{\alpha_{v,i}}$ $\alpha_{v,i} = \dfrac{v_{eb}}{v_s}$

From the input voltage divider $v_{eb} = \dfrac{R_E \,||\, R_{iBJT}}{R_s + R_E \,||\, R_{iBJT}} v_s$, so

$$\alpha_{v,i} = \frac{R_E \,||\, R_{iBJT}}{R_s + R_E \,||\, R_{iBJT}}$$

The overall voltage gain is then

$\boxed{A_v}$ $A_v = A_{vBJT}\alpha_{v,i}$

$$A_v = -\frac{h_{fb} R_{CL}}{h_{ib} + \left(h_{ib}h_{ob} - h_{fb}h_{rb}\right)R_{CL}} \frac{R_E \,||\, R_{iBJT}}{R_s + R_E \,||\, R_{iBJT}}$$

11.7.2.4 C.B.: Output Resistance

We want here to determine the value of the output resistance of the amplifier as already defined in Section 10.2.2 of Chapter 10. The output resistance R_o is the one that the amplifier presents to its load. Therefore, R_o acts as the source resistance for that load.

Similarly to what was already said for the input resistance, in general a resistance "seen" at two nodes is determined according to the Thévenin or Norton statements, that is, removing all power independent sources (voltage independent sources shorted, current independent sources open) and calculating the resistance between the open connection nodes. Here the nodes are those of the output port, and at that port only the AC voltage independent source v_s is present when looking towards the input.

With the load removed and the AC source killed ($v_s \overset{!}{=} 0$), the output resistances R_{oBJT} and R_o, as in Figure 11.49, looking backward, are as follows:

$$R_{oBJT} \overset{\text{def}}{=\joinrel=} \left.\frac{v_{cb}}{i_c}\right|_{R_L\text{ "open"}}$$

$$\left.\frac{v_{cb}}{i_c}\right|_{R_L\text{ "open"}}$$

$\boxed{R_{oBJT}}$ $R_{oBJT} \overset{\text{def}}{=\joinrel=} \dfrac{v_{cb}}{i_c}$

With the condition $v_s \overset{!}{=} 0$ for the input loop we obtain $h_{rb}v_{cb} = -\left(h_{ib} + R_S \,//\, R_E\right)i_e$, from which we can extract i, which can be substituted into Equation II C.E. obtaining

$$i_c = -h_{fb}\frac{h_{rb}}{h_{ib} + R_S \,||\, R_E}v_{cb} + h_{ob}v_{cb}$$

Therefore,

$$R_{oBJT} = \frac{1}{h_{ob} - \dfrac{h_{fb}h_{rb}}{h_{ib} + R_s \,||\, R_E}}$$

R_{oBJT} can be written in the equivalent form

$$R_{oBJT} = \frac{h_{ib} + R_s \,||\, R_E}{h_{ob}\left(R_s \,||\, R_E\right) + \left(h_{ib}h_{ob} - h_{rb}h_{fb}\right)} = \frac{h_{ib} + R_s \,||\, R_E}{h_{ob}\left(R_s \,||\, R_E\right) + \Delta h}$$

where $\Delta h = h_{ib}h_{ob} - h_{fb}h_{rb}$.

Observation

When we have $h_{rb} \cong 0$, we get the result $R_{oBJT} \cong \dfrac{1}{h_{ob}}$.

The overall output resistance is then

$$\boxed{R_O} \qquad R_o \overset{\text{def}}{=\joinrel=} \left.\frac{v_{cb}}{i_l}\right|_{R_L \text{ "open"}}$$

$$R_o = R_C \,||\, R_{oBJT}$$

11.8 C.E., C.B., C.C. COMPARISONS

Table 11.6 summarizes and compares the input/output resistances and current/voltage gains of the three BJT amplifier configurations.

TABLE 11.6

Comparisons among Different Amplifier Topologies

Amplifier	C.E.	C.C.	C.B.
Voltage gain A_{vBJT}	High $-\dfrac{h_{fe}R_{CL}}{h_{ie} + \Delta h * R_{CL}}$ degraded by R_{E1}	Low ($\cong 1$) $-\dfrac{h_{fc}R_{EL}}{h_{ic} + \Delta h * R_{EL}}$	High $-\dfrac{h_{fb}R_{CL}}{h_{ib} + \Delta h * R_{CL}}$

TABLE 11.6

Comparisons among Different Amplifier Topologies (*Continued*)

Amplifier	C.E.	C.C.	C.B.
Current gain A_{iBJT}	High $$\dfrac{h_{fe}}{1+h_{oe}R_{CL}}$$	High (≈ 100–200) $$\dfrac{-h_{fc}}{1+h_{oc}R_{EL}}$$	Low ($\cong 1$) $$\dfrac{h_{fb}}{1+h_{ob}R_{CL}}$$
Input res. R_{iBJT}	Moderate ($\approx 10^3\,\Omega$) $$\dfrac{h_{ie}+\Delta h * R_{CL}}{1+h_{oe}R_{CL}}$$	Very high ($\approx 10^6\,\Omega$) $$\dfrac{h_{ic}+\Delta h * R_{EL}}{1+h_{oc}R_{EL}}$$	Very low ($\approx 10^1$–$10^2\,\Omega$) $$\dfrac{h_{ib}+\Delta h * R_{CL}}{1+h_{ob}R_{CL}}$$
Output res. R_{oBJT}	Moderate ($\approx 10^4\,\Omega$) $$\dfrac{h_{ie}+R_s\,\|\,R_{12}}{h_{oe}\,(R_s\,\|\,R_{12})+\Delta h}$$	Very low ($\approx 10^1\,\Omega$) $$\dfrac{h_{ic}+R_s\,\|\,R_{12}}{h_{oc}\,(R_s\,\|\,R_{12})+\Delta h}$$	High ($\approx 10^3$–$10^7\,\Omega$) $$\dfrac{h_{ib}+R_s\,\|\,R_E}{h_{ob}\,(R_s\,\|\,R_E)+\Delta h}$$
Phase differ.	π or $(2n+1)\,\pi$	*none* or 2π	*none* or 2π
Key function	Transcond. ampl. Voltage ampl.	Voltage buffer Imped. match.	Current buffer HF amplifier

Note: Pay close attention to Δh, which is different for different configurations.

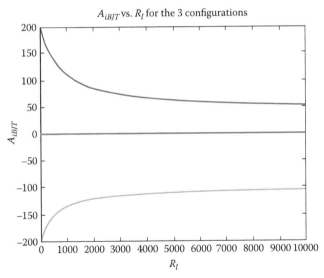

A_{iBJT} vs. R_l for the 3 configurations

FIGURE 11.52 A_{iBJT} vs. R_l for the three configurations: blue for C.E., green for C.C., red for C.B. (the latter constant and practically always near "1").

With MATLAB® we can visualize the current gains versus the load resistance for the three configurations (Figure 11.52):

```
Rg = 500; Rc = 4000; Re = 1000;
hfe = 200; hoe = 0.001;
hfb = -hfe/(1+hfe); hob = hoe/(1+hfe);
for Rl = 10:10000;
```

```
x(Rl) = Rl;
% y1 = AiBJTCE
y1(Rl) = hfe/(1+hoe*((Rc*Rl)/(Rc+Rl)));
% y2 = AiBJTCC
y2(Rl) = -(1+hfe)/(1+hoe*((Re*Rl)/(Re+Rl)));
% y3 = AiBJTCB
y3(Rl) = hfb/(1+hob*((Rc*Rl)/(Rc+Rl)));
end
plot(x,y1,x,y2,x,y3)
xlabel('Rl') ylabel('AiBJT') title('AiBJT vs Rl per le 3 configurazioni')
```

11.9 BJT SIMPLIFIED HYBRID MODEL

We have utilized the model given by the hybrid parameters for the BJT until now. It was evident how sometimes the mathematics can be quite difficult to solve. Therefore, we wonder if it is possible to reduce the complexity, accepting a negligible approximation on results.

To this aim, let us recall the hybrid model, for example, of the BJT in C.E. configuration that was shown in Figure 11.8 (Figure 11.53).

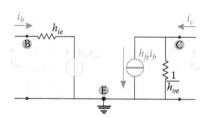

FIGURE 11.53   Hybrid model for the BJT in C.E. configuration.

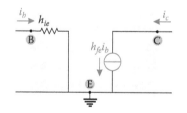

FIGURE 11.54   BJT simplified hybrid model for C.E. configuration.

In the complete *C.E.* amplifier we have the resistance $1/h_{oe}$ in parallel with the resistance R_{CL} (see Figure 11.30). Therefore, when $1/h_{oe} \gg R_{CL}$, or equally $h_{oe}R_{CL} < 0.1$, we can admit to neglect the term h_{oe}. As a consequence, the output KCL equation $i_c = h_{fe}i_b + h_{oe}v_{ce}$ becomes $i_c \cong h_{fe}i_b$. This result inplies that the voltage controlled voltage source $h_{re}v_{ce}$ can be written as $h_{re}v_{ce} = h_{re}h_{fe}i_bR_{CL}$ which, since practically we always have $h_{re}h_{fe} \approx 10^{-2}$, becomes a voltage value negligible respect to the one across the resistance h_{ie} in the input loop.

The conclusion is that when the load resistance R_{CL} can be considered quite "small," we can neglect the hybrid parameters h_{oe} and h_{re}, so that the BJT model can be simplified as represented in Figure 11.54.

Similar results exist for the other two amplifier configurations, so that for C.C. we can neglect the parameters h_{oc} and h_{rc} when $h_{oc}R_{EL} < 0.1$, and for C.B. we can neglect the parameters h_{rb} when $h_{ob}R_{CL} < 0.1$.

It is fundamental to understand when we can adopt the described approximation. This is because if we admit as acceptable the model schematized in the last figure, we admit that the BJT exhibits no feedback, so that the effect of the output to the input side will be ignored. If so, the BJT is said to be unilateral, along with the related amplifier. The consequence is that amplifier's input resistance value will be not dependent on the resistance's load one.

For the C.E. configuration for which we determined the input resistance to be

$$R_{iBJT} = h_{ie} - \frac{h_{re}h_{fe}R_{CL}}{1 + h_{oe}R_{CL}}$$

with $h_{oe}R_{CL} < 0.1$ it becomes simply

$$R_{iBJT-CE} = h_{ie}$$

For the swamped C.E. configuration for which we determined the input resistance to be

$$R_{iBJT} = \frac{h_{ie} + \Delta h * R_{CL}}{1 + h_{oe}(R_{CL} + R_{E1})} + \frac{[h_{oe}R_{CL} + \Delta h + 1 + h_{fe} - h_{re}]R_{E1}}{1 + h_{oe}(R_{CL} + R_{E1})}$$

with the condition $h_{oe}R_{CL} < 0.1$ it becomes simply

$$R_{iBJT-sCE} = h_{ie} + (1 + h_{fe})R_{E1}$$

The latter result can be justified by looking at Figure 11.55, which schematizes the simplified model of the BJT including the emitter resistor.

The current that flows through the resistance, h_{ie}, is i_b, while the current through the resistance R_{E1} is i_e. Now $i_e = i_b + i_c = i_b + h_{fe}i_b = (1 + h_{fe})i_b$. As a result, the equivalent resistance "seen" at the input port is h_{ie} plus not just R_{E1}, but $(1 + h_{fe})$ times R_{E1}. This relevant result is practically used very often to solve circuits with a BJT in a swamped C.E. configuration or in a C.C. configuration when C.E. hybrid parameters are used. This is because the simplified equation of the input resistance is very easy to remember and to apply. In any case, sometimes this simplification is unacceptable, so please be aware the mathematics can be more complicated, but that is what professionals are paid for.

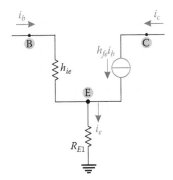

FIGURE 11.55 Simplified hybrid model of the BJT including the emitter resistor.

11.10 BIAS STABILITY

It is necessary that the quiescent Q-point, determined in DC conditions by the bias network, be as stable as possible to guarantee that the BJT is always working in the active region, regardless of the boundary conditions (for example, variations of the parameters of the components caused by changes in temperature). If not, that point could "move" from Q_1 to Q_2 for instance, as represented in Figure 11.56, and the output signal could be clipped.

The bias stability can be represented by the request to maintain a fixed DC collector current I_{CQ}. However, this current can change quite rapidly when the temperature changes and can change when the transistor is replaced by another of the same type even from the same manufacturer. In fact, when the temperature changes, the number of carriers changes accordingly, so that each single current (I_{BQ}, I_{EQ} and, I_{CQ} as well) changes. In addition, if the BJT is replaced by another one that is ostensibly the same, the parameters are not identically the same, because the technological processes are impossible to replicate perfectly.

The current I_{CQ} depends on all the circuit and transistor parameters: $I_{CQ} = f\left(V_{BEQ}, I_{BQ}, I_{CBO}, \beta, V_{CC}, R_E, ..\right)$

The overall changes of I_{CQ} can be written as

$$\Delta I_{CQ} = \frac{\delta I_{CQ}}{\delta V_{BEQ}} \Delta V_{BEQ} + \frac{\delta I_{CQ}}{\delta I_{BQ}} \Delta I_{BQ} + \frac{\delta I_{CQ}}{\delta I_{CBO}} \Delta I_{CBO} + \frac{\delta I_{CQ}}{\delta \beta} \Delta \beta + ...$$

in addition, since the parameters cannot change very much, we can approximate ($\delta \Leftrightarrow \Delta$):

$$\Delta I_{CQ} = S_{V_{BEQ}} \Delta V_{BEQ} + S_{I_{BQ}} \Delta I_{BQ} + S_{I_{CBO}} \Delta I_{CBO} + S_\beta \Delta \beta + ...$$

where

$$S_{V_{BEQ}} \stackrel{def}{=} \frac{\Delta I_{CQ}}{\Delta V_{BEQ}}\bigg|_{I_{BO}=cost,\ I_{CBO}=cost,\ \beta=cost,...} \qquad \left[\frac{A}{V}\right]$$

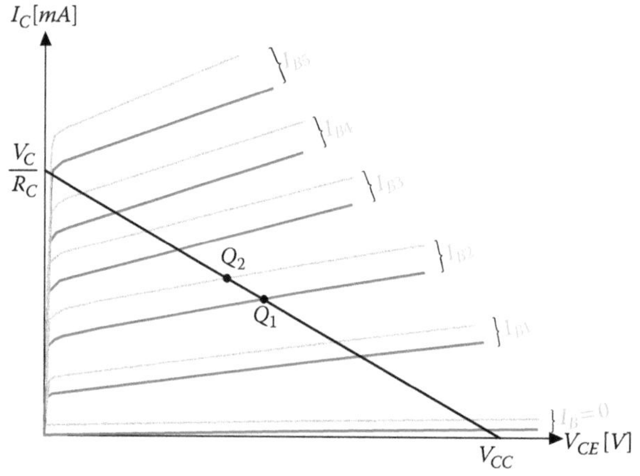

FIGURE 11.56 The Q-point could move according to boundary conditions.

$$S_{I_{BQ}} \overset{\text{def}}{=} \frac{\Delta I_{CQ}}{\Delta I_{BQ}}\Bigg|_{V_{BEQ}=cost,\ I_{CBO}=cost,\ \beta=cost,...}$$

$$S_{I_{CBO}} \overset{\text{def}}{=} \frac{\Delta I_{CQ}}{\Delta I_{CBO}}\Bigg|_{V_{BEQ}=cost,\ I_{BQ}=cost,\ \beta=cost,...}$$

$$S_{\beta} \overset{\text{def}}{=} \frac{\Delta I_{CQ}}{\Delta\beta}\Bigg|_{V_{BEQ}=cost,\ I_{BQ}=cost,\ I_{CBO}=cost,...} \qquad\qquad [\text{A}]$$

are called stability factors. Therefore, stabilization refers to the process of making the operating Q-point independent of temperature or parameter changes, making the stability factors as low as possible.

Obviously, these factors affect the stability of the Q-point differently. We have to pay attention mostly to $S_{I_{CBO}}$, $S_{V_{BEQ}}$, and S_{β}:

→ I_{CBO} rapidly increases with temperature so that, within the typical temperature range of electronic devices, I_{CBO} roughly doubles every $18°F$ or $10°C$, in particular

$$I_{CBO}\Big|_{T_2} = I_{CBO}\Big|_{T_1} * 2^{\frac{T_2-T_1}{10}}$$

→ V_{BE} decreases $2.5\,mV$ for every temperature degree, $\Delta V_{BE} = -2.5*(T_2 - T_1)$.
→ S_{β} changes quite linearly with temperature and, within the typical temperature working range of electronic devices, it can change on the order of 6:1. However, we have to worry about β because manufacturers cannot assure the same value for twin BJTs, so our design must guarantee that the amplifier works for a wide variation of β.

Curiosities

Most devices are not certified to function properly beyond temperature limits of $122–176°F$ or $50–80°C$.

For the most part, USB (Universal Serial Bus) flash memory drives can be considered stable storage devices, but only when operated in the range $32–160°F$ ($0–70°C$); otherwise malfunctions are possible.

11.10.1 Fixed Bias

The fixed bias network (Section 11.4.1.1) is the simplest one but has the least DC stability, as can be proved by the following analysis (Figure 11.57).

From BE-KVL,

$$V_{CC} = R_1 I_B + V_{BE}$$

so

$$I_B = \frac{V_{CC} - V_{BE}}{R_1}$$

However, the collector current is (Section 8.6.3 in Chapter 8)

$$I_C = \beta I_B + (1 + \beta) I_{CBO}$$

Therefore,

$$I_C = \beta \frac{V_{CC} - V_{BE}}{R_1} + (1 + \beta) I_{CBO}$$

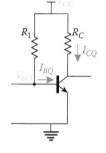

FIGURE 11.57 Fixed bias network.

We can then determine the stability factors as

$$S_{I_{CBO}} \overset{\text{def}}{=} \frac{\Delta I_{CQ}}{\Delta I_{CBO}} = 1 + \beta$$

$$S_{V_{BEQ}} \overset{\text{def}}{=} \frac{\Delta I_{CQ}}{\Delta V_{BEQ}} = -\frac{\beta}{R_1}$$

$$S_\beta \overset{\text{def}}{=} \frac{\Delta I_{CQ}}{\Delta \beta} = \frac{V_{CC} - V_{BE}}{R_1} + I_{CBO}$$

The expressions of $S_{V_{BEQ}}$ and S_β suggest better stability for higher values of R_1 which, on the other hand, cannot be as high as we desire. The parameters $S_{I_{CBO}}$ and S_β are proportional to β, and this is not good since β is of the order of 10^2.

11.10.2 Collector Feedback Bias

The fixed bias configuration has low stability. To reduce the problem we can adopt a collector feedback resistor, forming a collector-to-base bias circuit, to obtain a self-stabilizing bias.

The bias resistor, once named R_1, is now across the collector and base terminals. So it is renamed R_F, since it provides feedback of the output current I_C flowing through it into the transistor's base (Figure 11.58).

This configuration has an effect on stability. In fact, a higher transistor β would mean a higher collector current I_C, and a higher voltage drop across R_C. Therefore, the collector voltage V_{CE} would fall, so that the base current I_B would be reduced and, consequently, the collector current I_C reduces proportionally. It is the opposite with a lower transistor β.

The voltages across the BJT are

$$V_{CE} = V_{CB} + V_{BE} = R_F I_B + V_{BE}$$

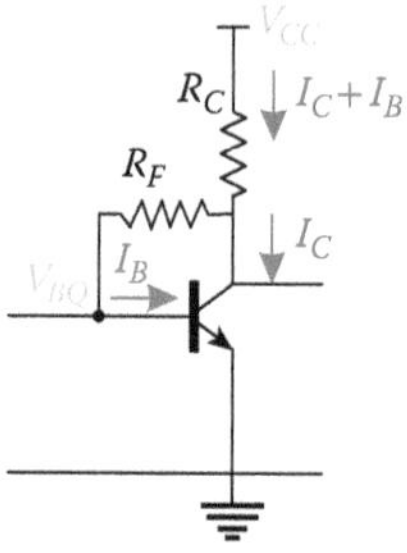

and according to CE-KVL

$$V_{CC} = R_C (I_B + I_C) + V_{CE}$$

So, mixing the previous equations

$$V_{CC} = R_C (I_B + I_C) + R_F I_B + V_{BE}$$

FIGURE 11.58 Collector feedback bias.

from which

$$I_B = \frac{V_{CC} - V_{BE} - I_C R_C}{R_F + R_C}$$

Therefore the expression for the collector current (Section 8.6.3 in Chapter 8)

$$I_C = \beta I_B + (1 + \beta) I_{CBO}$$

becomes

$$I_C = \beta \frac{V_{CC} - V_{BE} - I_C R_C}{R_F + R_C} + (1 + \beta) I_{CBO}$$

$$I_C \left(1 + \frac{\beta R_C}{R_F + R_C}\right) = \beta \frac{(V_{CC} - V_{BE})}{R_F + R_C} + (1 + \beta) I_{CBO}$$

The stability factors are then

$$S_{I_{CBO}} \stackrel{\text{def}}{=} \frac{\Delta I_{CQ}}{\Delta I_{CBO}} = \frac{1 + \beta}{1 + \beta \dfrac{R_C}{R_F + R_C}}$$

$$S_{V_{BEQ}} \stackrel{\text{def}}{=} \frac{\Delta I_{CQ}}{\Delta V_{BEQ}} = -\frac{\beta}{R_F + (1 + \beta) R_C}$$

$$S_\beta \stackrel{\text{def}}{=} \frac{\Delta I_{CQ}}{\Delta \beta} = \frac{V_{CC} - V_{BE} - R_C I_C + (R_F + R_C) I_{CBO}}{R_F + (1 + \beta) R_C}$$

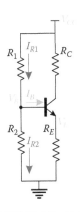

FIGURE 11.59 Voltage divider bias network.

which, compared to the previous ones calculated for the fixed bias network (Section 11.10.1) and assuming R_F in place of R_1, present the extra factors here highlighted in red in the formulas. It is then evident that the stability increases since the stability factors decrease. However, the advantages vanish if $R_F \gg R_C$, as it is obvious since the feedback effect would be practically cancelled.

11.10.3 Voltage Divider Bias

The voltage divider bias network (treated in Section 11.4.1.5) presents greater stability with respect to the collector fixed bias (Section 11.10.2) and the fixed bias (Sections 11.4.1.1 and 11.10.1) networks (Figure 11.59).

We perform here the exact analysis, approaching the problem with the Thévenin equivalent model of the input loop (Figure 11.60).

The BE-KVL gives us

$$V_{TH} = R_{TH}I_B + V_{BE} + R_E\left(I_B + I_C\right)$$

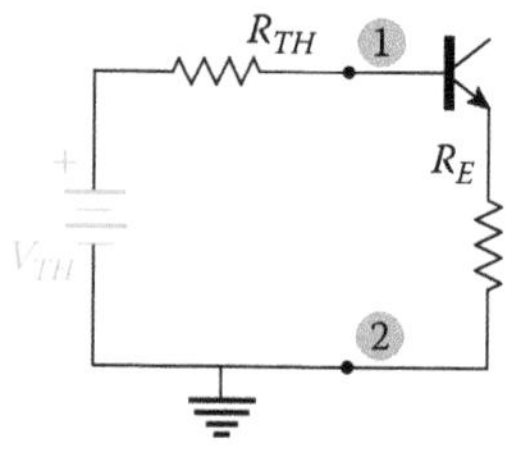

so

$$I_B = \frac{V_{TH} - V_{BE} - R_E I_C}{R_{TH} + R_E}$$

FIGURE 11.60 Thévenin equivalent model of the input loop.

which, inserted into the expression of the collector current $I_C = \beta I_B + (1 + \beta) I_{CBO}$ gives

$$I_C = \beta \frac{V_{TH} - V_{BE} - R_E I_C}{R_{TH} + R_E} + (1 + \beta) I_{CBO}$$

$$I_C\left(1 + \frac{\beta R_E}{R_{TH} + R_E}\right) = \frac{\beta V_{TH}}{R_{TH} + R_E} - \frac{\beta V_{BE}}{R_{TH} + R_E} + (1 + \beta) I_{CBO}$$

11.10.3.1 Stability versus I_{CBO}

The stability factor $S_{I_{CBO}}$, according to its definition, can then be calculated as

$$S_{I_{CBO}} \overset{\text{def}}{=} \frac{\Delta I_{CQ}}{\Delta I_{CBO}} = \frac{1 + \beta}{1 + \dfrac{\beta R_E}{R_{TH} + R_E}} = (1 + \beta)\frac{\left(1 + \dfrac{R_{TH}}{R_E}\right)}{\left(1 + \dfrac{R_{TH}}{R_E}\right) + \beta}$$

So *SICBO* results in the range (1) – (1 + β) with R_{TH}/R_E ranging from very low to very high values. To assure a good stability we must have *RTH* ≪ *RE*. Since typically, β ≫ 1 we can approximate

$$S_{I_{CBO}} \cong \frac{1 + \dfrac{R_{TH}}{R_E}}{1 + \dfrac{1}{\beta}\dfrac{R_{TH}}{R_E}} = \frac{1 + \dfrac{R_{TH}}{R_E}}{K_\beta}$$

where the parameter

$$K_\beta \overset{\text{def}}{=} \left(1 + \frac{1}{\beta}\frac{R_{TH}}{R_E}\right)$$

which can be considered practically about "1," so it does not add information to the equation.

11.10.3.2 Stability versus V_{BEQ}

The other stability factor $S_{V_{BEQ}}$, according to its definition, can be then calculated as

$$S_{V_{BEQ}} \overset{\text{def}}{=\!=} \frac{\Delta I_{CQ}}{\Delta V_{BEQ}} = -\frac{\beta}{R_{TH} + (1+\beta)R_E}$$

In addition, since, typically, $\beta \gg 1$ we can approximate

$$S_{V_{BEQ}} \cong -\frac{\beta}{R_{TH} + \beta R_E} \cong -\frac{1}{1 + \frac{1}{\beta}\frac{R_{TH}}{R_E}} \frac{1}{R_E} = -\frac{1}{K_\beta R_E}$$

In this case, it is useful to have a high value of R_E to guarantee stability versus V_{BEQ} variations.

Observation

Note the similarity of the expressions for $S_{I_{CBO}}$ and $S_{V_{BEQ}}$ with the ones already found for the collector feedback bias network, now with R_{TH} replacing R_F.

11.10.3.3 Stability versus β

Let us determine now the stability factor S_β.
  We know that (Section 8.6.3 in Chapter 8)

$$I_C = \beta I_B + (1+\beta) I_{CBO}$$

from which

$$I_B = \frac{I_C - (1+\beta) I_{CBO}}{\beta}$$

We will replace this expression of I_B in the BE-KVL equation of Figure 11.60, for which

$$V_{TH} = R_{TH} I_B + V_{BE} + R_E (I_B + I_C)$$

$$V_{TH} = R_{TH} \frac{I_C}{\beta} + V_{BE} + R_E \left(\frac{I_C}{\beta} + I_C \right)$$

$$V_{TH} - V_{BE} = \left[\frac{R_{TH}}{\beta} + R_E \left(1 + \frac{1}{\beta} \right) \right] I_C$$

$$V_{TH} - V_{BE} = \left[R_{TH} + (1+\beta) R_E \right] \frac{I_C}{\beta}$$

Now, the latter is an identity and since its left side does not depend on β we can consider the same for its right side. More conveniently we can affirm that, if for a certain β, let's call it β_1, there corresponds a certain value of the collector current, let's call it I_{C1}, and for another value of β, let's call it β_2, there corresponds another value of the collector current, let's call it I_{C2}, we can write the identity

$$\left[R_{TH} + (1+\beta_2) R_E \right] \frac{I_{C2}}{\beta_2} = \left[R_{TH} + (1+\beta_1) R_E \right] \frac{I_{C1}}{\beta_1}$$

from which

$$\frac{I_{C2}}{I_{C1}} = \frac{\left[R_{TH} + (1+\beta_1) R_E \right]}{\left[R_{TH} + (1+\beta_2) R_E \right]} \frac{\beta_2}{\beta_1}$$

and

$$\frac{I_{C2}}{I_{C1}} - 1 = \frac{\left[R_{TH} + (1+\beta_1) R_E \right]}{\left[R_{TH} + (1+\beta_2) R_E \right]} \frac{\beta_2}{\beta_1} - 1$$

$$\frac{I_{C2} - I_{C1}}{I_{C1}} = \frac{\left[R_{TH} + (1+\beta_1) R_E \right] \beta_2 - \left[R_{TH} + (1+\beta_2) R_E \right] \beta_1}{\left[R_{TH} + (1+\beta_2) R_E \right] \beta_1}$$

$$\frac{I_{C2} - I_{C1}}{I_{C1}} = \frac{(R_{TH} + R_E) \Delta\beta}{\left[R_{TH} + (1+\beta_2) R_E \right] \beta_1}$$

As a conclusion the stability factor S_β, according to its definition, is

$$\frac{\Delta I_{CQ}}{\Delta\beta} = \frac{(R_{TH} + R_E) I_{C1}}{\left[R_{TH} + (1+\beta_2) R_E \right] \beta_1} \quad \underline{\text{def}} \; S_\beta$$

in addition, the related I_{CQ} variation

$$\frac{\Delta I_{CQ}}{I_{C1}} = \frac{\left(1 + \dfrac{R_{TH}}{R_E} \right)}{\left(1 + \dfrac{1}{\beta_2} \dfrac{R_{TH}}{R_E} \right)} \frac{\Delta\beta}{\beta_1 \beta_2} = \frac{\left(1 + \dfrac{R_{TH}}{R_E} \right)}{K_{\beta2}} \frac{\Delta\beta}{\beta_1 \beta_2}$$

gives us a function of the ratio R_{TH}/R_E. It follows that a greater value of RE guarantees a higher stability versus β variations. Again, as previously mentioned,

$$K_{\beta2} \; \overset{\text{def}}{=\!=} \; \left(1 + \frac{1}{\beta_2} \frac{R_{TH}}{R_E} \right)$$

approaches "1," so it does not add relevant information to the equation.

11.11 KEY POINTS, JARGON, AND TERMS

→ Amplifier analysis splits into DC and AC analysis. The DC operation is to bias the transistor to determine the operating quiescent point Q about which the AC output signal can change corresponding to an AC input one.

→ In AC small signal analysis, we consider the signal's oscillations around the Q-point to be limited within the transistor's forward active region. In such a manner the amplifier operates in a condition of linearity, the superposition theorem can be adopted, and results from DC so that AC analysis can be summed.

→ Bias stability is when the Q-point does not move even with changes of parameters (temperature, current, beta, and even age of components)

→ A class A amplifier operating in C.E. configuration works so that the AC voltage source signal v_s moves above and below the base DC bias voltage level V_{BQ}. Accordingly the AC base current i_b moves around the DC bias level I_{BQ} along the AC load line. The base current change results in a larger change in collector current thanks to the transistor current gain (Figure 11.61).

→ The h–parameter network represents a model for the transistor when it works in its forward active region and within its frequency bandwidth.

→ A BJT can be adopted in an amplification network in three different configurations: C.E. (common emitter), C.C. (common collector), and C.B. (common base), depending on which terminal of the BJT is shared for both input and output loops in AC conditions. The C.E. configuration shows both high voltage and high current gain, has medium value for both input and output resistances, and the output voltage is 180° out-of-phase with the input voltage. The C.C. configuration shows current but no voltage gain, and has a high input resistance and a low output resistance. The C.B. configuration shows voltage but no current gain, and has a low input resistance and a high output resistance.

→ A capacitor is an AC short at high frequencies and a DC open at low frequencies, with respect to the amplifier's bandwidth.

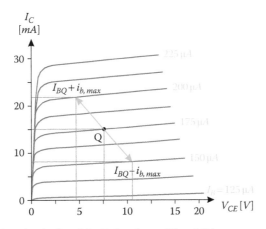

FIGURE 11.61 The Q-point is fixed by DC values. The AC base current swings around it.

→ Temperature stabilization refers to the process of minimizing undesired changes in an amplifier (or, in general, in any arbitrary circuit) caused by heat.

→ Small signal analysis refers to the methodology of obtaining the parameters of an amplifier when the BJT works in linear conditions, that is, only within its active region. The methodology consists of analyzing DC and AC conditions separately (thanks to the superposition theorem). For DC analysis the Q-point is calculated, considering that signal sources are removed, and capacitors behave as open circuits. In AC analysis the characteristics (gains, resistances) of the amplifier can be obtained, considering that the solutions in DC conditions are utilized, DC sources are killed, and capacitors behave as short circuits.

→ Stability refers to the fact that the quiescent Q-point, determined in DC conditions, should be as stable as possible, regardless of the boundary conditions.

11.12 EXERCISES

EXERCISE 1

A sinusoidal voltage source v_s drives a C.E. amplifier in its simplest configuration (Figure 11.62).

Assume $V_{cc} = 20\,V$, $\beta = 50$, and that the capacitive reactance is negligible at the frequencies of operation.

Find R_1 and R_C so that the Q-point is fixed in $I_{CQ} = 30\,mA$ and $V_{CEQ} = 10\,V$.

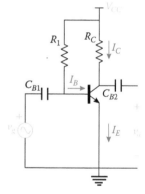

FIGURE 11.62 Simplest C.E. configuration.

ANSWER

$R_1 \cong 32.3\,k\Omega$; $R_C \cong 330\,\Omega$

EXERCISE 2

For a common emitter amplifier in its simplest version as in Figure 11.62 of the previous exercise, we have a DC source $V_{cc} = 12\,V$, a BJT Q-point such as $V_{BEQ} = 0.7\,V$; $I_{BQ} = 60\,\mu A$; $V_{CEQ} = 6\,V$; $I_{CQ} = 3\,mA$ and, correspondingly, $h_{re} = 800\,\Omega$; $h_{re} = 4 \times 10^{-4}$; $h_{fe} = 60$; $h_{oe} = 50 \times 10^{-6}\,\Omega^{-1}$.

Derive the overall values of current and voltage gains, A_i and A_v, and those of the input and output resistances, R_i and R_C.

ANSWER

$R_1 = 188\,k\Omega$; $R_C = 2\,k\Omega$.
$R_{iBJT} \cong 756\,\Omega$; $R_i \cong 753\,\Omega$. $A_{iBJT} \cong 54.5$; $A_i \cong -54.3$. $A_{vBJT} \cong A_v \cong -144.2$. $R_{oq} = 50\,k\Omega$; $R_O \cong 1923\,\Omega$.

(R_{iBJT} is close to R_i and A_{iBJT} is close to A_i since $R_1 \gg R_{iBJT}$, $188\,k\Omega \gg 756\,\Omega$, so R_1 does not contribute so much to the parallel configuration).

EXERCISE 3

If in the previous exercise the h_{re} value is ignored, what is the error percentage?

ANSWER

Taking A_i with no error at all, for the other parameters $E_{\%Av} \cong 5.45$; $E_{\%Ri} \cong 5.74\%$; $E_{\%Ro} \cong 5.45\%$.

These errors are so low that they can be in the range of variations of parameters due to thermal drifts.

For C.E. and C.B. configurations, if we consider $h_{re} \cong 0$ the relative error on the determined parameters can be quite low, and we have the advantage of easier mathematics, but pay close attention to the C.C. configuration for which $h_{re} \cong 1$.

EXERCISE 4

The Q-point of the simplest C.E. configuration, already represented in Figure 11.62, is fixed by the parameters $V_{CEQ} = 5\,V$; $I_{CQ} = 2\,mA$; $I_{BQ} = 5\,\mu A$; $V_{BEQ} = 0.7\,V$.

For the amplifiers, a BJT series BC107A is used that, in correspondence with the Q-point, has the following set of h-parameters: $h_{re} = 3\,k\Omega$; $h_{ie} = 1.7*10^{-4}$; $h_{fe} = 190$; $h_{oe} = 13*10^{-6}\,\Omega^{-1}$. In order to obtain a current gain for a BJT that is equal to $A_{iBJT} = 180$, which values are necessary for

- The collector resistance R_C (here it works as a load too)
- The bias resistor R_1
- The DC voltage supply V_{CC}

With these parameters, also determine the overall current gain A_i.

ANSWER

$R_C = 4.27\,k\Omega$, $V_{CC} = 13.55\,V$, $R_1 = 2.57\,M\Omega$, $A_i = -179.8$.

Pay close attention to the A_{iBJT} and A_i values. Why are they so similar?

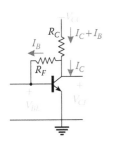

FIGURE 11.63 Collector-feedback bias topology.

EXERCISE 5

Consider the collector-feedback bias amplifier in Figure 11.63.

The circuit parameters are $V_{CC} = 12\,V$; $R_C = 2\,k\Omega$; $R_F = 120\,k\Omega$, and the BJT has $\beta = 80$.

Considering (as usual) $V_{BEQ} = 0.7\,V$ and I_{CBO} negligible, determine the Q-point.

ANSWER

$I_{BQ} \cong 40\,\mu A$, $I_{CQ} \cong 3.2\,mA$, $V_{CEQ} \cong 5.5\,V$.

EXERCISE 6

Determine the stability factors $S_{I_{CBO}}$, $S_{V_{BEQ}}$, S_β of the amplifier of the previous exercise.

ANSWER

$S_{I_{CBO}} \cong 35$, $S_{V_{BEQ}} \cong -0.28\,\dfrac{A}{V}$, $S_\beta \cong 17.3\,\mu A$.

EXERCISE 7

Consider the collector-feedback bias amplifier as already schematized in Figure 11.63.
Find the I_{CQ}–V_{CEQ} pairs in quiescent conditions when $V_{CC} = 12\,V$, $R_C = 2\,k\Omega$, and $R_F = 47\,k\Omega$, for the two occurrences $\beta = 80$ and $\beta = 120$.

ANSWERS

$I_{CQ}\big|_{\beta=80} \cong 4.38\,mA; \; V_{CEQ}\big|_{\beta=80} \cong 3.13\,V$

$I_{CQ}\big|_{\beta=120} \cong 4.738\,mA; \; V_{CEQ}\big|_{\beta=120} \cong 2.46\,V$

Observation

Higher β corresponds to higher I_{CQ} but lower V_{CEQ}, and thus a higher drop on R_C.

EXERCISE 8

A four-resistor bias for a BJT is given as in Figure 11.64.
Determine the values of all the resistors when $V_{CC} = 12\,V$, $I_{CQ} = 10\,mA$, the output voltage drops 50% on V_{CEQ}, 40% on R_C, and 10% on R_E, and a maximum I_{CQ} swing of $\Delta I_{CQ}/I_{CQ} = 5\%$ is requested with β varying.

Determine the dissipated powers in each resistor as well.

For the BJT we are using type 2N2222 for which $w\beta_{min} = 100$, $\beta_{max} = 300$, $\beta_{AVG} = 175$.

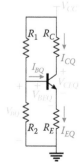

ANSWER

$R_C = \dfrac{V_{RC}}{I_C} = 480\,\Omega$, $R_E = \dfrac{V_{RE}}{I_E} \cong 119\,\Omega$

$P_{R_C} = 48\,mW$, $P_{R_E} = 12.1\,mW$

$R_1 \cong 5507\,\Omega$, $R_2 \cong 1096\,\Omega$

FIGURE 11.64 Four-resistor bias.

$P_{R_1} \cong 18.5\,mW$, $P_{R_2} \cong 3.3\,mW$

EXERCISE 9

Design a voltage divider with emitter feedback bias as in Figure 11.64, so that the total variation of the bias current collector ΔI_{CQ} is within 12%, equally divided among the stability factors $S_{I_{CBO}}$, $S_{V_{BEQ}}$, S_β, for a temperature variation in the range 77–167°F (125–75°C).

Utilize a BJT of type BC107 for which $I_{CBO}|_{T=25°C} = 15\,nA$, and when $I_C = 2\,mA$, $V_{CE} = 5\,V$ we get $h_{FEmin} = 110$, $h_{FEmax} = 450$, $h_{FEtyp} = 120$; $V_{BEon\,min} = 0.55\,V$, $V_{BEon\,max} = 0.7\,V$, $V_{BEon\,typ} = 0.62\,V$.

ANSWER

$R_E \cong 3400\,\Omega$, $R_1 = 25.4\,k\Omega$, $R_2 = 46.4\,k\Omega$, $R_C \cong 99\,\Omega$

Observation

These values satisfy the DC conditions. However, the designer must take into account the AC conditions as well; in particular, the R_C value will be fundamental. Therefore, if the AC conditions suggest different values for R_C, we have to change it, return to DC analysis and iterate the AC analysis until a good compromise is obtained.

EXERCISE 10

Given the circuit in Figure 11.65, for which $V_{CC} = 12\,V$; $V_{BE} = 0.7\,V$; $\beta = 70$; and $R_{C1} = 6\,k\Omega$; $R_{C2} = 4.7\,k\Omega$; $R_{E1} = 2.2\,k\Omega$; $R'_{E2} = 1\,k\Omega$; $R''_{E2} = 2.2\,k\Omega$, determine R_F so that $V_{CE1} = 3\,V$ and $V_{CE1} = 2\,V$.

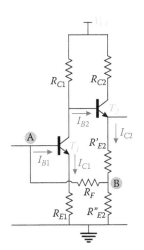

FIGURE 11.65 Voltage divider bias with feedback effect.

ANSWER

$R_F \cong 21\,k\Omega$

EXERCISE 11

Determine the h-parameter h_{fb} of a BJT with a C.B. configuration, as function of the h-parameters of the same BJT with a C.E. configuration, that is $h_{fb} = f(h_{ie}, h_{re}, h_{fe}, h_{oe})$.

ANSWER

$$h_{fb} = -\frac{h_{ie}h_{oe} + (1 - h_{re})h_{fe}}{h_{ie}h_{oe} + (1 - h_{re})(1 + h_{fe})}$$

EXERCISE 12

A four-resistor bias amplifier is supplied from a signal source with internal resistance $R_S = 1\,k\Omega$. The amplifier feeds a load resistor $R_L = 1\,k\Omega$ (Figure 11.66).

Considering that the values of resistors are $R_1 = 40\,k\Omega$, $R_2 = 10\,k\Omega$, $R_C = 5\,k\Omega$, $R_L = 1\,k\Omega$, and the BJT's hybrid parameters are $h_{ie} = 2\,k\Omega$, $h_{re} = 2{*}10^{-3}$, $h_{fe} = 120$, $h_{oe} = 2{*}10^{-4}\,S$, determine the following:

- The overall current amplifier $A_i = i_L / i_g$
- The overall input resistance $R_i = v_i / i_g$
- The overall voltage amplifier $A_v = v_L / v_i$
- The overall output resistance $R_o = v_L / i_L$

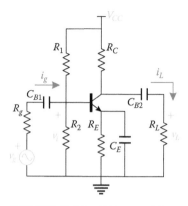

FIGURE 11.66   C.E. amplifier.

ANSWER

$A_{iBJT} \cong 103$; $R_{iBJT} \cong 1829\,\Omega$; $A_{vBJT} \cong -46.9$;
$R_{oBJT} \cong 8553\,\Omega$.
$A_i \cong -69.8$; $R_i \cong 1488\,\Omega$; $A_v \cong -28$; $R_o \cong 759\,\Omega$.

Observation

Pay close attention to the input and output resistance values, $R_i \cong 1488\,\Omega$ and $R_o \cong 759\,\Omega$. They are not suitable for an ideal voltage amplifier for which we require a high input resistance and a low output resistance. However, the obtained values can be of same interest for a transresistance amplifier, even if for the latter both resistances have to be as low as possible.

EXERCISE 13

Design a C.E. amplifier as in Figure 11.67 such that the BJT works in its active region, the overall voltage gain must be $A_v = -200$, and the DC supply $V_{CC} = 12\,V$.

Utilize an npn BJT with the following h-parameter set: $h_{fe} = h_{FE} = 100$, $h_{ie} = 1.5\,k\Omega$, but h_{re} and h_{oe} are negligible. Verify that the power collector $P_C = V_{CEQ}I_{CQ}$ is lower than $50\,mW$.

Suggestions:

- Choose a Q-point in the middle of the BJT's active region.
- V_{RC} and V_{RE} voltage drops 80% and 20% of $V_{CC}/2$ respectively.
- $I_{BQ} = 0.1 I_{R1}$

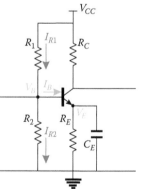

FIGURE 11.67   "Open" configuration of the CE amplifier.

ANSWER

$R_C = 3\,k\Omega$; $R_E = 750\,\Omega$; $R_1 \cong 63\,k\Omega$; $R_2 \cong 12\,k\Omega$;
$P_{CQ} \cong 9.6\,mW$.

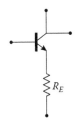

EXERCISE 14

Determine the h′-parameters h'_{ie}, h'_{re}, h'_{fe}, h'_{oe} of the scheme shown in Figure 11.68, as function of the h-parameters of the only BJT, that is,

$$h'_{ie} = f\left(h_{ie}, h_{re}, h_{fe}, h_{oe}\right)$$

$$h'_{re} = f\left(h_{ie}, h_{re}, h_{fe}, h_{oe}\right)$$

$$h'_{fe} = f\left(h_{ie}, h_{re}, h_{fe}, h_{oe}\right)$$

$$h'_{oe} = f\left(h_{ie}, h_{re}, h_{fe}, h_{oe}\right)$$

FIGURE 11.68 BJT with emitter resistor.

ANSWER

$$h'_{ie} = h_{ie} + \frac{(1-h_{re})(1+h_{fe})}{1+h_{oe}R_E}R_E, \; h'_{fe} = \frac{h_{fe}-h_{oe}R_E}{1+h_{oe}R_E}, \; h'_{re} = \frac{h_{re}+h_{oe}R_E}{1+h_{oe}R_E}, \; h'_{oe} = \frac{h_{oe}}{1+h_{oe}R_E}$$

So

$$[H'] = \begin{bmatrix} h'_{ie} & h'_{re} \\ h'_{fe} & h'_{oe} \end{bmatrix} = \begin{bmatrix} h_{ie} + \dfrac{h_{fe}R_E}{1+h_{oe}R_E} & \dfrac{h_{re}+h_{oe}R_E}{1+h_{oe}R_E} \\ \dfrac{h_{fe}-h_{oe}R_E}{1+h_{oe}R_E} & \dfrac{h_{oe}}{1+h_{oe}R_E} \end{bmatrix}$$

EXERCISE 15

Consider a four-resistor emitter feedback biased C.E. amplifier as represented in Figure 11.69.

Assume a DC voltage source $V_{CC} = 10\,V$, an AC voltage source with internal resistance $R_S = 500\,\Omega$, and a load resistor $R_L = 10\,k\Omega$. Assume a Q-point for which $I_{BQ} = 10\,\mu A$, $I_{CQ} = 1\,mA$, $V_{CEQ} = V_{CC}/2$, and $V_{R_CQ} = 0.8(V_{CC}/2)$. The voltage divider bias is such that $I_{R1} = 10I_{BQ}$. The BJT type at the Q-point presents the following set of h-parameters: $h_{ie} = 1.5\,k\Omega$, $h_{re} = 10^{-3}$, $h_{fe} = 200$, $h_{oe} = 10^{-3}\,\Omega^{-1}$.

Determine for the values of each resistor, R_C, R_E, R_1, R_2, in DC conditions.

Determine the values of the BJT and overall input/output resistances and gains in AC conditions.

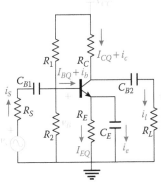

FIGURE 11.69 Four-resistor emitter feedback biased C.E. amplifier.

ANSWER

$R_C = 4\,k\Omega$, $R_E = 1\,k\Omega$, $R_1 = 83\,k\Omega$, $R_2 \cong 18.9\,k\Omega$
$R_{iBJT} \cong 1350\,\Omega$, $R_i \cong 1.24\,k\Omega$
$A_{iBJT} \cong 52$, $A_i \cong 47.7$
$R_{oBJT} \cong 1110\,\Omega$, $R_o \cong 800\,\Omega$
$A_{vBJT} \cong -110$, $A_v \cong -0.71$

EXERCISE 16

The C.B. configuration as represented in Figure 11.70 has BJT h-parameters equal to $h_{ib} = 20\,\Omega$; $h_{rb} = 30 \ast 10^{-3}$; $h_{fb} = -0.98$; $h_{ob} = 50 \ast 10^{-6}\,S$, and resistances $R_1 = 20\,k\Omega$; $R_2 = 20\,k\Omega$; $R_C = 10\,k\Omega$; $R_E = 7\,k\Omega$; $R_S = 1\,k\Omega$; $R_L = 50\,k\Omega$.
  Determine within the amplifier's bandwidth,

- The BJT and overall current gains, A_{iBJT} and A_i
- The BJT and overall input resistances, R_{iBJT} and R_i
- The BJT and overall voltage gains, A_{vBJT} and A_v
- The BJT and overall output resistances, R_{oBJT} and R_o

ANSWER

$R_{iBJT} \cong 22\,\Omega$; $R_i \cong 22\,\Omega$
$R_{oBJT} \cong 1.2\,M\Omega$; $R_o \cong 8276\,\Omega$
$A_{vBJT} \cong -362$; $A_v \cong -7.9$
$A_{iBJT} \cong -0.976$; $A_i \cong 0.162$

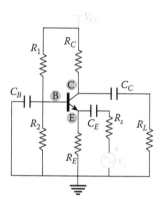

FIGURE 11.70 C.B. configuration.

EXERCISE 17

With respect to the previous exercise, with the same parameter values, determine the same gains and resistances but considering the absence of the base capacitor C_B, as represented in Figure 11.71.

ANSWER

$R_{iq} \cong 262\,\Omega$; $R_i \cong 252\,\Omega$
$R_{oq} \cong 202\,k\Omega$; $R_o \cong 8\,k\Omega$
$A_{vq} \cong 31$; $A_v \cong 6.3$
$A_{iq} \cong -0.976$; $A_i \cong 0.157$

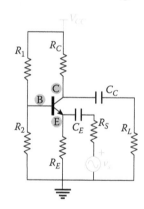

FIGURE 11.71 Configuration with the base terminal no longer grounded.

EXERCISE 18

The C.C. configuration as represented in Figure 11.72 has BJT h-parameters equal to (BC107 type) $h_{ie} = 4\,k\Omega$; $h_{re} = 2.2 \ast 10^{-4}$; $h_{fe} = 250$; $h_{oe} = 30 \ast 10^{-6}\,S$, and resistance equal to $R_1 = 140\,k\Omega$; $R_2 = 140\,k\Omega$; $R_E = 50\,k\Omega$; $R_L = 50\,k\Omega$; $R_S = 500\,\Omega$.
  Determine within the amplifier's bandwidth,
- The BJT and overall current gains, A_{iBJT} and A_i
- The BJT and overall input resistances, R_{iBJT} and R_i

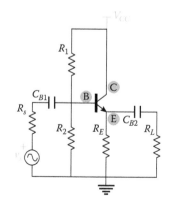

FIGURE 11.72 C.C. configuration.

- *The BJT and overall voltage gains, A_{vBJT} and A_v*
- *The BJT and overall output resistances, R_{oBJT} and R_o*

ANSWER

$A_{iBJT} = 143.4$, $A_i \cong 1.37$
$R_{iBJT} \cong 3.59\,M\Omega$, $R_i \cong 643\,\Omega$
$A_{vBJT} \cong 0.998$, $A_v \cong 0.22$
$R_{oBJT} \cong 31.35\,\Omega$, $R_o \cong 31.3\,\Omega$

EXERCISE 19

Let us consider a dual-output BJT amplifier that has two resistor loads, R_{L1} and R_{L2}, respectively, connected across the collector-ground and emitter-ground terminals (Figure 11.73).

Here none of the BJT's terminals are directly connected to the ground through AC.

Determine the expressions of

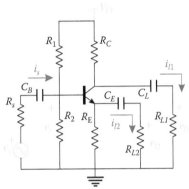

FIGURE 11.73 Dual-output BJT amplifier.

- *The BJT and overall input resistances, R_{iBJT} and R_i*
- *The BJT and overall voltage gains at the collector and emitter terminals, respectively $A_{vBJT,c}$, $A_{vBJT,e}$, $A_{v,c}$, $A_{v,e}$*
- *The BJT current gains at the collector and emitter terminals, $A_{iBJT,c}$ and $A_{iBJT,e}$*
- *The BJT and overall output resistances at the collector and emitter terminals, $R_{oBJT,c}$, $R_{oBJT,e}$, $R_{o,c}$, $R_{o,e}$*

For simplicity, consider $h_{rb} = 0$.

ANSWER

With the expressions $R_{L1C} = R_{L1} \mid\mid R_C$, $R_{L2E} = R_{L2} \mid\mid R_E$, $R_{12} = R_1 \mid\mid R_2$, $R_{12s} = R_s \mid\mid R_{12}$, the solutions can be written as

$$A_{iBJT,c} = \frac{h_{fe} - h_{oe}R_{L2E}}{1 + h_{oe}\left(R_{L1C} + R_{L2E}\right)}$$

$$R_{iBJT} = h_{ie} + \left(1 + A_{iBJT,c}\right)R_{L2E}, \; R_i = R_{12} \mid\mid R_{iq}$$

$$A_{vBJT,c} = -\frac{R_{L1C}}{R_{iBJT}} A_{iBJT,c}$$

$$A_{vBJT,e} = \frac{R_{L2E}}{R_{iq}}\left(1 + A_{iBJT,c}\right) \cong \frac{R_{L2E}}{R_{iq}} A_{iBJT,c}$$

$$A_{vBJT,e} = 1 - \frac{h_{ie}}{R_{iBJT}}$$

$$A_{v,c} = -\frac{R_{L1C}}{R_{iBJT}} A_{iBJT,c} \frac{R_{iBJT}}{R_{iBJT} + R_s}$$

$$A_{v,e} = \frac{R_{L2E}}{R_{iBJT}} A_{iBJT,c} \frac{R_{iBJT}}{R_{iBJT} + R_s}$$

Observation

The output collector and emitter voltages are out of phase by 180°. If we select a component such that $R_{L1C} = R_{L2E}$, the two voltages have the same module. The amplifier realizes a phase inverter, also known as phase splitter. This network can be useful to provide anti-phase input to another stage. Please pay close attention to the fact that the emitter resistor is not decoupled.

$$A_{iBJT,c} = \frac{R_{iq}}{R_{L1}} A_{vBJT,c}$$

$$A_{iBJT,e} = \frac{R_{iq}}{R_{L2}} A_{vq,e}$$

$$R_{oBJT,c} = \frac{1}{h_{oe}} + \frac{h_{fe}}{h_{oe}} \frac{R_{L2E}}{R_{L2E} + (h_{ie} + R_{12g})} + \left[R_{L2E} \,||\, (h_{ie} + R_{12s})\right] \cong \frac{1}{h_{oe}}(1 + h_{fe})$$

$$R_{o,c} = R_{oBJT,c} \,||\, R_{L1C}$$

$$R_{oBJT,e} = (R_{12s} + h_{ie}) \,||\, \frac{(R_{12s} + h_{ie})(1 + h_{oe}R_{L1C})}{h_{fe}} \,||\, \frac{(1 + h_{oe}R_{L1C})}{h_{oe}}$$

$$R_{o,e} = R_{L2E} \,||\, R_{oBJT,e}$$

EXERCISE 20

Given the amplifier shown in Figure 11.74, determine its ideal current gain A_i to realize a negative input resistance, $R_i < 0$.

ANSWER

$A_i > 1$

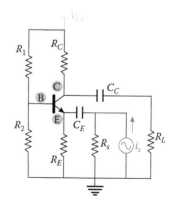

FIGURE 11.74 Current amplifier with current source.

12

Amplifiers: Basic MOSFET Configurations

12.1 MOSFET AS A SIGNAL AMPLIFIER

The MOSFET will be used here in a network to amplify a signal provided by an input source, which furnishes the input pairs v_s and i_s swinging around the zero value.

These signal pairs (directly or indirectly) fed into the MOSFET will produce in it the voltages and current $v_{gs}(t)$, $v_{ds}(t)$, $v_{dg}(t)$, $i_{ds}(t)$ which, if no bias is introduced, swing around zero as well. But this results in a problem, because if the MOSFET would "work" in the red circular area, depicted in Figure 12.1, around the origin O of the axes of its output characteristic curves, it does not produce in output the requested amplification of the input signal, but a "distorted" version of it.

To produce linear amplification, the MOSFET has to work in the green area of the figure, that is, in its saturation region (the name is different from that of the BJT), without entering the linear (yellow area) and cut-off (grey area) regions. In this manner, its voltages and current swing around the central fixed *quiescent* Q-point of that green area (Figure 12.3). To "translate" the O point of the origin to the Q-point of the saturation region, we have to bias the MOSFET, that is, apply DC voltage and current values V_{GSQ}, V_{DSQ}, V_{DGQ}, I_{DSQ} to it. As a consequence, the MOSFET experiences both DC and AC values (Figure 12.2):

$$v_{GS}(t) = V_{GSQ} + v_{gs}(t)$$

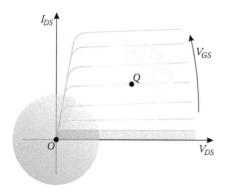

FIGURE 12.1 Output characteristics of the MOSFET and the working area around the point Q.

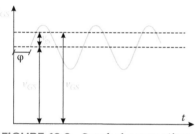

FIGURE 12.2 Symbol convention for the gate-source voltage

$$v_{DS}(t) = V_{DSQ} + v_{ds}(t)$$

$$v_{DG}(t) = V_{DGQ} + v_{dg}(t)$$

$$i_{DS}(t) = I_{DSQ} + i_{ds}(t)$$

The applied input signal is considered to be "small" when it swings around the Q-point but always remains "inside" the green area, so that the cutoff and linear regions are not entered.

The "small" input signal does not perturb the quiescent Q–point, which can then be computed separately, and the superposition theorem can be applied.

To remain in the green area of the saturation region allows us to schematize the MOSFET practically as an equivalent linear two-port network. In fact, with

$$i_{DS}(t) = f\{v_{GS}(t); v_{DS}(t)\}$$

and

$$I_{DS,Q} = f\{V_{GSQ}; V_{DSQ}\}$$

we can develop the following using a Taylor series limited to the first order:

$$i_{DS}(t) \cong I_{DSQ} + \frac{\delta i_{DS}}{\delta v_{GS}}\bigg|_{V_{GS,Q};V_{DS,Q}} v_{gs}(t) + \frac{\delta i_{DS}}{\delta v_{DS}}\bigg|_{V_{GS,Q};V_{DS,Q}} v_{ds}(t)$$

Since we will apply the superposition theorem, we will treat the DC quantities separately, focusing now only on the AC ones:

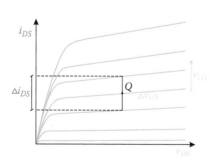

FIGURE 12.3 MOSFET's output characteristics, fixed Q-point, current vs. voltage variations.

$$i_{ds}(t) = \frac{\delta i_{DS}}{\delta v_{GS}}\bigg|_{V_{GSQ};V_{DSQ}} v_{gs}(t) + \frac{\delta i_{DS}}{\delta v_{DS}}\bigg|_{V_{GSQ};V_{DSQ}} v_{ds}(t)$$

Defining as *transconductance* g_m (the MOSFET's equivalent of the BJT's β) the value

$$g_m \underset{def}{=} \frac{\delta i_{DS}}{\delta v_{GS}}\bigg|_{V_{GSQ};V_{DSQ}}$$

and as drain (or channel or output) conductance g_d the value

$$g_d \overset{def}{=\!=} \left. \frac{\delta i_{DS}}{\delta v_{DS}} \right|_{V_{GSQ};V_{DSQ}}$$

We can then write the current $i_{ds}(t)$ as

$$i_{ds}(t) = g_m v_{gs}(t) + g_d v_{ds}(t)$$

which is a simple linear expression.

The MOSFET can be then represented by the admittance parameter equations (Section 5.4.1 in Chapter 5)

$$\begin{cases} i_1 = y_{11}i_1 + y_{12}v_2 \\ i_2 = y_{21}i_1 + y_{22}v_2 \end{cases}$$

with the particular assumptions

$$\begin{cases} i_g = f\left(v_{gs}, v_{ds}\right) = 0 \\ i_{ds}(t) = f\left(v_{gs}, v_{ds}\right) = y_{fs}v_{gs}(t) + y_{os}v_{ds}(t) = g_m v_{gs}(t) + g_d v_{ds}(t) \end{cases}$$

These equations can be represented by an equivalent circuit model with the gate terminal open (since the gate current is zero or its leakage value is considered negligible), and the output loop represented by a voltage controlled current source in parallel with its internal resistance (see Figure 12.7).

In saturation conditions we have

$$i_{DS} = K_n \left(v_{GS} - V_{th}\right)^2 \left(1 + \lambda v_{DS}\right)$$

which, for sake of simplicity, can be approximated as

$$i_{DS} \cong K_n \left(v_{GS} - V_{th}\right)^2$$

when we neglected the non-ideal factor proportional to λ, therefore the physical meaning of g_m, around Q, is

$$g_m \overset{def}{=\!=} \left. \frac{\delta i_{DS}}{\delta v_{GS}} \right|_{V_{GSQ};V_{DSQ}} = 2K_n\left(V_{GSQ} - V_{th}\right)$$

and, since $g_m = f\left(V_{GSQ}\right)$ the transconductance depends on the bias conditions, similarly to what happens with the h_f term for the BJT (Figure 12.3).

$$g_m \overset{def}{=\!=} \left. \frac{\delta i_{DS}}{\delta v_{GS}} \right|_{V_{GSQ};V_{DSQ}} = K_n(V_{GSQ} - V_{th})$$

With similar considerations it is possible to derive the physical meaning of g_d as well:

$$g_d \underset{=}{def} \left. \frac{\delta i_{DS}}{\delta v_{DS}} \right|_{V_{GSQ};V_{DSQ}} = K_n \left(V_{GSQ} - V_{th}\right)^2 \lambda$$

The quantities g_m and g_d are considered to be quite constant with respect to voltages and currents only when these make small excursions around the operating quiescent Q–point.

Accordingly, r_o is the MOSFET's output resistance due to the channel length modulation effect:

$$r_o \underset{=}{def} \left. \left(\frac{\delta i_{DS}}{\delta v_{DS}}\right)^{-1} \right|_{V_{GSQ};V_{DSQ}} = \frac{1}{\lambda K_n \left(v_{GS} - V_{th}\right)^2} \cong \frac{1}{\lambda I_{DS}}$$

In the following we will deal with a MOSFET used for analog electronic amplifiers, so it will be considered to act in its *saturation region*, for which

$$\begin{cases} condition: \quad 0 < v_{GS} - V_{th} < v_{DS} \\ current: \quad i_{DS} = K_n \left(v_{GS} - V_{th}\right)^2 \left(1 + \lambda v_{DS}\right) \end{cases}$$

so that the MOSFET behaves as a voltage controlled current source (VCCS).

12.1.1 Biasing Circuits

There are different methods to bias a MOSFET; the main ones are considered here.

Observation

Just as the BJT's parameters can vary with temperature, this also occurs with the MOSFET's parameters. In particular the *transconductance* K_n (Section 8.7.3 in Chapter 8) and the *threshold voltage* V_{th} can be subject to changes. So, the bias circuit should assure good stability.

12.1.1.1 Drain Feedback Bias

With the *drain feedback bias* a resistance R_G is connected across the drain-gate regions (Figure 12.4). In a practical sense, this is to have a gate voltage equal to the drain one, since no current flows into the gate because of the oxide.

From KVL analysis of the output mesh, we obtain

$$V_{DD} = R_D I_D + V_{DS}$$

while for the gate-source voltage we have

$$V_{GS} = V_{DS}$$

12.1.1.2 Voltage Divider Bias

Another bias is possible by means of a voltage divider network at the input port of the MOSFET, as in Figure 12.5.

Since no current flows into the gate terminal, we can apply the voltage divider formula, without admitting approximations, to determine the DC voltage at the gate port:

$$V_{GS} = \frac{R_2}{R_2 + R_1} V_{DD}$$

From KVL analysis of the output mesh

$$V_{DD} = R_D I_D + V_{DS}$$

12.1.1.3 Source Feedback Bias

In the previous biasing network we can add a source feedback bias. This is to increase the bias stability similarly as was done for the BJT with the emitter feedback bias (Section 11.4.1.4 in Chapter 11). Figure 12.6 shows the scheme with the complete biasing network.

The voltage across the gate-source terminal is

$$V_{GS} = V_G - V_S$$

where the gate voltage V_G is determined by the input voltage divider resistors

$$V_G = \frac{R_2}{R_2 + R_1} V_{DD}$$

The KVL equation of the output mesh is

$$V_{DD} = R_D I_D + V_{DS} + R_S I_D$$

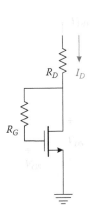

FIGURE 12.4 MOSFET with drain feedback bias.

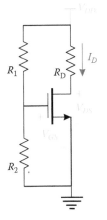

FIGURE 12.5 MOSFET with voltage divider bias.

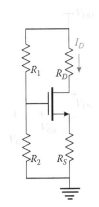

FIGURE 12.6 Four-resistor bias or voltage divider with source feedback bias.

12.1.2 Small-Signal Model

Similarly to what we have seen for the BJT, the AC response of the MOSFET also can be expressed in terms of two-port small-signal equivalent circuit. The choice can be among the six different topologies already discussed (Section 5.4.1 in Chapter 5), all of which are possible and useful. But, it is better to select the topology that better ties with the theory of the device. In particular, for the input loop, the presence of the oxide prevents any current flowing through the gate, which can be considered an open circuit. For the output loop *transconductance* g_m imposes the presence of a voltage controlled current source (VCCS), with the resistor r_o in parallel to take into account, by definition, the *output resistance* due to the channel length modulation effect. So, neglecting the gate capacitance within the bandwidth, the MOSFET small-signal equivalent model appears as in Figure 12.7.

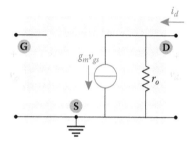

FIGURE 12.7 MOSFET small-signal equivalent circuit.

When $V_{DS} > V_{DS,sat}$, that is, below the pinch-off or linear region, we have

$$g_d = \frac{Z\mu_n C_{ox}}{L}\left(V_G - V_{th} - V_{DS}\right)$$

$$g_m = \frac{Z\mu_n C_{ox}}{L}V_{DS}$$

When $V_{DS} > V_{DS,sat}$, that is, above the pinch-off or saturation region, we have

$$g_d = 0$$

$$g_m = \frac{Z\mu_n C_{ox}}{L}\left(V_G - V_{th}\right)$$

Three different family configurations are possible for MOSFETs, depending on signal injection and extraction constraints: *common source (C.S.)*, *common-drain (C.D.)*, and *common-gate (C.G.)*.

12.2 C.S. CONFIGURATION

The *common source*, or C.S., configuration is the FET equivalent version of the *common emitter*, or C.E., configuration of the BJT amplifier.

A simple biasing network for a C.S. configuration is made of a drain resistor R_D, a source resistor R_S, and a gate resistor R_G. The gate-source bias drop is essentially provided by R_S, because the oxide's gate terminal does not allow a meaninful current to flow. But a high value resistor R_G is used anyway to connect the gate to the ground, to prevent any voltage drifts. The very high input resistance assures no loading effect on a AC signal source connected across gate-source terminals (Figure 12.8).

We will deal with the more commonly used voltage divider biasing network, for the C.S. configuration of a E-MOSFET amplifier. A voltage signal source is coupled into the gate through the capacitor C_{B1} for DC isolation. The bias is now provided by the resistors R_1 and R_2 and, since no current flows through the isolated gate, the resistors form a perfect voltage divider. The bias is for the MOSFET to work in the middle of its saturation region. The emitter resistor R_E provides a stabilizing DC effect, and it is AC shorted by the capacitor C_s. The resistor load R_L is coupled into the drain through the capacitor C_s for a DC isolation (Figure 12.9).

As the input voltage becomes more positive, the gate voltage becomes more positive accordingly, increasing from the quiescent condition. Proportionally, a signal current, flowing through the MOSFET's channel, produces a larger voltage drop across the drain resistor R_D, resulting in less voltage at the output drain terminal. The opposite occurs as the input signal goes more negative. This process produces a larger output signal, with respect the input signal, but 180° out of phase.

The circuit analysis is now performed separately for DC and AC conditions, respectively. This is because the amplifier is considered to work in linear conditions so that the superposition theorem can be applied.

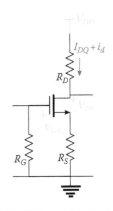

FIGURE 12.8 A simple biasing network for a C.S. configuration.

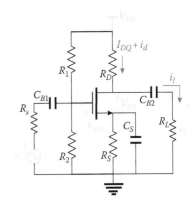

FIGURE 12.9 C.S. configuration with voltage divider and emitter feedback biasing.

12.2.1 DC Analysis

We can assume that all capacitors, for practical purposes, are short circuits to the signal within the amplifier's gain bandwidth, but open circuits in DC conditions. Accordingly, the equivalent DC model is represented in Figure 12.10.

The Q–point is fixed by the parameters $\{I_{DSQ}; V_{DSQ}; V_{GSQ}\}$.

To solve for I_{DSQ} it is crucial to calculate V_{GSQ}, which is

$$V_{GSQ} = V_{GQ} - V_{SQ}$$

From the input voltage divider equation $V_{GQ} = \dfrac{R_2}{R_2 + R_1} V_{DD}$, while $V_{SQ} = R_S I_{DSQ}$, therefore

$$V_{GSQ} = \frac{R_2}{R_2 + R_1} V_{DD} - R_S I_{DSQ}$$

Similarly to what we discussed for the BJT amplifiers, to guarantee linear amplification we have to operate within the MOSFET's saturation region, so we have to respect the DC condition

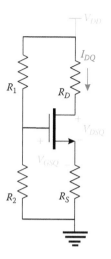

FIGURE 12.10   DC equivalent model of the C.S. configuration.

$$0 < V_{GS} - V_{th} < V_{DS}$$

From Section 8.7.6 of Chapter 8, we know that within the *saturation region* we must have the result that

$$I_{DSQ} = K_n \left(V_{GSQ} - V_{th} \right)^2 \left(1 + \lambda V_{DSQ} \right) \cong K_n \left(V_{GSQ} - V_{th} \right)^2$$

having assumed $\lambda V_{DSQ} \ll 1$.

Since K_n and V_{th} are normally given, it is sufficient to solve the system of two equations with a quadratic expression:

$$\begin{cases} V_{GSQ} = \dfrac{R_2}{R_2 + R_1} V_{DD} - R_S I_{DSQ} \\[2mm] I_{DSQ} \cong K_n \left(V_{GSQ} - V_{th} \right)^2 \end{cases}$$

so to obtain V_{GSQ} and I_{DSQ}.

In particular for V_{GSQ} from the previous system

$$R_s K_n V_{GSQ}^2 + \left(1 - 2 R_s K_n V_{th} \right) V_{GSQ} + \left(R_s K_n V_{th}^2 - V_{GQ} \right) = 0$$

This quadratic expression provides two solutions for V_{GSQ}, named V_{GSQ1} and V_{GSQ2}:

$$V_{GSQ1} = \frac{-\left(1 - 2 R_s K_n V_{th} \right) - \sqrt{\left(1 - 2 R_s K_n V_{th} \right)^2 - 4 R_s K_n \left(R_s K_n V_{th}^2 - V_{GQ} \right)}}{2 R_s K_n}$$

$$V_{GSQ2} = \frac{-\left(1 - 2 R_s K_n V_{th} \right) + \sqrt{\left(1 - 2 R_s K_n V_{th} \right)^2 - 4 R_s K_n \left(R_s K_n V_{th}^2 - V_{GQ} \right)}}{2 R_s K_n}$$

from which we can determine two values for the drain current

$$I_{DSQ1} \cong K_n \left(V_{GSQ1} - V_{th} \right)^2$$

$$I_{DSQ2} \cong K_n \left(V_{GSQ2} - V_{th} \right)^2$$

The last unknown V_{DSQ} is determined with the output KVL expression. So

$$V_{DSQ1} = V_{DD} - \left(R_D + R_S \right) I_{DSQ1}$$

$$V_{DSQ2} = V_{DD} - \left(R_D + R_S \right) I_{DSQ2}$$

Typically only one triplet between $\{I_{DSQ1};V_{DSQ1};V_{GSQ1}\}$ and $\{I_{DSQ2};V_{DSQ2};V_{GSQ2}\}$ satisfy the conditions for the MOSFET to work within its saturation region:

$$0 < V_{GSQ1} - V_{th} < V_{DSQ1} \text{ or } 0 < V_{GSQ2} - V_{th} < V_{DSQ2}$$

so the right one must be selected after the calculations.

To this aim it can be useful to refer to MATLAB®:

```
% here impose the values
R1 = ...; R2 = ...; Rd = ...; Rs = ...; Vdd =...;
Kn =...; Vth = ...;
% here the calculations
Vg = (R2/(R2+R1))*Vdd;
a = Rs*Kn; b = 1-2*Rs*Kn*Vth; c = Rs*Kn*Vth^2-Vg; delta = b^2-4*a*c;
Vgsq1 = (-b-sqrt(delta))/(2*a)
Vgsq2 = (-b+sqrt(delta))/(2*a)
Idsq1 = Kn*(Vgsq1-Vth)^2
Idsq2 = Kn*(Vgsq2-Vth)^2
Vdsq1 = Vdd-(Rd+Rs)*Idsq1
Vdsq2 = Vdd-(Rd+Rs)*Idsq2
if ((Vgsq1-Vth)>0) & ((Vgsq1-Vth)<Vdsq1) disp('saturation conditions
satisfied for Vgsq1'); end
if ((Vgsq2-Vth)>0) & ((Vgsq2-Vth)<Vdsq2) disp('saturation conditions
satisfied for Vgsq2'); end
```

Observation

A rule of thumb can sometimes be useful to choose the proper values for the biasing output loop resistors. We can select the source resistor R_S such that the voltage V_S across it is about one-fifth of the DC voltage source V_{DD}, and the drain resistor R_D such that the voltage V_D is placed in the middle of V_{DD} and V_S.

Of course, the following computations must assure that the related bias is correct.

12.2.2 AC Analysis

The AC analysis of a FET amplifier is very similar to that of a BJT amplifier. Again, the golden rules are to set all DC sources to zero (a zero voltage DC source is a short, while a zero current DC source is an open), to assume that all the capacitors are short circuits for the signal within the amplifier's gain bandwidth. Accordingly the AC equivalent model is represented in Figure 12.11.

For convenience, resistors R_1 and R_2 can be combined into $R_G = R_1 \| R_2$.

The AC analysis will be performed to determine the input and output resistances and the voltage gain. The current gain is not considered since no input current flows because of the gate oxide, so that the current gain is simply theorically infinite.

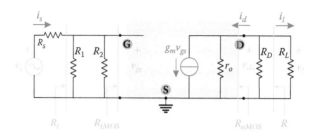

FIGURE 12.11   AC equivalent model
of the C.S. configuration.

12.2.2.1 C.S.: Input Resistance

We can distinguish two input resistances, that is, the one at the real input of the MOSFET

$$R_{iMOS} \overset{def}{=\!=} \frac{v_{gs}}{i_g}$$

and the one working as a load for the AC source

$$R_i \overset{def}{=\!=} \frac{v_{gs}}{i_s}.$$

We obtain the following result:

R_{iMOS}	$R_{iMOS} \overset{def}{=\!=} \dfrac{v_{gs}}{i_g}$

Since $i_g = 0$ we simply have

$$R_{iMOS} = \infty$$

R_i	$R_i \overset{def}{=\!=} \dfrac{v_{gs}}{i_s}$

It is easily determined that

$$R_i = R_G$$

So the overall input resistance is determined by the bias resistors only.

12.2.2.2 C.S.: Output Resistance

The *output resistance* R_o is the one that the amplifier presents to its load. As a consequence, R_o acts as the source resistance for that load. We recall here that, in general, a resistance "seen" at two nodes is determined according to the Thévenin or Norton statements, that is, removing all power independent sources (voltage independent sources shorted, current independent sources open) and calculating the resistance between the open connection nodes.

Here the nodes are the ones of the output port, and at that port only the AC voltage independent source v_s is present when looking toward the input. With the load removed and the AC source killed ($v_s \overset{!}{=} 0$), the output resistances R_{oBJT} and R_o, as represented in Figure 12.11, are as follows:

$$R_{oMOS} \underset{=}{def} \left. \frac{v_{ds}}{i_d} \right|_{R_L \text{ "open"}}$$

$$R_o \underset{=}{def} \left. \frac{v_{ds}}{i_l} \right|_{R_L \text{ "open"}}$$

R_{oMOS}	$R_{oMOS} \underset{=}{def} \left. \dfrac{v_{ds}}{i_d} \right|_{R_L \text{ "open"}}$

With the condition of $v_s \overset{!}{=} 0$, the current $g_m v_{gs} = 0$, so that the voltage controlled current source (VCCS) results open. As a consequence,

$$R_{oMOS} = r_o$$

R_o	$R_o \underset{=}{def} \left. \dfrac{v_{ds}}{i_l} \right|_{R_L \text{ "open"}}$

The overall output resistance is easily determined as

$$R_o = r_o \,||\, R_D$$

The output resistance is typically large, so that a voltage buffer is necessary to drive a low load resistor.

12.2.2.3 C.S.: Voltage Gain

The *voltage gain* A_v of the amplifier in the mid-band frequency regime refers to the ratio between the output voltage v_l, across the load resistance, and the input voltage v_s of the AC source ratio:

$$A_v \underset{=}{def} \frac{v_l}{v_s}$$

This equation can be rewritten as

$$A_v = \frac{v_l}{v_s} = \frac{v_{ds}}{v_s} = \frac{v_{ds}}{v_{gs}} \frac{v_{gs}}{v_s} = A_{vMOS} \alpha_{v,i}$$

So, it has been conveniently separated into two terms $A_{vMOS} = v_{ds}/v_{gs}$ and $\alpha_{v,i} = v_{gs}/v_s$.

The term A_{vMOS} can be considered the voltage gain due to the MOS itself, because of the ratio of its output and input voltages. The term $\alpha_{v,i}$ will reduce the overall voltage gain A_v because this takes into account the voltage divider effect owing to the input loop, as we will see, so this term furnishes an *attenuation effect*.
We obtain the following result:

A_{vMOS}	$A_{vMOS} \underset{=}{def} \dfrac{v_{ds}}{v_{gs}}$

From the output mesh $v_{ds} = -g_m v_{gs} \left(r_o \,||\, R_D \,||\, R_L \right)$, therefore,

$$A_{vMOS} = -g_m \left(r_o \,||\, R_D \,||\, R_L \right)$$

$\boxed{\alpha_{v,i}}$ $\alpha_{v,i} \underset{=}{def} \dfrac{v_{gs}}{v_s}$

From the input voltage divider equation $v_{gs} = \dfrac{R_G}{R_G + R_s} v_s$, therefore,

$\alpha_{v,i} = \dfrac{R_G}{R_G + R_s}$

The overall voltage gain is then

$\boxed{A_v}$ $A_v \underset{=}{def} A_{vMOS}\alpha_{v,i}$

$$A_v = -g_m \left(r_o \, || R_D || \, R_L \right) \dfrac{R_G}{R_G + R_s}$$

The output voltage is then 180° out of phase with respect to the input and the overall voltage gain can typically be high.

12.3 C.D. CONFIGURATION (SOURCE FOLLOWER)

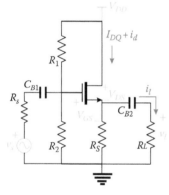

FIGURE 12.12 C.D. configuration with voltage divider biasing.

The *common drain*, or *C.D.* (Figure 12.12), amplifier is so called because the MOSFET shares its *drain* terminal for both input and output loops in AC conditions. In DC conditions this cannot be verified. Similarly to what we considered for the common collector of BJT configuration, this common drain topology is also referred to as a *source follower* since, as we will see, the magnitude of its voltage gain is less then unity.

For simplicity, the drain resistor is omitted here, but it can be present to offer a stability effect with respect to thermal runaway.

Observation

One of the many possible applications of a source follower is the realization of the readout circuits for *CMOS* (*Complementary MOS*) *image sensors*.

12.3.1 DC Analysis

The DC analysis of the C.D. configuration can be conducted similarly to that of the C.S. configuration. To determine the *Q*-point we consider the voltage divider expression

$$V_G = \dfrac{R_2}{R_2 + R_1} V_{DD}$$

the GS-KVL expression

$$V_{GS} = V_G - R_S I_{DS}$$

and the DS-KVL expression (when no drain resistance R_D is present)

$$V_{DD} = V_{DS} + R_S I_{DS}$$

12.3.2 AC Analysis

As usual, for the AC analysis we refer to the equivalent AC model, which is obtained considering the following:

- The capacitors act as short circuits at the working frequencies within the amplifier's bandwidth.
- The DC power supply is killed so the voltages are shorted.
- The MOSFET is replaced by its equivalent small-signal model.

The AC model of the *C.D.* configuration is represented in Figure 12.13.

For convenience, resistors R_1 and R_2 can be combined into $R_G = R_1 \,||\, R_2$, and the topology can be rearranged inverting the drain current direction (Figure 12.14).

The latter scheme, obtained with a mirror around the node "S," clearly visualizes that the input voltage of the MOSFET is across its gate-drain terminals, while the output voltage is across the source-drain terminals, exactly what we expected for a C.D. configuration.

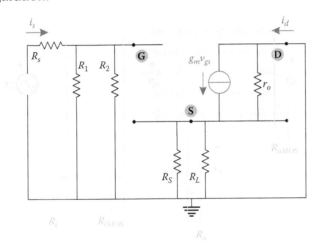

FIGURE 12.13   AC equivalent model
of the C.D. configuration.

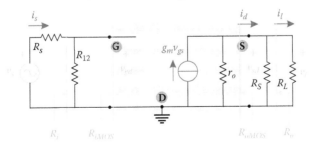

FIGURE 12.14   Rearranged AC equivalent
model of the C.D. configuration.

The current gain is not considered since no input current flows because of the gate oxide, so that the current gain is simply theoretically infinite.

12.3.2.1 C.D.: Input Resistance

We can distinguish two input resistances, that is, the one at the real input of the MOSFET $R_{iMOS} \overset{def}{=} \dfrac{v_{gs}}{i_g}$ and the one working as a load for the AC source $R_i \overset{def}{=} \dfrac{v_{gs}}{i_s}$.

As it was for the C.S. configuration, we obtain the result

$\boxed{R_{iMOS}}$ $R_{iMOS} \overset{def}{=} \dfrac{v_{gs}}{i_g}$

Since $i_g = 0$ we get

$$R_{iMOS} = \infty$$

$\boxed{R_i}$ $R_i \overset{def}{=} \dfrac{v_{gs}}{i_s}$

It is easily determined that
$$R_i = R_G$$

So the overall input resistance is determined by the bias resistors only.

Observation

An ideal voltage follower should have an infinite input resistance. From this point of view, since the gate current of a MOSFET is practically zero, the C.D. configuration should arrange a much better follower than the BJT's C.C. configuration. The ideal voltage follower *needs* a load, being a current amplifier, otherwise no current would flow.

12.3.2.2 C.D.: Output Resistance

The *output resistance* R_o is the one that the amplifier presents to its load. R_o is the resistance "seen" at the output nodes, determined by removing all power independent sources. Here no AC independent sources are present when looking toward the input, because the input is "isolated" by the gate oxide, as well modeled by the AC equivalent network.

To determine the output resistance in an easier way, we review the scheme shown in Figure 12.15.

The MOSFET is represented with two main resistances highlighted, that is, the reverse of its transconductance $1/g_m$ and the resistance r_o due to the channel length modulation effect. In this way, it is immediately clear by visual inspection that

$$R_{oMOS} = \frac{1}{g_m} \| r_o$$

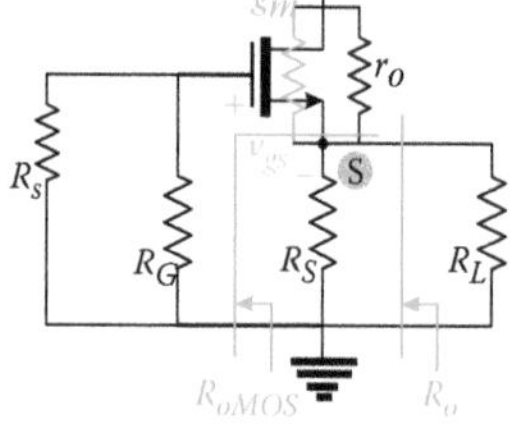

FIGURE 12.15 AC model to determine the output resistance of the C.D. configuration.

and

$$R_o = \frac{1}{g_m} \| r_o \| R_S$$

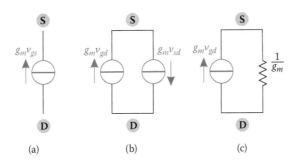

FIGURE 12.16 Steps to model another AC equivalent model of the C.D. configuration.

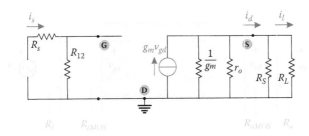

FIGURE 12.17 Another AC equivalent model of the C.D. configuration.

Another possibility to determine the output resistance is with another AC equivalent model from Figure 12.14. To do so, let's consider the voltage controlled current source $g_m v_{gs} = 0$ in Figure 12.16a, which can be swapped with the the sources $g_m v_{gd}$ and $g_m v_{sd}$ in Figure 12.16b, because $v_{gs} = v_{gd} - v_{sd}$ as can be easily verified. Now, thanks to the substitution theorem (Section 6.8 in Chapter 6) the $g_m v_{sd}$ current source, which depends on v_{sd}, can be simply replaced by the resistor $1/g_m$, as in Figure 12.16c.

The obtained new equivalent AC model is represented in Figure 12.17.

Now, a simple visual inspection is sufficient to obtain the same results for R_{oMOS} and R_o already reported.

The output impedance of a source follower is generally relatively low.

Observations

Thanks to the high input resistance and low output resistance this amplifier's configuration is suitable for a voltage buffer and to drive low-resistance loads.

The output resistance can be reduced by increasing g_m, which can be realized with a higher value of k_n that is obtained by making W/L as large as possible ($K_n = \dfrac{\mu_n C_S}{2} \dfrac{W}{L} \left[\dfrac{A}{V^2} \right]$, see Section 8.7.3 of Chapter 8).

12.3.2.3 C.D.: Voltage Gain

The MOSFET's KVL expression gives

$$v_{gd} = v_{gs} + v_{sd}$$

and the analysis of the AC equivalent model of Figure 12.14 gives

$$v_{sd} = g_m v_{gs} \left(r_o \mid\mid R_S \mid\mid R_L \right) = g_m \left(v_{gd} - v_{ds} \right) \left(r_o \mid\mid R_S \mid\mid R_L \right)$$

$$v_{sd} \left[1 + g_m \left(r_o \mid\mid R_S \mid\mid R_L \right) \right] = v_{gd} g_m \left(r_o \mid\mid R_S \mid\mid R_L \right)$$

Therefore,

$$A_{vMOS} \overset{def}{=} \frac{v_{sd}}{v_{gd}} = \frac{g_m \left(r_o \mid\mid R_S \mid\mid R_L \right)}{1 + g_m \left(r_o \mid\mid R_S \mid\mid R_L \right)}$$

Since the occurrence of $g_m \left(r_o \mid\mid R_S \mid\mid R_L \right) \gg 1$ is often verified, $A_{vMOS} \cong 1$, hence the name *source follower*.

Now, the overall voltage gain must take into account the attenuation factor $\alpha_{v,i}$ due to the input voltage divider:

$$\alpha_{v,i} = \frac{R_G}{R_G + R_S}$$

so, finally

$$A_v = \frac{g_m \left(r_o \mid\mid R_S \mid\mid R_L \right)}{1 + g_m \left(r_o \mid\mid R_S \mid\mid R_L \right)} \frac{R_G}{R_G + R_S}$$

12.4 C.G. CONFIGURATION

In a *common gate*, or C.G., configuration the input signal v_s is applied to the source and the output is taken from the drain, as in Figure 12.18.

12.4.1 DC Analysis

The DC analysis of the C.D. configuration can be conducted similarly to that of the C.S. configuration, so it is not reported here.

12.4.2 AC Analysis

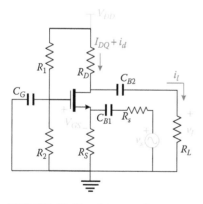

FIGURE 12.18 C.G. configuration with voltage divider biasing.

To perform the AC analysis we start by developing the AC equivalent model of the C.G. amplifier. To this aim, the capacitors can be represented by short circuits at the working frequencies and the MOSFET can be replaced by its equivalent network already described in Figure 12.7, to obtain the scheme of Figure 12.19.

It can be somewhat convenient to consider that the MOSFET's equivalent AC small-signal model represented in Figure 12.7 can be replaced by another model that considers the Norton equivalent of the output mesh rather than the Thévenin equivalent, as represented in Figure 12.20.

These two models are equivalent when $\gamma = g_m r_o$ (*substitution theorem*, Section 6.8 in Chapter 6). Now, the network represented in Figure 12.19 can be equivalently represented by considering the dependent source mirrored around the "S" node and replacing it with its Thévenin equivalent, as reported in Figure 12.21.

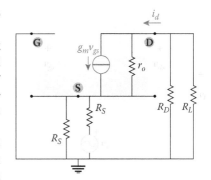

FIGURE 12.19   AC model of the C.G. configuration.

This new representaion is very useful to determine the AC resistances and gains of the amplifier.

12.4.2.1 C.G.: Input Resistance

By a simple visual inspection we obtain

$$v_{sg} = \gamma v_{gs} - \left[r_o + \left(R_D \mid\mid R_L\right)\right]i_d$$

$$-v_{gs} = \gamma v_{gs} - \left[r_o + \left(R_D \mid\mid R_L\right)\right]i_d$$

$$v_{gs}\left(1 + \gamma\right) = \left[r_o + \left(R_D \mid\mid R_L\right)\right]i_d$$

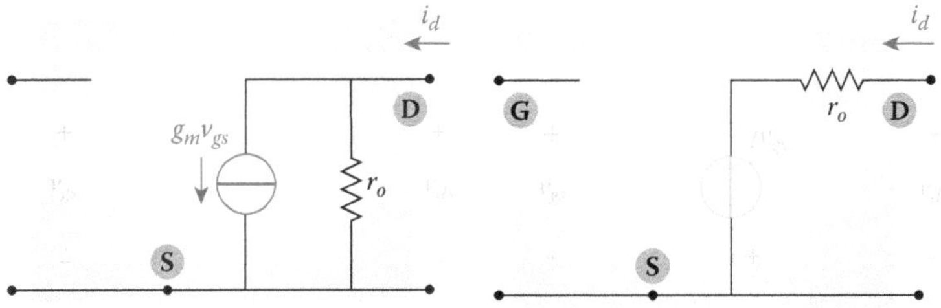

(a)   (b)

FIGURE 12.20   Two different models representing the MOSFET within its working frequencies.

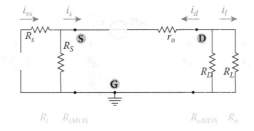

FIGURE 12.21   AC equivalent model of the C.G. configuration with Thévenin representation.

Now, we can distinguish two input resistances, that is, the one at the real input of the MOSFET

$$R_{iMOS} \underset{def}{=} \frac{v_{sg}}{i_s}$$

and the one working as a load for the AC source

$$R_l \underset{def}{=} \frac{v_s}{i_{vs}}$$

R_{iMOS}	$R_{iMOS} \underset{def}{=} \dfrac{v_{sg}}{i_s}$

$$R_{iMOS} = \frac{v_{sg}}{i_s} = \frac{-v_{gs}}{-i_d} = \frac{r_o + (R_D \mid\mid R_L)}{1 + \gamma} = \frac{r_o + R_D \mid\mid R_L}{1 + g_m r_o}$$

R_l	$R_l \underset{def}{=} \dfrac{v_s}{i_{vs}}$

$$R_l = R_S \mid\mid R_{iMOS} = R_S \mid\mid \frac{r_o + R_D \mid\mid R_L}{1 + g_m r_o}$$

12.4.2.2 C.G.: Output Resistance

KVL applied at the loop of the AC equivalent model delivers the following:

$$v_{sg} = (R_s \mid\mid R_S) i_d = -v_{gs}$$

$$v_{dg} = r_o i_d - \gamma v_{gs} + (R_s \mid\mid R_S) i_d$$

$$v_{dg} = \left\{ \left[r_o + (R_s \mid\mid R_S) \right] + \gamma (R_s \mid\mid R_S) \right\} i_d$$

Now, we can distinguish two output resistances, that is, the one at the real output of the MOSFET

$$R_{oMOS} \underset{def}{=} \frac{v_{dg}}{i_d}$$

and the one as "seen" by the load

$$R_o \underset{def}{=} \frac{v_{dg}}{i_l}$$

R_{oMOS}	$R_{oMOS} \underset{def}{=} \left. \dfrac{v_{dg}}{i_d} \right|_{R_L \text{"open"}}$

$$R_{oMOS} = \frac{v_{dg}}{i_d} = \left[r_o + (R_s \mid\mid R_S) \right] + \gamma (R_s \mid\mid R_S) = r_o + (1 + \gamma)(R_s \mid\mid R_S)$$

$$= r_o + (1 + g_m r_o)(R_s \mid\mid R_S)$$

$$\boxed{R_o} \quad R_o \underset{def}{=} \frac{v_{dg}}{i_l}$$

$$R_o = R_D \,||\, R_{oMOS} = R_D \,||\, \left[r_o + (1 + g_m r_o)(R_s \,||\, R_S) \right]$$

12.4.2.3 C.G.: Voltage and Current Gains

The *voltage gain* A_v of the amplifier in the mid-band frequency regime refers to the ratio between the output voltage v_l, across the load resistance and the input voltage v_s of the AC source:

$$A_v \underset{def}{=} \frac{v_l}{v_s}$$

This equation can be rewritten as

$$A_v = \frac{v_l}{v_s} = \frac{v_{dg}}{v_s} = \frac{v_{dg}}{v_{sg}} \frac{v_{sg}}{v_s} = A_{vMOS} \alpha_{v,l}$$

So, it has been conveniently separated into two terms $A_{vMOS} = \dfrac{v_{dg}}{v_{sg}}$ and $\alpha_{v,l} = \dfrac{v_{sg}}{v_s}$.

$$\boxed{A_{vMOS}} \quad A_{vMOS} \underset{def}{=} \frac{v_{dg}}{v_{sg}}$$

We obtain $v_{dg} = -(R_D \,||\, R_L)i_d$, and recalling that $v_{gs}(1+\gamma) = \left[r_o + (R_D \,||\, R_L) \right]i_d$, we can write

$$A_{vMOS} = \frac{v_{dg}}{v_{sg}} = \frac{v_{dg}}{-v_{gs}} = \frac{(R_D \,||\, R_L)i_d}{\dfrac{\left[r_o + (R_D \,||\, R_L) \right]i_d}{(1+\gamma)}} = (1+\gamma)\frac{R_D \,||\, R_L}{r_o + (R_D \,||\, R_L)}$$

$$= (1 + g_m r_o)\frac{R_D \,||\, R_L}{r_o + (R_D \,||\, R_L)}$$

$$\boxed{\alpha_{v,l}} \quad \alpha_{v,l} \underset{def}{=} \frac{v_{sg}}{v_s}$$

From the input voltage divider equation $v_{sg} = \dfrac{R_S}{R_S + R_s} v_s$, so

$$\alpha_{v,l} = \frac{R_S}{R_S + R_s}$$

The overall voltage gain is then

$$\boxed{A_v} \quad A_v \underset{def}{=} A_{vMOS} \alpha_{v,l}$$

$$A_v = (1 + g_m r_o)\frac{R_D \,||\, R_L}{r_o + (R_D \,||\, R_L)} \frac{R_S}{R_S + R_s}$$

We can consider the current gain A_{iMOS} practically equal to unity, since the drain and source currents are the same.

12.5 COMPARISONS AMONG C.S., C.D., C.G.

Table 12.1 shows a comparison among the three different MOSFET amplifier configurations.

The C.S. configuration is the only, quite interestingly, to provide a voltage gain, but the C.D. can buffer a poor voltage source into a nearly ideal one, and the C.G. can buffer a poor current source into a nearly ideal one.

12.6 BIAS STABILITY

We learned how the Q-point must be well defined in the center of the MOSFET's saturation region and must be as stable as possible, to assure *bias stability*. In that region the expression of the channel current is

$$i_{DS} = K_n \left(v_{GS} - V_{th}\right)^2$$

when, for simplicity, we assume $\lambda \cong 0$.

The *bias stability* is strictly related to the temperature, so that the previous formula can be empirically modified as

$$i_{DS} = K_n \left(v_{GS} - V_{th}\right)^2 \sqrt{\left(\frac{T_r}{T}\right)^3}$$

with $T_r = 300K$ being the room temperature.

TABLE 12.1

Comparison among C.S., C.D., and C.G. MOS Amplifier Topologies

Amplifier	C.S.	C.D. (S.F.)	C.G.
Voltage gain	> 1	$\cong 1$	> 1
A_{vMOS}	$-g_m \left(r_o \,||\, R_D \,||\, R_L\right)$	$\dfrac{g_m \left(r_o \,||\, R_S \,||\, R_L\right)}{1 + g_m \left(r_o \,||\, R_S \,||\, R_L\right)}$	$\left(1 + g_m r_o\right)\dfrac{R_D \,||\, R_L}{r_o + \left(R_D \,||\, R_L\right)}$
	Degraded by R_S		Degraded by R_S
Current gain	-	-	$\cong 1$
A_{iMOS}			
Input resistance	∞	∞	Low to moderately low
R_{iMOS}			$\dfrac{r_o + R_D \,||\, R_L}{1 + g_m r_o}$
			Decreased by R_S
Output resistance	Moderate to high	Low to moderately low	Moderate to high
R_{oMOS}	r_o	$\dfrac{1}{g_m} \,||\, r_o \,||\, R_S$	or very high
			$r_o + \left(1 + g_m r_o\right)\left(R_s \,||\, R_S\right)$
		Decreased by R_S	
Key function	Voltage amp.	Voltage buffer	Current buffer
	transconductance		
Phase difference	π or $(2n+1)\pi$	None or 2π	None or 2π

Differentiating,

$$di_{DS} = \frac{\delta i_{DS}}{\delta T} dT + \frac{\delta i_{DS}}{\delta v_{GS}} dv_{GS}$$

$$di_{DS} = -\frac{3}{2}\sqrt{\frac{T_r^3}{T^5}} K_n (v_{GS} - V_{th})^2 dT + 2K_n \sqrt{\left(\frac{T_r}{T}\right)^3} (v_{GS} - V_{th}) dv_{GS}$$

$$di_{DS} = -\frac{3}{2} K_n (v_{GS} - V_{th})^2 \sqrt{\left(\frac{T_r}{T}\right)^3} \frac{dT}{T} + 2K_n \sqrt{\left(\frac{T_r}{T}\right)^3} (v_{GS} - V_{th}) dv_{GS}$$

$$di_{DS} = -\frac{3}{2} i_{DS} \frac{dT}{T} + 2K_n \sqrt{\left(\frac{T_r}{T}\right)^3} (v_{GS} - V_{th}) dv_{GS}$$

Now, let's consider the four-resistor biasing network, as in Figure 12.6. Since from the input KVL we obtain

$$v_{GS} = V_G - R_S i_{DS} = \frac{R_2}{R_2 + R_1} V_{DD} - R_S i_{DS}$$

thus $dv_{GS} = -R_S di_{DS}$, and the previous equation becomes

$$\frac{di_{DS}}{i_{DS}} = \frac{-\frac{3}{2}}{1 + 2K_n \sqrt{\left(\frac{T_r}{T}\right)^3} (v_{GS} - V_{th}) R_S} \frac{dT}{T}$$

It is evident that the presence of the source resistor R_S improves the overall bias stability, so that the greater R_S is, the less current drifts with regard to thermal runaway effects.

12.7 MOSFET AS A SWITCH

Varying the drive voltage to the gate of a MOSFET, the drain-source channel resistance can be ranged from many hundreds of $k\Omega$s, to a few Ωs. This two extremes roughly correspond to "open-circuit" and "short-circuit" situations, for which we refer to the MOSFET as "OFF" and "ON."

In the circuit arrangement of Figure 12.22, an n–channel E-MOSFET can be used to switch a lightbulb ON and OFF, depending on the gate positive voltage value V_G.

With $V_G = 0$ (the switch open) the MOS is in its *cutoff region*; no channel will be formed so that no channel current can flow. As a consequence no current flows through the bulb, which is OFF, the voltage across the drain-source terminals is $V_{DS(off)}$ (the highest possible), and the output voltage is $v_o \cong V_{DD}$.

With $V_{GS} > V_{DS} + V_{th} > 0$ (the switch closed) the MOS is in its *ohmic region*; the channel is open so that a ohmic

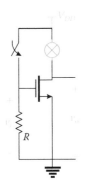

FIGURE 12.22 Circuit arrangement to use the MOSFET as a switch to turn the lightbulb ON and OFF.

current can flow (the highest possible). As a consequence, a current flows through the bulb that is ON, the voltage across the drain-source terminals is $V_{DS(on)}$ (the lowest possible), and the output voltage is $v_0 \cong 0$.

Practically, logic levels ("1"–"0" or "ON"–"OFF") work so that when the gate is connected to ground, the MOS is turned OFF (in its cutoff region) and when the gate is connected to V_{DD} the MOS is turned ON (in its ohmic region).

12.8 KEY POINTS, JARGON, AND TERMS

→ To produce linear amplification, the MOSFET has to work in its saturation region, without entering the linear and cutoff regions.

→ The *transconductance* is a key parameter of the MOSFET that expresses its performance as an amplifier. It is the ratio of the change in drain current to the change in gate voltage over a defined arbitrarily small region of the output saturation characteristics.

→ The channel modulation effect defines the MOSFET *output resistance.*

→ *Biasing* is the way to ensure that the signal always swings within the saturation region of the MOSFET.

→ Thanks to the superposition theorem it is possible to perform DC and AC analysis separately.

→ The three main configurations using a MOSFET in a network to provide amplifications are *common source, common drain,* and *common gate,* depending on which terminal of the MOSFET is shared for both input and output loops in AC conditions. The *common source* configuration is the only one, quite interestingly, to provide a voltage gain, but the *common drain* can buffer a poor voltage source into a nearly ideal one, and the *common ground* can buffer a poor current source into a nearly ideal one.

→ The source resistor increases the bias stability.

12.9 EXERCISES

EXERCISE 1
Demonstrate that for a common source stage with source degeneration, i.e., that includes an un-bypassed source resistor R_S, the MOSFET's voltage gain is

$$A_{vMOS} = -\frac{g_m r_o \left(R_D \mid\mid R_L\right)}{r_o + \left(1 + g_m r_o\right) R_S + \left(R_D \mid\mid R_L\right)}$$

EXERCISE 2
For the amplifier shown in Figure 12.23, the resistance values for the resistors are $R_D = 4.7\,k\Omega$, $R_F = 220\,k\Omega$, while the MOSFET's parameters are $g_m = 2\,mS$, $r_o = 133\,k\Omega$. Determinate the input and output resistances, R_i and R_o, and the voltage gain $A_v = v_o / v_i$.

ANSWERS

$R_i \cong 22.3\,k\Omega$, $R_o \cong 4.4\,k\Omega$, $A_v \cong -8.9$.

EXERCISE 3

An E-MOSFET C.S. amplifier is reported as in Figure 12.24.

The bias network has $R_1 = 5\,M\Omega$, $R_2 = 860\,k\Omega$, $R_D = 2\,k\Omega$, $V_{DD} = 12\,V$. The MOSFET has $V_{th} = 2\,V$ and a drain current $I_D = 0.1\,A$ when $V_{GS} = 5\,V$. Calculate the Q-point: V_{GSQ}, I_{DQ}, V_{DSQ}.

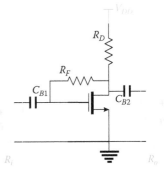

FIGURE 12.23 C.S. configuration with feedback resistor.

ANSWER

$V_{GSQ} \cong 1.76\,V$, $I_{DQ} \cong 63.4\,mA$, $V_{DSQ} \cong 10.7\,V$.

EXERCISE 4

An E-MOSFET is used in a common gate configuration for which the output signal is taken off the drain with respect to the ground, and the gate is AC connected directly to the ground (Figure 12.25).

The given parameters are $R_1 = 1\,M\Omega$, $R_2 = 1\,M\Omega$, $R_S = 2.2\,k\Omega$, $R_D = 4.7\,k\Omega$, $R_L = 500\,\Omega$, $R_s = 500\,\Omega$, $g_m = 0.01\,S$, $r_o = 50\,k\Omega$.

Calculate the following:

- The MOSFET and overall input resistances, R_{iMOS} and R_i
- The MOSFET and overall voltage gains, A_{vMOS} and A_v
- The MOSFET and overall output resistances, R_{oMOS} and R_o

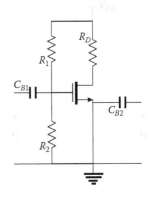

FIGURE 12.24 E-MOSFET C.S. amplifier with voltage divider bias.

ANSWER

$R_{iMOS} \cong 104.6\,\Omega$, $R_i \cong 99.9\,\Omega$

$A_{vMOS} \cong 23.16\,\Omega$, $R_o \cong 4.6\,k\Omega$

$A_{vMOS} \cong 23.16$, $A_v \cong 18.9$

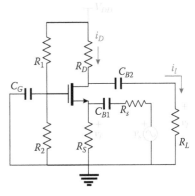

FIGURE 12.25 C.G. configuration.

13

Amplifiers: Variants

Here, we report some variants of the configurations already discussed and analyzed to improve the performance of some of the amplifiers.

In particular, we will deal with the methods used to increase the amplifier's *input resistance* value (Section 10.2.2 in Chapter 10), to increase the known low *efficiency* of the class A topology (Section 10.2.4), to increase the *bandwidth* (Section 10.2.6) and, finally, to increase the *signal to noise ratio* (Section 10.2.9) with the rejection of unwanted signals that can couple into the input.

13.1 INCREASED INPUT RESISTANCE

We already discussed and evaluated (particularly, but not exclusively, in Section 10.2.2 of Chapter 10) the importance of designing proper values of input and output resistances (or, in general, impedances) of an arbitrary amplifier. These resistance values can greatly affect the overall gain of an amplifier, so the best matching conditions (Section 6.9 in Chapter 6) have to be assured with respect to the type of amplifier (Section 10.3 in Chapter 10) to be realized. Here, we will deal in particular with the *input resistance* of the *C.C.* and *swamped C.E.* configurations.

We demonstred for C.C. (Section 11.6.3.2 in Chapter 11) that

$$R_{iBJT(C.C.)} = \frac{h_{ie}\left(1 + h_{oe}R_{EL}\right) + \left(1 - h_{re}\right)\left(1 + h_{fe}\right)R_{EL}}{1 + h_{oe}R_{EL}}$$

and in output open loop conditions considered when the load is unknown, the term R_{EL} can be replaced by R_E.

It was demonstrated for swamped C.E. (Section 11.5.2.1) that

$$R_{iBJT(s-C.E.)} = \frac{h_{ie} + \left(h_{ie}h_{oe} - h_{re}h_{fe}\right)R_{CL}}{1 + h_{oe}\left(R_{CL} + R_{E1}\right)} + \frac{\left[h_{oe}R_{CL} + \left(h_{ie}h_{oe} - h_{re}h_{fe}\right) + 1 + h_{fe} - h_{re}\right]R_{E1}}{1 + h_{oe}\left(R_{CL} + R_{E1}\right)}$$

and the emitter resistance R_{E1} can be renamed here for simplicity as R_E without losing generality.

As we have already seen (Section 11.5.2.2), in many practical cases the BJT's inner feedback effect can be considered zero, that is $h_{re} \cong 0$, and the Early effect as well, that is $h_{oe} \cong 0$. These conditions are quite often considered valid when

$$h_{fe}R_E \ll 0.1$$

which can be true but only admitting a certain approximation.

In these occurrences, the two previous equations become

$$R_{iBJT(C.C.)} = R_{iBJT(s-C.E.)} = R_{iBJT} = h_{ie} + \left(1 + h_{fe}\right)R_E$$

Therefore to increase the input resistance, proportional to the emitter resistance, we have to increase the value of the latter:

$$R_i \underset{R_E \to \infty}{\longrightarrow} \infty$$

But if we hypothetically admit that $R_E \to \infty$, the hypothesis $h_{fe}R_E \ll 0.1$ is no longer valid, but

$$R_{iBJT(C.C.)} \underset{R_E \to \infty}{\cong} \frac{1 + \left(h_{ie}h_{oe} - h_{re}h_{fe}\right) + h_{fe}}{h_{oe}} = \frac{1 + \Delta h + h_{fe}}{h_{oe}}$$

$$R_{iBJT(s-C.E.)} \underset{R_E \to \infty}{\cong} \frac{h_{oe}R_C + \left(h_{ie}h_{oe} - h_{re}h_{fe}\right) + 1 + h_{fe} - h_{re}}{h_{oe}} = \frac{h_{oe}R_C + \Delta h + 1 + h_{fe} - h_{re}}{h_{oe}}$$

where $\Delta h = h_{ie}h_{oe} - h_{re}h_{fe}$.

These two equations give an idea of the upper limits of the input resistances for the two configurations. Without considering the impracticable hypothesis $R_E \to \infty$, let's admit to "simply" having a very high value of the emitter resistance. In this case, the DC voltage drop across R_E would be very high so that, to keep the BJT working within its forward active region, the required DC voltage supply would be too large to be reasonable. Furthermore, the dissipated power due to the Joule effect (Section 4.4.8 in Chapter 4) would be excessive and the resulting temperature rise adds thermal drift problems. Last but not least, the overall input resistance R_i value as "seen" by the AC source would be decided by the bias resistors, R_1 and R_2 in parallel to R_E.

The problems coming from the voltage drop and the Joule effect can be solved by realizing R_E as an *equivalent* resistance rather than a *real* one. This is obtained by means of the following *Darlington pair* (Section 13.1.1), or *current source* (Section 13.1.2) configurations.

The effect due to the bias resistors can be reduced by *bootstrapping* (Section 13.1.3).

13.1.1 Darlington Pair

The *Darlington transistor pair*, or simply the *Darlington transistor* or *Darlington pair*, consists of two BJTs, T_1 and T_2, staked together as shown in Figure 13.1.

The input resistance of T_2 acts as emitter resistance for T_1, which "sees" an equivalent large resistance, which is due to a moderate emitter resistance of T_2 (not drawn in the previous scheme).

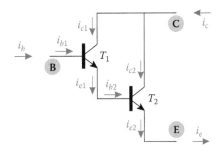

FIGURE 13.1   Darlington pair.

These two transistors are correlated to produce an equivalent single transistor having operating characteristics different in kind and/or quality from those of the components. So, the three B, C, and E terminals are equivalent to a single standard BJT, which can be turned on with a base-emitter voltage $V_{BE} \cong 1.4\,V$ rather than the usual $0.7\,V$ since it is across two base-emitter junctions rather than just one.

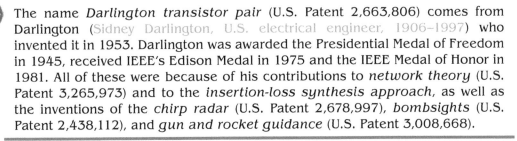

The equivalent AC model of the Darlington pair is schematized in Figure 13.2 where, for the sake of simplicity, we considered the h_{re} parameters negligible.

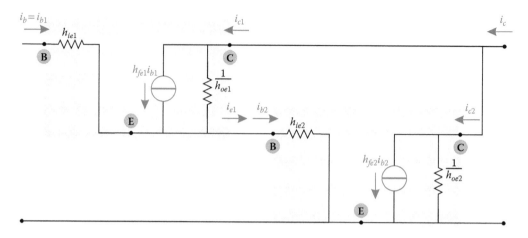

FIGURE 13.2 AC equivalent model of the Darlington pair.

Recalling the definitions, it is easy to determine the h-parameters of the entire pair as

$$h_{ie} = \left.\frac{v_{be}}{i_b}\right|_{v_{ce}=0} = h_{ie1} + \left(1 + h_{fe1}\right)h_{ie2}$$

$$h_{fe} = \left.\frac{i_c}{i_b}\right|_{v_{ce}=0} = \frac{h_{fe1} + \left(1 + h_{fe1}\right)h_{fe2} - h_{ie2}h_{oe1}}{1 + h_{oe1}h_{ie2}}$$

$$h_{re} = \left.\frac{v_{be}}{v_{ce}}\right|_{i_b=0} = \frac{h_{oe1}h_{ie2}}{1 + h_{oe1}h_{ie2}}$$

$$h_{oe} = \left.\frac{i_c}{v_{ce}}\right|_{i_b=0} = \frac{h_{oe1}\left(1 + h_{fe2}\right)}{1 + h_{oe1}h_{ie2}} + h_{oe2}$$

When the parameters h_{oe1} and h_{oe2} can both be considered negligible as well, these equations can be further simplified, becoming the commonly adopted and known forms

$$h_{ie} = h_{ie1} + \left(1 + h_{fe1}\right)h_{ie2} \cong h_{ie1} + h_{fe1}h_{ie2}$$

$$h_{fe} = h_{fe1} + \left(1 + h_{fe1}\right)h_{fe2} \cong h_{fe1}h_{fe2}$$

$$h_{re} = \frac{h_{oe1}h_{ie2}}{1 + h_{oe1}h_{ie2}} \cong h_{oe1}h_{ie2}$$

$$h_{oe} = \frac{h_{oe1}\left(1 + h_{fe2}\right)}{1 + h_{oe1}h_{ie2}} + h_{oe2} \cong h_{oe1}h_{fe2} + h_{oe2}$$

According to these results, the Darlington pair is very useful to realize an emitter follower configuration with a very high input resistance. In fact, let's analyze the simplified scheme shown in Figure 13.3.

We already know that the expressions of the current gains of the two BJTs in an emitter follower configuration can be respectively written as (Section 11.6.3.1 in Chapter 11)

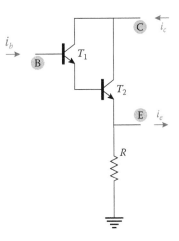

$$A_{iBJT2} = \frac{1 + h_{fe2}}{1 + h_{oe2}R_E}$$

$$A_{iBJT1} = \frac{1 + h_{fe1}}{1 + h_{oe2}R_{iBJT2}}$$

and their input resistance values (Section 11.6.3.2) as

$$R_{iBJT2} = \frac{h_{ie2} + (1 + h_{fe2})R_E}{1 + h_{oe2}R_E} = h_{ie2} + A_{iBJT2}R_E$$

$$R_{iBJT1} = \frac{h_{ie1} + (1 + h_{fe1})R_{iBJT2}}{1 + h_{oe1}R_{iBJT2}} = h_{ie1} + A_{iBJT1}R_{iBJT2}$$

FIGURE 13.3 Scheme of an emitter follower configuration adopting a Darlington pair.

so that combining the previous equations, the overall input resistance, coincident to the input resistance of the transistor T_1, results in

$$R_{iBJT1} = h_{ie1} + A_{iBJT1}(h_{ie2} + A_{iBJT2}R_E) \cong A_{iBJT1}A_{iBJT2}R_E$$

Accordingly, it is evident that the *Darlington pair* has very high current gain, the product $A_{iBJT1}A_{iBJT2}$, and a very high input resistance R_{iBJT1}, which is the emitter load resistance multiplied by the great overall current gain factor.

Observations

Very often the h-parameters of the two BJTs are approximated as equal. But the h-parameters depend on the operating Q-points, which are different for the two transistors, since the emitter current of T_1 is the base current for T_2. Therefore h_{fe1} is smaller than h_{fe2}.

A Darlington pair is commercially available as a single discrete unit in one package.

In addition to the *Darlington pair*, we can mention the *Sziklai pair* (George Clifford Sziklai, electronics engineer, 1909–1998), also known as the *complementary Darlington* since it uses a similar configuration to the Darlington one, but with one *npn* and one *pnp* transistor, connected in a way that their emitter terminals are in common rather than the collector terminals.

13.1.2 Current Source

An ideal DC *current source* has an infinite inner resistance (Section 4.5.3 in Chapter 4). This is one of the reasons why the current source can be used in place of a real emitter resistor since it satisfies the request of an equivalent high resistance value. This resistance being an *equivalent* and not a *real* one, no power dissipation problems arise. Furthermore, this DC source can provide the bias current so that the Q-point can be correctly located in the middle of the BJT's forward active region. This *current bias* method (see "Observation" in Section 11.4.1.5 of Chapter 11) is especially adopted in integrated circuit design, both for analog and digital applications. The simple *voltage bias* methods previously detailed (Section 11.4.1.1, Section 11.4.1.5, Section 11.10,2, and Section 12.1.1) can sometimes fail to produce a constant collector DC current when the supply of DC voltage or room temperature changes. This can happen, for example, with mobile phones when the DC voltage, supplied by the battery, changes with battery discharge, or when the user moves within areas with different temperatures.

A DC current source is not a "real" source, meaning that it is "simulated" by a network of transistors and resistor elements. The simplest way to realize the "simulation" is by means of two transistors forming the so-called *current mirror* or *current regulating circuit*, schematized in Figure 13.4.

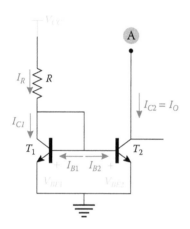

FIGURE 13.4   Basic BJT current mirror. It forms a current sinking network.

The transistor T_1 is said to form a *diode-connected BJT*, due to the direct connection with no voltage drop between its collector and base terminals. As a consequence T_1 never works in saturation conditions, but can only operate in active mode or be, simply, in cutoff mode.

The transistor T_1 "matches" the transistor T_2, in a sense that they have to be fabricated with the same procedures and on the same substrate, to guarantee twin characteristics (for example keeping the βs as similar as possible).

So, referring to the previous figure and adopting the subscripts "1" and "2" for the parameters of transistors T_1 and T_2 respectively, the starting conditions are

$$\beta_1 = \beta_2 = \beta$$

$$V_{BE1} = V_{BE2} = V_{BE}$$

$$I_{B1} = I_{B2} = I_B$$

and since $I_C \cong \beta I_B$ (Section 8.6.3 in Chapter 8), we also have

$$I_{C1} = I_{C2} = I_C$$

Now, from KCL at the collector of T_1

$$I_R = I_{C1} + I_{B1} + I_{B2} = I_C + 2I_B = I_C + 2\frac{I_C}{\beta}$$

but the output bias current I_o is equal to the collector current $I_{C2} = I_C$, therefore,

$$I_O = \frac{I_R}{1 + \dfrac{2}{\beta}}$$

Assuming typically $\beta \gg 1$, the equation becomes

$$I_o = I_R$$

hence the name *current mirror* since this network "copies" a current injected into the transistor T_1 and "creates" a current source into the collector of T_2. The value of the "copied" current I_R is established by

$$I_R = \frac{V_{CC} - V_{CE1}}{R} = \frac{V_{CC} - V_{BE}}{R}$$

so that it is sufficient to choose the proper values of V_{CC} and R.

At the node "A" in Figure 13.4 the load is applied, and the notable result is that I_o "mirrors" the current I_R regardless this load.

Observations

Examples of improved versions of the current source, because of the reduction of the dependence from β's finite value, are the *Wilson current mirror*, patented in 1971 (U.S. Patent 3,588,672) for which

$$I_O = \frac{I_R}{1 + \dfrac{2}{\beta(\beta + 2)}}$$

and the *cascode current mirror* for which

$$I_O = \frac{I_R}{1 + \dfrac{4}{\beta} + \dfrac{2}{\beta^2}}$$

The current source or current mirror is an example of a *translinear circuit*. This is because the diode-connected transistor T_1 converts a current to a logarithmically related voltage (see the *Shockley diode equation*, Section 8.5.2 in Chapter 8) which, in turn, is exponentially converted into a current by the transistor T_2 acting as a voltage-controlled current source.

The *LM134, LM234, LM334* ICs by National Semiconductors are three-terminal adjustable current sources. The value of the current is established with one external resistor, so that they can provide a current in the $1\,\mu A$–$10\,mA$ range.

13.1.3 Bootstrapping

Bootstrapping refers to the *bootstrap technique*, which involves a feedback effect (Section 11.3.2.2 in Chapter 11) so that the circuit is "pulled up as if by its own bootstraps." That is, it is a way of saying that the network can succeed on its own despite limited resources. The principal aim is to make the value of the AC input resistance R_i to "appear" much higher than the real one, to assure an input resistance value as high as possible when required. With this technique, the effect of the bias resistors, which reduce the input resistance value as "seen" by the AC source, can be reduced.

Curiosities

The term bootstrapping originates from *The Surprising Adventures of Baron Munchausen*, by Raspe (Rudolph Erich Raspe, German librarian, writer, and scientist, 1736–1794, Figure 13.5): "The Baron had fallen to the bottom of a deep lake. Just when it looked like all was lost, he thought to pick himself up by his own bootstraps."

Clearly, it refers to an impossible situation, but it illustrates a condition when somebody or something helps him/her/itself without any external aid.

In our network, the output influences the input to change the output itself. It is, again, a feedback effect.

FIGURE 13.5
R.E. Raspe

The *bootstrap* method is used in the statistical field too. In was introduced by Efron in 1979 (Bradley Efron, "Bootstrap methods: Another look at the jackknife." *Annals of Statistics*, 7, 1–26).

The bootstrap method was born for BJT devices, since the MOSFET ones already offer a very high input resistance due to their gate oxide, and it is especially useful for the *common collector* configuration.

A *bootstrap capacitor* C_F is connected across output-input loops of the C.C. amplifier shown in Figure 13.6 (here two equivalent schemes).

The capacitor C_F must behave as a short circuit in AC and an open circuit in DC. This is to allow an AC feedback effect of the resistor R_3, without affecting the bias conditions. So, the capacitor should have a reactance value $X_{C_F} = 1/2\pi C_F$ sufficiently lower than the resistance R_3 value at the lowest amplifier working frequency C_i:

$$C_F \gg \frac{1}{2\pi f_l R_3}$$

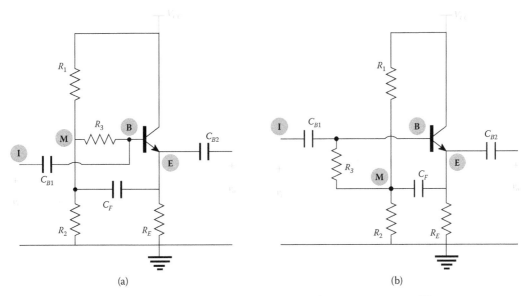

FIGURE 13.6 Two equivalent schemes of a bootstrap amplifier.

Defining the amplifier's voltage gain as $A_v = v_o/v_i$, the voltage V_{R3} across the resistor R_3 and the current i_{R3} flowing through it can be respectively written as

$$v_{R_3} = v_i - v_o = v_i - A_v v_i = (1 - A_v) v_i$$

$$i_{R_3} = \frac{v_{R_3}}{R_3} = \frac{(1 - A_v) v_i}{R_3}$$

As a consequence, the AC amplifier's input resistance $R_i = v_i/i_i$ results in

$$R_i = \frac{v_i}{i_{R_3}} = \frac{R_3}{(1 - A_v)}$$

where $i_i = i_{R3}$. But we know that the C.C. configuration has an A_v slightly less than unity, therefore the resistor R_3 AC behaves as if it is much larger than its real value!

Exactly the same result can be obtained adopting Miller's theorem (Section 6.6 in Chapter 6).

A bootstrap amplifier can increase the input resistance value by two orders of magnitude on average.

13.2 TRANSFORMER-COUPLED LOAD

The very high linearity of the class A amplifier is paid for in terms of very low efficiency. To partially solve the problem we can create a *transformer-coupled load*, which recalls a *class A power amplifier*, sometimes referred to as a *single ended power amplifier* (with only one transistor, different from the *push-pull amplifier* using two transistors).

For the sake of simplicity we can consider a C.E. amplifier with no emitter resistor and no load resistor (so an output open loop), as schematized in Figure 13.7.

The efficiency defined as $= P_o|_{AC}/P_i|_{DC}$ (Section 10.2.3 in Chapter 10) has to be solved by determining the RMS AC power $P_o|_{AC}$ delivered to load and the power supplied by the DC source $P_i|_{DC}$.

$$\boxed{P_i|_{DC}} \qquad P_i|_{DC} \underset{===}{def} V_{CC}I_{CC}$$

The DC current is equal to $I_{CC} = I_{CQ} + I_{R_1}$, but since $I_{CQ} \gg I_{R_1}$, we can approximate $I_{CC} \cong I_{CQ}$, so that $P_i|_{DC} \cong V_{CC}I_{CQ}$. In addition, to assure the Q-point is in the middle of the linear region we can state $V_{CEQ} = (1/2)V_{CC}$, so that

$$P_i|_{DC} = 2V_{CEQ}I_{CQ}$$

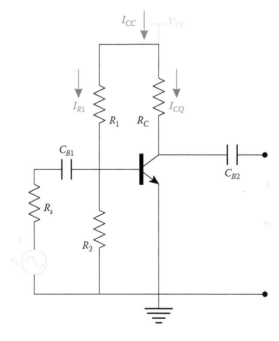

FIGURE 13.7 Simple C.E. configuration.

The quiescent current I_{CQ} flows through the collector resistive load R_C and causes a large waste of DC power that does not contribute to the useful AC output power.

$P_o|_{AC}$ $\qquad$ $P_o|_{AC} \overset{def}{=\!=} V_{rms,R_C} i_{rms,R_C}$

But since $v_{rms,R_C} = \dfrac{V_{CEQ}}{\sqrt{2}}$ and $i_{rms,R_C} = \dfrac{I_{CQ}}{\sqrt{2}}$ then

$$P_o|_{AC} = \frac{1}{2} V_{CEQ} I_{CQ}$$

Finally

$$\eta = \frac{P_o|_{AC}}{P_i|_{DC}} = \frac{\frac{1}{2} V_{CEQ} I_{CQ}}{2 V_{CEQ} I_{CQ}} = 0.25$$

that is,

$$\eta(\%) = 25\%$$

This value corresponds to the maximum efficiency of a class A amplifier, which is the lowest among all amplifier classes. Furthermore, this is a theoretical value but practically it is even lower, 15–20% in best cases, because actually we do not make the signal maximally swing up to the borders of the cutoff and saturation regions, to make sure we do not lose linearity.

Observation

This low efficiency is the reason class A is not adopted in large high-power amplifiers. In fact, when an amplifier is asked to produce, let's say, 100 W to a loudspeaker, it has to guarantee 400 W, 75% of which is wasted heat to be dissipated. Class A is typically used below 2 W, as in some radio or TV receivers or headphone systems.

The problem of poor efficiency arises practically because the collector resistive load R_c is essential only for AC conditions (Section 11.2.4 in Chapter 11), but results in a problem for DC conditions, because it dissipates DC power via the Joule effect. As an ideal situation, the resistance R_c would be present in AC and bypassed in DC. An inductance can help to solve the problem. Even better is if the inductance is part of a transformer, so that current and voltage levels can be increased or decreased according to the turns ratio n (Section 5.2 in Chapter 5). Figure 13.8 represents this occurrence, with a C.E. configuration for which the

AC load is represented by the collector resistor R_c itself (when an effective load resistor R_L is present the AC load would be $R_c \| R_L$).

Utilizing the transformer, the DC voltage drop across R_c is practically null, and if no emitter resistance R_E is adopted (as in our scheme) the load line becomes quite "vertical," as schematized in Figure 13.9, so that even the battery value can be reduced by half.

Regarding the AC load line, we determined its slope to be $-1/R_C$ (Figure 11.31 in Chapter 11), and with the utilization of a transfomer it can be varied, because the equivalent emitter load resistor R_{Ceq} results in

$$R_{Ceq} = \left(\frac{n_1}{n_2}\right)^2 R_C$$

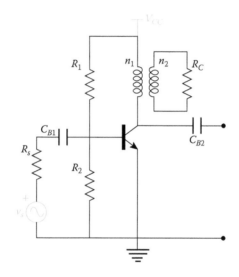

FIGURE 13.8   Transformer-coupled load.

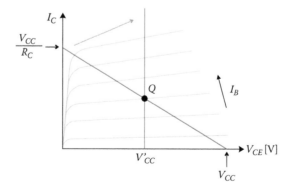

FIGURE 13.9   DC load line for $R_c = 0$.

But let's maintain the same battery value and see how the efficiency improves. Because of the presence of a transformed coupled load, no DC power will be dissipated by R_c, and thus there is no DC resistance between the DC source V_{CC} and the transistor T, so that the maximum V_{CEQ} and I_{CQ} will now be twice as large as without the transformer.

$P_i\big|_{DC}$ $\quad P_i\big|_{DC} \stackrel{def}{=\!=} V_{cc}I_{cc}$

Since no voltage drop is across transformer's primary, $V_{CEQ} = V_{CC}$ and, neglecting again the DC current in the bias resistors, $I_{CQ} = I_{cc}$; therefore,

$$P_i\big|_{DC} = V_{CEQ}I_{CQ}$$

$P_o\big|_{AC}$ $\quad P_o\big|_{AC} \stackrel{def}{=\!=} V_{rms,R_c}\, i_{rms,Rc}$

Now voltage swings in the range $V_{CEmax} - 0$ and collector current between $I_{Cmax} - 0$. Therefore,

$$v_{rms,R_c} = \frac{1}{\sqrt{2}}\frac{V_{CEmax} - 0}{2} = \frac{2V_{CC}}{2\sqrt{2}}$$

and

$$i_{rms,R_c} = \frac{1}{\sqrt{2}}\frac{I_{Cmax} - 0}{2} = \frac{2I_{CQ}}{2\sqrt{2}}$$

Finally,

$$P_o\big|_{AC} = \frac{V_{CC}}{\sqrt{2}}\frac{I_{CQ}}{\sqrt{2}} = \frac{1}{2}V_{CC}I_{CQ}$$

As a consequence

$$\eta = \frac{P_o\big|_{AC}}{P_i\big|_{DC}} = \frac{\frac{1}{2}V_{CC}I_{CQ}}{V_{CC}I_{CQ}} = 0.5$$

that is,

$$\eta(\%) = 50\%$$

so the maximum theoretical efficiency is doubled. In practical situations, this efficiency is reasonably about 30%–35%.

13.3 CASCODE AMPLIFIER

The *cascode* configuration combines a C.E. input with a C.B. output configuration with BJTs or C.S. and C.G. with FETs. The aim is to obtain: a large bandwidth, a reasonably high input resistance, a sufficient voltage gain, and very high output resistance. Combined results also offer high stability (Section 10.2.5 in Chapter 10), and high slew rate (Section 10.2.7) but the price is a relatively high DC supply voltage.

Curiosity

This configuration dates back to the age of the vaccum tubes. It realized the cascade of grounded cathode and grounded grid stages, hence the *cascode* name. The *cascode* configuration has one transistor on top of the other, different from the *cascade* (treated in Chapter 14), which has the output of one amplifier stage connected to the input of another amplifier stage.

Practically, the cascode configuration realizes a trade-off between output impedance and voltage gain limit, using a BJT in place of a tied resistor. This is schematized in Figure 13.10.

The three resistors named R_1, R_2, and R_3 are necessary to correctly bias the junctions of the two BJTs, the capacitor C_2 is to AC ground the base of T_2 so that it works in C.B. mode, while the transistor T_1 works in C.E. mode.

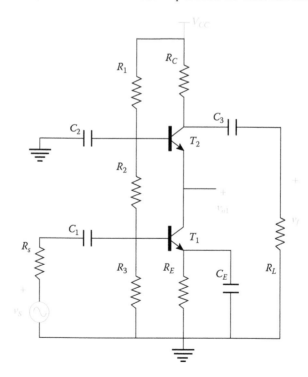

FIGURE 13.10 Cascode amplifier.

Up to now, we have considered the middle-band behavior of the amplifiers, without taking into account the effect of either the *real* capacitances or of the *parasitic* capacitances. In particular both the real and the parasistic capacitances were considered a short circuit at the working frequencies. But the capacitances have a fundamental role in determining the amplifier's bandwidth (Section 10.2.6 in Chapter 10).

We have already covered the real capacitances (Section 11.2.3 in Chapter 11), which determine the amplifier's lower cutoff frequency f_{c-l}, but not the parasitic ones, which are intrinsic to the doped semiconductor devices, and determine the higher cutoff frequency f_{c-h}. So, the small signal models adopted for middle-band behavior of the BJT and of the MOSFET are no longer valid for higher frequencies out of the amplifier's bandwith. Other high-frequency models are then necessary for BJTs and MOSFETs to include the capacitive effects of the junctions (see Section 8.5.4 of Chapter 8 as reference).

The model shown in Figure 13.11 is for the high-frequency behavior of a BJT. It takes into account the base-collector junction with a capacitance C_{BC} (sometimes referred to as C_μ) and the base-emitter junction with a capacitance C_{BE} (sometimes referred to as C_π). When the BJT works as an amplifier, the base-collector junction is in reverse-bias mode, so that C_{BC} is mainly a depletion capacitance (Section 8.5.4.1 in Chapter 8), while the base-emitter junction is in forward-bias mode so that C_{BE} is mainly a diffusion capacitance (Section 8.5.4.2). The most relevant is then C_{BC}, also because it determines a feedback effect, allowing an output signal to go back to the input loop.

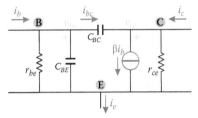

FIGURE 13.11 High-frequency BJT model.

The reason why the parasitic capacitances affect the behavior of the higher frequencies of the transistor is because when the transistor is working as an amplifier, it has to charge and discharge its inner capacitances as the signal changes. The current, performing this work, flows through the transistor inner resistances and through the source resistance of the driving stage. The combined action of capacitances and resistances just forms a low-pass filter.

Let's detail the effect of C_{BC}:

$$i_{bc} = C_{BC}\frac{d(v_{be}-v_{ce})}{dt} = C_{BC}\frac{d(v_{be}-A_{vBJT}v_{be})}{dt} = C_{BC}(1-A_{vBJT})\frac{dv_{be}}{dt}$$

This equation suggests that we can consider a new capacitance named $C_{BC}(1-A_{vBJT})$, which can be referred to as a *Miller capacitance* $C_M = C_{BC}(1-A_{vBJT})$, because in the previous model we can replace C_{BC} across the base-collector terminals with C_M across the base-emitter terminals just as Miller's theorem suggests (Section 6.6 in Chapter 6). The relevant result is that C_M can be much greater than the similar-capacitor C_{BC}, and its charging time can be relevant so that it causes a non-negligible delay before any collector current flows.

The *Miller effect* can be referred to here as the multiplication of the BJT's base-collector (or FET's drain-source) stray capacitance by the voltage gain. The increased input capacitance increases the lower cutoff frequency and so reduces the bandwidth.

Curiosity

The Miller effect is cited for the first time in John M. Miller's paper, "Dependence of the input impedance of a three-electrode vacuum tube upon the load in the plate circuit," *Scientific Papers of the Bureau of Standards*, vol. 15, no. 351, pages 367–385, 1920.

The *cascode* configuration combines a C.E. and a C.B. to reduce the problems coming from the Miller effect. In fact, C.E. provides a relatively low voltage gain because it is reduced by the low input impedance of C.B., which acts as a load for C.E. (see Figure 11.39 in Chapter 11), so that the Miller capacitance is reduced and the effects start at a higher frequency. As a consequence the amplifier's bandwidth is increased. The overall voltage gain can be considerable since it is a result of the poor voltage gain of C.E. but multiplied by the high voltage gain of C.B. The overall input resistance corresponds to the input resistance of C.E., which is known to be relatively high. The overall output resistance corresponds to the output resistance of C.B., which is known to be quite high.

13.4 DIFFERENTIAL AMPLIFIER

13.4.1 Introduction

A *differential amplifier* amplifies the difference between two input signals by some constant factor. This amplifier can also be referred to as a *differential pair*, since it is a two-transistor subtractor configuration, adopted to furnish a gain proportional to a *difference of signals* rather than to the particular signal as considered until now.

A classic example of application of differential amplifiers comes from telephone lines, which carry the voice signal between two (typically red and green) lines. These lines can run parallel to power lines, resulting in an induced $60\,Hz$, in the United States ($50\,Hz$ in Europe), voltage, with a peak of tens of volts from each wire to ground. The differential amplifiers are then utilized to amplify the voltage differences related to the voice signal, rejecting the voltage commonly related to the induced "interference."

Many electronic networks use an internal differential amplifier, as occurs, for instance, in the *operational amplifier,* which is one of the basic circuits of modern electronics. Many *integrated circuit* (*IC*) designs includes differential amplifiers as well.

13.4.2 Theory

A differentiated amplifier is a functional block with two input ports and two output ports, as represented in Figure 13.12.

The differential amplifier is made up of one *non-inverting* input (*NI*) and one *inverting* input (*I*), and one *main* output (*O1*) and one *secondary* output (*O2*). *NI* and *I* refer to the fact that signals at these input ports result 0° and 180° out of phase at the same output port, respectively.

A *single-ended* signal is taken across the main output port *O1* with respect to the ground (or a fixed potential anyway), while a *double-ended* signal is taken across the two output ports *O1–O2*.

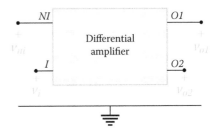

FIGURE 13.12   Differential amplifier.

Observation

The *single-ended* definition refers to any voltage across a certain node referenced to AC ground. A *differential mode* voltage refers to any voltage difference between two nodes. For the differential amplifier, it is preferred to take output not differentially but from one single end.

Here, we will deal with the common case of the single-ended output, and we will assume conditions of linearity, so that the superposition theorem can be applied.

We define v_{ni} as the voltage across the non-inverting input port and the ground, v_i as the voltage across the inverting input port and the ground, and A_D being the amplification value. Given the output sign inversion only for v_i, the amplified output signal at the port *O1* can be written as

$$v_{O1} = A_D \left(v_{ni} - v_i \right)$$

This expression, however, is true only for an *ideal* differential amplifier, because for a *real* one the signals present at the two input ports do not normally undergo the same amplification. The *real* model is then described by the expression

$$v_{O1} = A_1 v_{ni} + A_2 v_i$$

It is useful to decompose the expressions of the voltages as

$$v_{ni} = \frac{v_{ni} - v_i}{2} + \frac{v_{ni} + v_i}{2}; \quad v_i = \frac{v_i - v_{ni}}{2} + \frac{v_i + v_{ni}}{2}$$

defining

$$v_C = \frac{v_{ni} + v_i}{2}; \quad v_D = v_{ni} - v_i$$

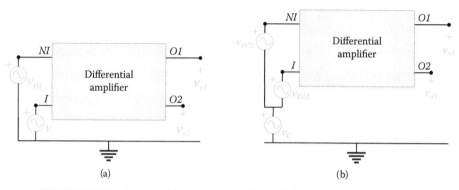

(a)   (b)

FIGURE 13.13 Schematic representation of the mathematical trick.

v_C is the so-called *common-mode* signal component (or *mean* or *semi-sum*), which represents what the two input signals have in common, and v_D the so-called *differential-mode* signal component, which represents the differences between the two input signals. This decomposition is schematically represented in Figure 13.13.

Therefore,

$$v_{O1} = A_1 v_{ni} + A_2 v_i = A_1 \left(\frac{v_D}{2} + v_C \right) + A_2 \left(-\frac{v_D}{2} + v_C \right)$$

$$= \frac{A_1 - A_2}{2} v_D + (A_1 + A_2) v_C = A_D v_D + A_C v_C$$

where $A_D = (A_1 + A_2)/2$ and $A_c = A_1 = A_2$.

We deduce that the *real* differential amplifier amplifies not only the difference of the input signals v_D, but also part of the common mode signal v_C.

Comparing the real versus ideal cases, we gather that they coincide only when $A_c v_C = 0$, i.e., basically when $A_c = 0$, which corresponds to ensuring that $A_1 = A_2$.

As a consequence, the cornerstone specification for a differential amplifier is named the *Common Mode Rejection Ratio* (*CMRR*) since it is a measure of the rejection of the common-mode input voltage in favor of the differential-input voltage:

$$CMRR = \left| \frac{A_D}{A_C} \right|$$

or

$$CMRR_{dB} = 20 \log_{10} CMRR = 20 \log_{10} \left| \frac{A_D}{A_C} \right|$$

Ideally, *CMRR* would be infinite with common mode signals being totally rejected.

Observation

To experimentally determine A_C, we can set the same signal v_{12} across the two inputs (so $v_{ni} = v_i \Leftrightarrow v_D = 0$) and measure the output signal ($v_{01} = A_D v_D \Leftrightarrow v_D = 0$). Similarly, to experimentally determine A_D we can set the same signal but out-phased across the two input ports.

Examples

The *electrocardiograph* is an instrument capable of measuring electrical voltages on the human body surface associated with heart muscle activity (see Figure 10.16 in Chapter 10). When used, because of the parasitic capacitances that connect the electrodes and the patient's body to the AC power supply (capacitive voltage divider), we have an unwanted $60\,Hz$ (or $50\,Hz$) common-mode signal. This can cause a problem, which can be avoided if we adopt a differential amplifier with a CMRR value as high as possible. Alternately, we could connect the patient directly to ground (thus bypassing the parasitic capacitances), but an accidental contact with the AC power supply would send a lethal current through the patient's body!

Everyone knows the problem of trying to hear music from headphones, or a voice from a smartphone when surrounded by noise or ambient sound. This problem can be "actively" reduced. In fact, there are some "active" noise-canceling headphones (by *Bose* with the *QuietConfort* model) and smartphones (by *Apple* with the iPhone 5) that are based on the technology of the differential amplifier. These devices detect the noise and replicate it but $180°$ out of phase, so that the original noise and the replicated one are erased in the output of a differential amplifier.

13.4.3 Basic Emitter-Coupled Pair

The heart of a differential amplifier can be realized by a differential pair composed of two emitter-coupled transistors, BJTs in our example (Figure 13.14).

We can distinguish two biased BJTs, T_1 and T_2, in common emitter configurations, sharing a unique emitter resistor R_E. As long as T_1 and T_2 are identical, with identical bias networks, the differential pair branches are indistinguishable from each other.

In a *differential* configuration, possible variations of the working point produced by very different causes have *the same* effects on the two transistors. Therefore the *differential* output signal will not be affected by these variations, being simply the amplified version of the *difference* between the two input signals, affected by *common* noises.

The two amplifiers presented in Figures 13.14a and b are similar, except that in the first the bias is made by a resistive network (as usual), while in the second the bias is made by two DC sources (for example two batteries). The two networks work to bias the junctions of the BJTs exactly in the same way. The first configuration is

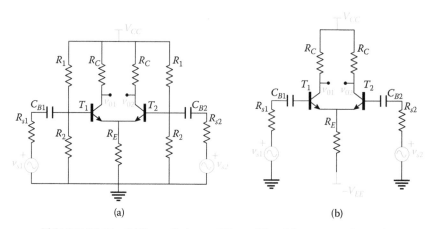

FIGURE 13.14   Differential amplifier with a bias network made by a (a) resistor divider and (b) two DC supplies.

termed an *unbalanced amplifier*, while the second is called a *balanced amplifier*. The terms unbalanced/balanced refer to a signal that is asymmetrically/symmetrically disposed about the reference potential. The reference is the ground in the first configuration, while it is the gate potential "around" which the signals swing as in a mirror in the second configuration. So, in the second case, unlike the first one, we have a symmetry of voltages with respect to zero. The bias realized by means of two batteries allows for a higher CMRR value.

For a qualitative analysis of this circuit, we assume to work in small-signal conditions (Section 10.2.4 in Chapter 10), so that the superposition theorem can be applied.

We start therefore considering $v_{s1} \neq 0$ and $v_{s2} = 0$. If so, v_{s1} will result out of the emitter of T_1, named v_{e1}, in phase and with a slightly lower value (T_1 behaves as a source follower). By visual inspection $v_{e1} = v_{e2}$, and

$$v_{be2} = v_{b2} - v_{e2} = 0 - v_{e2}$$

(the condition $v_{s2} = 0$ implies $v_{b2} = 0$). Therefore, signal v_{s1} goes across the base terminal of T_2 and the ground, named v_{b2}, 180° out of phase. Finally, this signal will become v_{o2} at the collector of T_2, amplified and again 180° out of phase (T_2 now behaves as a common emitter), so that the total phase shift will be zero.

Considering inverted input conditions now, i.e., $v_{s1} = 0$ and $v_{s2} \neq 0$, we have that the signal will simply result in output as v_{o2}, amplified and 180° out of phase (T_2 behaves again as a common emitter).

If we design the amplifier in such a way that both v_{s1} and v_{s2} are amplified the same amount, the signal v_{o2} is the amplified version of the difference $v_{s1} - v_{s2}$, exactly as we are looking for. When $v_{s1} = v_{s2}$ no output signal will be produced at all.

Given the different length in path for the two signals, their transit time will be different. The consequence is that the differential amplifier can find applications just below some *MHz* in frequency, otherwise phase shifts between the two signals would not be guaranteed to be 180°.

Serial data transmission can use the known protocols known as *RS422* or *RS485*. These protocols are based on differential techniques of transmission, to reject a considerable amount of common-mode noise.

The emitter resistor R_E, also called a *long-tail* resistor, can be usefully replaced by an active current source (Section 13.1.2) to improve the overall CMRR value.

13.5 KEY POINTS, JARGON, AND TERMS

→ The amplifier's input resistance can be increased utilizing a *Darlington pair*, a *current source*, or a *bootstrapping technique*.

→ A *Darlington pair* consists of two transistors connected in a way that the base of the second is directly connected with the emitter of the first, and both collectors are linked together.

→ A *current source* is the dual of a voltage source. It is realized by a network made of transistors and resistors elements, in a way that direct current flows into or out of a high-impedance output node.

→ *Bootstrapping* refers to a feedback effect used to increase the input resistance by means of an output signal.

→ A *transformer-coupled load* is used to increase the low *efficiency* of the class A topology.

→ A *cascode* configuration combines a C.E. input with a C.B. output with BJTs or C.S. and C.G. with FETs, to obtain a large bandwidth, a reasonably high input resistance and voltage gain, and a very high output resistance.

→ A *differential amplifier* consists of a particular amplifier whose output is not just proportional to the input, but to the difference between two input signals. This is to increase the *signal to noise ratio* with the rejection of unwanted signals that can couple into the input.

14

Amplifiers: Cascading Stages

A very weak signal cannot be sufficiently amplified by a *single stage amplifier*, like those we have dealt with so far—amplifiers made only of a single biased transistor forming a complete network. Single-stage transistor amplifiers can be inadequate for meeting most design requirements for any of the four amplifier types (voltage, current, transresistance, and transconductance). If a microphone produces about $20\,mV$ under normal conditions, you will need a single stage of amplification, but if it produces only $1\,mV$, you will need two or more stages of amplification, that is, a *multistage amplifier*, to provide greater amplification than a single stage could provide by itself.

In the previous chapters, we dealt with configurations with two transistors, such as the *Darlington pair* (Section 13.1.1 in Chapter 13), *current source* (Section 13.1.2), *cascode amplifiers* (Section 13.3), and *differential amplifiers* (Section 13.4), which can be considered "similar-multistage," since you cannot "isolate" a single transistor, with its own bias network, forming a complete isolated amplifier. Real multistage can be considered more properly as a *cascade of single stages*.

We treated single stages made of a single BJT configured as C.E. or C.C. or C.B., and single stages made of a single MOSFET configured as C.S. or C.D. or C.G.

When considering a multistage amplifier, the choice of the number of stages to be cascaded, and the choice of the type of transistors to be used and their relative configuration, strictly depends on the application, so no "golden rule" exists. But we can take into account the requested overall gain and the matching conditions that we need to satisfy. Useful rules of thumb can come from the general information furnished in Tables 11.6 and 11.7 of Chapter 11, and from the observation

TABLE 14.1

Characteristics of Some Double-Stage Configurations

Stages	Input resistance	Output resistance	Voltage gain
C.E.–C.E.	Medium	Medium	Very high
C.E.–C.B.	Medium	High	High
C.E.–C.C.	Medium	Low	Medium
C.C.–C.E.	High	Medium	Medium
C.C.–C.C.	Very high	Very low	< 1

that, in general, the MOSFET has a higher input resistance than a BJT, which, in turn, can provide lower output resistances.

Since a BJT type transistor can be configured as a C.E., C.B., or C.C. topology, it follows that two stages can be arranged according to nine different combinations: C.E.–C.E., C.E.–C.B. (see the *cascode*), C.E.–C.C., C.B.–C.E., C.B.–C.B., C.B.–C.C., C.C.–C.E., C.C.–C.B. (see the *differential amplifier*), and C.C.–C.C. (see the *Darlington pair*). But, of course, some combinations are more useful than others. This is because we have to ensure the required overall gain and, thus, the right resistance (or impedance) matching. So, the n-stage's output resistance $R_{o,n}$ and the $(n+1)$ - stage's input resistance $R_{i,n+1}$ must be in the proper ratio: $R_{i,n+1}/R_{o,n} \to \infty$ or $R_{i,n+1}/R_{o,n} \to 0$, depending on the type of requested gain.

Regarding the first three combinations with C.E. as the first stage, we have that C.E.–C.E. is useful to obtain a high voltage gain, C.E.–C.C. has a good voltage gain and a low output resistance, and C.E.–C.B. (*cascode*) has a good voltage gain and a large bandwidth. But a C.E. amplifier cannot drive a low resistance (or impedance) load directly, otherwise it will be overloaded, because of its typical of AC output resistance value. So C.C. as a second *buffer* stage between the C.E. amplifier and the load prevents the overloading effect.

A useful general reference table is shown in Table 14.1.

The designer sometimes prefers to realize double-stage amplifiers using *npn*- and *pnp*-type BJTs. This is because the temperature sensitivity of the amplifier can be greatly reduced. The reason relies on the subtraction of the voltage drifts with temperature variations of the two BJTs, since each transistor behaves in opposition, due to the opposite nature of *npn* versus *pnp*.

14.1 COUPLING

Each stage can be treated as a two-port network, so the stages can be hooked up according to different types of interconnections: *series, parallel,* and *cascade* (Section 5.5 in Chapter 5). Series and parallel types generally involve non-trivial mathematical calculations, so the *cascade* type is usually preferred.

The way the connection between two following stages is realized is called the *coupling* method.

The coupling can be realized *directly*, so that the output of the previous stage is connected to the input of the following stage without any additional circuitry, or can be realized by means of ad-hoc circuitry.

Coupling capacitors, inductors, and transformers causes low frequency gain loss in multistage amplifiers. This is for series coupling capacitors, because of their high reactance at low frequencies, and for parallel inductors and transformers, because of their low reactance at low frequencies.

14.1.1 RC, LC Coupling

RC coupling adopts a capacitor to couple the signal between stages. The capacitors ensure that the DC biasing of each stage is not affected by the bias of another stage or by the source and load. The RC coupling is the most common method of coupling two transistor stages of an amplifier.

LC coupling replaces the collector resistance R_C with an inductance L. This coupling allows for the amplification of higher frequencies, but produces poor voltage gain at low frequencies. The amplifier will peak at the resonant frequency determined by the LC couple.

In general, the RC coupling method is preferred because of the bulk, cost, and interference problems that are caused by the use of an inductance. The *RC coupling* is schematically represented in Figure 14.1.

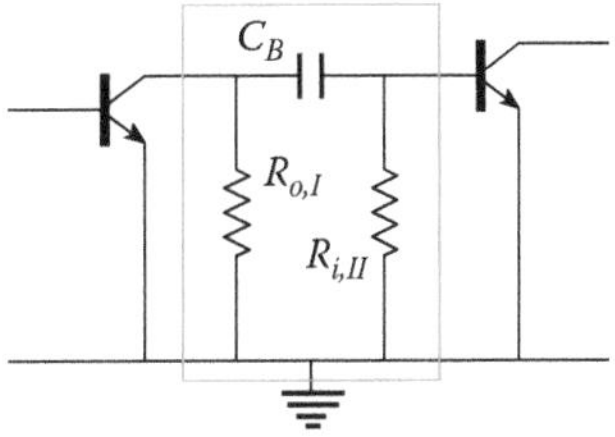

FIGURE 14.1 Schematic representation of RC coupling; "I" and "II" stand for first and second stage, respectively.

The output resistance $R_{o,I}$ of the first stage, the input resistance $R_{i,I}$ of the second stage, and the blocking capacitor C_B form the coupling RC network (enclosed in the green line in Figure 14.1). This network isolates the bias of each stage.

The main advantages of RC coupling are as follows:

→ The DC blocking capacitor prevents the bias conditions of the two stages from being influenced reciprocally. So for each stage the bias network can be designed without concern for the other(s).
→ A relatively flat bandwidth can be maintained, generally within a range between a few *Hz* and some *MHz*.
→ There is a relatively low harmonic distortion.

The main disadvantages are as follows:

→ The capacitor tends to limit the low-frequency response of the amplifier.
→ It is not useful for realizing impedance-matching conditions.

This coupling method is generally applied in audio amplifiers in radio and TV receivers.

Figure 14.2 represents an example of double stage C.E.–C.E. RC coupled amplifier.

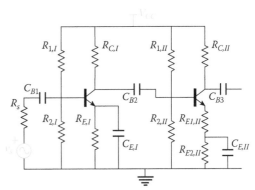

FIGURE 14.2 RC-coupled C.E.–C.E. configuration.

14.1.2 Transformer Coupling

Transformer coupling uses a transformer in a way that the primary winding is the output of one stage, and the secondary winding is the input of the next stage (Figure 14.3).

This coupling method allows us to DC isolate the adjacent stages, but signals are coupled from the first to the second stage. It allows low DC power dissipation, so that the amplifier's efficiency can increase (Section 13.2 in Chapter 13), and allows for impedance matching with proper design of the turns ratio.

The transformer coupling is often adopted for high-frequency amplifiers, such as those in RF (radio frequency) and IF (intermediate frequency) sections of radio and TV receivers, but not in LF (low frequency) applications, for which the transformer size would be prohibitive.

Figure 14.4 shows an example of a double-stage C.E.–C.E. configured with a transformer coupling method.

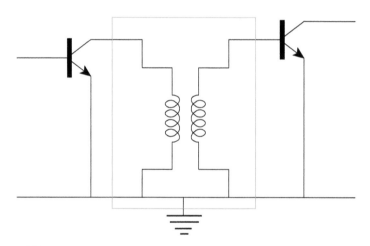

FIGURE 14.3 Schematization of the transformer coupling.

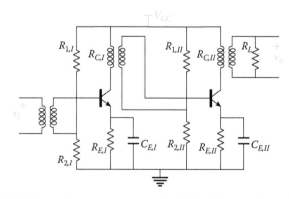

FIGURE 14.4 Transformer-coupled C.E.-C.E. configuration.

14.1.3 Impedance Coupling

Impedance coupling replaces the load resistor of the first stage with a coil, as shown in Figure 14.5.

In general, the amplifiers used for audio purposes need a large inductive value. The consequence is that the inductive reactance should be large, so that either inductance or frequency or both must be high. But load inductors with large amounts of inductance are most effective at high frequencies, so they are not suitable for audio applications.

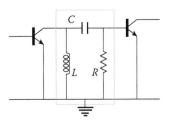

FIGURE 14.5
Schematization of the impedance coupling.

14.1.4 Direct Coupling

The *direct-coupling* method refers to the fact that two stages are directly connected without additional circuitry, as represented in Figure 14.6.

Because no DC blocking component is between the stages, the design requires the arrangement of the output quiescent conditions of one stage to be the same as the quiescent input conditions for the next stage. The consequence can be that any, even poor, instability of the quiescent Q-point of one stage becomes a signal to be amplified by the next stage, added as a "noise" to the real signal to be conditioned. In addition, temperature drifts, aging of components, and DC supply drifts can change the bias condition of the latter stage.

Nevertheless, *direct coupling* is adopted in different networks. This is in particularly true when the DC blocking components can seriously affect the low-frequency response of the amplifier, when the DC blocking components would be too large to be placed in an integrated circuit, and when the cost must be lowered.

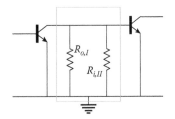

FIGURE 14.6 Schematic representation of direct coupling. Subscripts "I" and "II" stand for first and second stage, respectively.

With *direct coupling*, both DC and AC signals can be amplified down to $0\,Hz$, and the DC collector voltage of one stage provides base bias voltage of the following stage.

Figure 14.7 shows an example of a double-stage C.E.–C.E. configured with a direct coupling method.

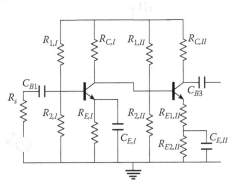

FIGURE 14.7 Direct coupled C.E.–C.E. configuration.

14.2 DC ANALYSIS

The study of multistage amplifiers will take into account linear conditions, and assume a small-signal regime, as already seen for single-stage amplifiers. The superposition theorem can then be applied, so that DC and AC analysis can be performed separately. In this view, the same arbitrary multistage amplifier can be analyzed with two different models, valid for DC and AC conditions, respectively.

The DC analysis must take into account that, basically, the coupling methods can be divided into two main groups, depending on the fact that the bias network can be considered separately or not for each amplifier's stage. So, the *RC*, *LC, transformer*, and *impedance* couplings need to consider the bias problem separately for each stage, while the *direct* coupling needs to establish the influence of the bias among stages. The two possibilities refer to AC- and DC-coupled multistage amplifiers, respectively.

14.2.1 AC-Coupled Multistage Amplifiers

In an AC-coupled multistage amplifier, the DC bias networks are *isolated* from each other. The amplifier is generally made of *n* stages and the isolation is provided by $n{-}1$ coupling capacitors C_B, C_B, .. $C_{B(n-1)}$ (Figure 14.8).

Thanks to the coupling capacitors, the DC analysis is *independent* for each stage. This is not true for AC analysis, because of the AC interdependence of the cascading.

14.2.2 DC-Coupled Multistage Amplifiers

In a DC-coupled multistage amplifier, the DC bias networks are *not isolated* from each other, so the DC analysis must take into account their interdependence. This is the same as we discussed for the AC analysis. Figure 14.9 schematizes

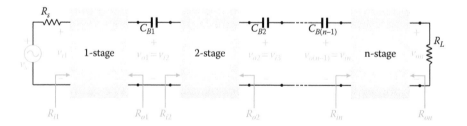

FIGURE 14.8 Schematization of an AC-coupled multistage amplifier.

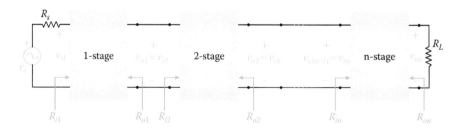

FIGURE 14.9 Schematization of a DC-coupled multistage amplifier.

a DC-coupled multistage amplifier without any frequency-dependent impedance between adjacent stages.

14.3 AC ANALYSIS

In Section 10.3 of Chapter 10, we modeled a one-stage amplifier with a linear two-port network, but it was a *particular* type of two-port network. In fact, in general, for many *passive* linear 2-ports, we can admit *reciprocity*, that is, input and output impedances are the same, $Z_i(\omega) = Z_o(\omega)$ (see Figure 5.10 in Chapter 5 for $R_1 = R_3$), so that $v_i = v_o$ when $i_i = i_o$ (see Figure 5.1). But this cannot be true for an amplifier. For an amplifier, a certain input current produces an output open-circuit voltage, but the same current on the output port cannot produce the equal open-circuit voltage across the input port. Furthermore, the amplifier cannot be considered to be *passive*, and its circuit symbol is not symmetric (Figure 10.1 in Chapter 10), contrary to the circuit symbols of resistors, capacitors, and inductors.

We adopted different models for a voltage amplifier (Figure 10.14 in Chapter 10), for a current amplifier (Figure 10.17), for a transresistance amplifier (Figure 10.19), and for a transconductance amplifier (Figure 10.22), all of them with the common aspects of *non-reciprocity* and *unilateral* characteristics. An amplifier with an output that exhibits no feedback to its input side is defined as *unilateral*, and the aforementioned amplifier's models lack feedback, because no output controlled source was at the input loop. The relevant consequence is that the source does not affect the output resistance, and the load does not affect the input resistance.

We will deal here with the AC analysis of amplifiers subjected to the hypothesis of unilateral stages, and to the hypothesis of non-unilateral, also termed *bilateral*, stages. Please pay close attention to the fact that the unilateral condition can result also from an approximation of a bilateral configuration, when the feedback effect is negligible. In any case, this feedback effect can be present for BJTs only, since MOSFETs always have unilateral behavior, because of the gate oxide that prevents any gate current to flow.

For the sake of simplicity, we will refer to the model of the *voltage amplifier* since it is the most common among all types of amplifiers.

14.3.1 Unilateral Stages

Figure 14.10 represents a cascading of n unilateral stages of voltage amplifiers.

To perform the AC analysis, let's consider the particular case of $n = 3$ stages. We can write

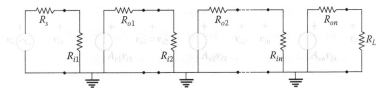

FIGURE 14.10 Cascading of unilateral stages.

$$A_v = \frac{v_l}{v_s} = \frac{v_{o3}}{v_s} = \frac{v_{o3}}{v_{o2}} \frac{v_{o2}}{v_{o1}} \frac{v_{o1}}{v_{i1}} \frac{v_{i1}}{v_s}$$

but

$$v_l = v_{o3} = \frac{R_L}{R_L + R_{o3}} A_{v3} v_{i3} = \frac{R_L}{R_L + R_{o3}} A_{v3} v_{o2}$$

$$v_{o2} = \frac{R_{i3}}{R_{i3} + R_{o2}} A_{v2} v_{i2} = \frac{R_{i3}}{R_{i3} + R_{o2}} A_{v2} v_{o1}$$

$$v_{o1} = \frac{R_{i2}}{R_{i2} + R_{o1}} A_{v1} v_{i1}$$

$$v_{i1} = \frac{R_{i1}}{R_{i1} + R_s} v_s$$

Therefore,

$$A_v = \frac{R_L}{R_L + R_{o3}} A_{v3} \frac{R_{i3}}{R_{i3} + R_{o2}} A_{v2} \frac{R_{i2}}{R_{i2} + R_{o1}} A_{v1} \frac{R_{i1}}{R_{i1} + R_s}$$

From this result, we can derive the observation that the overall voltage gain is generally given by the voltage gain of each stage, A_{vx} with $x = 1...n$, multiplied by attenuation factors due to voltage dividers between adjacent stages.

14.3.2 Non-Unilateral Stages

Figure 14.11 represents a cascading of n non-unilateral (or bilateral) stages of voltage amplifiers.

One or more non-unilateral stages make the analysis of cascading stages more tricky. This is because even the amplifier's total input resistance, as "seen" by the AC source, may depend on the final output load resistance, and the amplifier's total output resistance, as "seen" by the load, may depend on the source resistance. Any change of a single load implies the need to recalculate everything. This is not true for unilateral stages. Generally, the input impedance of each single stage depends on the input impedance of the next stage which, in turn, depends on the output impedance of the latter. In the same way, the output impedance of each single stage depends on the output impedance

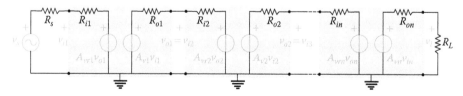

FIGURE 14.11   Cascading of bilateral stages.

of the preceding stage which, in turn, depends on the input impedance of the succeeding stage.

To solve an arbitrary amplifier made of cascading bilateral stages, a nested approach is commonly adopted from the last stage, on the very right side, to the first one, on the very left side. The goal is to simplify the problem, creating a new single two-port model of the entire amplifier. In any case, to solve the problem we have to deal with sets of simultaneous equations, and it is often preferred to recall the *transmission parameters* (see Table 5.1 in Chapter 5) which for cascading connections involves a matrix product (Section 5.5.4 of Chapter 5).

The non-unilateral stages are considered just for particular applications, so their mathematical computations are not here expressed.

14.4 DOUBLE-STAGE C.E.–C.E.

A double-stage amplifier made up of two C.E. networks, represented in Figure 14.12, can guarantee a high-voltage gain.

To perform the AC analysis, let's assume the BJT's simplified hybrid model is valid (Section 11.9 in Chapter 11), so that the AC equivalent model can be represented as in Figure 14.13.

For convenience, resistors $R_{1,I}$ and $R_{2,}$ can be combined into $R_{12,I} = R_{1,I} \parallel R_{2,I}$, and resistors $R_{1,II}$ and $R_{2,II}$ can be combined into $R_{12,II} = R_{1,II} \parallel R_{2,II}$.

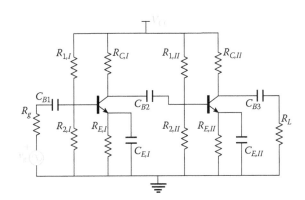

FIGURE 14.12   Double-stage C.E.-C.E.

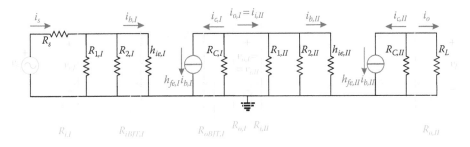

FIGURE 14.13   AC equivalent model of the double-stage C.E.-C.E.

14.4.1 AC Input Resistances

The AC input resistance of the first stage is defined as

$$R_{i,I} \underline{\mathrm{def}} \frac{v_{be,I}}{i_s}$$

It can be usefully rewritten as

$$R_{i,I} = \frac{v_{be,I}}{i_{b,I}} \frac{i_{b,I}}{i_s} = R_{iBJT,I} \alpha_{Ri,I}$$

with

$$R_{iBJT,I} = \frac{v_{be,I}}{i_{b,I}} \quad \text{and} \quad \alpha_{Ri,I} = \frac{i_{b,I}}{i_s}$$

But

$$R_{iBJT,I} = \frac{v_{be,I}}{i_{b,I}} = h_{ie,I}$$

and

$$\alpha_{Ri,I} = \frac{i_{b,I}}{i_g} = \frac{R_{12,I}}{R_{12,I} + h_{ie,I}}$$

therefore

$$R_{i,I} = R_{iBJT,I} \alpha_{Ri,I} = h_{ie,I} \frac{R_{12,I}}{R_{12,I} + h_{ie,I}}$$

that is,

$$R_{i,I} = h_{ie,I} \parallel R_{12,I}$$

The AC input resistance of the second stage is defined as $R_{i,II} \underline{\mathrm{def}} \frac{v_{i,II}}{i_{i,II}}$. Similarly to previous discussions, the result is that

$$R_{i,II} = R_{12,II} \parallel h_{ie,II}$$

14.4.2 AC Current Gain

The overall current gain is defined as $A_i \underline{\mathrm{def}} \frac{i_o}{i_e}$.
Since

$$i_o = -\frac{R_{C,II}}{R_{C,II} + R_L} i_{c,II}$$

$$i_{c,II} = h_{fe,II} i_{b,II}$$

$$i_{b,II} = \frac{R_{12,II}}{R_{12,II} + h_{ie,II}} i_{o,I}$$

$$i_{o,I} = -\frac{R_{C,I}}{R_{C,I} + R_{12,II} \parallel h_{ie,II}} i_{c,I}$$

$$i_{c,I} = h_{fe,I} i_{b,I}$$

$$i_{b,I} = \frac{R_{12,I}}{R_{12,I} + h_{ie,I}} i_g$$

we can finally write

$$A_i = \frac{R_{C,II}}{R_{C,II} + R_L} h_{fe,II} \frac{R_{12,II}}{R_{12,II} + h_{ie,II}} \frac{R_{C,I}}{R_{C,I} + R_{12,II} \parallel h_{ie,II}} h_{fe,I} \frac{R_{12,I}}{R_{12,I} + h_{ie,I}}$$

14.4.3 AC Voltage Gain

The overall voltage gain is defined as $A_{v,II} \stackrel{\text{def}}{=} \frac{v_l}{v_s}$. It depends on the voltage gain of each stage.

The voltage gain of the second stage is

$$A_{v,II} = \frac{v_{o,II}}{v_{be,II}} = \frac{v_{ce,II}}{v_{be,II}}$$

It is easy to demonstrate that

$$i_o = -\frac{R_{C,II}}{R_{C,II} + R_L} i_{c,II}$$

$$v_{ce,II} = v_o = -\left(R_{C,II} \parallel R_L\right) i_{c,II} = -\left(R_{C,II} \parallel R_L\right) h_{fe,II} i_{b,II}$$

$$v_{be,II} = h_{ie,II} i_{b,II}$$

The voltage gain of the second stage is then the ratio of hybrid parameters times the load resistance:

$$A_{v,II} = -\frac{h_{fe,II}}{h_{ie,II}} \left(R_{C,II} \parallel R_L\right)$$

The voltage gain of the first stage is

$$A_{v,I} \stackrel{\text{def}}{=} \frac{v_{o,I}}{v_{i,I}} = \frac{v_{ce,I}}{v_{be,I}}$$

Now, for $v_{ce,I}$,

$$v_{ce,I} = \left(R_{12,II} \parallel h_{ie,II}\right) i_{o,I}$$

but $i_{o,I} = -\dfrac{R_{C,I}}{R_{C,I} + \left(R_{12,II} + h_{ie,II}\right)} i_{c,I}$ and $i_{c,I} = h_{fe,I} i_{b,I}$ so

$$v_{ce,I} = -\left(R_{12,II} \parallel h_{ie,II}\right) h_{fe,I} i_{b,I} \dfrac{R_{C,I}}{R_{C,I} + \left(R_{12,II} + h_{ie,II}\right)}$$

Now, for $v_{be,I}$,

$$v_{be,I} = h_{ie} i_{b,I}$$

Therefore,

$$A_{v,I} = \dfrac{v_{ce,I}}{v_{be,I}} = -\dfrac{h_{fe,I}}{h_{ie,I}} \left(R_{12,II} \parallel h_{ie,II}\right) \dfrac{R_{C,I}}{R_{C,I} + \left(R_{12,II} + h_{ie,II}\right)}$$

which can be rewritten as

$$A_{v,I} = -\dfrac{h_{fe,I}}{h_{ie,I}} R_{C,I} \parallel \left(R_{12,II} \parallel h_{ie,II}\right)$$

The attenuation factor can be written according to the voltage divider equation

$$v_{be,I} = \dfrac{R_{12,I}}{R_{12,I} + R_s} v_s$$

therefore,

$$\alpha_v = \dfrac{R_{12,I}}{R_{12,I} + R_s}$$

Finally,

$$A_v = A_{v,II} A_{v,I} \alpha_v = \dfrac{h_{fe,II}}{h_{ie,II}} \left(R_{C,II} \parallel R_L\right) \dfrac{h_{fe,I}}{h_{ie,I}} R_{C,I} \parallel \left(R_{12,II} \parallel h_{ie,II}\right) \dfrac{R_{12,I}}{R_{12,I} + R_s}$$

Observation

Since the overall voltage gain of the C.E.–C.E. configuration can be really high, it is sometimes preferred to reduce and stabilize this gain, introducing a degeneration emitter resistor.

14.4.4 AC Output Resistance

The AC output resistance of the first stage is defined as

$$R_{o,I} \underline{\text{def}} \dfrac{v_{o,I}}{i_{o,I}} = R_{o,BJT,I} \parallel R_{C,I}$$

To determine it, we have to kill the independent sources, so $v_s \overset{!}{=} 0$. By visual inspection of the AC equivalent model we have that $i_{c,I} = h_{fe,I} i_{b,I}$, but $i_{b,I}$ is null since v_s is null, therefore $R_{oBJT,I} = \infty$, and

$$R_{o,I} = R_{C,I}$$

The AC output resistance of the second stage is defined as

$$R_{o,II} \underset{def}{=} \frac{v_{o,II}}{i_{o,II}} = R_{o,BJT,II} \,||\, R_{C,II}$$

Similarly to what was previously done for $R_{o,I}$ we have

$$R_{o,II} = R_{C,II}$$

The value of the output resistance of the second stage $R_{o,II}$ corresponds to that of the overall amplifier.

14.5 DOUBLE-STAGE C.E.–C.C.

An example of a double-stage C.E.–C.C. amplifier is shown in Figure 14.14.

Suppose all the resistance values are known, along with the DC source value, and the values of the relevant parameters of the BJTs (here indicated by the subscripts "1" and "2" for transistor T_1 and T_2, respectively):

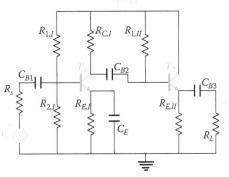

FIGURE 14.14 Example of a double-stage C.E.–C.C. amplifier.

→ β_1, h_{ie1}, h_{fe1}, β_2, h_{ie2}, h_{fe2}, considering $h_{re1} = h_{re2} \cong 0$ and $h_{oe1} = h_{ro2} \cong 0$

In forward active mode $V_{BEQ1} = V_{BEQ2} = 0.7\,V$.

The capacitors are assumed to be short circuits in the middle-band frequency.

We perform the DC analysis here to determine quiescent currents and voltages, and the AC analysis to determine the overall voltage gain.

14.5.1 DC Analysis

The equivalent DC model of the double stage C.E.–C.C. amplifier is represented in Figure 14.15.

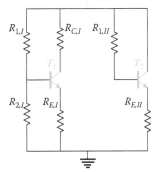

FIGURE 14.15 DC equivalent model of the double-stage C.E.–C.C.

It is evident how the two stages are independent so that we can perform their DC analysis separately.

Let's start by considering the C.C. stage.

From KVL-BE2,

$$V_{CC} = R_{1,II} I_{BQ2} + V_{BEQ2} + R_{E,II} I_{EQ2}$$

Recalling that when the BJT is in forward active mode, the relationship between the collector and base current of an arbitrary BJT is $I_C \cong \beta I_B$ (Section 8.6.3 in Chapter 8), and considering $I_C \cong I_E$, the previous equation can be rewritten as

$$V_{CC} = V_{BEQ2} + \left(\frac{R_{1,II}}{\beta} + R_{E,II} \right) I_{CQ2}$$

so that the value of the current I_{CQ2} is obtained.

From KVL-CE2,

$$V_{CC} = V_{CEQ2} + R_{E,II} I_{EQ2} \cong V_{CEQ2} + R_{E,II} I_{CQ2}$$

so that the value of the voltage V_{CEQ2} is obtained as well.

We can perform the DC analysis of the C.E. stage by considering its Thévenin equivalent at the input port as shown in Figure 14.16, with

$$V_{TH} = \frac{R_{2,I}}{R_{2,I} + R_{1,I}} ; \; R_{TH} = R_{2,I} \| R_{1,I}$$

Now, from KVL-BE1,

$$V_{TH} = I_{BQ1} R_{TH} + V_{BEQ1} + R_{E,I} I_{CQ1} \cong V_{BEQ1} + \left(\frac{R_{TH}}{\beta} + R_{E,I} \right) I_{CQ1}$$

so that the value of the current I_{CQ1} is obtained.

From KVL-CE1,

$$V_{CC} = R_{C,I} I_{CQ1} + V_{CEQ1} + R_{E,I} I_{EQ1} \cong V_{CEQ1} + \left(R_{C,I} + R_{E,I} \right) I_{CQ1}$$

so that the value of the voltage V_{CEQ1} is obtained as well.

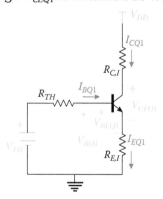

FIGURE 14.16 DC model of the C.E. stage with Thévenin equivalent at the input port.

14.5.2 AC Analysis

The AC equivalent model is represented in Figure 14.17.

For convenience, some of the resistors can be combined into $R_{12,I} = R_{1,I} \parallel R_{2,I}$, $R_{C,I1,II} = R_{C,I} \parallel R_{1,II}$, and $R_{E,IIL} = R_{E,II} \parallel R_L$.

The overall voltage gain is defined as

$$A_v \underline{\text{def}} \frac{v_l}{v_s}$$

but it can be conveniently expressed as

$$A_v = \frac{v_l}{v_{c1}} \frac{v_{c1}}{v_{b1}} \frac{v_{b1}}{v_s}$$

so we have to determine the three voltage ratios.

Regarding v_l/v_{c1}, we can write

$$\frac{v_l}{v_{c1}} = \frac{v_l}{v_{b2}} = \frac{\left(1 + h_{fe2}\right)R_{E,IIL}}{\left(1 + h_{fe2}\right)R_{E,IIL} + h_{ie2}}$$

which clearly represents an attenuation.

Regarding the term v_{c1}/v_{b1}, we can write

$$v_{c1} = -i_{c1}\left\{R_{C,I1,II} \parallel \left[h_{ie2} + \left(1 + h_{fe2}\right)R_{E,IIL}\right]\right\} = -h_{fe1}i_{b1}\left\{R_{C,I1,II} \parallel \left[h_{ie2} + \left(1 + h_{fe2}\right)R_{E,IIL}\right]\right\}$$

but

$$i_{b1} = \frac{v_{b1}}{h_{ie1}}$$

so

$$\frac{v_{c1}}{v_{b1}} = \frac{-h_{fe1}\left\{R_{C,I1,II} \parallel \left[h_{ie2} + \left(1 + h_{fe2}\right)R_{E,IIL}\right]\right\}}{h_{ie1}}$$

(regarding the resistance $h_{ie2} + \left(1 + h_{fe2}\right)R_{E,IIL}$, please refer to Section 11.9 of Chapter 11).

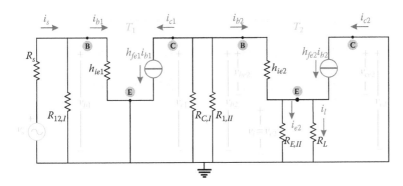

FIGURE 14.17   AC equivalent model of the double-stage C.E.–C.C.

Finally the term $\dfrac{v_l}{v_{c1}}$ can be expressed as

$$\frac{v_{b1}}{v_s} = \frac{R_{12,I}}{R_{12,I} + R_s}$$

which is another attenuation factor.

The overall voltage gain is therefore

$$A_v = -\frac{\left(1 + h_{fe2}\right) R_{E,IIL}}{\left(1 + h_{fe2}\right) R_{E,IIL} + h_{ie2}} \cdot \frac{h_{fe1}\left\{ R_{C,I1,II} \parallel \left[h_{ie2} + \left(1 + h_{fe2}\right) R_{E,IIL} \right] \right\}}{h_{ie1}} \cdot \frac{R_{12,I}}{R_{12,I} + R_s}$$

14.6 DOUBLE-STAGE C.C.–C.E.

Figure 14.18 shows an example of a double-stage C.C.–C.E. amplifier.

The AC analysis follows, with reference to the equivalent AC model (Figure 14.19).

For convenience, resistors $R_{1,I}$ and $R_{2,I}$ can be combined into $R_{12,I} = R_{1,I} \parallel R_{2,I}$, and resistors $R_{1,II}$ and $R_{2,II}$ can be combined into $R_{12,II} = R_{1,II} \parallel R_{2,II}$.

We want to perform the AC analysis first for the second stage and then for the first, according to the following definitions:

II stage

$$A_{iq,II} \;\underline{\text{def}}\; \frac{i_{c,II}}{i_{b,II}} = \frac{h_{fe}}{1 + h_{oe} R_{C,II}}$$

$$R_{iq,II} \;\underline{\text{def}}\; \frac{v_{b,II}}{i_{b,II}} = h_{ie}$$

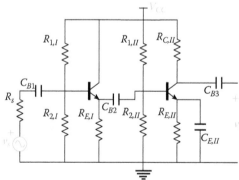

FIGURE 14.18   Example of a double-stage C.C.–C.E. amplifier.

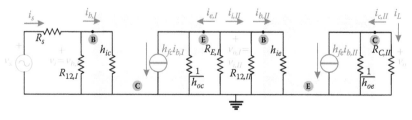

FIGURE 14.19   AC model of the double-stage C.C.–C.E. amplifier.

$$A_{vq,II} \underset{=}{\text{def}} \frac{v_{ce,II}}{v_{be,II}} = \frac{R_{c,II}}{R_{iq,II}} A_{iq,II}$$

$$R_{oq,II} \underset{=}{\text{def}} \frac{v_{ce,II}}{i_{c,II}} = \frac{1}{h_{oe}}$$

Calling the load resistance of the first stage $R_{L,I}$,

I stage

$$R_{L,I} = R_{E,I} \parallel R_{12,II} \parallel R_{iq,II}$$

we can define

$$A_{iq,I} \underset{=}{\text{def}} \frac{i_{c,I}}{i_{b,I}} = \frac{h_{fc}}{1 + h_{oc} R_{L,I}}$$

$$R_{iq,I} \underset{=}{\text{def}} \frac{v_{b,I}}{i_{b,I}} = h_{ic}$$

$$A_{vq,I} \underset{=}{\text{def}} \frac{v_{ce,I}}{v_{be,I}} = \frac{R_{L,I}}{R_{iq,I}} A_{iq,I}$$

$$R_{oq,I} \underset{=}{\text{def}} \frac{v_{ce,I}}{i_{c,I}} = \frac{1}{h_{oc}}$$

Taking into account the conversion among hybrid parameters as in Table 11.5 of Chapter 11, we have

$$A_{iq,I} = \frac{-\left(1 + h_{fe}\right)}{1 + h_{oe} R_{L,I}}$$

$$R_{iq,I} = h_{ie}$$

$$A_{vq,I} = \frac{R_{L,I}}{R_{iq,I}} A_{iq,I}$$

$$R_{oq,I} = \frac{1}{h_{oe}}$$

For the two stages,

$$A_{i,II} \underset{=}{\text{def}} \frac{i_L}{i_{i,II}} = \frac{-i_{c,II}}{i_{i,II}} = -\frac{i_{c,II}}{i_{b,II}} \frac{i_{b,II}}{i_{i,II}} = -A_{iq,II} \frac{R_{12,II}}{R_{12,II} + R_{iq,II}}$$

II stage

$$R_{i,II} = R_{12,II} \parallel R_{iq,II}$$

$$A_{v,II} \underset{=}{\text{def}} \frac{v_o}{v_{i,II}} = \frac{v_{ce,II}}{v_{i,II}} = \frac{v_{ce,II}}{v_{be,II}} \frac{v_{be,II}}{v_{i,II}} = A_{vq,II} \frac{R_{iq,II}}{R_{iq,II} + R_{12,II}}$$

$$R_{o,II} = R_{c,II} \parallel R_{oq,II}$$

$$A_{i,I} \underset{\text{def}}{=} \frac{i_{i,II}}{i_g} = \frac{i_{i,II}}{i_{e,I}}\frac{i_{e,I}}{i_{b,I}}\frac{i_{b,I}}{i_g} = -\frac{R_{E,I}}{R_{E,I}+R_{i,II}}A_{iq,I}\frac{R_{12,I}}{R_{12,I}+R_{iq,I}} \qquad \boxed{\text{I stage}}$$

$$R_{i,I} \underset{\text{def}}{=} \frac{v_i}{i_g} = R_{12,I} \parallel R_{iq,I}$$

$$A_{v,I} \underset{\text{def}}{=} \frac{v_{o,I}}{v_g} = \frac{v_{ec,I}}{v_g} = \frac{v_{ec,I}}{v_{bc,I}}\frac{v_{bc,I}}{v_g} = A_{vq,I}\frac{R_{12,I} \parallel R_{iq,I}}{\left(R_{12,I} \parallel R_{iq,I}\right)+R_g}$$

$$R_{o,I} = R_{E,I} \parallel R_{oq,I}$$

Finally,

$$A_i = A_{i,I}A_{i,II}$$

$$A_v = A_{v,I}A_{v,II}$$

$$R_i = R_{i,I}$$

$$R_o = R_{o,II}$$

14.7 DOUBLE-STAGE C.S.–C.E.

A double-stage C.S.–C.E. can be realized as in Figure 14.20.
The AC *unilateral* equivalent model is represented in Figure 14.21.

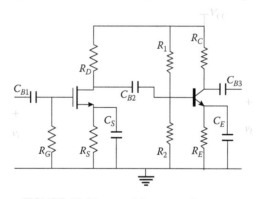

FIGURE 14.20 Double-stage C.S.–C.E.

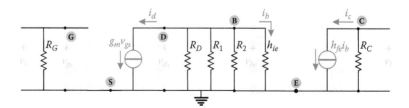

FIGURE 14.21 AC model of the double-stage C.S.–C.E.

14.7.1 AC Input Resistance

The AC input resistance R is the one that is "seen" by the AC input source (here absent), so R_G, in parallel with the MOSFET's gate resistance, which can be considered infinite, therefore,

$$R_i = R_G \parallel \infty = R_G$$

14.7.2 AC Output Resistance

To determine the output resistance R_o we have to set the AC input source $v_s = 0$. As a consequence the MOSFET is "off" and no drain current i_d flows, and there is also no flow from the base current i_b. Therefore,

$$R_o = R_C$$

14.7.3 AC Voltage Gain

The AC voltage gain of the first stage A_{v1} is that of a MOSFET in C.S. configuration, so from Section 12.2.2.3 of Chapter 12, we can write

$$A_{v1} = -g_m \left(R_D \parallel R_1 \parallel R_2 \parallel h_{ie} \right)$$

The AC voltage gain of the second stage $A_{v2} = v_l / v_{be}$ can be determined considering that $v_{be} = h_{ie}i_b$, $v_l = -h_{fe}i_b R_C$, so

$$A_{v2} = -\frac{h_{fe}}{h_{ie}} R_C$$

Thus, the overall voltage gain A_v can be expressed as

$$A_v = g_m \left(R_D \parallel R_1 \parallel R_2 \parallel h_{ie} \right) \frac{h_{fe}}{h_{ie}} R_C$$

14.8 BANDWIDTH

The *bandwidth* of an amplifier formed by cascading of n stages is

$$BW_{tot} = f'_{c-h} - f'_{c-l}$$

where, if all n stages have the same high cutoff frequency f_{c-h} and low cutoff frequency f_{c-l}, then

$$f'_{c-h} = f_{c-h} \sqrt{2^{\frac{1}{n}} - 1}$$

$$f'_{c-l} = \frac{f_{c-l}}{\sqrt{2^{\frac{1}{n}} - 1}}$$

while if each stage has different high and low cutoff frequencies then f'_{c-h} is due to the stage with the lowest f_{c-l} and f'_{c-l} is due to the stage with the highest f_{c-l}.

14.9 KEY POINTS, JARGON, AND TERMS

→ By definition, *one stage* refers to one single step of a process. Accordingly, a *one-stage amplifier* refers to an amplifier made up of one single transistor, arranged with its bias network, so as to produce amplification.

→ A *multistage amplifier* is an amplifier made with two or more stages arranged in a way that the output of one stage corresponds to the input for the next stage.

→ The *coupling method* refers to the way the connection between the two following stages is realized.

→ A stage is said to be *unilateral* when the output exhibits no feedback to its input side; *non-unilateral* or *bilateral* stages behave differently.

→ The analysis of a multistage amplifier must take into account the coupling method and the *unilateral* or *non-unilateral* characteristic of each single stage.

→ The *bandwidth* of a multistage amplifier must be determined considering the bandwidth of each single stage.

14.10 EXERCISES

EXERCISE 1

Given the network shown in Figure 14.22, we have, for the source $v_s = 10\,mV$, $R_s = 200\,\Omega$, for the amplifier $R_i = 1\,k\Omega$, $A_v = 20$, $R_o = 500\,\Omega$, and for the load $R_L = 100\,\Omega$.

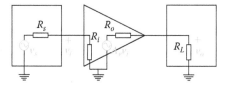

FIGURE 14.22 Scheme of a generic network including a voltage amplifier.

Determine the value of the output voltage v_o.

ANSWER

$v_o \cong 27.6\,mV$

EXERCISE 2

A voltage amplifier has a voltage gain of $A_v = 300$. It is used to supply a voltage of $v_L = 2\,V$ across a load $R_L = 5\,k\Omega$. The available AC voltage source has a value of $v_s = 10\,mV$ and an inner resistance of $R_s = 500\,\Omega$, and it is capable of a peak current $i_s = 1\,\mu A$.

Determine which input and output resistances are required by the amplifier.

ANSWER

$R_i = 9500\,\Omega$; $R_0 = 2125\,\Omega$.

EXERCISE 3

Consider the double C.E.-swamped C.E. configuration as represented in Figure 14.23. The parameters h_{re} and h_{oe} of both BJTs are negligible. Determine the expressions of the following:

- The input resistances of both stages and the overall input resistance
- The voltage gains of both stages and the overall voltage gain

EXERCISE 4

Consider the double C.S.–C.S. configuration as represented in Figure 14.24. Determine the expressions of the following:

- The overall input resistance
- The overall voltage gain
- The overall output resistance

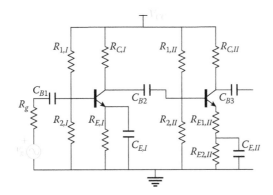

FIGURE 14.23   Double-stage C.E.-swamped C.E.

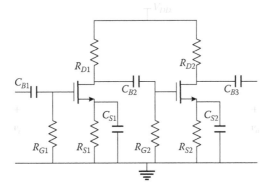

FIGURE 14.24   Double-stage C.S.–C.S.

Some Applications

15

We have used transistors to amplify signals. This fantastic property makes transistors suitable for a really large number of applications, and in this chapter we focus on some of the most relevant or curious ones.

15.1 AUDIO AMPLIFIER

An audio amplifier can be generally schematized with a cascade of black boxes each representing a single stage. As an example, we can represent an audio amplifier with a chain of an *input selection*, a *pre-amplifier section*, a *voltage amplifier*, a *tone control*, and a *power amplifier* (Figure 15.1).

The *input selection* is necessary to establish which source provides the signal to be amplified, and it realizes the best impedance matching condition (Section 6.9 in Chapter 6) with that source. This is because different sources can have very different impedances: microphones, radio tuners, antennas, etc., can offer impedances ranging from some hundred Ohms to several thousand Ohms.

The *pre-amplifier* amplifies the very weak signal from the AC source. It is required to add the lowest possible level of noise (Section 10.2.9 in Chapter 10). An example of a simple pre-amplifier schematic is shown in Figure 15.2.

The *voltage amplifier* (Section 10.3.1 in Chapter 10) boosts the signal voltage but its level is generally not adequate to drive a load.

The *tone control* is provided by filters, so that the signal can be treated with different frequency bandwith (generally "low," "medium," and "high" audible frequency ranges). The simplest way to realize it is by means of a *RC filter* (Section 7.3

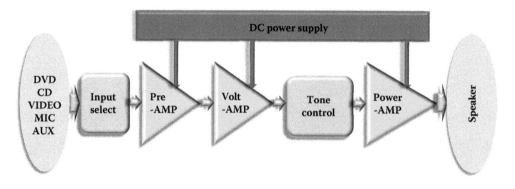

FIGURE 15.1   Chain of an arbitrary audio amplifier.

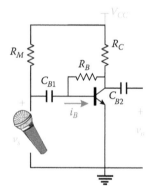

FIGURE 15.2   A simple
pre-amplifier.

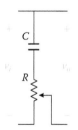

FIGURE 15.3   An RC filter
to realize the simplest
form of tone control.

in Chapter 7), with a variable resistor, so that the cutoff frequency, $f_c = 1/2\pi RC$ is established varying the value of the resistance R (Figure 15.3).

The *power amplifier* (Section 10.3.5 in Chapter 10) produces the relatively large power output (generally greater than $1\,W$) needed to drive a load such as a speaker. It should have good efficiency, so it is generally made of a class B or class AB amplifier (Section 10.4.2).

Observation

The *integrated circuit (IC)* LM386 by Texas Instruments is a Low Voltage Audio Power Amplifier. Practically, it is an entire radio amplifier on a single 8-pin package chip. It is battery operated, has a voltage gain ranging from 20 to 200, and has a low 2% distortion.

Another example is the *LM4871 Boomer®* by National Semiconductor. It is a Audio Power Amplifier on a 8-pin package *IC* capable of delivering $3\,W$ power into a $3\,\Omega$ resistance load with less than 10% THD.

A simplified version of audio amplifier is a *headphone amplifier*, as represented in Figure 15.4.

The first stage is a *swamped* C.E. configuration so to limit the gain to ensure linearity, while the second stage is a C.C configuration with a higher current gain thanks to the Darlington pair.

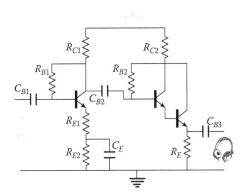

FIGURE 15.4   Headphone amplifier.

15.2 FM RADIO

Radio refers to anything associated with *radio electromagnetic waves* or *radiation* (Section 3.4 in Chapter 3). In particular, the tool with which we transmit and receive wireless radio signals is generally referred to as a *radio*.

Frequency modulation (FM) is a type of modulation (Section 1.8 in Chapter 1) used with radio, where the frequency of a high-frequency carrier is changed proportionally to the information signal.

The radio chain is made up of a transmitter and a receiver, detailed in Sections 15.2.1 and 15.2.2.

15.2.1 Transmitter

An *FM transmitter* is a network used to broadcast voice and/or music from an audio source to a radio receiver, based on *frequency modulation*.

The audio source can be a microphone, an iPod, a CD player, or similar devices. For instance, you can rebroadcast the sounds of your favorite TV program within your house.

Figure 15.5 shows two simple schematic circuits.

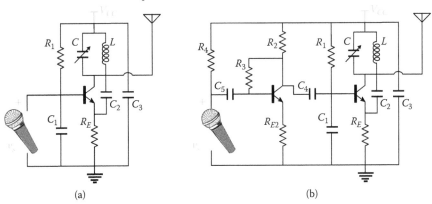

(a)   (b)

FIGURE 15.5   (a) Simple FM transmitter; (b) with increased gain.

The audio signal is introduced to a LC tank circuit (Section 7.3.3 in Chapter 7), which forms an oscillator vibrating at frequencies within the FM radio band. The oscillation voltage is across the base-collector junction, since the base is AC grounded by the capacitor C_1. The depth of the depletion layer, the reverse-biased base-collector junction, changes according to the modulated voltage v_{bc} across the same junction. The change in capacitance at the collector results in a change of the resonant frequency of the collector circuit.

Let's detail the single functional steps. At the very beginning, the capacitor C_1 starts to charge so that the BJT starts to turn "on." Consequently, an increasing current flows through the inductor's coil, across which a voltage drop rises, so reducing the voltage at the collector terminal. But the collector terminal is AC shorted with the emitter terminal by means of the capacitor C_2, so the voltage at the emitter terminal reduces as well. Now, since the base terminal is AC fixed (grounded) by C_1, the voltage across the base-emitter junction increases.

During these steps the Q-point moves from the cutoff region towards the saturation region, passing through the linear region of the BJT's input/output characteristics. This means that the BJT is turned more and more "on" until saturation, so that the collector current stabilizes and no longer varies. No voltage is now produced across the coil, and the voltage at the emitter junction is no longer lowered. The BJT goes toward an "off" condition, as the voltage drop across the resistor R_E increases more and more. This results in a current reduction through the coil, so that the voltage across it is now in the opposite direction. The voltage at the collector terminal increases and so does the one at the emitter terminal, thanks to the action of C_2, in a way that the BJT goes more and more "off." The Q-point is now reversing its direction, from the saturation region toward the cutoff one, and the BJT becomes fully "off."

The oscillation frequency of the overall process is due to the action of the inductor L, which charges and discharges, by means of its magnetic flux, its coupled capacitor C. So the oscillation frequency can be varied just by changing the value of the capacitance C. The modulated and amplified signal is passed to the antenna that propagated it.

The presence of the capacitor C_3 is to guarantee a rigid voltage drop across the DC source and the ground.

The two figures in Figure 15.5 represent the same FM transmitter but part (b) is a double-stage version to increase the overall gain. The first stage is a collector feedback bias common emitter amplifier (Section 11.10.2 in Chapter 11).

Curiosity

The inventor of modern FM radio transmission was Armstrong (Edwin Howard Armstrong, American electrical engineer and inventor, 1890–1954). It is strange to consider that Carson (John Renshaw Carson, degrees in Electrical Engineering and Science, 1886–1940), who invented the famous *single-sideband modulation*, wrote a paper that argued the absence of any advantages with FM transmission.

Armstrong, still an undergraduate, invented *regenerative feedback*, that is, amplification via *positive feedback* (mentioned in Section 15.4), followed by the *superheterodyne receiver*, still currently widely in use.

15.2.2 Receiver

The *radio receiver* is necessary to convert incoming modulated radio waves into a signal that produces sounds by means of a loudspeaker. The *FM radio receiver* works with electromagnetic waves, which are modulated in frequency. One possible simple version of an FM radio receiver is shown in Figure 15.6.

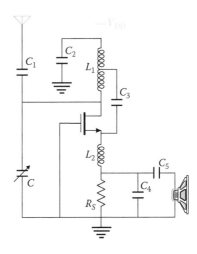

FIGURE 15.6 Simple FM radio receiver.

The working principle is based on a oscillating network, that is, a circuit with voltages and currents with repetitive variation. In particular, the proposed receiver is based on the so-called *Hartley oscillator* (Ralph Vinton Lyon Hartley, American electronics researcher, 1888–1970), well suited to the VHF range for FM broadcast, made of one transistor, two inductors, and one capacitor. Here, the transistor is a FET one, the two inductors are represented by L_1, which is divided in two parts by a centered tap connection, and the capacitor is C, which is a variable capacitor so as to "tune" the oscillation frequency. Please note that L_1 and C form a parallel for AC conditions, since C_2 represents an AC ground for L_1. The capacitor C_3 provides the feedback effect useful to the oscillation process. The function of L_2 is to block the RF components, and R_s together with C_4 smooths the signal. Finally, C_1 and C_5 are the DC blocking capacitors.

Curiosity

Different type of oscillators are named after their inventors. We can think of the Armstrong oscillator (Edwin Howard Armstrong, already mentioned), the Colpitts oscillator (Edwin Henry Colpitts, American electronic engineer, 1872–1949), the Vackář oscillator (Jiří Vackář, Czech electric engineer, 1919–2004), the Clapp oscillator (James Kilton Clapp, American electrical engineer, 1897–1965), the Franklin oscillator (Charles Samuel Franklin, British radio pioneer, 1879–1964), the Butler oscillator, the Seiler oscillator, and so on.

15.3 BATTERY CHARGER

We have the necessary information to realize a *battery charger*. Here, it is proposed in two versions: USB and AC mains powered.

15.3.1 USB-Powered Charger

We can charge the battery of cellular phones, portables, MP3 players, small toys, etc., by means of the USB port of our computer. But if we directly connect the battery to the USB port we will damage the battery itself. This is because over USB $5\,V$ are issued, while the average voltage of a phone battery can be, for instance,

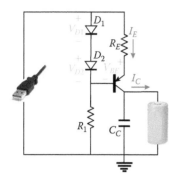

FIGURE 15.7 Example of a USB powered charger.

3.6 V. So we have to reduce the voltage but not the current, typically 100 mA for a USB socket, otherwise the battery will not be charged. The point is that an ad-hoc circuit is necessary to reduce the voltage, and to maintain it at a constant value, without reducing the current. It cannot be made by a simple resistor voltage divider, otherwise the current will heat up the resistors rather than charge the battery.

The circuit shown in Figure 15.7 can solve this.

The diodes have a negative temperature coefficient, that is, their voltage drop decreases increasing the temperature, typically $-2 (mV/°C)$ for silicon types. Please pay close attention to the BJT, which is a *pnp* type, so that its emitter has to be at a higher voltage than the collector to be correctly forward biased.

The two diodes and the transistor must be of the same silicon type, so that the voltage threshold of the diodes, and the voltage threshold of the base-emitter junction of the BJT can be as similar as possible. If so, the voltage drop across the branch made of D_1 and D_2 equals the voltage drop across the branch made of R_E and BJT's B-E junction. The consequence is that we can write the current I_E through R_E as $I_E = V_{D1}/R_E$, with V_{D1} the voltage drop across the diode D_1. The same current value applies to I_C flowing through the battery to be charged, since we admit $I_C \cong I_E$.

Now the battery typically increases its temperature during its charging process. This results in an increase of the overall temperature and, since the diodes have a negative temperature coefficient, their voltage drop decreases, in particular the value of V_{D1}. But $I_E = V_{D1}/R_E$ so I_E reduces as well. The effects are then balanced and the currents remain quite constant, as well as the voltage across the battery. We can consider that a sort of automatic adjustment is implemented.

15.3.2 AC Mains–Powered Charger

Figure 15.8 represents an AC mains powered charger for Nickel-Cadmium (*NiCd*) batteries. It is formed by a DC power supply (Section 9.2 in Chapter 9), with a regulator made of a BJT in emitter follower configuration.

NiCd batteries require a constant current to be charged. The emitter follower is capable of maintaining an almost constant output current, even when the load resistance of the battery varies. In fact, the reverse-biased zener diode (Section 8.5.5 in Chapter 8) maintains a constant voltage drop V_Z, irrespective of its current. The voltage V_Z is the sum of the one across the BJT's base-emitter junction V_{BE}, and the one across the emitter resistor V_{R_E}: $V_Z = V_{BE} + V_{R_E}$. Now, since the current I_E through R_E is given by

$$I_E = \frac{V_{R_E}}{R_E} = \frac{V_Z - V_{BE}}{R_E}$$

it can be considered practically constant, because we can admit V_Z, V_{BE}, and R_E are unchanged for a fixed temperature, or poorly varying with it. But for a BJT we

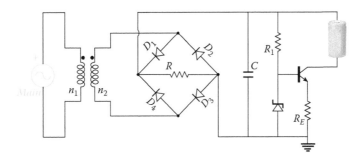

FIGURE 15.8 AC mains powered charger.

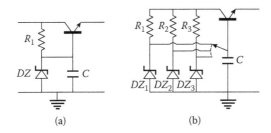

(a)   (b)

FIGURE 15.9 Regulators with
(a) unique voltage output or
(b) different voltage outputs.

can generally admit that $I_C \cong I_E$, so that the collector current flowing through the battery to be charged, is constant as well.

A regulator can be made with the possibility of choosing the value of the output voltage among different ones (Figure 15.9).

15.4 EARLY HEART PACEMAKER

The *pacemaker* is a small, battery-powered device, generally placed under the skin, used to help the heart beat in a regular rhythm. To this aim, when the heart rate becomes too slow or uneffective, this device paces, by transmitting a tiny electrical current to the heart muscle, causing it to restart its natural rhythm. Nowadays, the pacemaker is a very sophisticated programmable device, capable of adapting itself to different cardiac conditions, but the very early pacemaker was very simple. The first battery-operated, wearable artificial pacemaker is credited to Bakken (Earl E. Bakken, American engineer, businessman, and philanthropist, 1924–), and the working principle was on the basis of a circuit called a *blocking oscillator*. This circuit is a particular type of *wave generator* capable of

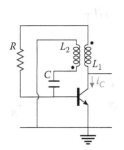

FIGURE 15.10
Simple form of a
blocking oscillator.

producing a narrow electric pulse. Its name derives from the fact that the transistor is cut off, or *blocked*, for most of its working cycle. The minimal configuration of a blocking oscillator consists of one transformer, one capacitor, one resistor, and one BJT, connected as schematized in Figure 15.10.

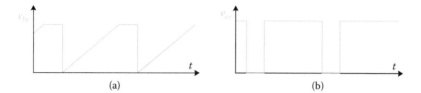

FIGURE 15.11   Waveforms of the voltages across the (a) base-
emitter and the (b) collector-emitter junctions during more cycles.

The working principle is as follows:

→ When this circuit is made "active," the DC source V_{CC}, through R, leads the BJT in forward bias conditions for a certain time.
→ With the BJT going "active," the current i_C flowing through the output loop increases more and more.
→ This increasing value of the current i_C produces, through L_1, an induced voltage across L_2.
→ The transformer is phased (see the two black points in Figure 15.10) in a way that any increase in i_C pulls the base up (the transformer is said to operate in *flyback mode*), increasing the base current (we talk about *regenerative* or *positive* feedback, that is, an output signal *increases* the input signal) so that the BJT saturates.
→ With the BJT saturated, there is no more change in i_C and, as a consequence, no coupling between L_1 and L_2.
→ The capacitor C now discharges through R, which reduces the voltage across the base-emitter junction and cuts off the transistor.

This process is iterated as represented in Figure 15.11.

The blocking oscillator can produce short pulses in the microsecond range.

The overall process can be conditioned and "triggered" by an input voltage v_i, which can be due to signals measured by electrocatheters implanted into the heart. The oscillator then generates pulses only when requested by the input signal.

15.5 OTHER TRANSISTOR APPLICATIONS

Transistors can be used to realize practically any circuital function. We can utilize transistors as switches (Section 12.7 in Chapter 12), and to design phase inverters (see "Observation" in Exercise 19 of Chapter 11), oscillator circuits, timers, relay controls, TV and radio receivers, active filters, frequency meters, clippers, and many other devices. In this section, we want to focus on some applications based on the possibility of changing the quiescent Q-point during the transistor's usage.

The voltage divider bias network analyzed so far established fixed values for quiescent currents and voltages, in order to guarantee the transistor would work in a state of linearity, when small signal conditions occur. But, for some applications, we can admit that those values are no longer set constant by a fixed resistor network.

The quiescent conditions can be varied by components capable of changing their resistance, or impedance, because of an external physical event. In this case we can utilize transistors to realize, for example, a liquid level indicator (Section 15.5.1), a battery voltage monitor (Section 15.5.2), a dark/light activated LED/relay (Section 15.5.3), or a measure of temperature (Section 15.5.4).

But the quiescent conditions can be varied by a capacitor too, which can be charged and discharged with an ad-hoc network. In this case we can utilize transistors to realize, for example, a simple timer (Section 15.5.5), or a cycling beep generator (Section 15.5.6).

15.5.1 Liquid Level Indicator

Two transistors can be arranged to form a *liquid level indicator*, as in Figure 15.12.

When the dipped electrodes detect the liquid (water for instance), the circuit is powered by the DC source V_{EE} (note that transistor T_2 is a *pnp* type), and the LED emits a warning light. The resistor R_C limits the current through the LED. The resistor R_1 can be adjusted to ensure the correct voltage divider bias, depending on the resistance offered by the liquid (to be previously measured). The two transistors need to obtain a sufficient gain, but just one transistor could be sufficient, depending on the type of liquid.

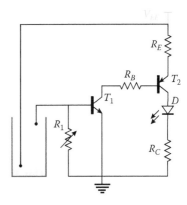

FIGURE 15.12   Liquid level indicator.

15.5.2 Battery Voltage Monitor

With two transistors we can design a *battery voltage monitor*, such that when the battery is fully charged a green diode lights, when the battery is partially discharged an orange diode lights, and when the battery has a charge below a threshold value a red diode lights.

An example of a voltage battery monitor circuit is shown in Figure 15.13.

With the lowest value of DC voltage supply, both the transistors are "off," and the only flowing current is I_3 through R_7 so that the red diode D_3 is "on" (diodes D_1 and D_2 are "off").

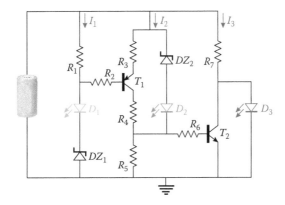

FIGURE 15.13   Battery voltage monitor.

When the value of DC voltage supply is increased, the zener diode DZ_2 goes "on" and the transistor T_2 saturates, so that $V_{CE} \cong 0.2\,V$. In this way, the orange diode D_2 lights and D_3 goes "off" because of the almost short circuit across the C-E junction.

When the value of DC voltage supply is increased more toward its highest value, the *pnp*-type transistor T_1 "saturates" so that the current I_2 flows through it and no longer through the diode D_2, which goes "off." The highest value of DC voltage makes the zener diode DZ_1 turn "on" and the green diode D_1 lights.

15.5.3 Dark/Light Activated LED/Relay

Currents flow through a BJT when its base-emitter junction is forward biased. But the bias depends on the voltage divider input loop, which can be made "light-sensitive," so that the light "establishes" when a BJT is in its "active" or "idle" mode.

The "light-sensitivity" comes from a component called a *light dependent resistor*, or *LDR*, which is essentially made of a semiconductor material capable of changing its resistance when illuminated. With incident light, photons are absorbed by the semiconductor (often *Cadmium Sulphide, CdS*, or *Cadmium Selenide, CdSe*; therefore electrons are excited from the valence into the conduction band and, consequently, the resistance lowers. The LDR's higest resistance is for dark conditions, so it is called *dark resistance*, and can be as high as mega-Ohms. With light, the value of the resistance drops dramatically to a few hundred Ohms. A LDR has a relation between irradiance and resistance that is not linear but actually exponential. This can result in a disadvantage, making this device unsuitable for precise measurements in analog applications, but it is preferred in digital, or "on-off," applications.

Example

LDRs are used in digital cameras to regulate the shutter speed as a function of the light intensity.

Figure 15.14 represents circuits that "turn on" an LED, depending on the room lighting (the LDR is drawn in blue). In particular, the circuit in Figure 15.14a

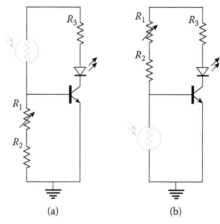

(a)   (b)

FIGURE 15.14 LED activated by (a) light or (b) darkness.

acitivates an LED by light; the circuit in Figure 15.14b activates the LED by darkness. The resistor R_3 limits the current flowing through the LED.

It can be more effective to activate another load, rather than a simple LED. The load can be even another circuit just "activated" by the light/darkness. For example, a lighting system can be switched on at nightfall. To this aim a *relay* can be helpful.

A *relay* is an electromechanical switch controlled by a current. It operates by means of an inductor that, when energized, generates a magnetic field acting on a metalic arm, which makes up a physical contact, so as to close or open a circuit loop.

Figures 15.15 adopts a relay in place of the LED. The relay is activated by light in Figure 15.15a, and by darkness in Figure 15.15b.

For all the circuits, the variable resistor R_1 allows a "fine-tuning" of the preset level of light/darkness required before the LED lights up or relay activates.

When more current is requested to activate the relay, the circuit can be rearranged with two transistors (Figure 15.16).

With a few changes, and with a microphone in place of the LDR, a similar circuit can be adopted for a sound activated LED/relay.

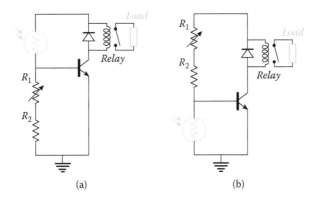

(a)   (b)

FIGURE 15.15   Relay activated by
(a) light or (b) darkness.

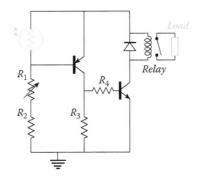

FIGURE 15.16   Relay activated
by darkness, with more
current flowing through it.

15.5.4 Measure of Temperature

A *thermal resistor*, commonly known as a *thermistor*, is a particular resistor with resistance that changes with temperature, so that a temperature measurement is possible. The thermistor translates the changes in temperature into electrical resistance, which can be used to drive an *ammeter*, that is, an instrument that measures an electric current. To this aim we can use a Wheatstone bridge (Section 4.4.13 in Chapter 4) with a thermistor in one of its arms (between nodes B and C), as represented in Figure 15.17.

The transistor is analog biased by the bridge and furnishes the current, proportional to the resistance of the thermistor, to drive the ammeter (in red).

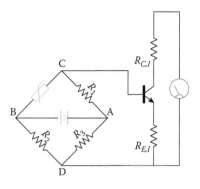

FIGURE 15.17 Circuit to measure the temperature.

15.5.5 Simple Timer

A simple *timer* can be designed with a few components. An example comes from the circuit shown in Figure 15.18.

When the switch is pressed, the capacitor will charge so that the BJT will be biased, and a collector current will flow through the LED, which will light. But when the switch is released, the LED will light just for the time the capacitor is discharged, so that the BJT is no longer biased. The discharging time represents the timer function.

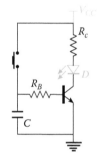

FIGURE 15.18 A simple timer.

15.5.6 Cyclic Beep Generator

It is possible to adopt a C.C. configuration to realize a simple *beep generator*, as schematized in Figure 15.19.

The working principle is quite simple. The DC source provides a DC current that, flowing through the resistor R_1, charges the capacitor C, which increases its voltage drop. With the right values of resistances of each component, the capacitor discharges and furnishes a current to forward bias the base-emitter junction of the BJT, which results in it being "activated." In this way, a current, mainly sourced by the battery via the output loop, flows through the loudspeaker, which beeps. The beep ends when the discharge of the capacitor lowers its voltage drop, so that the BJT can no longer be "active." But the overall process does not stop, because the capacitor is charged again by the current from the DC source, so that the BJT will be activated again. So this results in a nonstop process of cyclic beeps.

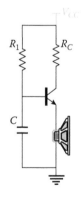

FIGURE 15.19 Cyclic beep generator.

CONCLUSIONS

What we have described so far is useful to give a solid basis to fully understand electronics. The circuital applications proposed are only a part of what electronics can offer.

You should therefore try and get curious and fascinated by what there is yet to know, in order to understand telecommunications, microelectronics, nanoelectronics, informatics, bioelectronics, etc. It is all about discovering what things like oscillators, power amplifiers, operational amplifiers, transmitters and receivers, mixers, and other things are and how they are made.

Good luck!

Index